METHODS IN MOLECULAR BIOLOGY™

John M. Walker, SERIES EDITOR

460. Essential Concepts in Toxicogenomics, edited by *Donna L. Mendrick and William B. Mattes, 2008*
459. Prion Protein Protocols, edited by *Andrew F. Hill, 2008*
458. Artificial Neural Networks: Methods and Applications, edited by *David S. Livingstone, 2008*
457. Membrane Trafficking, edited by *Ales Vancura, 2008*
456. Adipose Tissue Protocols, Second Edition, edited by *Kaiping Yang, 2008*
455. Osteoporosis, edited by *Jennifer J. Westendorf, 2008*
454. SARS- and Other Coronaviruses: *Laboratory Protocols,* edited by *Dave Cavanagh, 2008*
453. Bioinformatics, Volume II: *Structure, Function and Applications,* edited by *Jonathan M. Keith, 2008*
452. Bioinformatics, Volume I: *Data, Sequence Analysis and Evolution,* edited by *Jonathan M. Keith, 2008*
451. Plant Virology Protocols: *From Viral Sequence to Protein Function,* edited by *Gary Foster, Elisabeth Johansen, Yiguo Hong, and Peter Nagy, 2008*
450. Germline Stem Cells, edited by *Steven X. Hou and Shree Ram Singh, 2008*
449. Mesenchymal Stem Cells: *Methods and Protocols,* edited by *Darwin J. Prockop, Douglas G. Phinney, and Bruce A. Brunnell, 2008*
448. Pharmacogenomics in Drug Discovery and Development, edited by *Qing Yan, 2008*
447. Alcohol: *Methods and Protocols,* edited by *Laura E. Nagy, 2008*
446. Post-translational Modification of Proteins: *Tools for Functional Proteomics,* Second Edition, edited by *Christoph Kannicht, 2008*
445. Autophagosome and Phagosome, edited by *Vojo Deretic, 2008*
444. Prenatal Diagnosis, edited by *Sinhue Hahn and Laird G. Jackson, 2008*
443. Molecular Modeling of Proteins, edited by *Andreas Kukol, 2008.*
442. RNAi: Design and Application, edited by *Sailen Barik, 2008*
441. Tissue Proteomics: *Pathways, Biomarkers, and Drug Discovery,* edited by *Brian Liu, 2008*
440. Exocytosis and Endocytosis, edited by *Andrei I. Ivanov, 2008*
439. Genomics Protocols, Second Edition, edited by *Mike Starkey and Ramnanth Elaswarapu, 2008*
438. Neural Stem Cells: *Methods and Protocols,* Second Edition, edited by *Leslie P. Weiner, 2008*
437. Drug Delivery Systems, edited by *Kewal K. Jain, 2008*
436. Avian Influenza Virus, edited by *Erica Spackman, 2008*
435. Chromosomal Mutagenesis, edited by *Greg Davis and Kevin J. Kayser, 2008*
434. Gene Therapy Protocols: Volume II: *Design and Characterization of Gene Transfer Vectors,* edited by *Joseph M. LeDoux, 2008*
433. Gene Therapy Protocols: Volume I: *Production and In Vivo Applications of Gene Transfer Vectors,* edited by *Joseph M. LeDoux, 2008*

432. Organelle Proteomics, edited by *Delphine Pflieger and Jean Rossier, 2008*
431. Bacterial Pathogenesis: *Methods and Protocols,* edited by *Frank DeLeo and Michael Otto, 2008*
430. Hematopoietic *Stem Cell Protocols,* edited by *Kevin D. Bunting, 2008*
429. Molecular Beacons: *Signalling Nucleic Acid Probes, Methods and Protocols,* edited by *Andreas Marx and Oliver Seitz, 2008*
428. Clinical Proteomics: *Methods and Protocols,* edited by *Antonia Vlahou, 2008*
427. Plant Embryogenesis, edited by *Maria Fernanda Suarez and Peter Bozhkov, 2008*
426. Structural Proteomics: *High-Throughput Methods,* edited by *Bostjan Kobe, Mitchell Guss, and Huber Thomas, 2008*
425. 2D PAGE: *Sample Preparation and Fractionation,* Volume II, edited by *Anton Posch, 2008*
424. 2D PAGE: *Sample Preparation and Fractionation,* Volume I, edited by *Anton Posch, 2008*
423. Electroporation Protocols: *Preclinical and Clinical Gene Medicine,* edited by *Shulin Li, 2008*
422. Phylogenomics, edited by *William J. Murphy, 2008*
421. Affinity Chromatography: *Methods and Protocols,* Second Edition, edited by *Michael Zachariou, 2008*
420. Drosophila: *Methods and Protocols,* edited by *Christian Dahmann, 2008*
419. Post-Transcriptional Gene Regulation, edited by *Jeffrey Wilusz, 2008*
418. Avidin–Biotin Interactions: *Methods and Applications,* edited by *Robert J. McMahon, 2008*
417. Tissue Engineering, Second Edition, edited by *Hannsjörg Hauser and Martin Fussenegger, 2007*
416. Gene Essentiality: *Protocols and Bioinformatics,* edited by *Svetlana Gerdes* and *Andrei L. Osterman, 2008*
415. Innate Immunity, edited by *Jonathan Ewbank and Eric Vivier, 2007*
414. Apoptosis in Cancer: *Methods and Protocols,* edited by *Gil Mor and Ayesha Alvero, 2008*
413. Protein Structure Prediction, Second Edition, edited by *Mohammed Zaki and Chris Bystroff, 2008*
412. Neutrophil Methods and Protocols, edited by *Mark T. Quinn, Frank R. DeLeo, and Gary M. Bokoch, 2007*
411. Reporter Genes: *A Practical Guide,* edited by *Don Anson, 2007*
410. Environmental Genomics, edited by *Cristofre C. Martin, 2007*
409. Immunoinformatics: *Predicting Immunogenicity In Silico,* edited by *Darren R. Flower, 2007*
408. Gene Function Analysis, edited by *Michael Ochs, 2007*
407. Stem Cell Assays, edited by *Vemuri C. Mohan, 2007*
406. Plant Bioinformatics: *Methods and Protocols,* edited by *David Edwards, 2007*
405. Telomerase Inhibition: *Strategies and Protocols,* edited by *Lucy Andrews and Trygve O. Tollefsbol, 2007*

Bioinformatics

METHODS IN MOLECULAR BIOLOGY™

Bioinformatics

Volume I
Data, Sequence Analysis and Evolution

Edited by

Jonathan M. Keith, PhD

*School of Mathematical Sciences, Queensland University of Technology,
Brisbane, Queensland, Australia*

Humana Press

Editor
Jonathan M. Keith
School of Mathematical Sciences
Queensland University of Technology
Brisbane, Queensland, Australia
j.keith@qut.edu.au

Series Editor
John Walker
Hatfield, Hertfordshire AL10 9NP
UK

ISBN: 978-1-58829-707-5   e-ISBN: 978-1-60327-159-2
ISSN 1064-3745   e-ISSN: 1940-6029
DOI: 10.1007/978-1-60327-159-2

Library of Congress Control Number: 2007943036

© 2008 Humana Press, a part of Springer Science+Business Media, LLC
All rights reserved. This work may not be translated or copied in whole or in part without the written permission of the publisher (Humana Press, 999 Riverview Drive, Suite 208, Totowa, NJ 07512 USA), except for brief excerpts in connection with reviews or scholarly analysis. Use in connection with any form of information storage and retrieval, electronic adaptation, computer software, or by similar or dissimilar methodology now known or hereafter developed is forbidden.
The use in this publication of trade names, trademarks, service marks, and similar terms, even if they are not identified as such, is not to be taken as an expression of opinion as to whether or not they are subject to proprietary rights.
While the advice and information in this book are believed to be true and accurate at the date of going to press, neither the authors nor the editors nor the publisher can accept any legal responsibility for any errors or omissions that may be made. The publisher makes no warranty, express or implied, with respect to the material contained herein.

Cover illustration: Fig. 4, Chapter 19, "Inferring Ancestral Protein Interaction Networks," by José M. Peregrín-Alvarez

Printed on acid-free paper

9 8 7 6 5 4 3 2 1

springer.com

Preface

Bioinformatics is the management and analysis of data for the life sciences. As such, it is inherently interdisciplinary, drawing on techniques from Computer Science, Statistics, and Mathematics and bringing them to bear on problems in Biology. Moreover, its subject matter is as broad as Biology itself. Users and developers of Bioinformatics methods come from all of these fields. Molecular biologists are some of the major users of Bioinformatics, but its techniques are applicable across a range of life sciences. Other users include geneticists, microbiologists, biochemists, plant and agricultural scientists, medical researchers, and evolution researchers.

The ongoing exponential expansion of data for the life sciences is both the major challenge and the *raison d'être* for twenty-first century Bioinformatics. To give one example among many, the completion and success of the human genome sequencing project, far from being the end of the sequencing era, motivated a proliferation of new sequencing projects. And it is not only the quantity of data that is expanding; new types of biological data continue to be introduced as a result of technological development and a growing understanding of biological systems.

Bioinformatics describes a selection of methods from across this vast and expanding discipline. The methods are some of the most useful and widely applicable in the field. Most users and developers of Bioinformatics methods will find something of value to their own specialties here, and will benefit from the knowledge and experience of its 86 contributing authors. Developers will find them useful as components of larger methods, and as sources of inspiration for new methods. Volume I, Section IV in particular is aimed at developers; it describes some of the "meta-methods"—widely applicable mathematical and computational methods that inform and lie behind other more specialized methods—that have been successfully used by bioinformaticians. For users of Bioinformatics, this book provides methods that can be applied as is, or with minor variations to many specific problems. The Notes section in each chapter provides valuable insights into important variations and when to use them. It also discusses problems that can arise and how to fix them. This work is also intended to serve as an entry point for those who are just beginning to discover and use methods in Bioinformatics. As such, this book is also intended for students and early career researchers.

As with other volumes in the Methods in Molecular Biology™ series, the intention of this book is to provide the kind of detailed description and implementation advice that is crucial for getting optimal results out of any given method, yet which often is not incorporated into journal publications. Thus, this series provides a forum for the communication of accumulated practical experience.

The work is divided into two volumes, with data, sequence analysis, and evolution the subjects of the first volume, and structure, function, and application the subjects of the second. The second volume also presents a number of "meta-methods": techniques that will be of particular interest to developers of bioinformatic methods and tools.

Within Volume I, Section I deals with data and databases. It contains chapters on a selection of methods involving the generation and organization of data, including sequence data, RNA and protein structures, microarray expression data, and functional annotations.

Section II presents a selection of methods in sequence analysis, beginning with multiple sequence alignment. Most of the chapters in this section deal with methods for discovering the functional components of genomes, whether genes, alternative splice sites, non-coding RNAs, or regulatory motifs.

Section III presents several of the most useful and interesting methods in phylogenetics and evolution. The wide variety of topics treated in this section is indicative of the breadth of evolution research. It includes chapters on some of the most basic issues in phylogenetics: modelling of evolution and inferring trees. It also includes chapters on drawing inferences about various kinds of ancestral states, systems, and events, including gene order, recombination events and genome rearrangements, ancestral interaction networks, lateral gene transfers, and patterns of migration. It concludes with a chapter discussing some of the achievements and challenges of algorithm development in phylogenetics.

In Volume II, Section I, some methods pertinent to the prediction of protein and RNA structures are presented. Methods for the analysis and classification of structures are also discussed.

Methods for inferring the function of previously identified genomic elements (chiefly protein-coding genes) are presented in Volume II, Section II. This is another very diverse subject area, and the variety of methods presented reflects this. Some well-known techniques for identifying function, based on homology, "Rosetta stone" genes, gene neighbors, phylogenetic profiling, and phylogenetic shadowing are discussed, alongside methods for identifying regulatory sequences, patterns of expression, and participation in complexes. The section concludes with a discussion of a technique for integrating multiple data types to increase the confidence with which functional predictions can be made. This section, taken as a whole, highlights the opportunities for development in the area of functional inference.

Some medical applications, chiefly diagnostics and drug discovery, are described in Volume II, Section III. The importance of microarray expression data as a diagnostic tool is a theme of this section, as is the danger of over-interpreting such data. The case study presented in the final chapter highlights the need for computational diagnostics to be biologically informed.

The final section presents just a few of the "meta-methods" that developers of Bioinformatics methods have found useful. For the purpose of designing algorithms, it is as important for bioinformaticians to be aware of the concept of *fixed parameter tractability* as it is for them to understand NP-completeness, since these concepts often determine the types of algorithms appropriate to a particular problem. *Clustering* is a ubiquitous problem in Bioinformatics, as is the need to *visualize* data. The need to interact with massive data bases and multiple software entities makes the development of *computational pipelines* an important issue for many bioinformaticians. Finally, the chapter on *text mining* discusses techniques for addressing the special problems of interacting with and extracting information from the vast biological literature.

Jonathan M. Keith

Contents

Preface .... *v*
Contributors .... *ix*
Contents of Volume II .... *xi*

SECTION I: DATA AND DATABASES

1. Managing Sequence Data 3
   Ilene Karsch Mizrachi

2. RNA Structure Determination by NMR 29
   Lincoln G. Scott and Mirko Hennig

3. Protein Structure Determination by X-Ray Crystallography 63
   Andrea Ilari and Carmelinda Savino

4. Pre-Processing of Microarray Data and Analysis
   of Differential Expression 89
   Steffen Durinck

5. Developing an Ontology 111
   Midori A. Harris

6. Genome Annotation 125
   Hideya Kawaji and Yoshihide Hayashizaki

SECTION II: SEQUENCE ANALYSIS

7. Multiple Sequence Alignment 143
   Walter Pirovano and Jaap Heringa

8. Finding Genes in Genome Sequence 163
   Alice Carolyn McHardy

9. Bioinformatics Detection of Alternative Splicing 179
   Namshin Kim and Christopher Lee

10. Reconstruction of Full-Length Isoforms from Splice Graphs 199
    Yi Xing and Christopher Lee

11. Sequence Segmentation 207
    Jonathan M. Keith

12. Discovering Sequence Motifs 231
    Timothy L. Bailey

SECTION III: PHYLOGENETICS AND EVOLUTION

13. Modeling Sequence Evolution 255
    Pietro Liò and Martin Bishop

14. Inferring Trees 287
    Simon Whelan

15. Detecting the Presence and Location of Selection in Proteins................ 311
    Tim Massingham

16. Phylogenetic Model Evaluation... 331
    *Lars Sommer Jermiin, Vivek Jayaswal, Faisal Ababneh,
    and John Robinson*

17. Inferring Ancestral Gene Order.. 365
    Julian M. Catchen, John S. Conery, and John H. Postlethwait

18. Genome Rearrangement by the Double Cut and Join Operation............... 385
    Richard Friedberg, Aaron E. Darling, and Sophia Yancopoulos

19. Inferring Ancestral Protein Interaction Networks........................ 417
    José M. Peregrín-Alvarez

20. Computational Tools for the Analysis of Rearrangements
    in Mammalian Genomes... 431
    Guillaume Bourque and Glenn Tesler

21. Detecting Lateral Genetic Transfer: *A Phylogenetic Approach*........... 457
    Robert G. Beiko and Mark A. Ragan

22. Detecting Genetic Recombination.. 471
    Georg F. Weiller

23. Inferring Patterns of Migration.. 485
    Paul M.E. Bunje and Thierry Wirth

24. Fixed-Parameter Algorithms in Phylogenetics............................ 507
    Jens Gramm, Arfst Nickelsen, and Till Tantau

Index... 537

Evolution Index... 551

Contributors

FAISAL ABABNEH • *Department of Mathematics and Statistics, Al-Hussein Bin Talal University, Ma'an, Jordan*
TIMOTHY L. BAILEY • *ARC Centre of Excellence in Bioinformatics, and Institute for Molecular Bioscience, The University of Queensland, Brisbane, Queensland, Australia*
ROBERT G. BEIKO • *Faculty of Computer Science, Dalhousie University, Halifax, Nova Scotia, Canada*
MARTIN BISHOP • *CNR-ITB Institute of Biomedical Technologies, Segrate, Milano, Italy*
GUILLAUME BOURQUE • *Genome Institute of Singapore, Singapore, Republic of Singapore*
PAUL M.E. BUNJE • *Department of Biology, Lehrstuhl für Zoologie und Evolutionsbiologie, University of Konstanz, Konstanz, Germany*
JULIAN M. CATCHEN • *Department of Computer and Information Science and Institute of Neuroscience, University of Oregon, Eugene, OR*
JOHN S. CONERY • *Department of Computer and Information Science, University of Oregon, Eugene, OR*
AARON E. DARLING • *ARC Centre of Excellence in Bioinformatics, and Institute for Molecular Bioscience, The University of Queensland, Brisbane, Queensland, Australia*
STEFFEN DURINCK • *Katholieke Universiteit Leuven, Leuven, Belgium*
RICHARD FRIEDBERG • *Department of Physics, Columbia University, New York, NY*
JENS GRAMM • *Wilhelm-Schickard-Institut für Informatik, Universität Tübingen, Tübingen, Germany*
MIDORI A. HARRIS • *European Molecular Biology Laboratory – European Bioinformatics Institute, Hinxton, Cambridge, United Kingdom*
YOSHIHIDE HAYASHIZAKI • *Genome Exploration Research Group, RIKEN Yokohama Institute, Yokohama, Kanagawa, Japan; and Genome Science Laboratory, RIKEN Wako Institute, Wako, Saitama, Japan*
MIRKO HENNIG • *Department of Biochemistry and Molecular Biology, Medical University of South Carolina, Charleston, SC*
JAAP HERINGA • *Centre for Integrative Bioinformatics (IBIVU), VU University Amsterdam, Amsterdam, The Netherlands*
ANDREA ILARI • *CNR Institute of Molecular Biology and Pathology (IBPM), Department of Biochemical Sciences, University of Rome, "Sapienza," Roma, Italy*
VIVEK JAYASWAL • *School of Mathematics and Statistics, Sydney Bioinformatics and Centre for Mathematical Biology, University of Sydney, Sydney, New South Wales, Australia*
LARS SOMMER JERMIIN • *School of Biological Sciences, Sydney Bioinformatics and Centre for Mathematical Biology, University of Sydney, Sydney, New South Wales, Australia*
HIDEYA KAWAJI • *Functional RNA Research Program, Frontier Research System, RIKEN Wako Institute, Wako, Saitama, Japan*

JONATHAN M. KEITH • *School of Mathematical Sciences, Queensland University of Technology, Brisbane, Queensland, Australia*
NAMSHIN KIM • *Molecular Biology Institute, Institute for Genomics and Proteomics, Department of Chemistry and Biochemistry, University of California, Los Angeles, CA*
CHRISTOPHER LEE • *Molecular Biology Institute, Institute for Genomics and Proteomics, Department of Chemistry and Biochemistry, University of California, Los Angeles, CA*
PIETRO LIÒ • *Computer Laboratory, University of Cambridge, Cambridge, United Kingdom*
TIM MASSINGHAM • *European Molecular Biology Laboratory – European Bioinformatics Institute, Hinxton, Cambridge, United Kingdom*
ALICE CAROLYN MCHARDY • *IBM Thomas J. Watson Research Center, Yorktown Heights, NY*
ILENE KARSCH MIZRACHI • *National Center for Biotechnology Information, National Library of Medicine, National Institutes of Health, Bethesda, MD*
ARFST NICKELSEN • *Institut für Theoretische Informatik, Universität zu Lübeck, Lübeck, Germany*
JOSÉ M. PEREGRÍN-ALVAREZ • *SickKids Research Institute, Toronto, Ontario, Canada*
WALTER PIROVANO • *Centre for Integrative Bioinformatics (IBIVU), VU University Amsterdam, Amsterdam, The Netherlands*
JOHN H. POSTLETHWAIT • *Institute of Neuroscience, University of Oregon, Eugene, OR*
MARK A. RAGAN • *ARC Centre of Excellence in Bioinformatics, and Institute for Molecular Bioscience, The University of Queensland, Brisbane, Queensland, Australia*
JOHN ROBINSON • *School of Mathematics and Statistics and Centre for Mathematical Biology, University of Sydney, Sydney, New South Wales, Australia*
CARMELINDA SAVINO • *CNR-Institute of Molecular Biology and Pathology (IBPM), Department of Biochemical Sciences, University of Rome, "Sapienza," Roma, Italy*
LINCOLN G. SCOTT • *Cassia, LLC, San Diego, CA*
TILL TANTAU • *Institut für Theoretische Informatik, Universität zu Lübeck, Lübeck, Germany*
GLENN TESLER • *Department of Mathematics, University of California, San Diego, La Jolla, CA*
GEORG F. WEILLER • *Research School of Biological Sciences and ARC Centre of Excellence for Integrative Legume Research, The Australian National University, Canberra, Australian Capital Territory, Australia*
SIMON WHELAN • *Faculty of Life Sciences, University of Manchester, Manchester, United Kingdom*
THIERRY WIRTH • *Museum National d'Histoire Naturelle, Department of Systematics and Evolution, Herbier, Paris, France*
YI XING • *Department of Internal Medicine, Carver College of Medicine and Department of Biomedical Engineering, University of Iowa, Iowa City, IA*
SOPHIA YANCOPOULOS • *The Feinstein Institute for Medical Research, Manhasset, NY*

Contents of Volume II

SECTION I: STRUCTURES

1. UNAFold: *Software for Nucleic Acid Folding and Hybridization*
   Nicholas R. Markham and Michael Zuker

2. Protein Structure Prediction
   Bissan Al-Lazikani, Emma E. Hill, and Veronica Morea

3. An Introduction to Protein Contact Prediction
   Nicholas Hamilton and Thomas Huber

4. Analysis of Mass Spectrometry Data in Proteomics
   Rune Matthiesen and Ole N. Jensen

5. The Classification of Protein Domains
   Russell L. Marsden and Christine A. Orengo

SECTION II: INFERRING FUNCTION

6. Inferring Function from Homology
   Richard D. Emes

7. The Rosetta Stone Method
   Shailesh V. Date

8. Inferring Functional Relationships from Conservation of Gene Order
   Gabriel Moreno-Hagelsieb

9. Phylogenetic Profiling
   Shailesh V. Date and José M. Peregrín-Alvarez

10. Phylogenetic Shadowing: Sequence Comparisons of Multiple Primate Species
    Dario Boffelli

11. Prediction of Regulatory Elements
    Albin Sandelin

12. Expression and Microarrays
    Joaquín Dopazo and Fátima Al-Shahrour

13. Identifying Components of Complexes
    Nicolas Goffard and Georg Weiller

14. Integrating Functional Genomics Data
    Insuk Lee and Edward M. Marcotte

SECTION III: APPLICATIONS AND DISEASE

15. Computational Diagnostics with Gene Expression Profiles
    Claudio Lottaz, Dennis Kostka, Florian Markowetz, and Rainer Spang

16. Analysis of Quantitative Trait Loci
    Mario Falchi

17. Molecular Similarity Concepts and Search Calculations
    Jens Auer and Jürgen Bajorath

18. Optimization of the MAD Algorithm for Virtual Screening
    Hanna Eckert and Jürgen Bajorath
19. Combinatorial Optimization Models for Finding Genetic Signatures from Gene Expression Datasets
    Regina Berretta, Wagner Costa, and Pablo Moscato
20. Genetic Signatures for a Rodent Model of Parkinson's Disease Using Combinatorial Optimization Methods
    Mou'ath Hourani, Regina Berretta, Alexandre Mendes, and Pablo Moscato

SECTION IV: ANALYTICAL AND COMPUTATIONAL METHODS

21. Developing Fixed-Parameter Algorithms to Solve Combinatorially Explosive Biological Problems
    Falk Hüffner, Rolf Niedermeier, and Sebastian Wernicke
22. Clustering
    Geoffrey J. McLachlan, Richard W. Bean, and Shu-Kay Ng
23. Visualization
    Falk Schreiber
24. Constructing Computational Pipelines
    Mark Halling-Brown and Adrian J. Shepherd
25. Text Mining
    Andrew B. Clegg and Adrian J. Shepherd

Section I

Data and Databases

Chapter 1

Managing Sequence Data

Ilene Karsch Mizrachi

Abstract

Nucleotide and protein sequences are the foundation for all bioinformatics tools and resources. Researchers can analyze these sequences to discover genes or predict the function of their products. The INSD (International Nucleotide Sequence Database—DDBJ/EMBL/GenBank) is an international, centralized primary sequence resource that is freely available on the internet. This database contains all publicly available nucleotide and derived protein sequences. This chapter summarizes the nucleotide sequence database resources, provides information on how to submit sequences to the databases, and explains how to access the sequence data.

Key words: DNA sequence database, GenBank, EMBL, DDBJ, INSD.

1. Introduction

The International Nucleotide Sequence Database (INSD) is a centralized public sequence resource. As of August 2007, it contains over 101 million DNA sequences comprised of over 181 billion nucleotides, numbers that continue to increase exponentially. Scientists generate and submit their primary sequence data to INSD as part of the publication process. The database is archival and represents the results of scientists' experiments. The annotation, represented by annotated features such as coding regions, genes, and structural RNAs, on the sequence is based on the submitter's observations and conclusions rather than those of the database curators.

2. Sequence Databases

2.1. International Collaboration

The International Nucleotide Sequence Database (INSD) Collaboration is a partnership among GenBank (http://www.ncbi.nlm.nih.gov/Genbank/) NCBI, NLM, NIH, Bethesda, MD, *(1)* EMBL database (http://www.ebi.ac.uk/EMBL/) EBI, Hinxton, United Kingdom, *(2)* and DDBJ (http://www.ddbj.nig.ac.jp/) National Institute of Genetics, Mishima, Japan *(3)*. For over 25 years, GenBank, EMBL, and DDBJ have maintained this active, successful collaboration for building and maintaining nucleotide sequence databases. Representatives from the three databases meet annually to discuss technical and biological issues affecting the databases. Ensuring that sequence data from scientists worldwide is freely available to all is the primary mission of this group. As part of the publication process, scientists are required to deposit sequence data in a public repository; the INSD encourages publishers of scientific journals to enforce this policy to ensure that sequence data associated with a paper are freely available from an international resource. One advantage to depositing sequences in INSD is that they are available electronically, which is far more usable to the scientific community than having the sequences in a printed paper. Consequently, scientists can download a sequence from INSD and analyze the sequences as part of their research.

The three databases have built a shared web site, http://insdc.org (**Fig. 1.1**), which contains information, policies, and procedures that are important for the collaborators, submitters, and users. This site contains:

1. The data release policy
2. Standards for submission of sequences to the nucleotide databases
3. The Feature Table Document, outlining legal features and syntax to be included in the sequence record in order to standardize annotation across the databases (*see* **Note 1**)
4. Links to the three contributing databases' websites, where submitters can find information regarding submission and sequence retrieval that is specific to each database

Each of the three INSD contributors has its own set of submission (*see* **Section 4.1**) and retrieval tools (*see* **Note 2**). Although scientists may submit data to any of the three databases for inclusion in INSD, data processed at each site are exchanged daily so that a sequence submitted to any of the three databases will be retrievable from all three sites. In order to avoid entries that are out of sync at the three sites, submitters must update their submissions only at the site where they initially submitted the data. Updates are also propagated to the other two sites through the

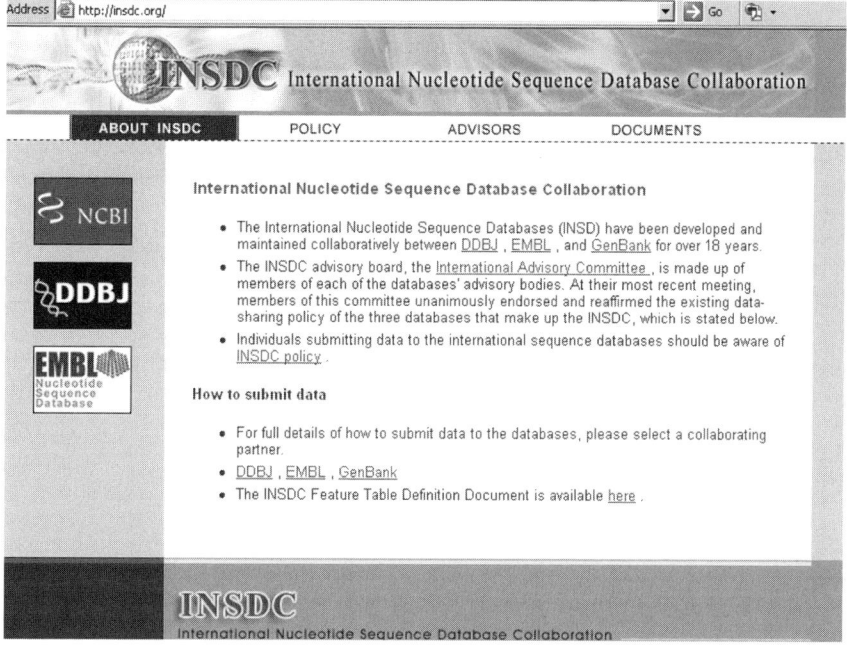

Fig. 1.1. Home page for International Nucleotide Sequence Database Collaborators (http://www.insdc.org).

daily exchange mechanism, just as are new submissions. Although the sequence data and all of the information contained within a sequence record are the same, each site presents the data in a slightly different format.

2.2. The NCBI RefSeq Project

The Reference Sequence (RefSeq) database *(4)* at NCBI (http://www.ncbi.nlm.nih.gov/RefSeq/) provides a curated non-redundant collection of genomic DNA, transcript (RNA), and protein sequences from a number of biologically significant organisms. RefSeq entries are not part of INSD but are derived from INSD records. RefSeq entries can be retrieved from Entrez (*see* **Section 10.2**) and the NCBI ftp site (*see* **Section 10.1**). They differ from primary INSD data in that each RefSeq record is a synthesis of information from a number of sources, so it is not a piece of primary research data itself. RefSeq curators use information from the literature and other sources to create a curated view of each sequence. At present, not all RefSeq records represent a fully curated "review" of a particular gene or transcript and are therefore labeled in the COMMENT (*see* **Section 7**) section of the flat file. Entries labeled as REVIEWED have been curated by a member of the NCBI staff or a collaborator who reviewed additional available sequence and the literature to expand the annotation and sequence of the record. This information is included as links to other sources: published articles, the Entrez Gene database, and organism-specific databases. VALIDATED RefSeq records have undergone an initial review and are still subject to a final review. The annotation in PROVISIONAL RefSeq records has not yet been subject to curator review, yet there is good evidence that the annotation represents valid transcripts and proteins. PREDICTED RefSeq records contain transcript and protein annotation that represents *ab initio* predictions, which may or may not be supported by transcript evidence. The RefSeq collection, especially the REVIEWED RefSeq records, provides an excellent view into an organism's genome.

3. Types of INSD Sequence Records

3.1. Direct Submissions

INSD processes thousands of new sequence submissions per month from scientists worldwide. The typical INSD submission consists of a single, contiguous stretch of DNA or RNA sequence with annotations which represent the biological information in the record. The submissions come from a variety of submitters, from small laboratories doing research on a particular gene or genes, to genome centers doing high-throughput sequencing. Many sequences that come from large sequencing centers undergo automated bulk submission processing with very little annotator review. (These types of submissions are described later.) The "small-scale" submissions may be a single sequence or sets of related sequences, and they usually contain annotation. Submissions include mRNA sequences with coding regions, fragments of genomic DNA with a single gene or multiple genes, or ribosomal RNA gene clusters. If part of the nucleotide sequence encodes a protein, a coding sequence (CDS) and resulting conceptual translation are annotated. A protein accession number (`/protein_id`) is assigned to the translation product.

Multiple sequences can be submitted and processed together for GenBank as a set. Groups of sequences may also be submitted to DDBJ and EMBL but they do not use the same set concept as does GenBank. With the Sequin submission tool (which is further described later), submitters can specify that the sequences are biologically related by classifying them as environmental sample, population, phylogenetic, or mutation sets. Environmental sample, population, phylogenetic, and mutation sets all contain a group of sequences that span the same gene or region of the genome. Members of environmental sets are unclassified or unknown organisms. Population sets contain sequences from different isolates of the same organism, whereas phylogenetic sets contain sequences from related organisms. The multiple mutations of a single gene from a single species can be submitted as a mutation set. Each sequence within a set is assigned its own accession number and can be viewed independently in Entrez. However, each set is also indexed within the PopSet division of Entrez, allowing scientists to view the relationship among the set's sequences through an alignment.

3.2. EST/STS/GSS

Expressed Sequence Tags (ESTs), Sequence Tagged Sites (STSs), and Genome Survey Sequences (GSSs) are generally submitted in a batch and are usually part of a large sequencing project devoted to a particular genome. These entries have a streamlined submission process and undergo minimal processing before being released to the public. EST sequences (http://www.ncbi.nlm.nih.gov/dbEST/) are single-pass cDNA sequences from a particular tissue and/or developmental stage and are commonly used by genome annotation groups to place predicted genes on the genome.

STS sequences (http://www.ncbi.nlm.nih.gov/dbSTS/) are short genomic landmark sequences that define a specific location on the genome and are, therefore, useful for mapping.

GSS sequences (http://www.ncbi.nlm.nih.gov/dbGSS/) are short sequences derived from genomic DNA and include, but are not limited to, single-pass genome sequences, BAC ends, and exon-trapped genomic sequences. Like EST sequence records, GSS sequence records do not contain annotation but can be useful for genome mapping or annotation. EST, STS, and GSS sequence records reside in their own respective divisions within INSD, rather than in the taxonomic division of the organism.

3.3. Mammalian Gene Collection

The Mammalian Gene Collection (MGC; http://mgc.nci.nih.gov/) is a trans-NIH initiative that provides full-length open reading frame (FL-ORF) clones for human, mouse, cow, and rat genes. The sequences for the MGC clones can be found in INSD, and the clones that correspond to each of the sequences can be purchased for further study. As of October 2006, there were over 24,000 human full ORF clones, which represent about 14,400 non-redundant genes. In Entrez, users can link from the GenBank view of the MGC sequence to other biological resources at NCBI: the Gene database to see information about the gene structure; MapViewer to see placement of the clone in the genome; dbSNP to see the polymorphisms that exist for the gene; and OMIM, the catalog of human genes and genetic disorders. Investigators can then obtain a clone that they have identified as interesting by computational study, to perform laboratory experiments and learn more about the gene in question.

3.4. Microbial Genomes

INSD has received and released more than 300 complete microbial genomes since 1996. These genomes are relatively small in size compared with their eukaryotic counterparts, ranging from 500,000 to 5 million bases. Nonetheless, these genomes contain thousands of genes, coding regions, and structural RNAs.

The database staff reviews the annotation on the genomes to confirm that it is consistent with the annotation standards. Many of the genes in microbial genomes have been identified only by similarity to genes in other genomes, and their gene products are often classified as hypothetical proteins. Each gene within a microbial genome is assigned a locus tag, a unique identifier for a particular gene in a particular genome. Since the function of many of the genes is unknown, locus tag names have become surrogate gene names. To ensure that a locus tag is unique for a particular gene, submitters must register with DDBJ, EMBL, or GenBank for a unique locus tag prefix for each genome being submitted.

3.5. High Throughput Genomic Sequence

The High Throughput Genomic (HTG) sequence division (http://www.ncbi.nlm.nih.gov/HTGS/) was created to accommodate a growing need to make unfinished, clone-based genomic sequence data rapidly available to the scientific community. HTG entries are submitted in bulk by genome centers, processed by an automated system, and then immediately released to the public database. HTG sequences are submitted at differing levels of completion (**Fig. 1.2**). Phase 0 sequences are one-to-few reads of a single clone and are not usually computationally assembled by overlap into larger contiguous sequences (contigs). Phase 1 entries are assembled into contigs that are separated by sequence gaps and whose relative order and orientation are not known. Phase 2 entries are also assembled unfinished sequences that may or may not contain sequence gaps. If there are gaps, then the contigs are in the correct order and orientation. Phase 3 sequences are of finished quality and have no gaps. HTG sequences may be annotated. The human genome was completed using the clone-based HTG sequencing method.

3.6. Whole Genome Shotgun

In 2001, a new approach for sequencing complete genomes was introduced: Whole Genome Shotgun (WGS) sequencing (http://www.ncbi.nih.gov/Genbank/wgs.html). This has become such a dominant sequencing technique that more nucleotides of WGS sequence have been added to INSD over the past 6 years than from all of the other divisions since the inception of the database. Rather than using traditional clone-based sequencing technology,

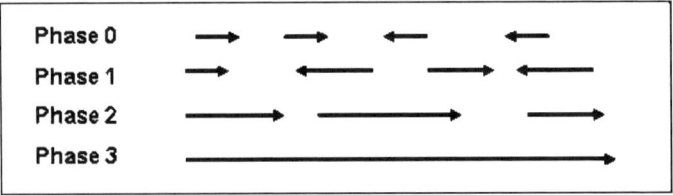

Fig. 1.2. Diagram showing the orientation and gaps that might be expected in high throughput sequence from phases 0, 1, 2, and 3.

as is done for HTG, WGS sequences are generated by breaking the genome into random fragments and sequencing them. Multiple overlapping reads of sequence are then computationally assembled to form a contig. Contig sequences can be assembled into larger structures called scaffolds. Scaffold records, which are in the CON division, contain information regarding the relative positions of contigs in the genome; they can be simply multiple contigs connected by gaps or they could be entire chromosomes built from contigs and gaps. Gaps between contigs may be of known or unknown length. All of the contig sequences from a single genome assembly project, along with the instructions for building scaffolds, are submitted together as a single WGS project. Like HTG, these sequences can be submitted with or without annotation. As sequencing progresses and new assemblies are computed, a new version is submitted to INSD that supersedes the previous version. Although there is no tracking of sequences between assemblies, sequences from prior assemblies are always retrievable by accession number or gi in Entrez.

A WGS project accession number has a different format from other INSD accession numbers. It is composed of a four-letter project ID code, a two-digit assembly version number, and a six- to eight-digit contig ID. For example, the *Neurospora crassa* WGS project was assigned project accession AABX00000000, the first version of the genome assembly is AABX01000000 and AABX01000111 is the 111th contig of the WGS project assembly version 1. For each project, a master record (**Fig. 1.3**) is created that contains information that is common among all the records of the sequencing projects, such as the biological source, submitter information, and publication information. Each master record includes links to the range of accession numbers for the individual contigs in the assembly and links to the range of accessions for the scaffolds from the project.

3.7. Third-Party Annotation

The vast amount of publicly available data from the human genome project and other genome sequencing efforts is a valuable resource for scientists throughout the world. Although a laboratory studying a particular gene or gene family has sequenced numerous cDNAs; it may have neither the resources nor inclination to sequence large genomic regions containing the genes, especially when the sequence is available in public databases. Instead, researchers might choose to download genomic sequences from INSD and perform analyses on these sequences. However, because the researchers did not perform the sequencing, the sequence with its new annotations cannot be submitted to INSD, excluding potentially important scientific information from the public databases. To address this problem, the INSD established a dataset for Third Party Annotation (TPA; www.ncbi.nlm.nih.gov/Genbank/tpa.html). All sequences in TPA are derived from the publicly available

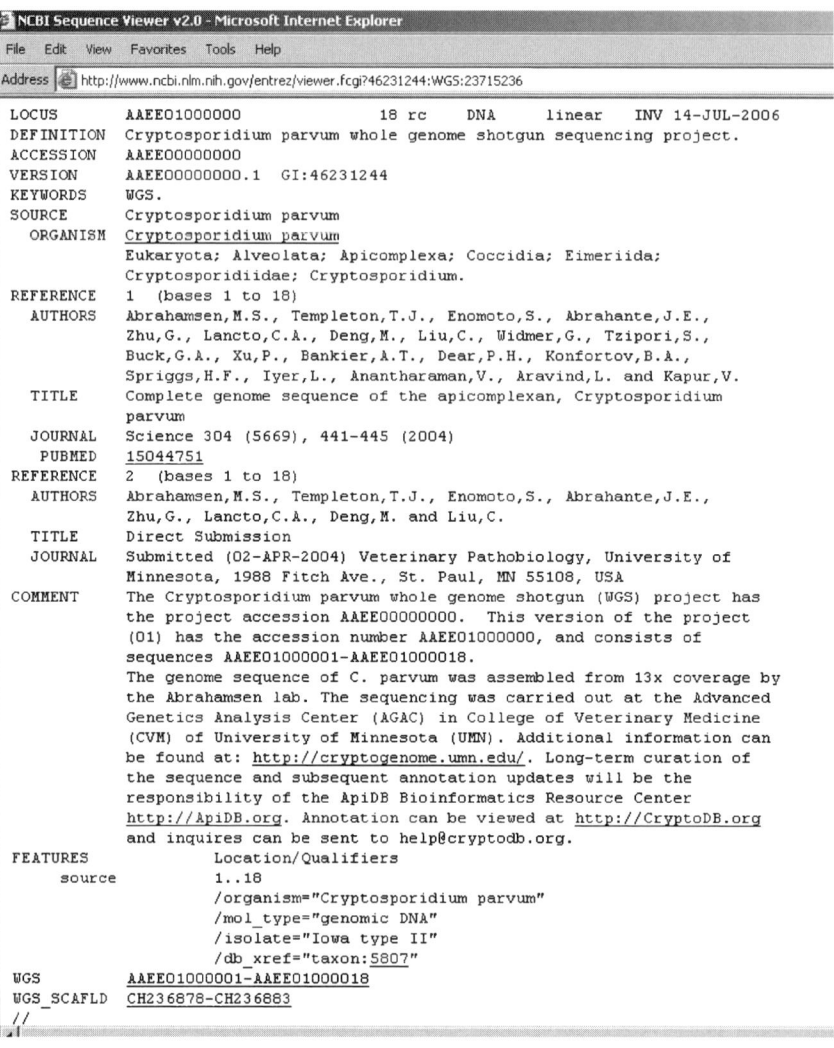

Fig. 1.3. WGS master sequence record that contains two new line types: WGS, which points to the component contig records, and WGS_SCAFLD, which points to the scaffold records.

collection of sequences in INSD and/or the Trace Archive (*see* **Note 3**). Researchers can submit both new and alternative annotation of genomic sequence to INSD as TPA records. Additionally, the TPA dataset contains an mRNA sequence created either by combining exonic sequences from genomic sequences or assembling overlapping EST or other mRNA sequences. TPA sequences are released to the public database only when their accession numbers and/or sequence data appear in a peer-reviewed publication in a biological journal.

There are two categories of data in the TPA repository, TPA:experimental and TPA:inferential. To be in the experimental category of TPA, a record's annotation must be supported by wet-lab experimental evidence. The annotation of sequences in inferential TPA is derived by inference rather than direct experimentation. Sequences in this category of TPA may be members of gene families in which the annotation of homologous family members was determined experimentally by the submitter, and the annotation of the TPA:inferential sequence is based on a comparison to this experimentally derived sequence. Additionally, a complete annotated organellar or viral genome, in which the annotated features may be a mix of experimentally and inferentially determined data, assembled from sequences in the primary databases may be submitted as a TPA:inferential record.

4. Submission of Sequence Data to INSD

4.1. Web-Based Submission Tools

DDBJ, EMBL, and GenBank each have a web-based submission tool in which submitters complete a series of forms with information about the sequences and annotation. DDBJ's web-based submission tool is called SAKURA (http://sakura.ddbj.nig.ac.jp/), EMBL's web-based submission tool is called Webin (http://www.ebi.ac.uk/embl/Submission/webin.html), and GenBank's web-based submission tool is called BankIt (http://www.ncbi.nlm.nih.gov/BankIt/). Each database also has additional submission systems for the submission of large datasets, batch submissions, and genome submissions.

Sakura, Webin, and BankIt are convenient ways to submit a small number of sequences with simple annotation. Users are prompted to enter submitter information, the nucleotide sequence, biological source information, and features and annotation pertinent to the submission. All of these tools have help documentation to guide submitters and sets of annotation examples that detail the information that is required for each type of submission.

These web-based tools have quality assurance and validation checks and report the results back to the submitter for resolution. For example, in BankIt, a BLAST similarity search compares the sequence to the UniVec (http://www.ncbi.nlm.nih.gov/VecScreen/) database to prevent the deposition of sequences that still contain cloning vector sequence. Before submitting their records to the database, submitters have the opportunity to review their entries to confirm that the information included is necessary and correct.

Webin and Sakura submitters can access previous submissions in order to resume partially completed submissions or copy information from a previous submission to a new submission by entering the ID and password from the previous submission. GenBank is developing the next generation of BankIt to allow submitters to set up a login so that they can access information from previous submissions, among other things.

5. Sequin

For submitters who prefer a non web-based tool or have large or complex submissions, Sequin (http://www.ncbi.nlm.nih.gov/Sequin/index.html) is a stand-alone application that can be used for the annotation and analysis of nucleotide sequences. It is available on NCBI's ftp site: ftp://ftp.ncbi.nih.gov/sequin/ (*see* **Note 4**) and is the preferred submission tool for complex submissions to GenBank that contain a significant amount of annotation or many sequences. Although DDBJ and EMBL accept submissions generated by Sequin, they prefer to receive submissions generated by Sakura or Webin, respectively. Sequin uses a series of wizards to guide submitters through preparing their files for submission. Such files include: nucleotide and amino acid sequences in FASTA format (*see* **Note 5**), spreadsheets for source organism information (*see* **Note 8a**) and tables for feature annotation (*see* **Note 8c**). For submitting multiple, related sequences (e.g., those in a phylogenetic or population study), Sequin accepts the output of many popular multiple sequence-alignment packages, including FASTA+GAP, PHYLIP, and NEXUS. With a sequence alignment, submitters can annotate features on a single record and then propagate these features to the other records instead of annotating each record individually. Prior to submission to the database, the submitter is encouraged to validate his or her submission and correct any error that may exist (*see* **Note 6**). Completed Sequin submissions are submitted by e-mailing them to DDBJ, EMBL, or GenBank or by uploading them directly to GenBank using SequinMacroSend (http:/www.ncbi.nlm.nih.gov/LargeDirSubs/dir_submit.cgi).

5.1. tbl2asn

tbl2asn (http://www.ncbi.nlm.nih.gov/Genbank/tbl2asn2.html) is a command-line program that automates the creation of sequence records for submission to GenBank using many of the same functions as Sequin. It is used primarily for submission of complete genomes and large batches of sequences in which the detailed tools in Sequin are not necessary. tbl2asn requires a template file, plus FASTA sequence files (*see* **Note 5**), and tab-delimited five-column table files (*see* **Note 8c**) with the annotation. The program reads these files to create a submission that can be validated (*see* **Note 6**) and submitted to GenBank. Tbl2asn is available by ftp from the NCBI ftp site (*see* **Note 4**): ftp://ftp.ncbi.nih.gov/toolbox/ncbi_tools/converters/by_program/tbl2asn/. This program, like Sequin, can be downloaded to be run in a number of different operating systems.

6. The GenBank Sequence Record

A GenBank sequence record is most familiarly viewed as a flat file with the information being presented in specific fields. DDBJ and GenBank flat files are similar in design. The EMBL flat file format, however, is different in design but identical in content to flat files from GenBank and DDBJ.

6.1. Definition, Accession, and Organism

Here is the top of the GenBank flat file, showing the top seven fields.

```
LOCUS       AF123456 1510bp mRNA linear VRT
            25-JUL-2000
DEFINITION  Gallus gallus doublesex and mab-3
            related transcription factor 1
            (DMRT1) mRNA, partial cds.
ACCESSION   AF123456
VERSION     AF123456.2 GI:6633795
KEYWORDS    .
SOURCE      Gallus gallus (chicken)
  ORGANISM  Gallus gallus
            Eukaryota; Metazoa; Chordata;
            Craniata; Vertebrata; Euteleostomi;
            Archosauria; Aves; Neognathae;
            Galliformes; Phasianidae;
            Phasianinae; Gallus
```

The first token of the LOCUS field is the locus name. At present, this locus name is the same as the accession number, but in the past, more descriptive names were used. For instance, HUMHBB is the locus name for the human beta-globin gene in the record with accession number U01317. With the increase in the number of redundant sequences over time, the generation of descriptive locus names was abandoned. Following the locus name is the length of the sequence, molecule type of the sequence, topology of the sequence (linear or circular), GenBank taxonomic or functional division, and date of the last modification. The DEFINITION line gives a brief description of the sequence including information about the source organism, gene(s), and molecule information. The ACCESSION is the database-assigned accession number, which has one of the following formats: two letters and six digits or one letter and five digits for INSD records; four letters and eight digits for WGS records; and two letters, an underscore, and six to eight digits for RefSeq records. The VERSION line contains the sequence version and the gi, a unique numerical identifier for that particular sequence. A GenBank record may or may not have KEYWORDS. Historically, the KEYWORD field in the GenBank record was used as a summary of the information present in the record. It was a free text field and may have contained gene name, protein name, tissue localization, and so on. Information placed on this line was more appropriately placed elsewhere in the sequence record. GenBank strongly discourages the use of keywords to describe attributes of the sequence. Instead, GenBank uses a controlled vocabulary on the KEYWORD line to describe different submission types or divisions (*see* **Note 7**).

The SOURCE and ORGANISM contain the taxonomic name and the taxonomic lineage, respectively, for that organism.

7. Reference Section

The next section of the GenBank flat file contains the bibliographic and submitter information:

```
REFERENCE   1 (bases 1-1,510)
AUTHORS     Nanda, I., Shan, Z.H.,
            Schartl, M., Burt, D.V., Koehler, M.,
            Nothwang, H.-G., Gruetzner, F.,
            Paton, I.R., Windsor, D., Dunn, I.,
            Engel, W., Staeheli, P.,
            Mizuno, S., Haaf, T., and Schmid, M.
TITLE       300 million years of conserved
            synteny between chicken Z and
            human chromosome 9.
JOURNAL     Nat Genet 21(3):258-259 (1999)
PUBMED      10080173
...
REFERENCE   3 (bases 1-1,510)
AUTHORS     Haaf, T. and Shan, Z.H.
TITLE       Direct Submission
JOURNAL     Submitted (25-JAN-1999) Max-Planck
            Institute for Molecular Genetics,
            Ihnestr. 73, Berlin 14195, Germany
COMMENT     On Dec 23, 1999 this sequence
            version replaced gi:4454562.
```

The REFERENCE section contains published and unpublished references. Many published references include a link to a PubMed ID number that allows users to view the abstract of the cited paper in Entrez PubMed. The last REFERENCE cited in a record reports the names of submitters of the sequence data and the location where the work was done. The COMMENT field may have submitter-provided comments about the sequence. In addition, if the sequence has been updated, then the COMMENT will have a link to the previous version.

7.1. Features and Sequence

The FEATURES section contains the source feature, which has additional information about the source of the sequence and the organism from which the DNA was isolated. There are approximately 50 different standard qualifiers that can be used to describe the source. Some examples are /strain, /chromosome, and /isolation_source.

```
FEATURES     Location/Qualifiers
  source     1..1510
             /organism="Gallus gallus"
             /mol_type="mRNA"
             /db_xref="taxon:9031"
             /chromosome="Z"
             /map="Zp21"
  gene       <1..1510
             /gene="DMRT1"
             /note="expressed in genital ridges
             of both males and
             females prior to the sexual
             determination, restricted to
             the male gonads after sex
             determination; DMRT1 is a
             candidate sex determination factor
             in vertebrates"
  CDS        <1..936
             /gene="DMRT1"
             /note="cDMRT1"
             /codon_start=1
             /product="doublesex and mab-3
             related transcription factor 1"
             /protein_id="AAF19666.1"
             /db_xref="GI:6633796"
             /translation="PAAGKKLPRLPKCARCRNHGYS
             SPLKGHKRFCMWRDCQCKKCSL
             IAERQRVMAVQVALRRQQAQEEELGISHPVPLP-
             SAPEPVVKKSSSSSSCLLQDSSSPA
             HSTSTVAAAAASAPPEGRMLIQDIPSIPSRGH-
             LESTSDLVVDSTYYSSFYQPSLYPYY
             NNLYNYSQYQMAVATESSSSETGGTFVGSAM-
             KNSLRSLPATYMSSQSGKQWQMKGMEN
             RHAMSSQYRMCSYYPPTSYLGQGVGSPTCVTQI-
             LASEDTPSYSESKARVFSPPSSQDS
             GLGCLSSSESTKGDLECEPHQEPGAFAVSPVLEGE"
ORIGIN
    1 ccggcggcgg gcaagaagct gccgcgtctg
      cccaagtgtg cccgctgccg caaccacggc
   61 tactcctcgc cgctgaaggg gcacaagcgg
      ttctgcatgt ggcgggactg ccagtgcaag
```

```
121 aagtgcagcc tgatcgccga gcggcagcgg
    gtgatggccg tgcaggttgc actgaggagg
```
...

Following the source feature are features that describe the sequence, such as gene, CDS (coding region), mRNA, rRNA, variation, and others. Like the source feature, other features can be further described with feature-specific qualifiers. For example, an important qualifier for the CDS feature is a `/translation` that contains the protein sequence. Following the Feature section is the nucleotide sequence itself.

7.2. GenBank Divisions

GenBank records are grouped into 20 divisions; the majority of these are taxonomic groupings. The other divisions group sequences by a specific technological approach, such as HTG or GSS. Sequences in the technique-based divisions often have a specific keyword in the record (*see* **Note 7**).

These GenBank divisions are:

1. PRI: Primate sequences
2. ROD: Rodent sequences
3. MAM: Other mammalian sequences
4. VRT: Other vertebrate sequences
5. INV: Invertebrate sequences
6. PLN: Plant, fungal, and algal sequences
7. BCT: Bacterial sequences
8. VRL: Viral sequences
9. PHG: Bacteriophage sequences
10. SYN: Synthetic sequences
11. UNA: Unannotated sequences
12. EST: Expressed sequence tags
13. PAT: Patent sequences
14. STS: Sequence tagged sites
15. GSS: Genome survey sequences
16. HTG: High throughput genomic sequences
17. HTC: High throughput cDNA sequences
18. ENV: Environmental sampling sequences
19. TPA: Third Party Annotation
20. CON: Constructed entries

8. INSD Sequence Processing

8.1. New Submissions

Direct submissions to INSD are analyzed and validated in a multiple-step process by the annotation staff at each of the three sites. The first level of review occurs before accession numbers are assigned to ensure that the submissions meet the minimal criteria to be accepted into INSD. Sequences should be >50 bp in length and be sequenced by, or on behalf of, the group submitting the sequence. They must also represent molecules that exist in nature (not a consensus sequence or a mix of genomic or mRNA sequence). Submissions are also confirmed to be new data rather than updates to sequences submitted previously by the same submitter.

Once sequences receive accession numbers, they are reviewed more extensively by the annotation staff, who use a variety of tools to annotate and validate the sequence and annotation data. At GenBank, a more robust version of the submission tool Sequin is used by the annotation staff to edit sequence records. Global editing allows the annotation staff to annotate large sets quickly and efficiently. Submissions are stored in a database that is accessed through a queue management tool, which automates some of the processing steps before staff review. These steps include: confirming taxonomic names and lineages, confirming bibliographic information for published references, starting BLAST searches, and running automatic validation checks. Hence, when an annotator is ready to work on a record, all of this information is immediately accessible. In addition, all of the correspondence between GenBank staff and submitters is stored in the database with the record.

The GenBank annotation staff checks all submissions for:

- Biological validity: Does the conceptual translation of a coding region match the amino acid sequence provided by the submitter? Is the source organism name present in NCBI's taxonomy database? Does the submitter's description of the sequence agree with the results of a BLAST similarity search against other sequences?

- Vector contamination: Sequences are screened against NCBI's UniVec to detect contaminating cloning vector sequence.

- Publication status: If there is a published reference, a PubMed ID is added to the record so that the sequence and publication records can be linked in Entrez.

- Formatting and spelling

Similar procedures are also carried out at DDBJ and EMBL.

If there are problems with the sequence or annotation, the annotator works with the submitter by e-mail to correct the problems. Using Sequin, the annotator can incorporate corrected sequence and/or annotation files from submitters during processing. Completed entries are sent to the submitter for a final review before their release into the public database. Submitters may request that INSD hold their sequence until a specified future release date. Records will be held until that date or when the accession number or the sequence is published, whichever is first.

The GenBank annotation staff currently processes about 2,200 submissions per month, corresponding to approximately 40,000 sequences.

8.2. Processing of Batch Submissions: HTG, WGS, and Genomes

In contrast to the processing of "direct submissions," genome submissions, including WGS and HTG, go through a more automated process. HTG submissions are deposited by the submitter in an FTP directory, and a number of automated validation checks are performed. If there are no problems with the sequences, then they are released directly into the public database without any manual review. If there are problems, the sequences are flagged so that a GenBank annotator can review the problem and advise the submitter to deposit a corrected version of the record. GenBank annotators do not make the modifications for the submitters; they require that the submitters make the change and resubmit the sequence. This strategy forces submitters to make the appropriate changes in their own database, so that problems are corrected when they next update these records.

Genome submissions and WGS submissions are processed by GenBank annotators using a number of scripts and tools to review the annotation in bulk. Often these types of submissions contain thousands of features, and manual review of each feature is not feasible.

8.3. Updates and Maintenance of the Database

An INSD record can be updated by the submitter any time new information is acquired. Updates can include: adding a new sequence, correcting an existing sequence, adding a publication, or adding a new annotation (*see* **Note 8**). The new, updated record replaces the older one in the database and retrieval and analysis tools. However, because INSD is archival, a copy of the older record is maintained in the database. If the sequence has changed as part of the update, the version (accession.version) gets incremented and the gi changes. For example, the sequence in AF123456.1 was updated on December 23, 1999 and became AF123456.2. A COMMENT is added to the updated GenBank flat file that indicates when the sequence is updated and provides a link to the older version of the sequence.

```
COMMENT  On Dec 23, 1999 this sequence
         version replaced gi: 4454562.
```

However, if the sequence does not change during the update, the accession.version and the gi do not change. Regardless of the actual update, users can retrieve older versions of the record using the Sequence Revision History tool (**Fig. 1.4**) from Entrez (http://www.ncbi.nlm.nih.gov/entrez/sutils/girevhist.cgi).

If sequence records are released to the database and then later found to include misidentified source organisms or contaminated

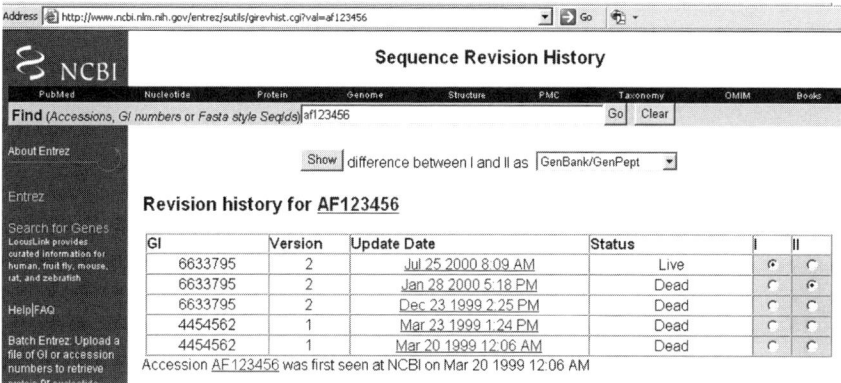

Fig. 1.4. Sequence Revision History page allows users to retrieve older versions of a sequence record prior to it being updated. Sequence changes are indicated by incrementing the version number. This page also allows users to show differences between two different versions of a sequence record by choosing a version in columns I and II and then clicking the Show button.

sequence data, the records can be removed from distribution to the public databases. When removed, the records are no longer available for text-based queries in the retrieval tools of the INSD databases. However, they still remain retrievable by accession number so that scientists who previously worked with this sequence data still may access them. In Entrez, a comment is added to the record to indicate why it was removed from distribution so that any scientist who has used its data is alerted that information in it may be compromised. This removal mechanism is also used for submitters who request that we hold their sequences confidential after they have been released. This situation can occur when sensitive data has been released publicly before a corresponding paper has been published.

9. Pitfalls of an Archival Primary Sequence Database

9.1. Bad Annotation and Propagation

Because INSD is an archival primary sequence database, submitters "own" their annotation and are primarily responsible for ensuring that it is correct. Database staff review records for accuracy and inform submitters of any problems. However, if submitters do not agree that this analysis is accurate, they do not need to make any changes to their records. If entries with poor annotation appear in the database, they may be used by other scientists to validate their own data and possibly prepare a submission to the database. This may lead to the propagation of bad data to subsequent entries in the database.

Other submitter-based sources of error result from segments of the database that are not manually reviewed, as are the direct submissions. These segments are some of the bulk sequences such as HTG, GSS, EST, and WGS, which undergo automated processing. The sheer volume of data that is submitted to these data streams precludes a manual check of all the sequences. The automated checks in this bulk processing are intended to catch the most egregious problems. However, mis-annotation of genes, for instance, could slip through this processing undetected.

The NCBI RefSeq project was created in part to deal with this problem. Entries in RefSeq (especially those that are flagged REVIEWED) have undergone some curation to assure that the annotation of the sequence in question is correct. In addition to curation by the database staff, there is a mechanism for users and experts in the scientific community to add annotation to these records using GeneRif (http://www.ncbi.nlm.nih.gov/projects/GeneRIF/GeneRIFhelp.html). Scientists can add comments and links to published scientific papers that enhance or confirm the annotation.

9.2. Vector Contamination

The GenBank staff actively removes vector contamination from sequence submissions when it is discovered. Our method for detecting and correcting such data is to use a specialized BLAST database that has been developed to check for vector contamination. We screen all direct submissions against the UniVec (**Fig. 1.5**) http://www.ncbi.nlm.nih.gov/VecScreen/VecScreen.html database, remove the vector contamination from any submission in which it is detected, and report the vector's discovery and removal to the submitters. Although the GenBank submissions staff screens all incoming submissions against the UniVec database, there are a number of new cloning vector sequences and linkers that are not yet represented in the database. Therefore, it is still possible that sequences in the database contain some vector contamination. UniVec is being updated to address this problem.

9.3. Annotation on the Wrong Strand

When sequences contain coding regions, it is often easy to confirm the strandedness of the gene on the nucleotide sequence by determining the conceptual translation. This translation can then be run through a BLAST similarity search to determine the correct reading frame or sequence strand for the coding region. Unfortunately, if the gene is a structural RNA, that determination cannot be made as easily. Because ribosomal RNA sequences have been submitted to INSD on both strands, BLAST similarity search results cannot clearly determine the strandedness of the ribosomal RNA gene on the sequence. To overcome this problem, INSD checks sequences against a ribosomal RNA BLAST

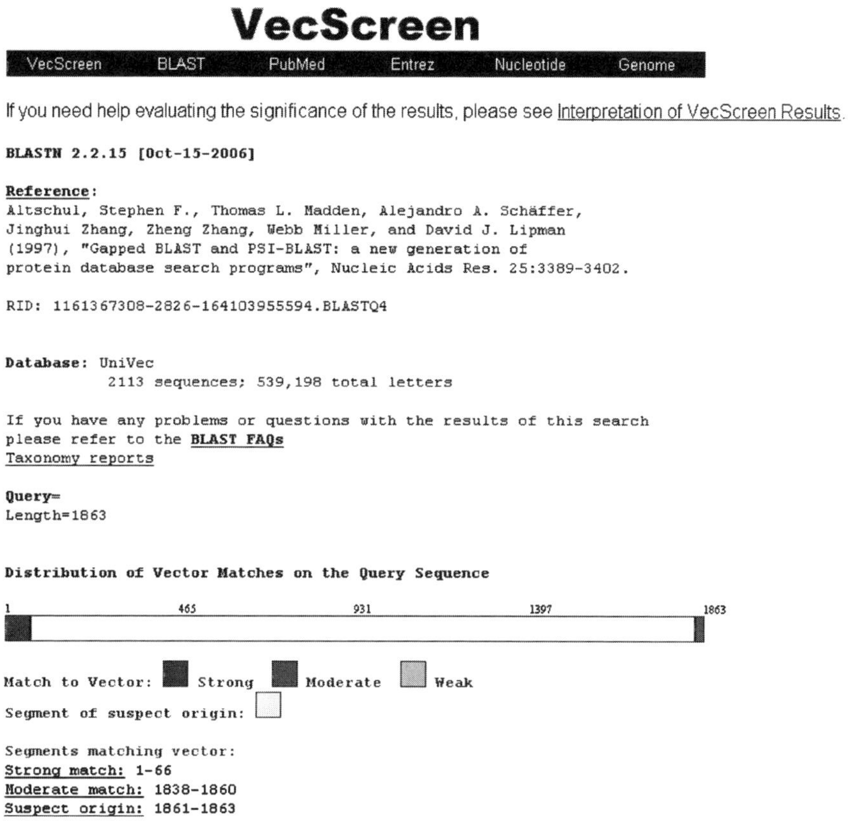

```
                              VecScreen
   VecScreen     BLAST      PubMed      Entrez     Nucleotide     Genome

If you need help evaluating the significance of the results, please see Interpretation of VecScreen Results.

BLASTN 2.2.15 [Oct-15-2006]

Reference:
Altschul, Stephen F., Thomas L. Madden, Alejandro A. Schäffer,
Jinghui Zhang, Zheng Zhang, Webb Miller, and David J. Lipman
(1997), "Gapped BLAST and PSI-BLAST: a new generation of
protein database search programs", Nucleic Acids Res. 25:3389-3402.

RID: 1161367308-2826-164103955594.BLASTQ4


Database: UniVec
          2113 sequences; 539,198 total letters

If you have any problems or questions with the results of this search
please refer to the BLAST FAQs
Taxonomy reports

Query=
Length=1863


Distribution of Vector Matches on the Query Sequence

1              465           931           1397        1863


Match to Vector:  ■ Strong  ■ Moderate  □ Weak
Segment of suspect origin: □

Segments matching vector:
Strong match: 1-66
Moderate match: 1838-1860
Suspect origin: 1861-1863
```

Fig. 1.5. Graphical display of results for VecScreen BLAST analysis that shows that the query sequence has cloning vector sequence spanning nucleotides 1–66 and 1838–1860. The sequence alignments showing the similarity are also present on this output page (not shown).

database in which all of the RNA genes are present on the plus strand. Thus, the annotators can be sure that new ribosomal RNA submissions are released to INSD with the RNA on the plus strand.

10. Accessing Sequence Data

10.1. FTP

DDBJ, EMBL, and GenBank all have periodic releases of the sequences in the database. The sequences are available for download by ftp from each of the three sites (ftp://ftp.ddbj.nig.ac.jp/database/ddbj/, ftp://ftp.ebi.ac.uk/pub/databases/embl/, and ftp://ftp.ncbi.nih.gov/genbank/). In addition, NCBI makes available a complete RefSeq release (ftp://ftp.ncbi.nih.gov/refseq/release/). Users can download the complete release and search the databases at their own site.

10.2. Entrez Retrieval System

NCBI has a single search and retrieval system for all of the databases that it maintains. This system is called Entrez *(4)* (http://www.ncbi.nlm.nih.gov/gquery/gquery.fcgi). Users can query an individual database or all of the databases with a single query. The number of sequences in Entrez Nucleotide and Entrez Protein continue to grow at an exponential rate. The Entrez Nucleotide database is a collection of sequences from several sources, including INSD and RefSeq. The Entrez Nucleotide database has divided into three subsets due to the rapid growth of the database. Recently, the EST and GSS divisions of GenBank have been separated from the rest of the sequences that are contained in the CoreNucleotide subset. The Entrez Protein database is a collection of sequences from a variety of sources, including SwissProt, PIR, PRF, PDB, and translations from annotated coding regions in INSD and RefSeq. One significant advantage to the Entrez retrieval system is the network of links that join entries from each of the databases. A nucleotide sequence record can have links (**Fig. 1.6**) to the Taxonomy, PubMed, PubMed Central, Protein, Popset, Unigene, Gene, and Genome databases. In addition to those links, a protein record will also have links to Related Structures, Conserved Domains, and Related Domains. All of these links can be accessed through a pull-down Links menu. In addition, the GenBank flat file also displays hyperlinks to the Taxonomy and PubMed databases. Links to external databases can be made by LinkOut or by db_xrefs within the entry. By taking advantage of these links, users

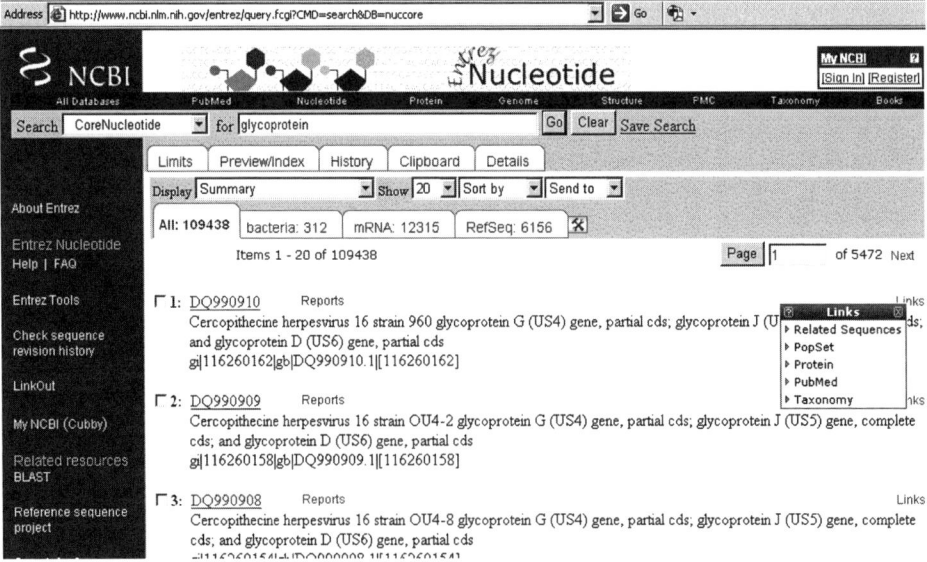

Fig. 1.6. Results page from Entrez Nucleotide Database query. The Links menu shows list of links to Related Squences, Entrez PopSet, Entrez Protein, PubMed, and Entrez Taxonomy from DQ988166, the first sequence retrieved by this query. This Links assist with the navigation through the NCBI site for additional information related to sequences.

can make important scientific discoveries. These links are critical to discovering the relationship between a single piece of data and the information available in other databases.

In Entrez, sequence data can be viewed in a number of different formats. The default and most readable format is the GenBank flat file view. The graphical view, which eliminates most of the text and displays just sequence and biological features, is another display option. Other displays of the data, for instance XML, ASN.1, or FASTA formats, are intended to be more computer readable.

11. Conclusion

Managing the constantly increasing amount of nucleotide and protein sequence data determined from different sources at different institutions requires the ability to accept, process, and display a standard, computable form of this data. To achieve this, DDBJ, EMBL, and GenBank work in collaboration (INSD) to collect and validate sequence data and make it easily available through public web sites. INSD defines standard elements for the sequence data records to ensure that the information can be submitted to and obtained from any of the collaborators' web sites consistently, even when different tools are used to collect or display the data. The accuracy of the sequence data is confirmed by validation steps at both the submission and processing stages. This consistency and accuracy are maintained not only for typical nucleotide sequence data, but also for specialized data, such as EST, GSS, STS, HTG, complete genome, and Third Party Annotation sequences. In addition, because sequence data are not static information, INSD provides methods for updating, correcting, or adding additional information to existing sequence records, either before or after they become publicly available. Finally, sequence data is not useful unless it is easily available to researchers in their labs and at their desks. Each INSD collaborator provides these public users with multiple tools to search, retrieve, and analyze sequence data in a variety of formats to allow such use.

12. Notes

1. The DDBJ/EMBL/GenBank Feature Table: Definition, which can be found at http://www.insdc.org/files/documents/feature_table.html, lists all allowable features and qualifiers

for a DDBJ/EMBL/GenBank record. This document gives information about the format and conventions, as well as examples, for the usage of features in the sequence record. Value formats for qualifiers are indicated in this document. Qualifiers may be:

a. Free text

b. Controlled vocabulary or enumerated values

c. Citation or reference numbers

d. Sequence

Other syntax related to the flat file is described in this document. The document also contains reference lists for the following controlled vocabularies:

e. Nucleotide base codes (IUPAC)

f. Modified base abbreviations

g. Amino acid abbreviations

h. Modified and unusual Amino Acids

i. Genetic Code Tables

j. Country Names

2. The web search and retrieval tool for the three members of INSD are:
getentry (http://getentry.ddbj.nig.ac.jp/) for DDBJ
SRS (http://srs.ebi.ac.uk/) for EMBL
Entrez (http://www.ncbi.nlm.nih.gov/entrez/query.fcgi?db=Nucleotide) for NCBI.

3. The NCBI Trace Archive (http://www.ncbi.nlm.nih.gov/Traces/) is a permanent repository of DNA sequence chromatograms (traces), base calls, and quality estimates for single-pass reads from various large-scale sequencing projects. Data are exchanged regularly between the NCBI Trace Archive and the Ensembl Trace Server at the EBI and Sanger Institute in the United Kingdom.

4. The Sequin and tbl2asn files can also be obtained by command line ftp. To get the latest version of Sequin or tbl2asn, ftp to ftp.ncbi.nih.gov, use "anonymous" as the login and your e-mail address as the password. Change directories to asn1-converters/by_program/tbl2asn for tbl2asn and sequin/CURRENT for Sequin. Set bin mode, and download the appropriate version of the program for your operating system.

5. When preparing a submission by Sequin or tbl2asn, information about the sequence can be incorporated into the Definition Line of the fasta formatted sequence. FASTA format is simply the raw sequence preceded by a definition line. The definition line begins with a > sign and is followed immediately

by the sequence identifier and a title. Information can be embedded into the title that Sequin and tbl2asn use to construct a submission. Specifically, you can enter organism and strain or clone information in the nucleotide definition line and gene and protein information in the protein definition line using name-value pairs surrounded by square brackets. Example:

>myID [organism=Drosophila melanogaster] [strain=Oregon R] [clone=abc1].

6. The Sequin and tbl2asn submission utilities contain validation software which will check for problems associated with your submission. The validator can be accessed in Sequin from the Search->Validate menu item or in tbl2asn using –v in the command line. If the submission has no problems, there will be a message that the validation test succeeded. If not, a window will pop up in Sequin listing the validation errors and warnings. In tbl2asn, validator messages are stored in a file with a .val suffix. In Sequin, the appropriate editor for making corrections can be launched by double-clicking on the error message in the validator window. The validator includes checks for such things as missing organism information, incorrect coding region lengths, internal stop codons in coding regions, inconsistent genetic codes, mismatched amino acids, and non-consensus splice sites. If is important that submissions are checked for problems prior to submission to the database.

7. Sequences in many of the functional divisions have keywords that are related to that division. For example, EST sequences have an EST keyword. Similarly, GSS, STS, HTC, WGS, and ENV sequence records have keywords that are indicative of the GenBank division in which they belong. Furthermore, TPA sequences have keywords that identify the category, experimental or inferential, to which the TPA record belongs. For instance, TPA inferential records have three keywords: "Third Party Annotation; TPA; TPA:inferential". TPA sequence records that have experimental evidence are flagged with TPA:experimental. There are a number of sequencing projects that utilize keywords to flag sequences that belong to these projects. MGC, FLIcDNA, and BARCODE are three examples. Phase one and phase two HTG sequence records are in the HTG division of GenBank. These sequence records must contain two keywords, HTG and HTGS_PHASE1 or HTGS_PHASE2. Other keywords have been adopted by the HTGS submitting genome centers to indicate the status of the sequence. Phase 3 HTG sequences are considered finished and move to the appropriate taxonomic division. These sequences still retain the HTG keyword.

8. Updates to GenBank records can be submitted by e-mail to gb-admin@ncbi.nlm.nih.gov, through the BankIt update form (http://www.ncbi.nlm.nih.gov/BankIt/) or by Sequin-MacroSend (http://www.ncbi.nlm.nih.gov/LargeDirSubs/dir_submit.cgi) for large Sequin files. GenBank requests that update files are properly formatted. For publication or other general information, updates should be submitted as text in an e-mail.

   a. Updates to source information (i.e., strain, cultivar, country, specimen_voucher) in a two-column tab-delimited table, for example:

   ```
   acc. num.    strain
   AYxxxxxx     82
   AYxxxxxy     ABC
   ```

   b. Updates to nucleotide sequence should be submitted with the complete new sequence(s) in fasta format.

   ```
   >AYxxxxxx
   cggtaataatggaccttggaccccggcaaagcggagagac
   >AYxxxxxy
   ggaccttgga ccccggcaaagcggagagaccggtaataat
   ```

   c. Updates to feature annotation should be submitted as a tab-delimited five-column feature table. The first line of the table should read:
   >Feature SeqId

   The sequence identifier (SeqId) must be the same as that used on the sequence. The table_name is optional. Subsequent lines of the table list the features. Each feature is on a separate line. Qualifiers describing that feature are on the line below. Columns are separated by tabs.

   Column 1: Start location of feature

   Column 2: Stop location of feature

   Column 3: Feature key

   Line2:

   Column 4: Qualifier key

   Column 5: Qualifier value

   For example:

   ```
   >Feature gb|AYxxxxxx|AYxxxxxx
   <1       1050   gene
                   gene    ATH1
                   gene_syn    YPR026W
   <1       1009   CDS
                   product acid trehalase
   ```

```
                 note          role in sugar
                               metabolism
                               codon_start     2
        <1       1050  mRNA
                 product acid trehalase
        1626     1590  tRNA
        1570     1535
                 product tRNA-Phe
        1626     1535  gene
                 gene          trnF
```

Acknowledgment

This research was supported by the Intramural Research Program of the NIH, NLM, NCBI.

References

1. Benson, D. A., Karsch-Mizrachi, I., Lipman, D. J., Ostell, J., Wheeler, D. L. (2008) GenBank. *Nucleic Acids Res* 36(Database issue), D25–D30.

2. Cochrane, G., Akhtar, R., Aldebert, P., Althorpe, N., Baldwin, A., Bates, K., Bhattacharyya, S., Bonfield, J., Bower, L., Browne, P., Castro, M., Cox, T., Demiralp, F., Eberhardt, R., Faruque, N., Hoad, G., Jang, M., Kulikova, T., Labarga, A., Leinonen, R., Leonard, S., Lin, Q., Lopex, R., Lorenc, D., McWilliam, H., Mukherjee, G., Nardone, F., Plaister, S., Robinson, S., Sobhany, S., Vaughan, R., Wu, D., Zhu, W., Apweiler, R., Hubbard, T., Birney, E. (2008) Priorities for nucleotide trace, sequence and annotation data capture at the Ensembl Trace Archive and the EMBL Nucleotide Sequence Database. *Nucleic Acids Res* 36 (Database issue), D5–D12.

3. Sugawara, H., Ogasawara, O., Okubo, K., Gojobori, T., Tateno, Y. (2006). DDBJ with new system and face. *Nucleic Acids Res* 36(Database issue), D22–D24.

4. Wheeler, D. L., Barrett, T., Benson, D. A., Bryant, S. H., Canese, K., Chetvernin, V., Church, D. M., Dicuccio, M., Edgar, R., Frderhen, S., Feolo, M., Geer, L. Y., Helmberg, W., Kapustin, Y., Khovayko, O., Landsman, D., Lipman, D. J., Madden, T. L., Maglott, D. R., Miller, V., Ostell, J., Pruitt, K. D., Schuler, G. D., Shumway, M., Sequeira, E., Sherry, S. T., Sirotkin, K., Souvorov, A., Starchenko, G., Tatusov, R. L., Tausova, T. A., Wagner, L., Yaschenko, E. (2008) Database resources of the National Center for Biotechnology Information. *Nucleic Acids Res.* 36(Database issue), D13–D21.

Chapter 2

RNA Structure Determination by NMR

Lincoln G. Scott and Mirko Hennig

Abstract

This chapter reviews the methodologies for RNA structure determination by liquid-state nuclear magnetic resonance (NMR). The routine production of milligram quantities of isotopically labeled RNA remains critical to the success of NMR-based structure studies. The standard method for the preparation of isotopically labeled RNA for structural studies in solution is *in vitro* transcription from DNA oligonucleotide templates using T7 RNA polymerase and unlabeled or isotopically labeled nucleotide triphosphates (NTPs). The purification of the desired RNA can be performed by either denaturing polyacrylamide gel electrophoresis (PAGE) or anion-exchange chromatography. Our basic strategy for studying RNA in solution by NMR is outlined. The topics covered include RNA resonance assignment, restraint collection, and the structure calculation process. Selected examples of NMR spectra are given for a correctly folded 30 nucleotide-containing RNA.

Key words: RNA, RNA synthesis, RNA purification, NMR, resonance assignment, structure determination.

1. Introduction

RNA continues to surprise the scientific community with its rich structural diversity and unanticipated biological functions, including catalysis and the regulation of gene expression. Knowledge of the three-dimensional structure of biological macromolecules is indispensable for describing and understanding the underlying determinants of molecular recognition. RNA-ligand recognition generally occurs by "induced-fit" rather than by rigid "lock-and-key" docking (1, 2). These recognition processes apparently necessitate conformational flexibility for which liquid state NMR spectroscopy is uniquely suited to answer important questions in this area by looking at dynamic ensembles of structures.

Large quantities of RNA can be routinely prepared from either DNA template-directed *in vitro* transcription using T7 RNA polymerase (as well as T3 or SP6), or phosphoramidite-based chemical synthesis. This chapter focuses on the *in vitro* transcription method using T7 RNA polymerase, which is both more efficient, and cost effective (especially for RNAs >50 nucleotides) *(3, 4)*. However, the disadvantages of *in vitro* transcription include difficulties associated with the selective incorporation of isotopically labeled nucleotides or modified nucleotides, which are often functionally important.

The proliferation of RNA structure determinations using NMR spectroscopy is the combined result of:

1. The availability of efficient methods for isotopic labeling of RNA molecules, which permits heteronuclear experiments to be performed that resolve the severe spectral overlap inherent in proton spectra of RNAs

2. The rapid development of pulse sequences tailored for RNA spin systems facilitating many structure determinations

Severe spectral overlap in unlabeled RNA seriously limits the application of solution studies by NMR (**Fig. 2.1**). In contrast to the abundant ^{1}H isotope, the naturally occurring nuclei ^{12}C and ^{14}N cannot be readily studied with high-resolution NMR techniques. The production of isotopically labeled RNA remains critical to the success of these NMR-based structure studies *(5)* and a variety of synthetic methods have been developed for the routine production of isotopically labeled nucleotides. Labeled NTPs for *in vitro* transcription reactions can be readily produced

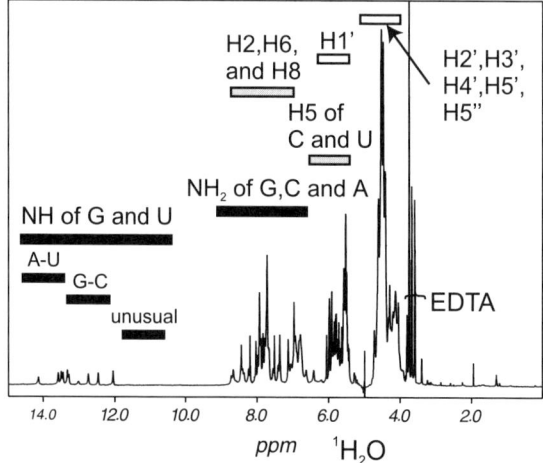

Fig. 2.1. 1D ^{1}H spectrum of the 30 nucleotide HIV-2 TAR RNA recorded in H$_2$O. Typical ^{1}H chemical shift ranges are indicated; solid black bars highlight exchangeable imino and amino protons, gray bars non-exchangeable base, and open bars non-exchangeable ribose protons.

by enzymatic phosphorylation of ribonucleoside monophosphates (NMPs) isolated from bacteria such as *Methylophilus methylotrophus* or *E. coli* grown on ^{13}C- and/or ^{15}N enriched media. Optimized and detailed protocols for the preparation of labeled NTPs are published elsewhere and are not covered in this chapter *(6–9)*. Alternatively, a variety of isotopically labeled NTPs are commercially available (e.g., Cambridge Isotope Labs, Sigma-Aldrich, Spectra Gases). Through the use of ^{13}C and ^{15}N isotopic labeling and multidimensional heteronuclear NMR experiments (**Fig. 2.2**), studies of 15-kDa RNAs are commonplace and recent methodological developments have been reviewed *(10–14)*.

New experiments to measure RNA orientation dependent dipolar couplings *(15–17)* and cross-correlated relaxation rates *(18, 19)* have been developed, providing additional structural information. Furthermore, NMR experiments have been introduced that allow the direct identification of donor and acceptor nitrogen atoms involved in hydrogen bonds *(20, 21)*. The unambiguous identification of hydrogen bonds is important in nucleic acid structure determination, particularly for tertiary structural interactions; in the absence of such direct measurements, hydrogen-bonding partners can be misassigned, which will subsequently impact the precision of the resulting structure. All these recently introduced parameters are especially important for structure determination of RNA due to the low proton density, and because a significant number of protons are potentially involved in exchange processes.

We have applied most of the reviewed methods to the 30-nucleotide human immunodeficiency virus (HIV)-2 transactivation response element (TAR) RNA, one of the best-characterized

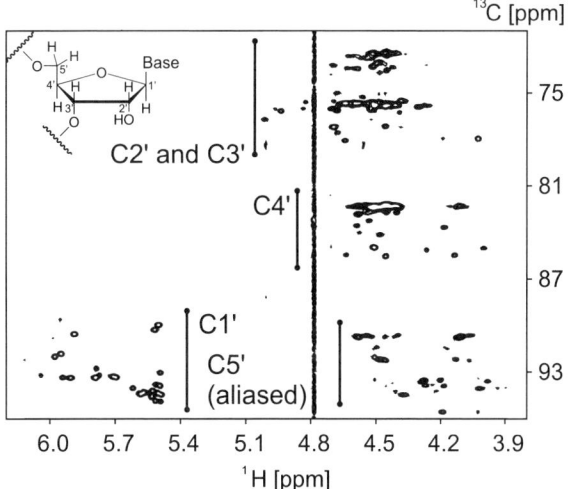

Fig. 2.2. 2D ^{1}H,^{13}C CT-HSQC spectrum of the TAR RNA. Typical ribose ^{13}C chemical shift ranges are indicated. The spectrum was acquired such that the ^{13}C5′ resonances are aliased in ω_1 (−1*spectral width) to improve digital resolution.

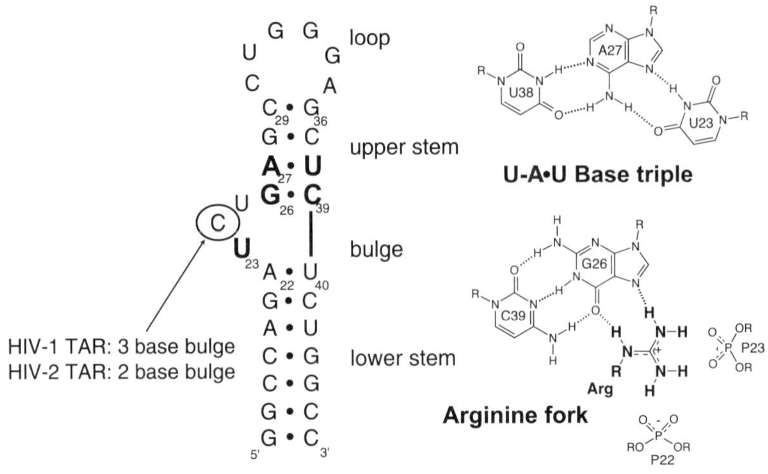

Fig. 2.3. Sequence and secondary structure of the TAR RNA where the bold typeface highlights nucleotides important for Tat recognition. Upon binding of argininamide, the TAR RNA undergoes a conformational change in the bulge region where the essential nucleotides, U38, A27 and U23, form a base triple.

medium-sized RNA molecules. The TAR RNA hairpin loop interacts with Tat, one of the regulatory proteins encoded by HIV. Tat contains an arginine-rich motif responsible for binding to its target *(22, 23)*. Formation of the Tat–TAR interaction is critical for viral replication. Peptides from the basic region of Tat retain the specificity of RNA binding and the amide derivative of arginine also binds specifically to TAR, although with greatly reduced affinity *(24, 25)*. The nucleotides on TAR important for Tat binding are clustered around a three-nucleotide bulge, shown in **Fig. 2.3**. Upon binding of Tat, Tat peptides or argininamide, the TAR RNA undergoes a major conformational change in the bulge region. In the bound form, the essential nucleotides, U38, A27 and U23, form a base triple, shown in **Fig. 2.3**, which results in an opening of the major groove for peptide recognition *(26–29)*. Additional solution studies of the TAR RNA in the absence of ligands have been performed *(30, 31)*.

2. Materials

2.1. In Vitro Transcription

1. Oligonucleotide Transcription Buffer (10×): 800 mM N-(2-hydroxyethyl)-piperazine-N′-2-ethanesulfonic acid potassium salt (K-HEPES), pH 8.1, 10 mM spermidine, and 0.1% (w/v) Triton X-100 prepared in water (*see* **Note 1**).

2. Plasmid Transcription Buffer (10×): 400 mM Tris-HCl, pH 8.1, 10 mM spermidine, 0.1% (w/v) Triton X-100 prepared in water.

3. Solution of 50% (w/v) polyethylene glycol 8000 (PEG-8000) prepared in water.
4. 0.5 M ethylene diamine tetraacetic acid disodium salt (EDTA), pH 8.0, prepared in water *(32)*.
5. TE Buffer: 10 mM Tris-HCl, 1 mM EDTA, pH 8.0, prepared in water.
6. DNA Transcription Promoter Oligonucleotide (60 µM) prepared in water (*see* **Note 2**).
7. DNA Transcription Template Oligonucleotide (60 µM) prepared in water (*see* **Notes 3** and **4**).
8. Linearized Double-Stranded Plasmid DNA Template (≥3 mg/mL) prepared in TE Buffer (*see* **Note 5**).
9. Solutions of 100 mM nucleotide-5′-triphosphates (ATP, UTP, GTP, and CTP) prepared in pH 7.0 water (*see* **Note 6**).
10. Solution of 1 M dithiothreitol (DTT) prepared in water.
11. Solution of 1 M magnesium chloride ($MgCl_2$) prepared in water.
12. Bacteriophage T7 RNA polymerase (*see* **Note 7**).
13. Phenol/chloroform (1:1, v/v; Fisher Scientific) equilibrated with TE buffer.
14. Chloroform/*i*-amyl alcohol (29:1, v/v).
15. Solution of 3 M Sodium Acetate, pH 5.3, prepared in water.
16. 100% Ethanol.
17. 80% Formamide Stop/Loading Buffer (2×): 80% (v/v) formamide, 20% (v/v) 0.5 M (EDTA), pH 8.0, 0.02% (w/v) Bromophenol blue, and 0.02% (w/v) Xylene cyanol prepared in water.
18. 8 M Urea prepared in water.

2.2. Polyacrylamide Gel Electrophoresis (PAGE)

1. Twenty Percent Acrylamide/Bisacrylamide Solution: 29:1 (w/w) acrylamide/bisacrylamide, 8 M urea, 90 mM Trisborate (TBE), 2 mM EDTA, pH 8.1 (*see* **Note 8**).
2. TBE Running Buffer: 90 mM TBE, 2 mM EDTA, pH 8.1, prepared in water.
3. N,N,N′,N′-Tetramethyl-ethylenediamine (TEMED, Bio-Rad) (*see* **Note 9**).
4. Ammonium Persulfate Solution (APS): 30% (w/v) solution in water (*see* **Note 10**).
5. Elutrap Electroelution System (with BT1 and BT2 membranes, Schleicher & Schuell BioScience).
6. CentriPrep concentrator with appropriate molecular weight cut-off (Millipore).

7. NMR Buffer (e.g., 10 mM sodium phosphate, pH 6.5, 50 mM NaCl, 0.1 mM EDTA, 0.02% NaN_3, prepared in 90% H_2O/10% D_2O).

2.3. Anion-Exchange Chromatography

1. Low salt loading buffer, e.g., 20 mM potassium phosphate, pH 6.5, 0.5 mM EDTA, 0.02% sodium azide (NaN_3), and 100 mM KCl.
2. High salt elution buffer, e.g., 20 mM potassium phosphate, pH 6.5, 0.5 mM EDTA, 0.02% NaN_3, and 2 M KCl.
3. Two HiTrap Q columns (Amersham Pharmacia).
4. NAP25 column (Amersham Pharmacia).
5. CentriPrep concentrator with appropriate molecular weight cut-off (Millipore).
6. NMR Buffer (e.g. 10 mM sodium phosphate, pH 6.5, 50 mM NaCl, 0.1 mM EDTA, 0.02% NaN_3, prepared in 90% H_2O/10% D_2O).

3. Methods

3.1. RNA Sample Preparation and Purification

The yield of *in vitro* transcribed RNA can depend on a variety of factors, many of which are not fully understood. The rational sparse matrix method of duplicate 40–60 conditions in small-scale (10–50 μL) transcription reactions can be easily employed to find the optimal reaction conditions. Trace amounts of α-^{32}P-labeled nucleotide (typically 5.0×10^5:1 [mol/mol] GTP:α-^{32}P-GTP (800 Ci/mmol), Perkin Elmer, Wellesley, MA) can be included to permit later radioanalytic quantitation of the transcription products. After four hours of incubation at 37°C the reactions are quenched with stop/loading buffer, and loaded directly to a 20% (29:1) denaturing polyacrylamide electrophoresis gel. The dried gel is phosphorimaged and the optimal conditions for transcription are chosen. The conditions can be chosen to either maximize the total yield of RNA, or in the case of isotopically labeled nucleotides, to maximize the yield of RNA per mole of input nucleotides. In addition, computational methods can assist in the interpretation of the experimental transcription optimization data *(33)*. Typically, before embarking on a large-scale synthesis, a pilot 1 mL transcription reaction is carried out to verify the isolated yield. Transcription reactions are carried out on a scale of 1–40 mL, and typical isolated yields are 1–10 nmol RNA per mL of transcription.

Two strategies are available for preparing large quantities of RNA by *in vitro* run-off transcription *(3, 4)*. Transcriptions for short RNAs (<50 nucleotides) are carried out from synthetic

DNA templates. The non-coding (top) strand and the template strand are purchased and gel purified on at least a 1-μmol scale for large-scale preparations. The preparation of RNA by standard *in vitro* run-off transcriptions from synthetic DNA templates using T7 polymerase becomes inefficient if the RNA transcript is longer than ~60 nucleotides. Thus, larger RNA transcripts are typically synthesized using linearized plasmid DNA containing the target RNA coding sequence under a T7 promoter (**Fig. 2.4**).

RNA structural studies in solution by NMR require milligram amounts of the desired RNA of specific length and sequence. Traditionally, the purification of RNA transcripts is achieved by preparative denaturing (8 M urea) PAGE and subsequent electroelution from the polyacrylamide gel matrix *(34)*. This method separates large quantities of the desired RNA from unincorporated nucleotides and short, abortive transcripts with single nucleotide resolution, but tends to be laborious and time consuming. Additional disadvantages are the co-purification of water-soluble acrylamide impurities that are a result of incomplete polymerization along with the RNA transcript. These impurities show a high affinity for RNA, and their complete removal by dialysis is difficult, necessitating either additional purification steps or extensive rinsing of the RNA transcript with water using an appropriate CentriPrep concentrator.

An alternative purification protocol employs anion-exchange chromatography *(35, 36)*. Using this fast chromatography purification approach, the most time-consuming step for preparing large quantities of RNA for structural studies – PAGE purification followed by electroelution – can be eliminated and sample contamination with acrylamide is circumvented. It should be noted that this technique also preserves the co-transcriptionally adopted folding state of the desired RNA. This is in marked contrast to denaturing PAGE purification, which is typically accompanied by several precipitation steps, and represents an important advantage in cases in which annealing procedures fail to reproduce a natively folded RNA target.

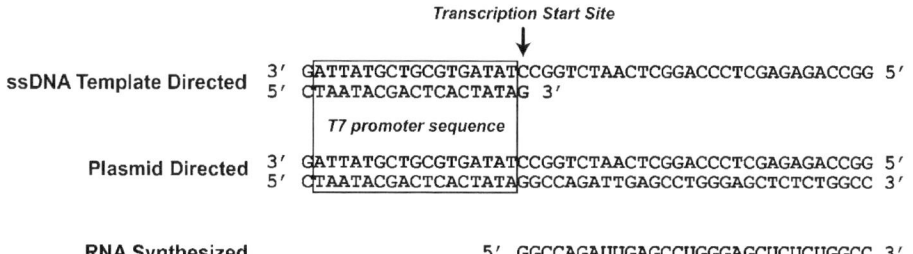

Fig. 2.4. Single and double-stranded DNA template sequences for the *in vitro* transcription of the HIV-2 TAR RNA (*see* **Note 2**).

3.1.1. Large-Scale DNA Template Directed In Vitro RNA Transcription

1. Transcription reactions are carried out under the following conditions: 80 mM K-HEPES (pH 8.1), 1 mM spermidine, 10 mM DTT, 0.01% Triton X-100, 80 mg/mL PEG-8000, 8–48 mM $MgCl_2$, 2–6 mM each NTP, 0.3 µM template oligonucleotide DNA, 0.3 µM promoter oligonucleotide DNA, and ~2,000–4,000 units/mL T7 RNA polymerase (*see* **Notes 11, 12,** and **13**).
2. The reactions are incubated for 4 hours in a water bath at 37°C.
3. The reactions are quenched with the addition of 0.1 volume of 0.5 M EDTA (pH 8.0) (*see* **Note 14**).
4. The reactions are extracted with an equal volume of phenol/chloroform (1:1, v/v) equilibrated with TE buffer to remove T7 RNA polymerase prior to purification. The organic layer is further extracted with an equal volume of water to ensure all the RNA is removed from the reaction.
5. The aqueous layers are combined and back extracted with an equal volume of chloroform/*i*-amyl alcohol (29:1, v/v) to remove any traces of phenol.
6. The aqueous layer is ethanol precipitated with the addition of 0.1 vol of 3 M sodium acetate and 3.5 volumes cold 100% ethanol at −20°C.
7. The crude RNA precipitate is collected by centrifugation, and resuspended in equal volumes of 80% Formamide Stop/Loading Buffer and 8M urea.
8. The sample is suitable for loading to a denaturing polyacrylamide gel.

3.1.2. RNA Purification by Denaturing Polyacrylamide Electrophoresis

1. These instructions are general and are easily adaptable to other formats, and reaction scales, including minigels. It is critical that the glass plates for the gels are extensively cleaned with detergent (e.g., Alconox, Alconox, New York, NY), ammonium-based glass cleaner (e.g., Wendex, S.C. Johnson), and finally 95% ethanol.
2. Prepare a polyacrylamide gel of the appropriate percentage, size, and thickness by mixing acrylamide/bisacrylamide solution, 1 µL APS and 1 µL TEMED per mL acrylamide/bisacrylamide solution *(32)*. The gel should polymerize in about 30 minutes.
3. Once the gel polymerizes, carefully remove the comb and wash the wells with TBE running buffer.
4. Place the gel into the appropriate gel running apparatus and add TBE running buffer to the upper and lower chambers of the gel unit.

5. Complete the assembly of the gel unit by connecting the power supply. The gel should be pre-run for at least 30 minutes at the appropriate voltage to allow thermal equilibration of the gel plates, prior to loading your samples.

6. Run the RNA sample a sufficient time to resolve the *n-1* nucleotide transcription product, typically two-thirds of the gel if the correct percentage polyacrylamide gel was used.

7. Take the gel off the apparatus and carefully remove the gel from the plates, placing the gel on clear cellophane.

8. The RNA can be easily visualized by UV_{256} shadowing, and excised from the gel with a clean razor blade or scalpel.

9. Place the gel pieces into an Elutrap Electroelution System in TBE running buffer at 4°C, and the RNA is extracted from the gel in a manner outlined by the vendor. Typically, removing four fractions over a period of 6 hours at 200V is sufficient to extract RNA from even a twenty percent polyacrylamide gel.

10. The RNA containing fractions are combined and precipitated by adding one-tenth the volume of 3M sodium acetate, followed by 3.5 volumes of cold 100% ethanol. Place the solution at –20°C.

11. The desired RNA is collected by centrifugation.

12. RNA samples are desalted using an appropriate CentriPrep concentrator and lyophilized.

13. The lyophilized RNA is dissolved in desired final volume (e.g., 500 µL for a standard 5-mm NMR sample tube) of NMR buffer.

14. The NMR sample is annealed in a manner appropriate to the specific RNA to form native structure (*see* **Note 15**).

3.1.3. RNA Purification by Anion-Exchange Chromatography

1. The transcription reaction is clarified by centrifugation (14,000*g*) to remove traces of precipitated pyrophosphate (*see* **Note 16**).

2. Equilibrate two HiTrap Q columns (Amersham Pharmacia) in low salt loading buffer at room temperature.

3. The transcription reaction mixture is applied to the equilibrated columns.

4. To separate the desired RNA from unincorporated nucleotides and plasmid DNA template, the loaded sample is typically eluted at a low flow rate of 1 mL/minute, in 3-mL fractions, with an increasing KCl gradient created by simultaneously decreasing the percentage of low salt loading buffer and increasing the percentage of high salt elution buffer being passed through the columns.

5. A common gradient for anion-exchange purification after transcription starts with 100% loading buffer to wash the columns, continues through a gradual climb from 0% to 60% elution buffer for the first 2 hours, then to 100% elution buffer over another 5 minutes. A typical elution profile, detected at 260 nm, generally shows three major peaks: the first one containing unincorporated nucleotides and short, abortive transcripts; the second peak containing the desired RNA; and the last fractions containing the plasmid DNA template (*see* **Note 17**).

6. Pure fractions are combined and concentrated using an appropriate CentriPrep (Millipore) concentrator.

7. Concentrated fractions are desalted and buffer exchanged by passage through a NAP25 gel filtration column (Amersham Pharmacia) equilibrated with an NMR buffer. Alternatively, pure fractions can also be dialyzed into NMR buffer, and then concentrated with a CentriPrep concentrator.

8. NMR samples are concentrated to desired final volumes (e.g., 500 µL for a standard 5-mm NMR sample tube) using an appropriate CentriPrep concentrator.

3.2. NMR Resonance Assignment and Restraint Collection

3.2.1. Resonance Assignment Strategy

Assignment of RNA resonances is commonly achieved through identification of sequential base to ribose nuclear Overhauser effect (NOE) patterns seen in helical regions of nucleic acid structure (**Fig. 2.5**), in analogy to the procedure originally utilized for DNA studies in the 1980s *(37)*. With the advent of isotopic labeling for RNA, the basic NOE assignment approach was initially expanded to include multi-dimensional (3D and 4D) versions of the standard nuclear Overhauser effect spectroscopy (NOESY), which simplified assignment and identification of NOEs *(38, 39)*.

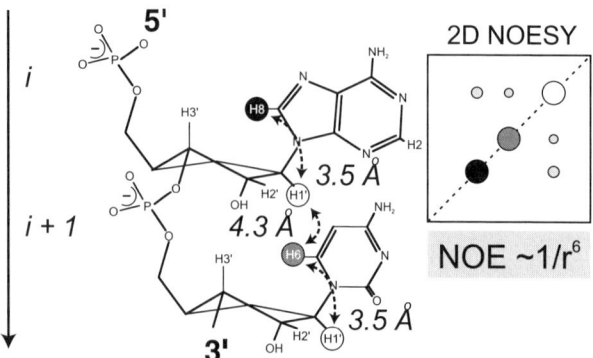

Fig. 2.5. Schematic representation of a 5'-pApC-3' dinucleotide with arrows indicating the intra- and interresidual distances used for NOE based sequential assignments of A-form helical conformations. A schematic 2D NOESY with cross-peaks correlating H1'(i)- H8(i), H1'(i)-H6(i+1), and H1' (i+1)-H6(i+1) is shown.

The NOE-based approach, however, relies on assumptions about structure and assignments, and is susceptible to errors from structural bias; methodology that achieves sequential assignment via unambiguous through-bond correlation experiments, as is the case for proteins, would be more ideal. Unfortunately, complete sequential assignments of even medium-sized RNA molecules using through-bond experiments such as HCP (*see* **Note 18**), HCP-TOCSY (*To*tal *C*orrelation *S*pectroscop*y*) and HP-HET-COR (*Het*eronuclear *Cor*relation) are hampered by notoriously overlapped resonances and modest sensitivity. Thus, through-bond assignment using HCP-like experiments is not feasible for larger RNA target molecules (~20 kDa). A hybrid approach with HCN and NOESY experiments is the optimal compromise to achieve unambiguous assignments. The HCN experiments can determine intranucleotide correlations within and between the base and ribose resonances, which will significantly reduce the ambiguity present in the NOESY-based assignment procedure. HCCH-based experiments are used to unambiguously assign crowded ribose, pyrimidine H5/H6, and adenosine spin systems. A variety of through-bond correlation experiments facilitate the assignments of exchangeable imino- and amino proton resonances linked to non-exchangeable base H6 and H8 protons *(12, 40)*.

3.2.2. NMR Restraints for Structure Determination

After sequence-specific assignments of RNAs are obtained, the structure determination is based on collecting sufficient numbers of proton-proton distance restraints utilizing NOESY experiments. The structural analysis of the RNA backbone conformation is complicated by the lack of useful ^{1}H-^{1}H NOE distance restraints available that define the backbone torsions (**Fig. 2.6**).

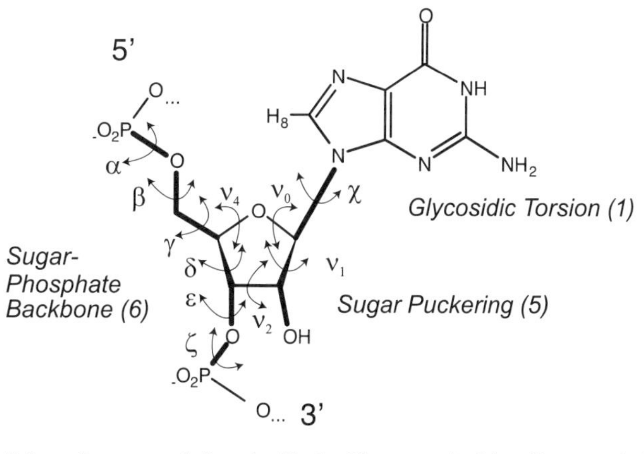

Fig. 2.6. Schematic representation of a 5′-pGp-3′ mononucleotide with arrows indicating the various torsional degrees of freedom in the sugar-phosphate backbone, the pentose ring, and the glycosidic torsion.

Potentially, the short distance restraints between pairs of protons (<6Å) can be complemented with torsion angle information accessible through J-coupling constants. Vicinal ^3J scalar coupling constants can provide useful structural information about the sugar pucker, the β and ε backbone torsion angle conformations, as well as the glycosidic torsion χ, which defines the orientation of nucleobases with respect to the sugar moiety. In addition, NMR experiments have been introduced that allow the direct identification of donor and acceptor nitrogen atoms involved in hydrogen bonds. These recently introduced parameters are especially important for structure determination of RNA due to the low proton density.

However, there is a practical difficulty in defining RNA structures precisely by NMR because NOE and J-coupling–based structure calculation relies on either short range distance (<6Å) or local torsion angle information. RNAs often are elongated structures, which are better approximated as cylindrical rather than globular shapes. There is a lack of NOE information between distant ends of the molecule; as a result, the relative orientations of helical segments at opposite ends of the molecule are poorly defined. Recent advances in methodology help to alleviate or overcome this shortcoming *(15, 41)*.

New experiments to measure orientational, rather than distance-dependent, dipolar couplings and cross-correlated relaxation rates have been developed, providing additional structural information. Methods have been developed to create a slightly anisotropic environment for molecules tumbling in solution. This results in a small degree of alignment of the molecule, and the dipolar couplings no longer average to zero, while retaining the quality of high-resolution NMR spectra. The most promising system for NMR studies of partially aligned RNA is a Pf1 bacteriophage solution *(16, 42)*. There is a narrow useful range of alignments suitable for high-resolution NMR studies. Higher phage concentrations are associated with stronger alignments and produce larger residual dipolar couplings, whereas lower concentrations correspond to lower degrees of ordering, reflected in smaller dipolar couplings. Too much alignment gives larger dipolar couplings, but also results in line broadening to such an extent that high-resolution NMR is not possible.

Residual dipolar couplings (RDC) data should be combined with the traditionally used NOE distance restraints and torsion angles derived from scalar J-couplings. The RDC data do not only provide additional information for a better definition of the global orientation of the three stems with respect to each other, but also carry valuable information on the dynamic properties of the RNA studied *(43–45)*.

3.2.3. NOE Distance Restraints

1. The main source of structural data will still be obtained from NOEs, which provide distance restraints for pairs of hydrogen atoms. Only short proton-proton distances in the range <6Å are accessible through NOESY-type experiments. Identification of NOEs will be facilitated by resolving the ^{1}H,^{1}H NOE connectivities that are essential for determining the structure into three and four dimensions through detection of the heteronuclear (^{13}C/^{15}N) chemical shifts of the proton-attached nuclei.

2. NOESY-type experiments should be recorded with varying mixing times (50–300 ms). NOE cross peaks obtained with long mixing times (>100 ms) are harder to quantitate and should be used with caution in structure calculations; however, they can tremendously help during the assignment process (*see* **Notes 19, 20, 21,** and **22**).

3. Imino proton resonances should be assigned sequence specifically at an early stage from water flip-back, WATERGATE-2D NOESY *(46)* spectra (τ_{mix} = 200 ms) to verify the construct integrity and secondary structure predictions (**Fig. 2.7**).

4. The identification of NOEs can be further facilitated by utilizing isotope filtered/edited NOESY experiments in combination with nucleotide-specific isotopically labeled RNA *(47)*.

3.2.4. Torsion Angle Restraints

1. The ribose sugar geometry is defined by five alternating torsion angles (v_0 through v_4). Usually, the ribose sugar adopts one of the energetically preferred C2′-endo (South) or C3′-endo (North) conformations. A number of ^{1}H,^{1}H and ^{1}H,^{13}C scalar couplings are available to determine the sugar pucker

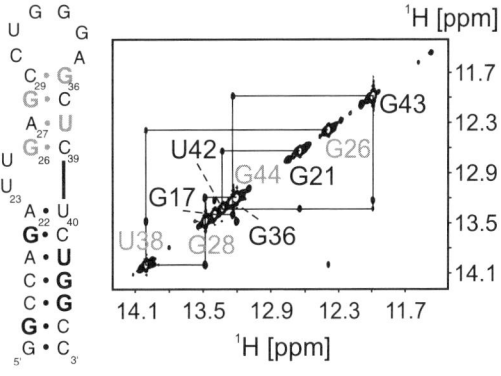

Fig. 2.7. 2D ^{1}H,^{1}H water flip-back, WATERGATE NOESY *(46)* spectrum of the TAR RNA (τ_{mix} = 200 ms). Sequential assignments of the imino proton resonances by NOE connectivities are indicated. The observable upper stem G and U residues are shown in bold gray and lower stem G and U residues in bold black.

qualitatively with the combination of H1'-H2' and H3'-H4' coupling constants being the most useful for smaller RNAs. The $^3J(\text{H1}',\text{H2}')$ vicinal coupling is >8 Hz for C2'-endo puckers and ~1 Hz for C3'-endo puckers (**Fig. 2.8**), typically found in A-form helices *(48–50)*. The opposite behavior is expected for the $^3J(\text{H3}',\text{H4}')$ coupling constant with C2'-endo puckers associated with small and C3'-endo puckers with relatively large coupling constant values (*see* **Note 23**).

2. Measurement of the γ torsion is difficult due to the need for stereospecific assignments of the H5' and H5' proton resonances. The two-bond C4',H5'/H5" couplings can be used in conjunction with the vicinal H4',H5'/H5" couplings to define γ *(50, 51)*.

3. Two heteronuclear vicinal $^1\text{H},^{13}\text{C}$ couplings contain useful information about the glycosidic torsion angle χ. The $^3J(\text{H1}',\text{C})$ couplings involving the C4,C8 carbons in purines and the C2,C6 carbons in pyrimidines, respectively, all depend on the χ torsion *(50, 52)*. The preferred orientation around χ in A-form helix is *anti*, which makes the base accessible for commonly found hydrogen bonding interaction.

4. The ε and β torsions can be determined by measuring a variety of $^{13}\text{C},^{31}\text{P}$ and $^1\text{H},^{31}\text{P}$ scalar couplings. Some of these torsions may be measured directly in 2D $^1\text{H},^{31}\text{P}$ heteronuclear COSY (or HETCOR) experiments *(53, 54)* and non-refocused $^1\text{H},^{31}\text{P}$

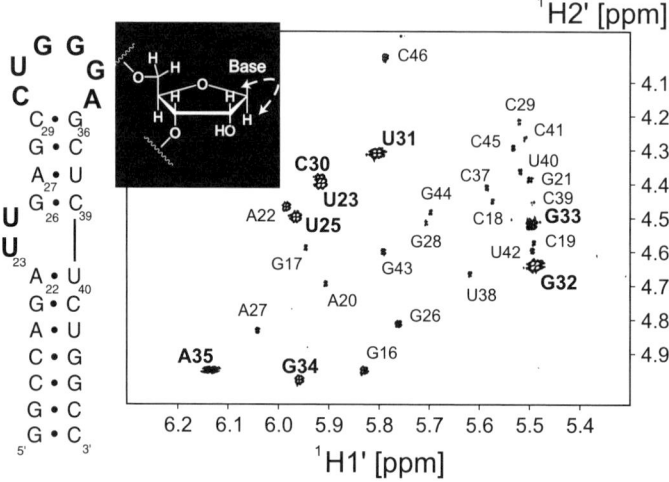

Fig. 2.8. Ribose H1'-H2' region of a 2D $^1\text{H},^1\text{H}$ DQF (*Double Quantum Filtered*): COSY *(129)* spectrum of the TAR RNA. Assignments for the H1'-H2' cross-peaks probing the individual sugar puckers are indicated. Residues shown in bold adopt either C2'-endo or mixed C2'-endo/C3'-endo sugar puckers, resulting in more efficient magnetization transfer due to larger $^3J(\text{H1}',\text{H2}')$ couplings. Weaker cross-peaks are associated with residues adopting C3'-endo sugar puckers, typically found in A-form helices. The inset shows a ribose ring; the arrow highlights the H1'- H2' connection.

HSQCs (*H*eteronuclear *S*ingle *Q*uantum *C*oherences) if the phosphorus and proton resonances are sufficiently resolved (**Fig. 2.9**). However, both the ribose proton and phosphorus resonances involved are generally overlapped for even moderate size RNAs. Accurate measurements for $^{13}C,^{31}P$ and $^{1}H,^{31}P$ couplings can be obtained from both phosphorus-fitting of doublets from singlets (so-called P-FIDS) *(55)* or spin echo difference experiments *(56–60)*. J-HMBC techniques can be applied to determine $^{3}J(H,P)$ couplings *(61)*. A quantitative version of the HCP experiment allows for quantitation of $^{3}J(C4',P)$ *(62)*.

5. The α and ζ torsions are not accessible by J-coupling measurements because the involved ^{16}O nuclei have no magnetic moment. Some groups have used ^{31}P chemical shifts as a guide for loose constraints on these torsions *(63)*; however, the correlation between ^{31}P chemical shifts and the phosphodiester backbone conformation is not well understood in RNA.

6. Cross-correlated relaxation rates have been introduced to high-resolution NMR as a novel parameter for structure determination *(18, 19)*. Such methods have been employed to gain information on the α and ζ torsions. The cross-correlated relaxation between a ribose ^{13}C-^{1}H dipole and the ^{31}P chemical shift anisotropy (CSA) carries valuable structural information about the phosphodiester conformation *(64)*. Additionally, applications have been published where the cross-correlated relaxation between a ^{13}C-^{1}H dipole

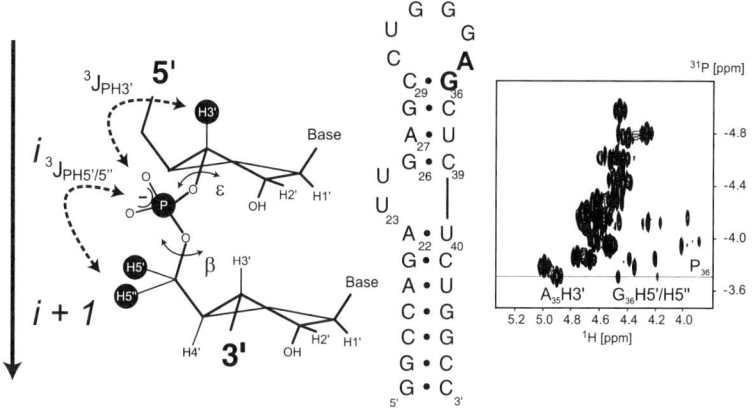

Fig. 2.9. Schematic representation of a 5'-NpN-3' dinucleotide with arrows highlighting the $^{3}J(H3',P)$ and $^{3}J(H5'/5'',P)$ couplings associated with the torsion angles ε and β, respectively. 2D $^{1}H,^{31}P$-HETCOR spectrum of the TAR RNA *(53)*. Assignments for the H3'/H5'/H5'',^{31}P-correlations along the $^{31}P36$ resonance are indicated in the spectrum and shown in bold in the secondary structure representation.

and the glycosidic ^{15}N CSA is utilized to collect information about the glycosidic torsion angle χ *(65, 66)*. Another example is the measurement of cross-correlated relaxation rates between neighboring ^{13}C-^{1}H dipoles within the ribose ring that can be used to define the sugar pucker. For RNAs, cross-correlated relaxation rates can be measured using an experiment that belongs to the HCCH class, and precisely determine the ribose sugar pucker without the need of any empirical Karplus parameterization *(67)*. The resolution of this experiment can be further enhanced by a combination with a CC-TOCSY transfer *(68)* (*see* **Note 24**).

3.2.5. Residual Dipolar Coupling Restraints

1. One-bond dipolar couplings on the order of ±10–30 Hz can be introduced using ~15 mg/mL filamentous Pf1-phages as co-solutes, which creates an anisotropic environment for the RNA target molecule (*see* **Note 25**).

2. For a directly bonded pair of nuclei with known distance, such as ^{1}H-^{13}C or ^{1}H-^{15}N in labeled RNA, angular restraints can be extracted from dipolar coupling data and incorporated during the structure calculation. Such one-bond dipolar couplings can be measured in a straightforward and sensitive manner. The difference between scalar J coupling constant values measured in isotropic and anisotropic media gives the residual dipolar coupling. Two NMR experiments are commonly performed to measure one bond ^{1}D(H,C) RDC constants. Base C2-H2, C5-H5, C6-H6, and C8-H8 dipolar couplings are typically derived from analyzing peak positions in CT-TROSY and CT-antiTROSY experiments *(69)*. For the ribose 1′–4′ one-bond dipolar couplings, a J-modulated CT-HSQC should be acquired *(70)*. A J-modulated ^{1}H,^{15}N-HSQC provides additional one bond ^{1}D(H,N) RDC restraints *(71)*.

3. The determination of the phosphate backbone conformation in solution remains an experimentally intriguing problem. New parameters based on incomplete averaging in partially aligned RNA samples such as dipolar ^{1}H,^{31}P couplings *(53, 72, 73)* or ^{31}P CSA *(74)* hold the promise to significantly impact on the precision of RNA structure determination in solution.

3.2.6. Hydrogen Bond Restraints

1. Canonical base-pair hydrogen bonding of the Watson-Crick type is fundamental in all biological processes in which nucleic acids are involved. The partially covalent character of hydrogen bonds gives rise to measurable scalar spin-spin couplings of, for example, the type ^{h2}J(N,N) and ^{h1}J(H,N) that represent important additional NMR parameters for the structure determination of nucleic acids in solution *(20, 21)*. In addition to the unambiguous determination of donor D

and acceptor A nuclei involved in hydrogen bond formation, the magnitude of the $^hJ(D,A)$ couplings reports on the hydrogen bond geometry and could potentially provide more precise distance information for structure calculations. The simultaneous identification of nuclei involved in hydrogen bonds and quantification of corresponding $^{h2}J(N,N)$ scalar couplings is accomplished using a HNN-COSY (*Co*rrelation *S*pectroscop*y*) experiment or one of its variants (**Fig. 2.10**).

2. Several groups have also reported measuring scalar couplings across hydrogen bonds in non-canonical base pairs and in tertiary structural interactions *(75–81)*.

3. The large two-bond $^2J(H,N)$ scalar couplings within the purine bases allow reasonably efficient magnetization transfer during INEPT (*I*nsensitive *N*uclei *E*nhancement by *P*olarization *T*ransfer) delays *(82)*. The independent assignments of potential nitrogen hydrogen bond acceptor sites using the intra-residue $^2J(H2,N1)$, $^2J(H2,N3)$, and $^2J(H8,N7)$ correlations for the purine residues in the RNA molecule can be obtained from a two-bond $^2J(H,N)$ $^1H,^{15}N$-HSQC experiment.

4. The 2′-hydroxyl group plays fundamental roles in both the structure and function of RNA and is the major determinant of the conformational and thermodynamic differences

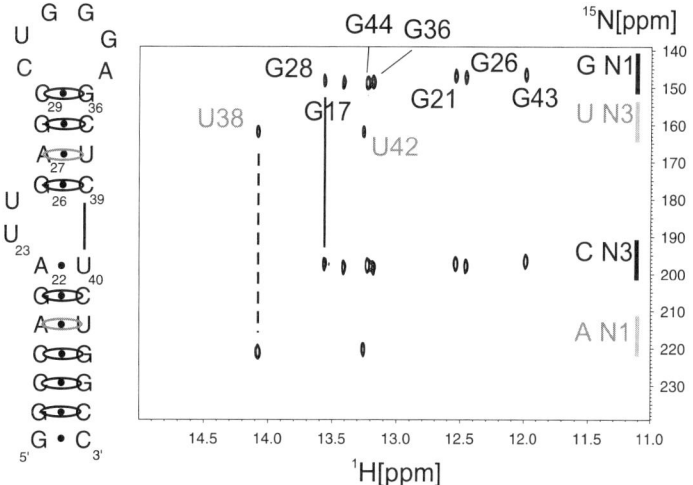

Fig. 2.10. 2D $^1H,^{15}N$ HNN-COSY of the TAR RNA. Assignments of the imino proton nitrogen correlations are indicated with the observable G and U residues shown in black and gray, respectively. Canonical base-pair hydrogen bonding of the Watson-Crick type correlates U N3 nitrogen donor sites with A N1 nitrogen acceptor sites (dashed line) and G N1 nitrogen donor sites with C N3 nitrogen acceptor sites (solid line). Typical ^{15}N chemical shift ranges are indicated; solid black bars highlight G N1 and C N3 nitrogens, while gray bars highlight A N1 and U N3 nitrogen chemical shift ranges.

between RNA and DNA. In aqueous solution the rapid exchange of the hydroxyl proton with the solvent typically prevents its observation in RNA at room temperature by NMR. Most recently, a conformational analysis of 2′-OH hydroxyl groups of the HIV-2 TAR RNA by means of NMR scalar coupling measurements in solution at low temperature has been reported *(83, 84)*. Cross hydrogen bond scalar couplings involving two slowly exchanging 2′-OH hydroxyl protons were observed and analyzed in a frame shifting mRNA pseudoknot *(85)*.

3.2.7. Base-to-Base H(C/N)H-Type Correlation Experiments

1. A set of HNCCH- and HCCNH-TOCSY experiments have been developed that correlate the exchangeable imino and amino proton resonances with the non-exchangeable base resonances for the complicated spin systems of all four nucleotides as shown in **Fig. 2.11** *(81, 86–91)*.

2. Complementary HCCH-COSY *(92)* and HCCH-TOCSY *(93, 94)* experiments are used to unambiguously assign pyrimidine H5/H6 and adenosine spin systems.

3.2.8. Base-to-Ribose HCN-Type Correlation Experiments

1. Optimized HCN-type pulse schemes for the through-bond correlation of ribose and base resonances utilizing MQ (multi-quantum)- instead of SQ (single-quantum)-evolution periods have been proposed and show significant sensitivity gains, essential for successful investigations of larger RNA systems *(95, 96)*. Also, TROSY (*Transverse Relaxation-Optimized Spectroscopy*) versions of HCN experiments have been successfully applied to RNA *(97, 98)*.

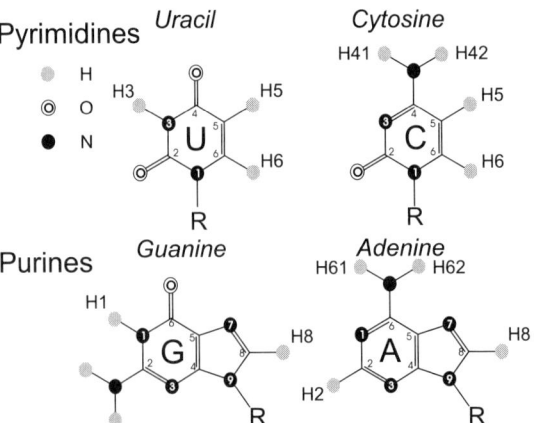

Fig. 2.11. The four different aromatic nucleobases, uracil, cytosine, guanine, and adenine. Exchangeable imino and amino proton resonances and non-exchangeable aromatic proton resonances that are correlated in HNCCH- and HCCNH-TOCSY experiments are shown as gray circles.

2. In favorable cases, magnetization can be transferred all the way through from the anomeric H1′ proton to the base H6/8 protons circumventing assignments through joint glycosidic N1/9 nitrogen chemical shift *(99, 100)*.

3.2.9. Ribose-to-Ribose HCCH-Type Correlation Experiments

1. The magnetization transfer through the ribose proton spin systems is hampered due to the small $^3J(H1',H2')$ vicinal coupling, present in most commonly populated A-form RNA, correlating the H1′ and H2′ resonances. Ribose proton spin system assignments from homonuclear ^{1}H, ^{1}H - COSY- and TOCSY experiments can be obtained more readily using HCCH-COSY and -TOCSY experiments on ribose rings uniformly labeled with ^{13}C, which allows magnetization transfer and chemical shift evolution on the C1′ to C5′ carbons *(39, 101–104)*.

2. The powerful hybrid HCCH-COSY-TOCSY *(105, 106)* experiment can also be employed to unambiguously assign crowded ribose spin systems.

3.2.10. Ribose-to-Phosphate Backbone H(C)P-Type Correlation Experiments

1. For unlabeled RNAs, a number of relatively efficient ^{1}H,^{31}P-multi-dimensional correlation schemes are available for sequential assignment of ^{31}P and ribose ^{1}H resonances. Magnetization can be transferred from excited ^{31}P resonances to the $^3J(H,P)$ scalar coupled ribose protons for detection using either COSY- *(54)* or heteronuclear TOCSY-type *(107)* transfer steps. The resulting two-dimensional H3′/H5′/H5″,^{31}P-correlations can be concatenated with homonuclear ^{1}H,^{1}H NOESY or TOCSY experiments to transfer magnetization to potentially better resolved resonances like H1′ or aromatic H8/H6 resonances *(108, 109)*.

2. A straightforward extend approach for ^{13}C labeled RNAs is HCP correlation via sequential INEPT transfers (^{1}H → ^{13}C → ^{31}P → ^{13}C → ^{1}H) *(110, 111)* correlating nuclei of adjacent nucleotides i and i + 1 (**Fig. 2.12**). Subsequent experiments, HCP-CCH-TOCSY (112) and P(CC)H-TOCSY *(113)* combine the HCP and HCCH-TOCSY experiments and thus resolve relevant correlations on the better dispersed C1′/H1′ resonances.

3.3. Structure Calculation

3.3.1. Generation of Restraint File

1. The intensity of NOESY cross peaks is approximately proportional to the inverse of the averaged distance to the power of six, $<1/r_{ij}^6>$, assuming an isolated pair of proton spins i and j. For RNA NMR studies, NOE-derived distance restraints are often determined semi-quantitatively and placed into four categories: strong, medium, weak, and very weak NOEs. A conservative approach sets all the lower bounds to 1.8 Å

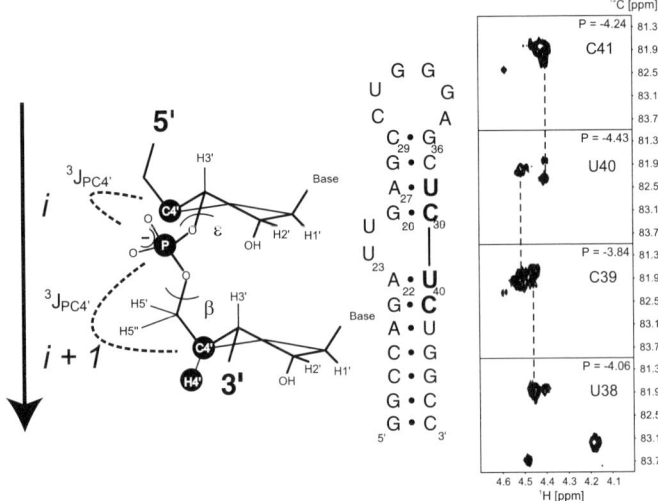

Fig. 2.12. Schematic representation of a 5′-NpN-3′ dinucleotide with arrows highlighting the $^3J(C4',P)$ couplings associated with the torsion angles ε and β, respectively. 2D $^1H4',^{13}C4'$ planes of a 3D HCP spectrum of the TAR RNA. Sequential assignments for the $^1H4',^{13}C4'$-correlations along the ^{31}P resonance frequencies of C41, U40, C39, and U38 are indicated (dashed lines) and shown in bold in the secondary structure representation.

(van der Waals radius) with upper bounds ranging from 3.0 Å for the most intense NOEs to 7.0 Å for the weakest NOEs found in H_2O experiments.

CNS/Xplor syntax as compiled in distance restraint table (e.g., noe.tbl):

assign <1st atom-sel.> <2nd atom-sel.> <distance> $<d_{minus}>$ $<d_{plus}>$

Example: proton H1 of residue 4 and proton H4′ of residue 30 are separated by 2.4 ± 0.6 Å

assign (resid 4 and name H1) (resid 30 and name H4′) 2.4 0.6 0.6

2. J-coupling restraints can be implemented in two different ways during the structure determination. They can be introduced qualitatively by restricting a torsion angle in a loose manner (±30°) to one of the three staggered rotamers along the phosphodiester backbone, or defining the preferred ribose sugar pucker such as C2′-endo or C3′-endo. Alternatively, vicinal J-couplings can be quantitatively related to a certain torsion angle using semi-empirical Karplus relations of the form: $^3J = A \cos^2\theta + B \cos\theta + C$, where θ is the intervening torsion angle *(40, 114)*. CNS/Xplor syntax as compiled in torsion angle restraint table (e.g. torsion.tbl):

assign <1st atom-sel.> <2nd atom-sel.>
<3rd atom-sel.> <4th atom-sel.> <real> <real> <real> <integer>

The four numbers are, respectively:

1. Force constant in kcal/(mole radians exp{exponent})
2. Equilibrium torsion angle in degrees
3. Range around the equilibrium value in degrees
4. Exponent for restraint calculation

Example: restrict residue 2 to North or C3'-endo sugar pucker

```
The sugar pucker can be defined with the
following ribose torsion angles: ν1 = O4'-
C1'-C2'-C3' and ν2 = C1'-C2'-C3'-C4'.
assign (resid 2 and name O4') (resid 2 and
name C1')
     (resid 2 and name C2') (resid 2 and
     name C3')
     1.0 -20.0 10.0 2
assign (resid 2 and name C1') (resid 2 and
name C2')
     (resid 2 and name C3') (resid 2 and
     name C4') 1.0 35.0 5.0 2
```

3. The size of dipolar couplings for an axially symmetric RNA molecule depends on the average value of an orientational function, $\frac{1}{2}(3\cos^2\theta - 1)$, and the inverse cubic distance, $1/r^3$, between the coupled nuclei. Here, the angle θ characterizes the axial orientation of the internuclear vector that connects the coupled nuclei with respect to the principal axis system of the molecular alignment tensor.

```
CNS/Xplor syntax as compiled in RDC table
(e.g., dipolar.tbl):
A pseudomolecule OXYZ is defined with orthogo-
nal vectors OX, OY, and OZ. OXYZ reorients it-
self during the refinement process to satisfy
the experimentally measured RDC data against
an energy penalty with its origin fixed in
space away from the target RNA molecule.
assign <external origin-sel.>
     <z-unit vector-sel.>
     <x-unit vector-sel.>
     <y-unit vector-sel.>
     <1st atom-sel.> <2nd atom-sel.> <RDC>
     <RDC_error>
Example: An RDC value of 15.6 ± 0.6 Hz is
measured for the one-bond interaction bet-
ween C1' and H1' of residue 2:
assign (resid 500 and name OO)
     (resid 500 and name Z)
     (resid 500 and name X)
     (resid 500 and name Y)
     (resid 2 and name C1')
     (resid 2 and name H1') 15.6000 0.6000
```

3.3.2. Molecular Dynamics Simulation

1. Most commonly, starting structures are calculated from randomized RNA coordinates using solely energy terms from holonomic constraints such as geometric and non-bonded terms using restrained molecular dynamics calculations.

2. To generate a family of structures consistent with the NMR data, the second step refines against the experimentally derived NOE distance and torsion restraints. We typically follow widely used approaches using restrained molecular dynamics in torsion angle space. Families of structures are generated from random extended structures in Xplor *(115)* or CNS *(116)* using *ab initio* simulated annealing. Torsion angle dynamics (TAD) as implemented into, e.g., Xplor or CNS proved to be robust and have a higher convergence rate with respect to molecular dynamics in Cartesian coordinate space *(117)*.

3. The generated structures are further refined against RDC data in a series of molecular dynamic runs with increasing dipolar force constants. Xplor and CNS provide modules for refinements against novel NMR parameters, for example, chemical shifts and anisotropic interactions such as RDCs and phosphorus chemical shift anisotropies.

4. The lowest energy structures after simulated annealing and subsequent refinement against sets of RDCs collected are minimized using the AMBER module Sander *(118)*. Due to more adapted force fields, AMBER yields better and more consistent results for nucleic acids *(119)*.

3.3.3. Structural Statistics

1. In evaluating the quality of a family of RNA NMR structures, a number of statistics can be evaluated: Root Mean Square Deviation (RMSD), number of NOE, RDC, and torsion restraints; residual distance, dipolar coupling, and torsion violations; and the largest distance, dipolar coupling, and torsion violations. Typically, the distance restraints are further dissected into the number of inter-residue, intra-residue, and inter-molecular NOEs.

2. Useful RMSDs to consider include only regions of interest and are usually a more accurate descriptor of the quality of the structure than the overall global RMSD. Local RMSDs are given because the overall global RMSD can easily be in the 2.0–3.0 Å range, which might otherwise be indicative of poor convergence. Almost every RNA structure studied includes a region that is poorly defined, such as a disordered loop, terminal base pair, or a nucleotide without any inter-nucleotide NOEs. This situation is comparable to protein NMR studies, which often neglect the N and C terminal ends of proteins because of the lack of structural data from these regions (*see* **Note 26**).

3. In contrast to crystallographic B-factors, a general measure for the uncertainty in NMR-derived structures is not available. The commonly used RMSD, which is a measure for the precision of the data, tends to overestimate the accuracy of NMR structure ensembles and therefore is a problematic measure for the uncertainty in the atomic coordinates. However, the measurement of a large set of RDCs permits cross-validation to assess the accuracy of NMR-derived atomic coordinates. Structure calculations should be carried out omitting a randomly chosen subset of the RDC data while refining against the remaining RDCs. The accuracy of a family of RNA NMR structures is cross-validated by the agreement between the structures (which are used to back-calculate the RDCs) and the omitted RDC subset *(120, 121)*. Alternatively, a comparison between calculated and observed ^{1}H chemical shifts represents another possibility for cross-validation of structures derived from NMR restraints *(122)*.

4. Notes

1. Unless stated otherwise, all solutions should be prepared in water that has a resistivity of 18.2 MΩ•cm and total organic content of less than five parts per billion. This standard is referred to as "water" in the text.

2. The T7 promoter DNA strand used for oligonucleotide-based *in vitro* transcription should be of the following sequence: 5'-C TAA TAC GAC TCA CTA TAG-3'. The addition of a cytidine nucleotide 5' of the T7 promoter sequence increases stability of the dsDNA and increases yields of product RNA *(123)*.

3. When designing the template strand of ssDNA, care should be taken at both the 5', as well as 3' end to insure optimal yields of RNA *(3, 4)*. If the RNA product contains unacceptable 3'-end inhomogeneity, the template ssDNA can be prepared with a 5' non-hydrogen bonding nucleoside such as 4-methylindole *(124)*. Alternatively, the desired RNA can be transcribed with a 3'-end flanking sequence that folds into a hammerhead ribozyme that cleaves co-transcriptionally to yield a homogenous 3'-end with a 2'-, 3'-cyclic phosphate group *(125, 126)*.

4. 5'-YpA-3' steps in single-stranded regions constitute hot spots for RNA hydrolysis and thus can contribute to long-term chemical instability of an NMR sample. In favorable cases, these dinucleotide steps can be eliminated without compromising the RNA structure.

5. In addition to Notes 2, 3, and 4, care should also be taken when designing a restriction enzyme site at the 3′-end of the plasmid for linearization. The remaining nucleotides should not only reduce 3′-end inhomogeneity, but also should be compatible with secondary or tertiary interactions that may be present in the RNA.

6. It is recommended that nucleotide-5′-triphosphates should be prepared *(127)* or purchased (Sigma-Aldrich, St. Louis, MO; Cambridge Isotope Labs, Andover, MA) as the free acid, ammonium, or sodium salt whenever possible. In our hands, lower transcription yields can result when lithium, magnesium, triethylammonium, and cyclohexylammonium salts are used.

7. T7 RNA polymerase is commercially available (e.g., New England Biolabs, Beverly, MA) but expensive. We prepare T7 RNA polymerase for transcriptions from an *E. coli* overexpressing strain, several million Units at a time, approximately every 6 months.

8. Unpolymerized acrylamide/bisacrylamide is a neurotoxin; therefore, care should be taken to avoid direct exposure.

9. N,N,N,N-Tetramethyl-ethylenediamine (TEMED, Bio-Rad) is best stored at room temperature in a desiccator. Quality of gels and rate of polymerization decline after opening; therefore, purchasing small amounts of TEMED is recommended.

10. Ammonium persulfate (APS) is best stored at 4°C. Quality of gels and rate of polymerization decline over time; therefore, it is recommended that stocks should be prepared frequently.

11. If plasmid DNA is used, one should substitute for plasmid Transcription buffer and omit the PEG-8000.

12. During the transcription reaction, there is a buildup of pyrophosphate that may slow down and in extreme cases inhibit the polymerase reaction by sequestering Mg^{2+}. Transcription yields may be improved with the addition of 1 unit of inorganic pyrophosphatase (IPP, Sigma-Aldrich) per milliliter of transcription. IPP hydrolyzes (insoluble) pyrophosphates. Care should be taken to optimize the transcription in the presence of IPP.

13. Transcription yields may also be improved with the addition of 10 units of RNAase Inhibitor (RNAsin, Promega Corp.) per milliliter of transcription. Care should be taken to optimize the transcription in the presence of inhibitor.

14. When transcription optimizations are being performed, one can directly bring each reaction up in loading buffer and apply directly to the polyacrylamide gel.

15. No general procedures for annealing can be given as conditions can vary between RNAs. Typically, simple stem-loop structures such as the 30 nucleotide-containing TAR RNA can be properly annealed by heat denaturation (95°C for 2 minutes) followed by a snap-cooling step (4°C for 10 minutes) under low to moderate salt conditions.

16. Transcription reactions can be extracted with an equal volume of phenol/chloroform (Fisher Scientific) equilibrated with TE buffer (10 mM Tris-HCl, 1 mM EDTA, pH 8.0) to remove enzymes prior to anion-exchange chromatography.

17. Anion-exchange FPLC gradient conditions should be optimized to increase the resolution for each desired RNA target.

18. Names given to the through bond correlation experiments are derived from the series of nuclei through which magnetization is transferred during the experiment.

19. Before embarking on a detailed and time-consuming NMR investigation of a chosen RNA, it is extremely important to optimize the sample conditions for acquisition of the various required NMR experiments. It is critical to determine at the outset if the system is suitable for a high-resolution NMR structure elucidation. Considerations include: the RNA construct, salt concentrations, pH, and buffer type and concentration.

20. The imino proton region of the proton NMR spectrum of an unlabeled RNA sample in H_2O provides a sensitive diagnostic for this purpose. An example imino proton 1D spectrum for a correctly folded 30mer RNA is shown in **Fig. 2.13**. One peak should be observed for each Watson-Crick base pair in the molecule. Since the imino protons exchange

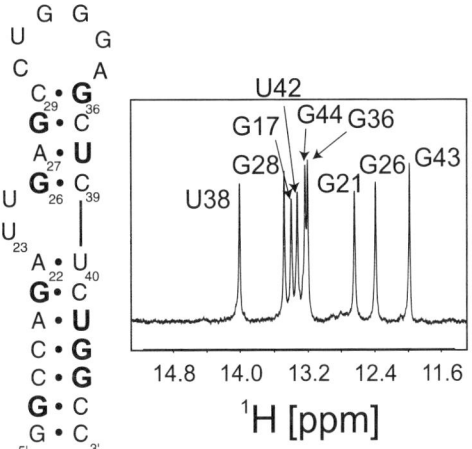

Fig. 2.13. 1D jump-return echo ¹H imino spectrum of the TAR RNA recorded in H_2O. Assignments for observable G and U imino protons are indicated and shown in bold in the secondary structure representation.

rapidly with the bulk H_2O, the spectrum was recorded with a jump-return echo sequence that avoids presaturation, while providing most efficient water suppression *(128)*. The pyrimidine base protons can provide a valuable alternative, circumventing problems related to solvent exchange. H5-H6 cross-peaks can be conveniently monitored in 2D TOCSY or COSY spectra; an example is given in **Fig. 2.14**.

21. The sample conditions are surveyed directly by NMR spectroscopy as a function of RNA and Mg^{2+} concentration in a phosphate buffer (10 mM Na- or K-phosphate, pH 6–7) with moderate monovalent salt (typically 50–100 mM NaCl or KCl) in order to identify constructs and solution conditions suitable for a subsequent structure determination. The goal is to obtain the narrowest line width and best chemical shift dispersion for the observable imino and/or H5-H6 base protons that report on secondary structure formation.

22. Potential problems with interpretation of obtained NOESY cross-peak intensities in terms of ^1H-^1H distances in structure calculations arise mainly from the phenomenon called spin diffusion. Spin diffusion causes a breakdown of the isolated spin pair approximation because other nearby protons provide competing indirect pathways for observing the direct NOE between the two protons. Spin diffusion effects play a role, especially when longer NOESY mixing times (>100 ms) are used. This usually leads to damped NOESY cross-peak intensities that build up through the direct pathway, resulting in underestimated interproton distances. Additionally, multistep transfer pathways can occur, resulting in false NOE assignments. For example, the imino protons of guanines might show spin diffusion mediated NOEs to the

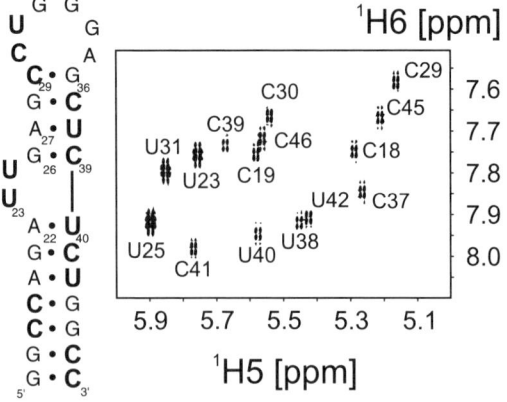

Fig. 2.14. H5-H6 region of a 2D ^1H,^1H DQF-COSY (129) spectrum of the TAR RNA. Assignments for the pyrimidine H5-H6 cross peaks are indicated and shown in bold in the secondary structure representation.

non-exchangeable aromatic H5 and H6 protons of cytidines in Watson-Crick base pairs through the cytidine amino protons. However, in an early stage of the assignment procedure based on NOESY correlations, spin diffusion pathways can aid the identification of spin systems. Thus, for assignments it is recommended to analyze NOESY spectra acquired with shorter (~50 ms) and longer (~150 ms) mixing times.

23. Often ribose puckers are found with homonuclear H1′,H2′/H3′,H4′ coupling constants in the 3–6 Hz, indicative of conformational exchange between the C2′- and C3′-endo puckers. This mixed conformation is typically left unrestrained.

24. The quantitative analysis of scalar J-couplings, especially in the case of homonuclear ^{3}J(H,H) couplings related to the ribose sugar pucker, becomes more and more difficult with increasing molecular weight. In contrast, the efficiency of cross-correlated relaxation pathways scales linearly with the overall correlation time of the molecule, which is related to its size. These new methods that exploit cross-correlated relaxation as a tool for structure determination should allow the characterization of conformations for larger RNA molecules, for which purpose J-coupling analysis is not feasible anymore.

25. Pf1-phage is commercially available (ASLA Biotech Ltd., Riga, Latvia). The phage solution can be exchanged into the NMR buffer by pelleting the phage in an ultracentrifuge (50K for 6 hours) and resuspending in NMR buffer multiple times. Prior to use, the phage should be spun down and resuspended by gently shaking for 6 hours with the RNA sample used in the isotropic experiments.

26. It does not appear that there will be a simple and quick procedure for NMR assignment of RNA molecules. Neglecting the problems with sensitivity or overlap, complete assignment requires a large number of experiments, if all of the optimized sequences are performed (~4 experiments for the bases, ~3 experiments to correlate the base resonances to the ribose, and ~2–3 experiments to correlate the ribose resonances). This results in a very rough estimate of about 20 days measurement time (assuming on average 2 days measurement time per experiment) for a RNA sample, with sample concentrations in the mM range and a molecular weight between 10 and 25 kDa, carried out on spectrometers with at least 500 MHz (proton resonance frequency). The subsequent data analysis and structure elucidation tends to be even more time consuming due to the absence of robust, automated procedures so that a complete RNA structure analysis using procedures reviewed here can not be accomplished in less than 2 month.

Acknowledgments

The authors thank S. Daudenarde and G. Pérez-Alvarado for invaluable discussions. This work was supported by the National Institutes of Health (AI040187 and GM66669, to M.H.).

References

1. Leulliot, N., Varani, G. (2001) Current topics in RNA-protein recognition: control of specificity and biological function through induced fit and conformational capture. *Biochemistry* 40, 7947–7956.
2. Williamson, J. R. (2000) Induced fit in RNA-protein recognition. *Nat Struct Biol* 7, 834–837.
3. Milligan, J. F., Groebe, D. R., Witherell, G. W., et al. (1987) Oligoribonucleotide synthesis using T7 RNA polymerase and synthetic DNA templates. *Nucleic Acids Res* 15, 8783–8798.
4. Milligan, J. F., Uhlenbeck, O. C. (1989) Synthesis of small RNAs using T7 RNA polymerase. *Methods Enzymol* 180, 51–62.
5. Perez-Canadillas, J. M., Varani, G. (2001) Recent advances in RNA-protein recognition. *Curr Opin Struct Biol* 11, 53–58.
6. Batey, R. T., Battiste, J. L., Williamson, J. R. (1995) Preparation of isotopically enriched RNAs for heteronuclear NMR. *Methods Enzymol* 261, 300–322.
7. Batey, R. T., Cloutier, N., Mao, H., et al. (1996) Improved large scale culture of *Methylophilus methylotrophus* for 13C/15N labeling and random fractional deuteration of ribonucleotides. *Nucleic Acids Res* 24, 4836–4837.
8. Batey, R. T., Inada, M., Kujawinski, E., et al. (1992) Preparation of isotopically labeled ribonucleotides for multidimensional NMR spectroscopy of RNA. *Nucleic Acids Res* 20, 4515–4523.
9. Nikonowicz, E. P., Sirr, A., Legault, P., et al. (1992) Preparation of 13C and 15N labelled RNAs for heteronuclear multidimensional NMR studies. *Nucleic Acids Res* 20, 4507–4513.
10. Zidek, L., Stefl, R., Sklenar, V. (2001) NMR methodology for the study of nucleic acids. *Curr Opin Struct Biol* 11, 275–281.
11. Cromsigt, J., van Buuren, B., Schleucher, J., et al. (2001) Resonance assignment and structure determination for RNA. *Methods Enzymol* 338, 371–399.
12. Furtig, B., Richter, C., Wohnert, J., et al. (2003) NMR spectroscopy of RNA. *Chembiochem* 4, 936–962.
13. Latham, M. P., Brown, D. J., McCallum, S. A., et al. (2005) NMR methods for studying the structure and dynamics of RNA. *Chembiochemistry* 6, 1492–1505.
14. Wu, H., Finger, L. D., Feigon, J. (2005) Structure determination of protein/RNA complexes by NMR. *Methods Enzymol* 394, 525–545.
15. Bax, A., Kontaxis, G., Tjandra, N. (2001) Dipolar couplings in macromolecular structure determination. *Methods Enzymol* 339, 127–174.
16. Hansen, M. R., Mueller, L., Pardi, A. (1998) Tunable alignment of macromolecules by filamentous phage yields dipolar coupling interactions. *Nat Struct Biol* 5, 1065–1074.
17. Tjandra, N., Bax, A. (1997) Direct measurement of distances and angles in biomolecules by NMR in a dilute liquid crystalline medium. *Science* 278, 1111–1114.
18. Reif, B., Hennig, M., Griesinger, C. (1997) Direct measurement of angles between bond vectors in high-resolution NMR. *Science* 276, 1230–1233.
19. Schwalbe, H., Carlomagno, T., Hennig, M., et al. (2001) Cross-correlated relaxation for measurement of angles between tensorial interactions. *Methods Enzymol* 338, 35–81.
20. Dingley, A. J., Grzesiek, S. (1998) Direct observation of hydrogen bonds in nucleic acid base pairs by internucleotide (2)J(NN) couplings. *J Am Chem Soc* 120, 8293–8297.
21. Pervushin, K., Ono, A., Fernandez, C., et al. (1998) NMR scaler couplings across Watson-Crick base pair hydrogen bonds in DNA observed by transverse relaxation optimized spectroscopy. *Proc Natl Acad Sci U S A* 95, 14147–14151.
22. Churcher, M. J., Lamont, C., Hamy, F., et al. (1993) High affinity binding of TAR RNA by the human immunodeficiency virus type-1 tat protein requires base-pairs in the RNA stem and amino acid residues

flanking the basic region. *J Mol Biol* 230, 90–110.

23. Long, K. S., Crothers, D. M. (1995) Interaction of human immunodeficiency virus type 1 Tat-derived peptides with TAR RNA. *Biochemistry* 34, 8885–8895.

24. Tao, J., Frankel, A. D. (1992) Specific binding of arginine to TAR RNA. *Proc Natl Acad Sci U S A* 89, 2723–2726.

25. Tao, J., Frankel, A. D. (1993) Electrostatic interactions modulate the RNA-binding and transactivation specificities of the human immunodeficiency virus and simian immunodeficiency virus Tat proteins. *Proc Natl Acad Sci U S A* 90, 1571–1575.

26. Puglisi, J. D., Chen, L., Frankel, A. D., et al. (1993) Role of RNA structure in arginine recognition of TAR RNA. *Proc Natl Acad Sci U S A* 90, 3680–3684.

27. Puglisi, J. D., Tan, R., Calnan, B. J., et al. (1992) Conformation of the TAR RNA-arginine complex by NMR spectroscopy. *Science* 257, 76–80.

28. Aboul-ela, F., Karn, J., Varani, G. (1995) The structure of the human immunodeficiency virus type-1 TAR RNA reveals principles of RNA recognition by Tat protein. *J Mol Biol* 253, 313–332.

29. Brodsky, A. S., Williamson, J. R. (1997) Solution structure of the HIV-2 TAR-argininamide complex. *J Mol Biol* 267, 624–639.

30. Aboul-ela, F., Karn, J., Varani, G. (1996) Structure of HIV-1 TAR RNA in the absence of ligands reveals a novel conformation of the trinucleotide bulge. *Nucleic Acids Res* 24, 3974–3781.

31. Long, K. S., Crothers, D. M. (1999) Characterization of the solution conformations of unbound and Tat peptide-bound forms of HIV-1 TAR RNA. *Biochemistry* 38, 10059–10069.

32. Sambrook, J., Fritsch, E. F., Maniatis, T. (1989) *Molecular Cloning: A Laboratory Manual*, Cold Spring Harbor Laboratory Press, Cold Spring Harbor, NY.

33. Yin, Y., Carter, C. W., Jr. (1996) Incomplete factorial and response surface methods in experimental design: yield optimization of tRNA(Trp) from in vitro T7 RNA polymerase transcription. *Nucleic Acids Res* 24, 1279–1286.

34. Wyatt, J. R., Chastain, M., Puglisi, J. D. (1991) Synthesis and purification of large amounts of RNA oligonucleotides. *Biotechniques* 11, 764–769.

35. Anderson, A. C., Scaringe, S. A., Earp, B. E., et al. (1996) HPLC purification of RNA for crystallography and NMR. *RNA* 2, 110–117.

36. Shields, T. P., Mollova, E., Ste Marie, L., et al. (1999) High-performance liquid chromatography purification of homogenous-length RNA produced by trans cleavage with a hammerhead ribozyme. *RNA* 5, 1259–1267.

37. Wuthrich, K. (1986) *NMR of Proteins and Nucleic Acids*, Wiley, New York.

38. Nikonowicz, E. P., Pardi, A. (1992) Three-dimensional heteronuclear NMR studies of RNA. *Nature* 355, 184–186.

39. Nikonowicz, E. P., Pardi, A. (1993) An efficient procedure for assignment of the proton, carbon and nitrogen resonances in 13C/15N labeled nucleic acids. *J Mol Biol* 232, 1141–1156.

40. Wijmenga, S. S., van Buuren, B. N. M. (1998) The use of NMR methods for conformational studies of nucleic acids. *Prog Nucl Magn Reson Spectrosc* 32, 287–387.

41. Zhou, H., Vermeulen, A., Jucker, F. M., et al. (1999) Incorporating residual dipolar couplings into the NMR solution structure determination of nucleic acids. *Biopolymers* 52, 168–180.

42. Hansen, M. R., Hanson, P., Pardi, A. (2000) Filamentous bacteriophage for aligning RNA, DNA, and proteins for measurement of nuclear magnetic resonance dipolar coupling interactions. *Methods Enzymol* 317, 220–240.

43. Meiler, J., Prompers, J. J., Peti, W., et al. (2001) Model-free approach to the dynamic interpretation of residual dipolar couplings in globular proteins. *J Am Chem Soc* 123, 6098–6107.

44. Peti, W., Meiler, J., Bruschweiler, R., et al. (2002) Model-free analysis of protein backbone motion from residual dipolar couplings. *J Am Chem Soc* 124, 5822–5833.

45. Tolman, J. R. (2002) A novel approach to the retrieval of structural and dynamic information from residual dipolar couplings using several oriented media in biomolecular NMR spectroscopy. *J Am Chem Soc* 124, 12020–12030.

46. Lippens, G., Dhalluin, C., Wieruszeski, J. M. (1995) Use of a Water Flip-Back Pulse in the Homonuclear Noesy Experiment. *J Biomol Nmr* 5, 327–331.

47. Peterson, R. D., Theimer, C. A., Wu, H. H., et al. (2004) New applications of 2D filtered/edited NOESY for assignment and

structure elucidation of RNA and RNA-protein complexes. *J Biomol Nmr* 28, 59–67.

48. Duchardt, E., Richter, C., Reif, B., et al. (2001) Measurement of 2J(H,C)- and 3J(H,C)-coupling constants by alpha/beta selective HC(C)H-TOCSY. *J Biomol NMR* 21, 117–126.

49. Schwalbe, H., Marino, J. P., Glaser, S. J., et al. (1995) Measurement of H,H-Coupling Constants Associated with nu1, nu2, and nu3 in Uniformly 13C-Labeled RNA by HCC-TOCSY-CCH-E.COSY *J Am Chem Soc* 117, 7251–7252.

50. Schwalbe, H., Marino, J. P., King, G. C., et al. (1994) Determination of a complete set of coupling constants in 13C-labeled oligonucleotides. *J Biomol NMR* 4, 631–644.

51. Hines, J. V., Varani, G., Landry, S. M., et al. (1993) The stereospecific assignment of H5′ and H5″ in RNA using the sign of two-bond carbon-proton scalar coupling. *J Am Chem Soc* 115, 11002–11003.

52. Trantirek, L., Stefl, R., Masse, J. E., et al. (2002) Determination of the glycosidic torsion angles in uniformly C-13-labeled nucleic acids from vicinal coupling constants (3)J(C2/4-H1′) and (3)J(C6/8-H1′). *J Biomol NMR* 23, 1–12.

53. Carlomagno, T., Hennig, M., Williamson, J. R. (2002) A novel PH-cT-COSY methodology for measuring JPH coupling constants in unlabeled nucleic acids. application to HIV-2 TAR RNA. *J Biomol NMR* 22, 65–81.

54. Sklenar, V., Miyashiro, H., Zon, G., et al. (1986) Assignment of the 31P and 1H resonances in oligonucleotides by two-dimensional NMR spectroscopy. *FEBS Lett* 208, 94–98.

55. Schwalbe, H., Samstag, W., Engels, J. W., et al. (1993) Determination of 3J(C,P) and 3J(H,P) coupling constants in nucleotide oligomers with FIDS-HSQC. *J Biomol NMR* 3, 479–486.

56. Hoogstraten, C. G., Pardi, A. (1998) Measurement of carbon-phosphorus J coupling constants in RNA using spin-echo difference constant-time HCCH-COSY. *J Magn Reson* 133, 236–240.

57. Legault, P., Jucker, F. M., Pardi, A. (1995) Improved measurement of 13C, 31P J coupling constants in isotopically labeled RNA. *FEBS Lett* 362, 156–160.

58. Szyperski, T., Fernandez, C., Ono, A., et al. (1999) The 2D [31P] spin-echo-difference constant-time [13C, 1H]-HMQC experiment for simultaneous determination of 3J(H3′P) and 3J(C4′P) in 13C-labeled nucleic acids and their protein complexes. *J Magn Reson* 140, 491–494.

59. Hu, W., Bouaziz, S., Skripkin, E., et al. (1999) Determination of 3J(H3i, Pi+1) and 3J(H5i/5i, Pi) coupling constants in 13C-labeled nucleic acids using constant-time HMQC. *J Magn Reson* 139, 181–185.

60. Clore, G. M., Murphy, E. C., Gronenborn, A. M., et al. (1998) Determination of three-bond 1H3′-31P couplings in nucleic acids and protein-nucleic acid complexes by quantitative J correlation spectroscopy. *J Magn Reson* 134, 164–167.

61. Gotfredsen, C. H., Meissner, A., Duus, J. O., et al. (2000) New methods for measuring 1H-31P coupling constants in nucleic acids. *Magn Reson Chem* 38, 692–695.

62. Richter, C., Reif, B., Worner, K., et al. (1998) A new experiment for the measurement of nJ(C,P) coupling constants including 3J(C4′i,Pi) and 3J(C4′i,Pi+1) in oligonucleotides. *J Biomol NMR* 12, 223–230.

63. Legault, P., Pardi, A. (1994) 31P chemical shift as a probe of structural motifs in RNA. *J Magn Reson B* 103, 82–86.

64. Richter, C., Reif, B., Griesinger, C., et al. (2000) NMR spectroscopic determination of angles and in RNA from CH-dipolar coupling, P-CSA cross-correlated relaxation. *J Am Chem Soc* 122, 12728–12731.

65. Duchardt, E., Richter, C., Ohlenschlager, O., et al. (2004) Determination of the glycosidic bond angle chi in RNA from cross-correlated relaxation of CH dipolar coupling and N chemical shift anisotropy. *J Am Chem Soc* 126, 1962–1970.

66. Ravindranathan, S., Kim, C. H., Bodenhausen, G. (2003) Cross correlations between 13C-1H dipolar interactions and 15N chemical shift anisotropy in nucleic acids. *J Biomol NMR* 27, 365–375.

67. Felli, I. C., Richter, C., Griesinger, C., et al. (1999) Determination of RNA sugar pucker mode from cross-correlated relaxation in solution NMR spectroscopy. *J Am Chem Soc* 121, 1956–1957.

68. Richter, C., Griesinger, C., Felli, I., et al. (1999) Determination of sugar conformation in large RNA oligonucleotides from analysis of dipole-dipole cross correlated relaxation by solution NMR spectroscopy. *J Biomol NMR* 15, 241–250.

69. Andersson, P., Weigelt, J., Otting, G. (1998) Spin-state selection filters for the measurement of heteronuclear one-bond coupling constants. *J Biomol NMR* 12, 435–441.

70. Tjandra, N., Bax, A. (1997) Measurement of dipolar contributions to 1JCH splittings from magnetic-field dependence of J modulation in two-dimensional NMR spectra. *J Magn Reson* 124, 512–515.
71. Tjandra, N., Grzesiek, S., Bax, A. (1996) Magnetic field dependence of nitrogen-proton J splittings in N-15-enriched human ubiquitin resulting from relaxation interference and residual dipolar coupling. *J Am Chem Soc* 118, 6264–6272.
72. Hennig, M., Carlomagno, T., Williamson, J. R. (2001) Residual dipolar coupling TOCSY for direct through space correlations of base protons and phosphorus nuclei in RNA. *J Am Chem Soc* 123, 3395–3396.
73. Wu, Z., Tjandra, N., Bax, A. (2001) Measurement of 1H3′-31P dipolar couplings in a DNA oligonucleotide by constant-time NOESY difference spectroscopy. *J Biomol NMR* 19, 367–370.
74. Wu, Z., Tjandra, N., Bax, A. (2001) 31P chemical shift anisotropy as an aid in determining nucleic acid structure in liquid crystals. *J Am Chem Soc* 123, 3617–3618.
75. Dingley, A. J., Masse, J. E., Feigon, J., et al. (2000) Characterization of the hydrogen bond network in guanosine quartets by internucleotide 3hJ(NC)′ and 2hJ(NN) scalar couplings. *J Biomol NMR* 16, 279–289.
76. Dingley, A. J., Masse, J. E., Peterson, R. D., et al. (1999) Internucleotide scalar couplings across hydrogen bonds in Watson-Crick and Hoogsteen base pairs of a DNA triplex. *J Am Chem Soc* 121, 6019–6027.
77. Hennig, M., Williamson, J. R. (2000) Detection of N-H...N hydrogen bonding in RNA via scalar couplings in the absence of observable imino proton resonances. *Nucleic Acids Res* 28, 1585–1593.
78. Liu, A. Z., Majumdar, A., Hu, W. D., et al. (2000) NMR detection of N-H...O=C hydrogen bonds in C-13,N-15-labeled nucleic acids. *J Am Chem Soc* 122, 3206–3210.
79. Majumdar, A., Kettani, A., Skripkin, E. (1999) Observation and measurement of internucleotide 2JNN coupling constants between 15N nuclei with widely separated chemical shifts. *J Biomol NMR* 14, 67–70.
80. Majumdar, A., Kettani, A., Skripkin, E., et al. (1999) Observation of internucleotide NH...N hydrogen bonds in the absence of directly detectable protons. *J Biomol NMR* 15, 207-2-11.
81. Wohnert, J., Ramachandran, R., Gorlach, M., et al. (1999) Triple-resonance experiments for correlation of H5 and exchangeable pyrimidine base hydrogens in (13)C,(15)N-labeled RNA. *J Magn Reson* 139, 430–433.
82. Sklenar, V., Peterson, R. D., Rejante, M. R., et al. (1994) Correlation of nucleotide base and sugar protons in a N-15-labeled Hiv-1 RNA oligonucleotide by H-1-N-15 Hsqc experiments. *J Biomol NMR* 4, 117–122.
83. Fohrer, J., Hennig, M., Carlomagno, T. (2006) Influence of the 2′-hydroxyl group conformation on the stability of A-form helices in RNA. *J Mol Biol* 356, 280–287.
84. Hennig, M., Fohrer, J., Carlomagno, T. (2005) Assignment and NOE analysis of 2′-hydroxyl protons in RNA: implications for stabilization of RNA A-form duplexes. *J Am Chem Soc* 127, 2028–2029.
85. Giedroc, D. P., Cornish, P. V., Hennig, M. (2003) Detection of scalar couplings involving 2′-hydroxyl protons across hydrogen bonds in a frameshifting mRNA pseudoknot. *J Am Chem Soc* 125, 4676–4677.
86. Simorre, J. P., Zimmermann, G. R., Mueller, L., et al. (1996) Correlation of the guanosine exchangeable and nonexchangeable base protons in 13C-/15N-labeled RNA with an HNC-TOCSY-CH experiment. *J Biomol NMR* 7, 153–156.
87. Simorre, J. P., Zimmermann, G. R., Mueller, L., et al. (1996) Triple-resonance experiments for assignment of adenine base resonances in C-13/N-15-labeled RNA. *J Am Chem Soc* 118, 5316–5317.
88. Simorre, J. P., Zimmermann, G. R., Pardi, A., et al. (1995) Triple resonance HNCCCH experiments for correlating exchangeable and nonexchangeable cytidine and uridine base protons in RNA. *J Biomol NMR* 6, 427–432.
89. Sklenar, V., Dieckmann, T., Butcher, S. E., et al. (1996) Through-bond correlation of imino and aromatic resonances in C-13-,N-15-labeled RNA via heteronuclear TOCSY. *J Biomol Nmr* 7, 83–87.
90. Wohnert, J., Gorlach, M., Schwalbe, H. (2003) Triple resonance experiments for the simultaneous correlation of H6/H5 and exchangeable protons of pyrimidine nucleotides in C-13, N-15-labeled RNA applicable to larger RNA molecules. *J Biomol Nmr* 26, 79–83.
91. Fiala, R., Jiang, F., Patel, D. J. (1996) Direct correlation of exchangeable and non-exchangeable protons on purine bases in 13C,15N-labeled RNA using a HCCNH-TOCSY experiment. *J Am Chem Soc* 118, 689–690.
92. Simon, B., Zanier, K., Sattler, M. (2001) A TROSY relayed HCCH-COSY experiment

for correlating adenine H2/H8 resonances in uniformly 13C-labeled RNA molecules. *J Biomol NMR* 20, 173–176.
93. Legault, P., Farmer, B. T., Mueller, L., et al. (1994) Through-bond correlation of adenine protons in a C-13-labeled ribozyme. *J Am Chem Soc* 116, 2203–2204.
94. Marino, J. P., Prestegard, J. H., Crothers, D. M. (1994) Correlation of adenine H2/H8 resonances in uniformly C-13 labeled Rnas by 2d Hcch-Tocsy: a new tool for H-1 assignment. *J Am Chem Soc* 116, 2205–2206.
95. Marino, J. P., Diener, J. L., Moore, P. B., et al. (1997) Multiple-quantum coherence dramatically enhances the sensitivity of CH and CH2 correlations in uniformly 13C-labeled RNA. *J Am Chem Soc* 119, 7361–7366.
96. Sklenar, V., Dieckmann, T., Butcher, S. E., et al. (1998) Optimization of triple-resonance HCN experiments for application to larger RNA oligonucleotides. *J Magn Reson* 130, 119–124.
97. Fiala, R., Czernek, J., Sklenar, V. (2000) Transverse relaxation optimized triple-resonance NMR experiments for nucleic acids. *J Biomol NMR* 16, 291–302.
98. Riek, R., Pervushin, K., Fernandez, C., et al. (2001) [(13)C,(13)C]- and [(13)C,(1)H]-TROSY in a triple resonance experiment for ribose-base and intrabase correlations in nucleic acids. *J Am Chem Soc* 123, 658–664.
99. Farmer, B. T., Muller, L., Nikonowicz, E. P., et al. (1993) Unambiguous resonance assignments in carbon-13, nitrogen-15-labeled nucleic acids by 3D triple-resonance NMR. *J Am Chem Soc* 115, 11040–11041.
100. Sklenar, V., Rejante, M. R., Peterson, R. D., et al. (1993) Two-dimensional triple-resonance HCNCH experiment for direct correlation of ribose H1′ and base H8, H6 protons in 13C,15N-labeled RNA oligonucleotides. *J Am Chem Soc* 115, 12181–12182.
101. Fesik, S. W., Eaton, H. L., Olejniczak, E. T., et al. (1990) 2D and 3D NMR spectroscopy employing carbon-13/carbon-13 magnetization transfer by isotropic mixing. Spin system identification in large proteins. *J Am Chem Soc* 112, 886–888.
102. Kay, L. E., Ikura, M., Bax, A. (1990) Proton-proton correlation via carbon-carbon couplings: a three-dimensional NMR approach for the assignment of aliphatic resonances in proteins labeled with carbon-13. *J Am Chem Soc* 112, 888–889.
103. Pardi, A. (1995) Multidimensional heteronuclear NMR experiments for structure determination of isotopically labeled RNA. *Methods Enzymol* 261, 350–380.
104. Pardi, A., Nikonowicz, E. P. (1992) Simple procedure for resonance assignment of the sugar protons in carbon-13 labeled RNAs. *J Am Chem Soc* 114, 9202–9203.
105. Hu, W., Kakalis, L. T., Jiang, L., et al. (1998) 3D HCCH-COSY-TOCSY experiment for the assignment of ribose and amino acid side chains in 13C labeled RNA and protein. *J Biomol NMR* 12, 559–564.
106. Glaser, S. J., Schwalbe, H., Marino, J. P., et al. (1996) Directed TOCSY, a method for selection of directed correlations by optimal combinations of isotropic and longitudinal mixing. *J Magn Reson B* 112, 160–180.
107. Kellogg, G. W. (1992) Proton-detected hetero-TOCSY experiments with application to nucleic acids. *J Magn Reson* 98, 176–182.
108. Kellogg, G. W., Szewczak, A. A., Moore, P. B. (1992) Two-dimensional hetero-TOCSY-NOESY. Correlation of phosphorus-31 resonances with anomeric and aromatic proton resonances in RNA. *J Am Chem Soc* 114, 2727–2728.
109. Kellogg, G. W., Schweitzer, B. I. (1993) Two- and three-dimensional 31P-driven NMR procedures for complete assignment of backbone resonances in oligodeoxyribonucleotides. *J Biomol NMR* 3, 577–595.
110. Heus, H. A., Wijmenga, S. S., Vandeven, F. J. M., et al. (1994) Sequential backbone assignment in C-13-labeled Rna via through-bond coherence transfer using 3-dimensional triple-resonance spectroscopy (H-1, C-13, P-31) and 2-dimensional hetero Tocsy. *J Am Chem Soc* 116, 4983–4984.
111. Marino, J. P., Schwalbe, H., Anklin, C., et al. (1994) A 3-dimensional triple-resonance H-1,C-13,P-31 experiment: sequential through-bond correlation of ribose protons and intervening phosphorus along the Rna oligonucleotide backbone. *J Am Chem Soc* 116, 6472–6473.
112. Marino, J. P., Schwalbe, H., Anklin, C., et al. (1995) Sequential correlation of anomeric ribose protons and intervening phosphorus in RNA oligonucleotides by a 1H, 13C, 31P triple resonance experiment: HCP-CCH-TOCSY. *J Biomol NMR* 5, 87–92.
113. Wijmenga, S. S., Heus, H. A., Leeuw, H. A., et al. (1995) Sequential backbone assignment of uniformly 13C-labeled RNAs by a two-dimensional P(CC)H-TOCSY triple resonance NMR experiment. *J Biomol NMR* 5, 82–86.

114. Marino, J. P., Schwalbe, H., Griesinger, C. (1999) J-coupling restraints in RNA structure determination. *Acc Chem Res* 32, 614–623.

115. Schwieters, C. D., Kuszewski, J. J., Tjandra, N., et al. (2003) The Xplor-NIH NMR molecular structure determination package. *J Magn Reson* 160, 65–73.

116. Brunger, A. T., Adams, P. D., Clore, G. M., et al. (1998) Crystallography & NMR system: A new software suite for macromolecular structure determination. *Acta Crystallographica Section D-Biological Crystallography* 54, 905–921.

117. Stein, E. G., Rice, L. M., Brunger, A. T. (1997) Torsion-angle molecular dynamics as a new efficient tool for NMR structure calculation. *J Magn Reson* 124, 154–164.

118. Pearlman, D. A., Case, D. A., Caldwell, J. W., et al. (1995) AMBER, a computer program for applying molecular mechanics, normal mode analysis, molecular dynamics and free energy calculations to elucidate the structures and energies of molecules. *Computer Physics Communications* 91, 1–41.

119. Tsui, V., Case, D. A. (2000) Molecular dynamics simulations of nucleic acids with a generalized born solvation model. *J Am Chem Soc* 122, 2489–2498.

120. Clore, G. M., Garrett, D. S. (1999) R-factor, free R, and complete cross-validation for dipolar coupling refinement of NMR structures. *J Am Chem Soc* 121, 9008–9012.

121. Clore, G. M., Kuszewski, J. (2003) Improving the accuracy of NMR structures of RNA by means of conformational database potentials of mean force as assessed by complete dipolar coupling cross-validation. *J Am Chem Soc* 125, 1518–1525.

122. Cromsigt, J. A., Hilbers, C. W., Wijmenga, S. S. (2001) Prediction of proton chemical shifts in RNA. Their use in structure refinement and validation. *J Biomol NMR* 21, 11–29.

123. Baklanov, M. M., Golikova, L. N., Malygin, E. G. (1996) Effect on DNA transcription of nucleotide sequences upstream to T7 promoter. *Nucleic Acids Res* 24, 3659–3660.

124. Moran, S., Ren, R. X., Sheils, C. J., et al. (1996) Non-hydrogen bonding 'terminator' nucleosides increase the 3'-end homogeneity of enzymatic RNA and DNA synthesis. *Nucleic Acids Res* 24, 2044–2052.

125. Ferre-D'Amare, A. R., Doudna, J. A. (1996) Use of cis- and trans-ribozymes to remove 5' and 3' heterogeneities from milligrams of in vitro transcribed RNA. *Nucleic Acids Res* 24, 977–978.

126. Price, S. R., Ito, N., Oubridge, C., et al. (1995) Crystallization of RNA-protein complexes. I. Methods for the large-scale preparation of RNA suitable for crystallographic studies. *J Mol Biol* 249, 398–408.

127. Scott, L. G., Tolbert, T. J., Williamson, J. R. (2000) Preparation of specifically 2H- and 13C-labeled ribonucleotides. *Methods Enzymol* 317, 18–38.

128. Sklenar, V., Brooks, B. R., Zon, G., et al. (1987) Absorption mode two-dimensional NOE spectroscopy of exchangeable protons in oligonucleotides. *FEBS Lett* 216, 249–252.

129. Rance, M., Sorensen, O. W., Bodenhausen, G., et al. (1983) Improved spectral resolution in cosy 1H NMR spectra of proteins via double quantum filtering. *Biochem Biophys Res Commun* 117, 479–485.

Chapter 3

Protein Structure Determination by X-Ray Crystallography

Andrea Ilari and Carmelinda Savino

Abstract

X-ray biocrystallography is the most powerful method to obtain a macromolecular structure. The improvement of computational technologies in recent years and the development of new and powerful computer programs together with the enormous increment in the number of protein structures deposited in the Protein Data Bank, render the resolution of new structures easier than in the past. The aim of this chapter is to provide practical procedures useful for solving a new structure. It is impossible to give more than a flavor of what the x-ray crystallographic technique entails in one brief chapter; therefore, this chapter focuses its attention on the Molecular Replacement method. Whenever applicable, this method allows the resolution of macromolecular structures starting from a single data set and a search model downloaded from the PDB, with the aid only of computer work.

Key words: X-ray crystallography, protein crystallization, molecular replacement, coordinates refinement, model building.

1. Introduction

1.1. Protein Crystallization

The first requirement for protein structure determination by x-ray crystallography is to obtain protein crystals diffracting at high resolution. Protein crystallization is mainly a "trial and error" procedure in which the protein is slowly precipitated from its solution. As a general rule, the purer the protein, the better the chances to grow crystals. Growth of protein crystals starts from a supersaturated solution of the macromolecule, and evolves toward a thermodynamically stable state in which the protein is partitioned between a solid phase and the solution. The time required before the equilibrium is reached has a great influence on the final result, which can go from an amorphous or microcrystalline precipitate to large single crystals. The supersaturation conditions can be obtained by addition of precipitating agents (salts, organic solvents, and polyethylene glycol polymers) and/or by modifying some of the internal parameters of the solution, such as pH, temperature and protein concentration. Since proteins are labile molecules, extreme conditions of precipitation, pH and temperature should be avoided. Protein crystallization involves three main steps *(1)*:

1. Determination of protein degree of purity. If the protein is not at least 90–95% pure, further purification will have to be carried out to achieve crystallization.
2. The protein is dissolved in a suitable solvent from which it must be precipitated in crystalline form. The solvent is usually a water buffer solution.
3. The solution is brought to supersaturation. In this step, small aggregates are formed, which are the nuclei for crystal growth. Once nuclei have been formed, actual crystal growth begins.

1.2. Crystal Preparation and Data Collection

1.2.1. Protein Crystals

A crystal is a periodic arrangement of molecules in three-dimensional space. Molecules precipitating from a solution tend to reach the lowest free energy state. This is often accomplished by packing in a regular way. Regular packing involves the repetition in the three space dimensions of the unit cell, defined by three vectors, a, b, and c, and three angles α, β, and γ between them. The unit cell contains a number of asymmetric units, which coincide with our macromolecule or more copies of it, related by symmetry operations such as rotations with or without translations. There are 230 different ways to combine the symmetry operations in a crystal, leading to 230 space groups, a list of which can be found in the *International Table of Crystallography (2)*. Nevertheless, only 65 space groups are allowed in protein crystals, because the application of mirror planes and inversion points would change the configuration of amino acids from L to D, and D-amino acids are never found in natural protein.

Macromolecule crystals are loosely packed and contain large solvent-filled holes and channels, which normally occupy 40–60% of the crystal volume. For this reason, protein crystals are very fragile and have to be handled with care. In order to maintain its water content unaltered, protein crystals should always be kept in their mother liquor or in the saturated vapor of their mother liquor *(3, 4)*. During data collection (see the following) x-rays may cause crystal damage due to the formation of free radicals. The best way to avoid damage is crystal freezing. In cryo-crystallography protein crystals are soaked in a solution called "cryoprotectant" so that, when frozen, vitrified water, rather then crystalline ice, is formed. In these conditions, crystals exposed to x-rays undergo negligible radiation damage. Cryo-crystallography usually allows a complete data set to be collected from a single crystal and results in generally higher quality and higher resolution diffraction data, while providing more accurate structural information. Normally, all measurements—both in house and using synchrotron radiation—are performed at 100 K.

1.2.2. X-Ray Diffraction

X-ray scattering or diffraction is a phenomenon involving both interference and coherent scattering. Two mathematical descriptions of the interference effect were put forward by Max von Laue and W. L. Bragg *(5, 6)*. The simplest description, known as Bragg's law, is presented here. According to Bragg, x-ray diffraction can be viewed as a process similar to reflection by planes of atoms in the crystal. Incident x-rays are scattered by crystal planes, identified by the Miller indices hkl (*see* **Note 1**) with an angle of reflection θ. Constructive interference only occurs when the path-length difference between rays diffracting from parallel crystal planes is an integral number of wavelengths. When the crystal planes are separated by a distance d, the path length difference is 2d·sinθ. Thus, for constructive interference to occur the following relation must hold true: nλ = 2d·sinθ. As a consequence of Bragg's law, to "see" the individual atoms in a structure, the radiation wavelength must be similar to interatomic distances (typically 0.15 nm or 1.5 Å).

1.2.3. X-Ray Sources

X-rays are produced in the laboratory by accelerating a beam of electrons emitted by a cathode into an anode, the metal of which dictates the wavelength of the resulting x-ray. Monochromatization is carried out either by using a thin metal foil that absorbs much of the unwanted radiation or by using the intense low-order diffraction from a graphite crystal. To obtain a brighter source, the anode can be made to revolve (rotating anode generator) and is water-cooled to prevent it from melting. An alternative source of x-rays is obtained when a beam of electrons is bent by a magnet. This is the principle behind the synchrotron radiation sources that are capable of producing x-ray beams some thousand times more intense than a rotating anode generator. A consequence of this high-intensity radiation source is that data collection times have been drastically reduced. A further advantage is that the x-ray spectrum is continuous from around 0.05 to 0.3 nm (*see* **Note 2**).

1.2.4. X-Ray Detector

In an x-ray diffraction experiment, a diffraction pattern is observed that could be regarded as a three-dimensional lattice, reciprocal to the actual crystal lattice (*see* **Note 3** and **Fig. 3.1**). For a crystal structure determination the intensities of all diffracted reflections must be measured. To do so, all corresponding reciprocal lattice points must be brought to diffracting conditions by rotating the

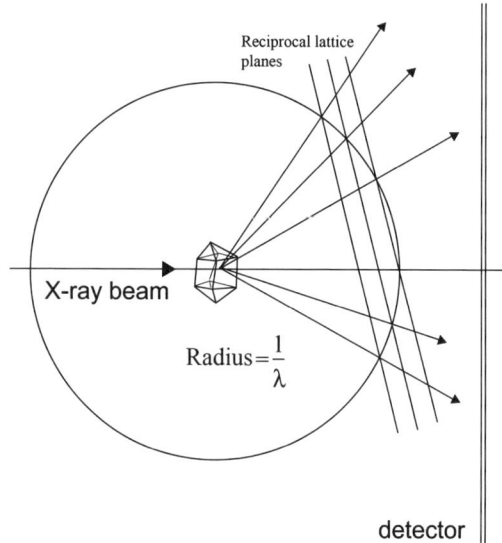

Fig. 3.1. Ewald sphere. A diagrammatic representation of the generation of an x-ray diffraction pattern (*see* **Note 3**).

lattice (i.e., by rotating the crystal) until the required reciprocal lattice points are on a sphere with radius $1/\lambda$. It follows that an x-ray diffraction instrument consists of two parts:

1. A mechanical part for rotating the crystal
2. A detecting device to measure the position and the intensity of the diffracted reflections

For a protein structure determination the number of diffracted beams to be recorded is extremely high (of the order of 10^4–10^6) and requires highly efficient hardware. The most efficient and fastest devices for data collection in protein crystallography are known as the image plate, and the CCD camera (*see* **Notes 4** and **5**). These instruments are much more sensitive and faster than an x-ray film, reducing considerably the time for exposure and data processing and solving the time-consuming data collection problem.

1.2.5. Data Measurement and Data Processing

Successful data integration depends on the choice of the experimental parameters during data collection. Therefore, it is crucial that the diffraction experiment is correctly designed and executed. The essence of the data collection strategy is to collect every unique reflection at least once. The most important issues that have to be considered are:

1. The crystal must be single.
2. In order to have a good signal-to-noise ratio, it is recommended to measure crystal diffraction at the detector edge.
3. The exposure time has to be chosen carefully: it has to be long enough to allow collection of high resolution data (*see* **Note 6**),

but not so long as to cause overload reflections at low resolution and radiation damage.

4. The rotation angle per image should be optimized: an angle too large will result in spatial overlap of spots, an angle too small will give too many partial spots (*see* **Note 7**).

5. High data multiplicity will improve the overall quality of the data by reducing random errors and facilitating outlier identification.

Data analysis, performed with modern data reduction programs, is normally performed in three stages:

1. Autoindexing of one image. The program deduces the lattice type, the crystal unit cell parameters and crystal orientation parameters from a single oscillation image.

2. Indexing of all images. The program compares the diffraction measurements to the spots predicted on the basis of the autoindexing parameters, assigns the hkl indices and calculates the diffraction intensities for each spot in all the collected images.

3. Scaling. The program scales together the data of all the collected images and calculates the structure factor amplitudes for each reflection (identified by the indices hkl).

1.3. Structure Determination

The goal of x-ray crystallography is to obtain the distribution of the electron density which is related to the atomic positions in the unit cell, starting from the diffraction data. The electronic density function has the following expression:

$$\rho(x, y, z) = (1/V) \sum_{hkl} F_{hkl} \, e^{-2\pi i(hx+ky+lz)} \quad [1]$$

where F_{hkl} are the structure factors, V is the cell volume and h,k,l are the Miller indices. F is a complex number and can be represented as a vector with a module and a phase. It is possible to easily calculate the amplitude of F directly from the x-ray scattering measurements but the information on the phase value would be lost. Different experimental techniques can be used to solve the "phase problem," allowing the building of the protein three-dimensional structure: multiple isomorphous replacement (MIR), multiple anomalous diffraction (MAD), and Molecular Replacement (MR). The last one can be performed by computational calculations using only the native data set.

1.3.1. Molecular Replacement

The Molecular Replacement method consists in fitting a "probe structure" into the experimental unit cell. The probe structure is an initial approximate atomic model, from which estimates of the phases can be computed. Such a model can be the structure of a protein evolutionarily related to the unknown one, or even of the same protein from a different crystal form, if available. It is well known that the level of resemblance of two protein structures correlates well with the level of sequence identity *(7)*. If the starting model has at least 40% sequence identity with the protein of which the structure is to be determined, the structures are expected to be very similar and Molecular Replacement will have a high probability of being successful. The chance of success of this procedure progressively decreases with a decrease in structural similarity between the two proteins. It is not uncommon that Molecular Replacement based on one search probe only fails. If this happens, the use of alternative search probes is recommended. Alternatively, a single pdb file containing superimposed structures of homologous proteins can be used. Lastly, if conventional Molecular Replacement is unsuccessful, models provided by protein structure prediction methods can be used as probes in place of the structure of homologous proteins.

The Molecular Replacement method is applicable to a large fraction of new structures since the Protein Data Bank (http://www.rscb.org) *(8)* is becoming ever larger and therefore the probability of finding a good model is ever increasing.

Molecular replacement involves the determination of the orientation and position of the known structure with respect to the crystallographic axes of the unknown structure; therefore, the problem has to be solved in six dimensions. If we call X the set of vectors representing the position of the atoms in the probe and X′ the transformed set, the transformation can be described as:

$$X' = [R] X + T \qquad [2]$$

where R represents the rotation matrix and T the translation vector. In the traditional Molecular Replacement method, Patterson functions (*see* **Note 8**), calculated for the model and for the experimental data, are compared. The Patterson function has the advantage that it can be calculated without phase information. The maps calculated through the two Patterson functions can be superimposed with a good agreement only when the model is correctly oriented and placed in the right position in the unit cell. The calculation of the six variables, defining the orientation and the position of the model, is a computationally expensive problem that requires an enormous amount of calculation. However, the Patterson function properties allow the problem to be divided into two smaller problems: the determination of: *(1)* the rotation matrix, and *(2)* the translation vector. This is possible because the Patterson map is a vector map, with peaks corresponding to the positions of vectors between atoms in the unit cell. The Patterson map vectors can be divided into two categories: intramolecular vectors (self-vectors) and intermolecular vectors (cross-vectors). Self-vectors (from one atom in the molecule to another atom in the same molecule) depend only on the orientation of the molecule, and not on its position in the cell; therefore, they can be exploited in the rotation function. Cross-vectors depend both on the orientation of the molecule and on its position in the

1.3.2. Rotation Function

As mentioned, the rotation function is based on the observation that the self-vectors depend only on the orientation of the molecule and not on its position in the unit cell. Thus, the rotation matrix can be found by rotating and superimposing the model Patterson (calculated as the self-convolution function of the electron density, see **Note 8**) on the observed Patterson (calculated from the experimental intensity). Mathematically, the rotation function can be expressed as a sum of the product of the two Patterson functions at each point:

$$F(R) = \int_v P_{cryst}(u) P_{self}(Ru) du \qquad [3]$$

where P_{cryst} and P_{self} are the experimental and the calculated Patterson functions, respectively, R is the rotation matrix and r is the integration radius. In the integration, the volume around the origin where the Patterson map has a large peak is omitted. The radius of integration has a value of the same order of magnitude as the molecule dimensions because the self-vectors are more concentrated near the origin. The programs most frequently used to solve x-ray structures by Molecular Replacement implement the fast rotation function developed by Tony Crowther, who realized that the rotation function can be computed more quickly using the Fast Fourier Transform, expressing the Patterson maps as spherical harmonics *(9)*.

1.3.3. Translation Function

Once the orientation matrix of the molecule in the experimental cell is found, the next step is the determination of the translation vector. This operation is equivalent to finding the absolute position of the molecule. When the molecule (assuming it is correctly rotated in the cell) is translated, all the intermolecular vectors change. Therefore, the Patterson functions' cross-vectors, calculated using the observed data and the model, superimpose with good agreement only when the molecules in the crystal are in the correct position. The translation function can be described as:

$$T(t) = \int_v P_{cryst}(u) P_{cross}(ut) du \qquad [4]$$

where P_{cryst} is the experimental Patterson function, whereas P_{cross} is the Patterson function calculated from the probe oriented in the experimental crystal, t is the translation vector and u is the intermolecular vector between two symmetry-related molecules.

1.4. Structure Refinement

Once the phase has been determined (e.g., with the Molecular Replacement method) an electron density map can be calculated and interpreted in terms of the polypeptide chain. If the major part of the model backbone can be fitted successfully in the electronic density map, the structure refinement phase can begin. Refinement is performed by adjusting the model in order to find a closer agreement between the calculated and the observed structure factors. The adjustment of the model consists in changing the three positional parameters (x, y, z) and the isotropic temperature factors B (*see* **Note 9**) for all the atoms in the structure except the hydrogen atoms. Refinement techniques in protein x-ray crystallography are based on the least squares minimization and depend greatly on the ratio of the number of independent observations to variable parameters. Since the protein crystals diffract very weakly, the errors in the data are often very high and more than five intensity measurements for each parameter are necessary to refine protein structures. Generally, the problem is poorly over-determined (the ratio is around 2) or sometimes under-determined (the ratio below 1.0). Different methods are available to solve this problem. One of the most commonly used is the Stereochemically Restrained Least Squares Refinement, which increases the number of the observations by adding stereochemical restraints *(10)*. The function to minimize consists in a crystallographic term and several stereochemical terms:

$$Q = \sum w_{hkl}\{|F_{obs}|-|F_{cal}|\}^2 + \sum w_D(d_{ideal}-d_{model})^2$$
$$+ \sum w_T(X_{ideal}-X_{model})^2 + \sum w_P(P_{ideal}-P_{model})^2 \quad [5]$$
$$+ \sum w_{NB}(E_{min}-E_{model})^2 + \sum w_C(V_{ideal}-V_{model})^2$$

where w terms indicate weighting parameters: "w_{hkl}" is the usual x-ray restraint, "w_D" restrains the distance (d) between atoms (thus defining bond length, bond angles, and dihedral angles), "w_T" restrains torsion angles (X), "w_P" imposes the planarity of the aromatic rings (P), "w_{NB}" introduces restraints for non-bonded and Van der Waals contacts (E), and finally "w_C" restrains the configuration to the correct enantiomer (V). The crystallographic term is calculated from the difference between the experimental structure factor amplitudes F_{obs} and the structure factor amplitudes calculated from the model F_{calc}. The stereochemical terms are calculated as the difference between the values calculated from the model and the corresponding ideal values. The ideal values for the geometrical parameters are those measured for small molecules and peptides. The refinement program minimizes the overall function by calculating the shifts in coordinates that will give its minimum value by the least squares fitting method. The classical least square method can produce overfitting artefacts by moving faster toward agreement with structure factor amplitudes than toward correctness of the phases, because its shift directions assume the current model phases to be error-free constants. The refinement programs, developed more recently, use the Maximum Likelihood method, which allows a representation of the uncertainty in the phases so that the latter can be used with more caution (*see* **Note 10**).

Another popular refinement method is known as "simulated annealing" in which an energy function that combines the x-ray term with a potential energy function comprising terms for bond stretching, bond angle bending, torsion potentials, and van der Waals interactions, is minimized *(11)*.

The parameter used for estimating the correctness of a model in the refinement process is the crystallographic R factor (R_{cryst}), which is usually the sum of the absolute differences between observed and calculated structure factor amplitudes divided by the sum of the observed structure factor amplitudes:

$$R_{cryst} = \Sigma |F_{obs} - F_{calc}| / \Sigma |F_{obs}| \quad [6]$$

Using R_{cryst} as a guide in the refinement process could be dangerous because it often leads to over-fitting the model. For this reason, it is recommended to also use the so-called R_{free} parameter, which is similar to R_{cryst} except for the fact that it is calculated from a fraction of the collected data that has been randomly chosen to be excluded from refinement and maps calculation. In this way, the R_{free} calculation is independent from the refinement process and "phase bias" is not introduced. During the refinement process both R factors should decrease reaching a value in the 10 to 20% range.

1.4.1. Model Building

A key stage in the crystallographic investigation of an unknown structure is the creation of an atomic model. In macromolecular crystallography, the resolution of experimentally phased maps is rarely high enough so that the atoms are visible. However, the development of modern data collection techniques (cryo-crystallography, synchrotron sources) has resulted in remarkable improvement in the map quality, which, in turn, has made atomic model building easier. Two types of maps are used to build the model: the "2Fo-Fc" map and the "Fo-Fc" map (where Fo indicates the observed structure factor amplitudes and Fc the calculated ones). The first one is used to build the protein model backbone and is obtained by substituting the term $|2Fo-Fc| \exp(-i\varphi_{calc})$ (where φ_{calc} is the phase calculated from the model) to the structure factor term in the equation of the electronic density *(1)*. The "Fo-Fc" map helps the biocrystallographer to build difficult parts of the model and to find the correct conformation for the side chains. Moreover, it is used to add solvent and ligand molecules to the model. The "Fo-Fc" map is obtained by substituting the term $|Fo-Fc| \exp(-i\varphi_{calc})$ to the structure factor term in the equation of the electronic density.

2. Materials

2.1. Crystallization and Crystal Preparation

1. Hampton Research crystallization and cryoprotection kits.
2. Protein more than 95% pure, at a concentration between 5 and 15 mg/mL.
3. VDX plates to set up crystallization trials by hanging drop method.
4. Siliconized glass cover slides and vacuum grease.
5. Magnetic crystal caps, and mounted cryo loops.
6. Cryo tong and crystal wand.
7. Dewars to conserve and transport crystals at nitrogen liquid temperature.

2.2. Data Measurement and Data Processing

1. Goniometer head.
2. HKL suite (XDisplay, Denzo, and Scalepack) for macromolecular crystallography.

2.3. Molecular Replacement

1. Linux boxes or Silicon Graphics computers.
2. One of the following programs: AMoRe (freely available), MolRep (freely available), Xplor, CNS.

2.4. Refinement and Model Building

1. Collaborative computational Project no. 4 interactive (CCP4i) suite containing programs to manipulate the datasets, solve the structure and refine the model and calculate the maps (freely available).
2. One of the following programs: QUANTA (Molecular Structure, Inc.), Xfit, Coot, O to build the model. The last three are freely available.
3. Refmac5 (CCP4i package), Xplor, CNS, programs to refine the model.

3. Methods

3.1. Crystallization and Crystal Preparation

Precise rules to obtain suitable single-protein crystals have not been defined yet. For this reason, protein crystallization is mostly a trial and error procedure. This can be summarized in three steps:

1. Check protein sample purity, which has to be around 90–95%.
2. Slowly increase the precipitating agent concentration (PEGs, salts, or organic solvents) in order to favor protein aggregation.
3. Change pH and/or temperature.

It is usually necessary to carry out a large number of experiments to determine the best crystallization conditions, whereas using a minimum amount of protein per experiment. The protein concentration should be about 10 mg/mL; therefore, 1 mg of purified protein is sufficient to perform about 100 crystallization experiments. Crystallization can be carried out using different techniques, the most common of which are: liquid-liquid diffusion methods, crystallization under dialysis and vapor diffusion technique. The latter is described in detail since it is easy to set up and allows the biocrystallographer to utilize a minimum protein amount. The vapor diffusion technique can be performed in two ways: the "hanging drop" and the "sitting drop" methods.

1. In the "hanging drop" method, drops are prepared on a siliconized microscope glass cover slip by mixing 1–5 µL of protein solution with the same volume of precipitant solution. The slip is placed upside-down over a depression in a tray; the depression is partly filled (about 1ml) with the required precipitant solution (reservoir solution). The chamber is sealed by applying grease to the circumference of depression before the cover slip is put into place (**Fig. 3.2A**).

2. The "sitting drop" method is preferable when the protein solution has a low surface tension and the equilibration rate between drop solution and reservoir solution needs to be slowed down. A schematic diagram of a sitting drop vessel is shown in **Fig. 3.2B**.

The parameters that can be varied include: nature and concentration of the precipitating agent; buffers to explore the entire pH range; additional salts and detergents; and others.

3.2. Crystal Cryoprotection

The most widely used cryomounting method consists of the suspension of the crystal in a film of an "antifreeze" solution, held by surface tension across a small diameter loop of fiber, and followed by rapid insertion into a gaseous nitrogen stream. The cryoprotected solution is obtained by adding cryo protectant agents such as glycerol, ethylene glycol, MPD (2-methyl-2,4-pentandiol), or low molecular weight PEG (polyethylene glycol) to the precipitant solution. The crystal is immersed in this solution for a few seconds prior to being flash-frozen. This method places little mechanical stress on the crystal, so it is excellent for fragile samples. Loops are made from very fine (~10 µm diameter) fibers of nylon. As some crystals degrade in growth and harvest solutions, liquid nitrogen storage is an excellent way to stabilize crystals for long periods *(12)*. This system is particularly useful when preparing samples for data collection at synchrotron radiation sources, in that by minimizing the time required by sample preparation, it allows the biocrystalographer to use the limited time available at these facilities to collect data.

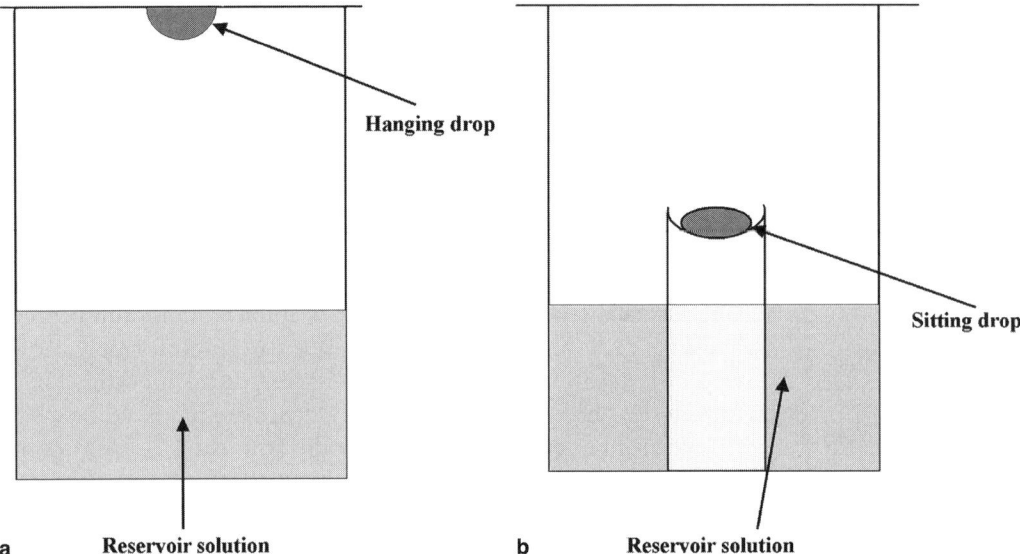

Fig. 3.2. (**A**) "Hanging drop" crystallization method. A drop of protein solution is suspended from a glass cover slip above a reservoir solution, containing the precipitant agent. The glass slip is siliconized to prevent spreading of the drop. (**B**) "Sitting drop" crystallization method. A drop of protein solution is placed in a plastic support above the reservoir solution.

3.3. Data Measurement

Once the crystal is placed in the fiber loop, the latter must be attached to a goniometer head. This device has two perpendicular arcs that allow rotation of the crystal along two perpendicular axes. Additionally, its upper part can be moved along two perpendicular sledges for further adjustment and centering of the crystal. The goniometer head must be screwed on to a detector, making sure that the crystal is in the x-ray beam. In agreement with Bragg's law, the crystal-to-detector distance should be as low as possible to obtain the maximum resolution together with a good separation between diffraction spots. Generally, a distance of 150 mm allows collection of high quality data sets with a good resolution (i.e., <2.0 Å) for protein crystals with unit cell dimensions around 60–80 Å. Long unit cell dimensions (a, b, and/or c longer than 150 Å), large mosaicism (>1.0 degree) (*see* **Note 11**), and large oscillation range (>1.0 degree), are all factors affecting spot separations and causing potential reflection overlaps.

Data collection is best performed interactively, with immediate data processing to get a fast feedback during data collection. This strategy avoids gross inefficiencies in the setup of the experiment; for example, incomplete data sets and/or reflection overlaps and/or large percentages of overloaded reflections.

3.4. Data Processing

The basic principles involved in the integrating diffraction data from macromolecules are common to many data integration programs currently in use. This section describes the data processing performed by the HKL2000 suite *(13)*. The currently used data processing methods exploit automated subroutines for indexing the x-ray crystal data collection, which means assigning correct hkl index to each spot on a diffraction image (**Fig. 3.3**).

1. Peak search. The first automatic step is the peak search, which chooses the most intense spots to be used by the autoindexing subroutine. Peaks are measured in a single oscillation image, which for protein crystals, requires 0.2–1.0 oscillation degrees.

2. Autoindexing of one image. If autoindexing succeeds a good match between the observed diffraction pattern and predictions is obtained. The auto-indexing permits the identification of the space group and the determination of cell parameters (*see* **Note 12** and **Table 3.1**). Other parameters also have to be refined. The most important are the crystal

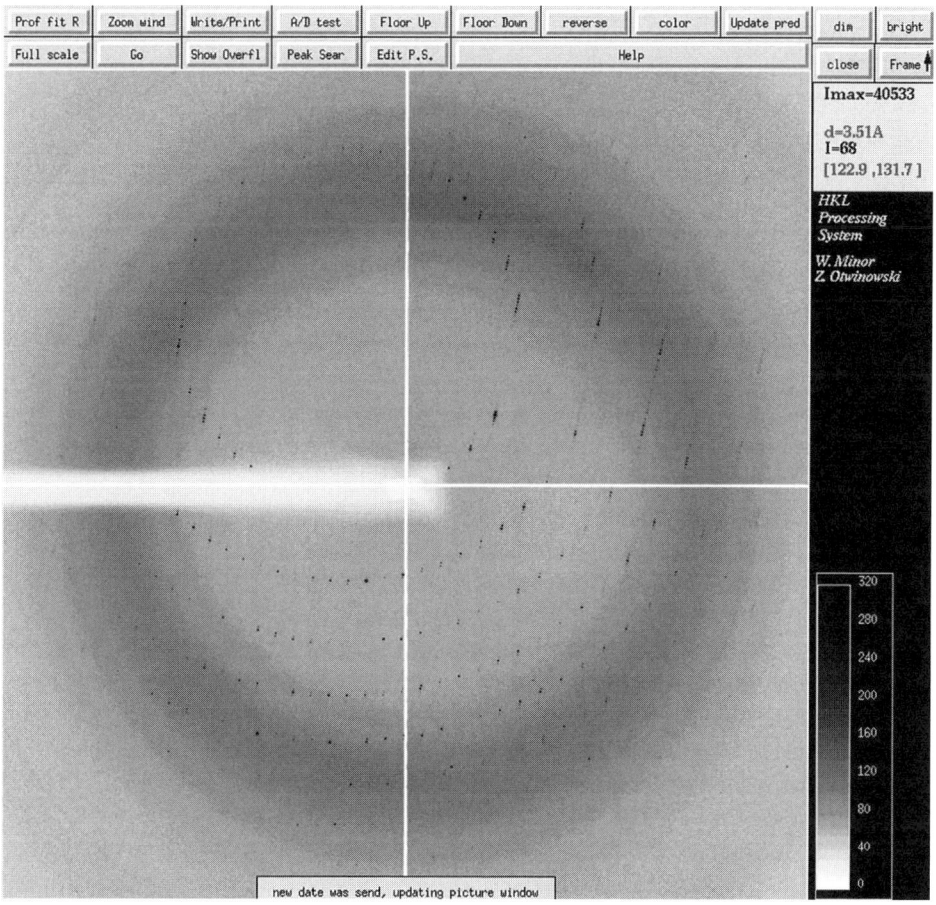

Fig. 3.3. Diffraction oscillation image visualized with the program Xdisp (HKL2000 suite) of the whole human sorcin collected at the ESRF synchrotron radiation source (Grenoble, France). The spot distances from the image centre are proportional to the resolution, so the spots at the image edge are the highest resolution spots.

Table 3.1
Output of the Denzo autoindexing routine

Lattice	Metric tensor distortion index	Best cell (symmetrized) Best cell (without symmetry restrains)					
Primitive cubic	60.04%	64.84 148.55	64.96 148.55	315.85 148.55	90.09 90.00	89.92 90.00	60.43 90.00
I centred cubic	74.97%	65.32 236.68	322.35 236.68	322.36 236.68	20.04 90.00	84.33 90.00	84.38 90.00
F centred cubic	79.51%	335.11 326.68	322.36 326.68	322.56 326.68	23.24 90.00	157.19 90.00	157.36 90.00
Primitive rhombohedral	2.75%	322.35 320.18 64.90	322.34 320.18 64.90	315.85 320.18 953.96	11.69 11.62 90.00	11.60 11.62 90.00	11.57 11.62 120.00
Primitive hexagonal	0.22%	65.32 65.08	64.84 65.08	315.85 315.85	89.92 90.00	90.17 90.00	120.12 120.00
Primitive tetragonal	13.37%	64.84 64.90	64.96 64.90	315.85 315.85	90.09 90.00	89.92 90.00	60.43 90.00
I centred tetragonal	13.68%	64.84 64.90	64.96 64.90	634.88 634.88	87.11 90.00	92.88 90.00	60.43 90.00
Primitive orthorhombic	13.37%	64.84 64.84	64.96 64.96	315.85 315.85	90.09 90.00	89.92 90.00	60.43 90.00
C centred orthorhombic	0.09%	65.32 65.32	112.17 112.17	315.85 315.85	90.01 90.00	89.83 90.00	90.13 90.00
I centred orthorhombic	13.68%	64.84 64.84	64.96 64.96	634.88 634.88	87.11 90.00	92.88 90.00	60.43 90.00
F centred orthorhombic	2.37%	65.32 65.32	112.17 112.17	634.88 634.88	89.99 90.00	95.73 90.00	90.13 90.00
Primitive monoclinic	0.07%	64.84 64.84	315.85 315.85	64.96 64.96	90.09 90.00	119.57 119.57	90.08 90.00
C centred monoclinic	0.05%	65.32 65.32	112.17 112.17	315.85 315.85	89.99 90.00	90.17 90.17	90.13 90.00
Primitive triclinic	0.00%	64.84	64.96	315.85	90.09	90.08	119.57
Autoindex unit cell		65.24	65.24	315.85	90.00	90.00	120.00
Crystal rotx, roty, rotz	−8.400	55.089	70.885				
Autoindex Xbeam, Ybeam	94.28	94.90					

The lattice and unit cell distortion table, and the crystal orientation parameters are shown. These results were obtained for the F112L human mutant sorcin (soluble Resistance related calcium binding protein) (25).

and detector orientation parameters, the center of the direct beam and the crystal-to-detector distance.

3. Autoindexing of all the images. The autoindexing procedure, together with refinement, is repeated for all diffraction images.

Data are processed using a component program of the HKL2000 suite called Denzo. The scaling and merging of indexed data, as well as the global refinement of crystal parameters, is performed with the program Scalepack, which is another HKL2000 suite component. The values of unit-cell parameters refined from a single image may be quite imprecise. Therefore a post-refinement procedure is implemented in the program to allow for separate refinements of the orientation of each image while using the same unit cell for the whole data set. The quality of x-ray data is first assessed by statistical parameters reported in the scale.log file. The first important parameter is the I/σ (I: intensity of the signal, σ the standard deviation), that is the signal-to-noise ratio, which is also used to estimate the maximum resolution. The second parameter is χ^2, which is closely related to I/σ (*see* **Note 13**). The program tries to bring χ^2 close to 1.0 by manipulating the error model. Another important parameter is R_{sym}, which is a disagreement index between symmetry related reflections and of which the average value should be below 10% (*see* **Note 14**). The output of the data processing procedure is a file with suffix .hkl, containing all the measured intensities with their relative σ values and the corresponding hkl indices. Using the program Truncate implemented in the CCP4i suite *(14)*, it is possible to calculate: the structure factor amplitudes from the intensities by the French and Wilson method *(15)*, the Wilson plot to estimate an overall B factor (*see* **Note 9**), and an absolute scale factor, and intensity statistics to evaluate the correctness of the data reduction procedure. The truncated output data are stored in a file that usually has the extension.mtz.

3.5. Molecular Replacement

1. Search model. The first operation is searching the databases for a probe structure similar to the structure to be solved. Since we do not know the structural identity of our protein with homologous proteins we use sequence identity as a guide. Proteins showing a high degree of sequence similarity with our "query" protein can be identified in protein sequence databases using sequence comparison methods such as BLAST *(16)*. The protein of known three-dimensional structure showing the highest sequence identity with our query protein is generally used as the search model.

2. Files preparation. The Pdb file of the search probe has to be downloaded from the Protein Data Bank. The file has to be manipulated before molecular replacement is performed. The water molecules as well as the ligand molecules have to be removed from the file. The structure can be transformed

into a polyalanine search probe to avoid model bias during the Molecular Replacement procedure (*see* **Note 15**). The other file needed to perform the Molecular Replacement is the file with extension .mtz resulting from earlier data processing (*see* **Section 3.4.**), containing information about crystal space group, cell dimensions, molecules per unit cell and a list of the collected experimental reflections.

3. Molecular Replacement. The Molecular Replacement procedure consists in Rotation and Translation searches to put the probe structure in the correct position in the experimental cell. This operation can be done using different programs, the most common of which is AMoRe *(17)*. This chapter describes briefly the use of MolRep *(18)*, one of the most recent programs to solving macromolecular structures by Molecular Replacement. This program belongs to the CCP4i suite and is automated and user friendly. The program performs rotation searches followed by translation searches. The only input files to upload are the .mtz and the .pdb files. The values of two parameters have to be chosen: the integration radius and the resolution range to be used for Patterson calculation. In the rotation function, only intramolecular vectors need to be considered. Since all vectors in a Patterson function start at the unit cell axes origin, the vectors closest to the origin will in general be intramolecular. By judiciously choosing a maximum Patterson radius, we can improve the chances of finding a strong rotation hit. Usually a value of the same order of magnitude as the search probe dimensions is chosen. Regarding the second parameter, high-resolution reflections (>3.5 Å) will differ substantially because they are related to the residue conformations. On the other hand, low-resolution reflections (<10Å) are influenced by crystal packing and solvent arrangement. Thus, the resolution range that should be used is usually within 10–3.5Å.

4. Output files. The output files to check are: the file with extension .log that lists all the operations performed by the program, and the coordinates file representing the MR solution, that is the model rotated and translated in the real cell, in pdb format. As shown in Table 3.2, after the rotational and translational searches are performed, the program lists all the possible solutions (*see* **Note 16**) followed by the rotation angles and the translation shifts necessary to position the model in the real cell, the crystallographic R factors and finally the correlation coefficients (*see* **Note 17**). The first line of **Table 3.2** represents a clear solution for a MR problem. In fact the crystallographic R factor is below 0.5, and the correlation coefficient is very high (75.9%). Moreover, there is a jump between the first possible solution (first line) and the second possible solution (second line).

Table 3.2
Output of the MolRep program after rotation and translation searches

		alpha	beta	gamma	Xfrac	Yfrac	Zfrac	TF/sig	R-fac	Corr
Sol_TF_7	1	32.27	84.87	78.84	0.824	0.498	0.091	65.34	0.333	0.759
Sol_TF_7	2	32.27	84.87	78.84	0.324	0.041	0.092	25.76	0.482	0.478
Sol_TF_7	3	32.27	84.87	78.84	0.324	0.454	0.091	24.18	0.477	0.481
Sol_TF_7	4	32.27	84.87	78.84	0.324	0.498	0.016	23.57	0.483	0.467
Sol_TF_7	5	32.27	84.87	78.84	0.422	0.498	0.091	23.37	0.479	0.478
Sol_TF_7	6	32.27	84.87	78.84	0.324	0.498	0.325	23.12	0.482	0.471
Sol_TF_7	7	32.27	84.87	78.84	0.238	0.498	0.092	23.01	0.481	0.473
Sol_TF_7	8	32.27	84.87	78.84	0.324	0.498	0.372	22.99	0.479	0.475
Sol_TF_7	9	32.27	84.87	78.84	0.324	0.498	0.400	22.97	0.480	0.473
Sol_TF_7	10	32.27	84.87	78.84	0.324	0.000	0.196	22.93	0.490	0.456

The present results (data not published) have been obtained for the protein Dps (Dna binding proteins, from starved cells from *Listeria monocytogenes* using as search model the Dps from *Listeria innocua* (Pdb code 1QHG).

3.6. Structure Refinement

Several programs can be used to perform structure refinement. The most common are: CNS written by Brünger *(19)*, which uses conventional least square refinement as well as simulated annealing to refine the structure; and REFMAC5 (CCP4i suite) written by Murshudov *(20)* that uses maximum likelihood refinement. Although CNS and many other programs have been used with success, this chapter illustrates the use of REFMAC5 implemented in CCP4i because it provides a graphic interface to compile the input files; this feature is particularly helpful for beginners.

1. Rigid body refinement. First, the initial positions of the molecules in the unit cell and in the crystal cell provided by MR procedures have to be refined. For this purpose, rigid body refinement should be performed. This method assigns a rigid geometry to parts of the structure and the parameters of these constrained parts are refined rather than individual atomic parameters. The input files to be uploaded are the MR solution and the .mtz file containing the experimental reflections. The resolution at which to run rigid body refinement has to be specified (in general the rigid body refinement should start at the lowest resolution range) and the rigid entity should be defined (this can be an entire protein, a protein subunit or a protein domain). To define the rigid entities in REFMAC5, simply select the chain and protein regions that are to be fixed.

2. Output files. The output files are: (a) the .log file that contains a list of all the operations performed, statistics about the geometrical parameters after each refinement cycle, crystallographic R factor and R free factor values, and finally the figure of merit (*see* **Note 18** and **Table 3.3**); (b) the .pdb file containing the refined coordinates of the model; (c) the .mtz file containing the observed structure factors (F_{obs}), the structure factor amplitudes calculated from the model (F_{calc}) and the phase angles calculated from the model.

3. Coordinates and B factors refinement. The program REFMAC 5 refines the x, y, z and B parameters using the maximum likelihood method. As for the rigid body refinement, the input files are the .mtz file containing the F_{obs} and the .pdb file containing the coordinates of the model. It is also necessary to restrain the stereochemical parameters using the maximum likelihood method. It is possible to choose a numerical value for the relative weighting terms or, more easily, to choose a single value for the so-called "weight matrix" that allows the program to restrain all the stereochemical parameters together. The value of the "weight matrix" should be between 0.5, indicating loose stereochemical restraints, and 0, indicating strong stereochemical restraints which keep geometrical parameters of the macromolecules

Table 3.3
Summary of 10 cycles of DpsTe (*see* Note 22) coordinate refinement using REFMAC5. The R_{fact}, R_{free}, Figures of Merits (FOM) and root mean square deviation values of some stereo-chemical parameters are shown

Ncyc	Rfact	Rfree	FOM	LLG	rmsBOND	rmsANGLE	rmsCHIRAL
0	0.213	0.213	0.862	1165259.2	0.004	0.734	0.055
1	0.196	0.210	0.865	1151022.5	0.010	1.022	0.074
2	0.191	0.209	0.867	1146576.9	0.011	1.106	0.080
3	0.188	0.209	0.868	1144297.8	0.011	1.144	0.083
4	0.187	0.209	0.869	1142920.2	0.011	1.166	0.085
5	0.186	0.209	0.870	1142088.8	0.011	1.178	0.086
6	0.185	0.209	0.870	1141496.4	0.011	1.186	0.087
7	0.184	0.209	0.870	1141031.5	0.011	1.190	0.088
8	0.184	0.209	0.871	1140743.6	0.011	1.192	0.088
9	0.184	0.209	0.871	1140461.8	0.011	1.195	0.088
10	0.183	0.209	0.871	1140311.0	0.011	1.196	0.088

near the ideal values. In REFMAC5 NCS (*see* **Note 19**) restraints can also be used for refinement.

3.7. Model Building

After the Molecular Replacement and the first cycles of coordinates refinement, only a partial model has been obtained. In this model, the side chains are absent, and often parts of the model do not match the electronic density map. Therefore, the building of the first structural elements is followed by refinement cycles that should lead to an improvement on the statistics (i.e., the R factor has to decrease and the figure of merit has to increase). The most common programs used for model building are QUANTA, COOT *(21)*, O *(22)*, and XFIT (XTALVIEW PACKAGE) *(23)*. XFIT and COOT permit direct calculation of density maps. Two maps are necessary to build a model: the 2Fo-Fc map contoured at 1σ which is used to trace the model and the Fo-Fc map contoured at 3σ, which is necessary to observe the differences between the model and the experimental data.

1. Starting point. First find a match between protein sequence and the 2Fo-Fc density map. If the phases are good, this operation should not be too difficult. The electron density map should be clear (especially if it has been calculated from high-resolution data) and should allow the identification of the amino acids (*see* **Note 20**) (**Fig. 3.4**).

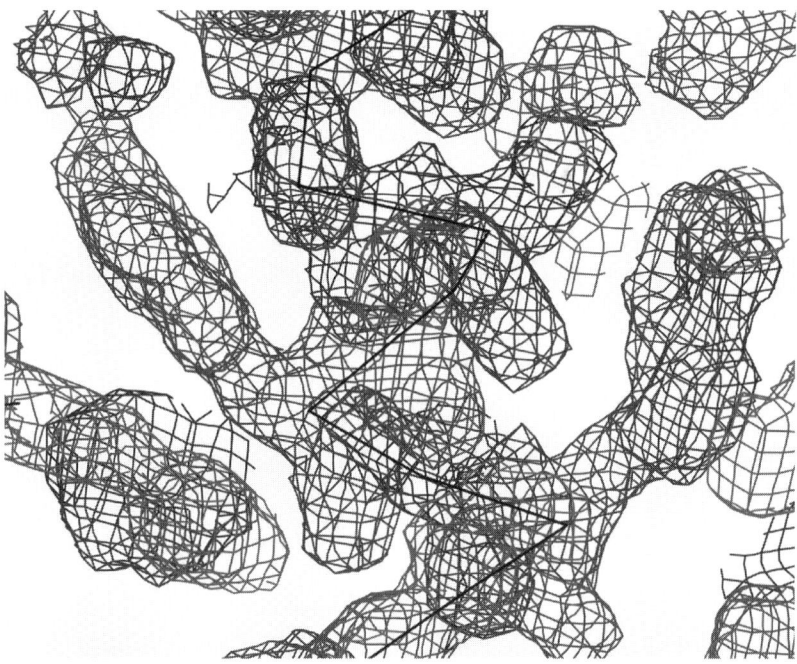

Fig. 3.4. Initial Electronic density map of Dps from *Thermosynechococcus elongatus* (*see* **Note 22**) calculated after Molecular Replacement. Cα trace of the model is superimposed on the map. The electronic density of a Trp residue and a Tyr residue are easily recognizable in the map.

2. Initial building. Once the first residue has been identified and fitted into the electron density map, model building can be performed by fitting the whole protein sequence residue by residue in the map.

3. Building of the unfitted structure elements. If the initial model does not contain all the protein residues, it is possible to build the main chain of the protein region missing from the model "ab initio," using one of the programs cited above. As an example, with XFIT it is possible to add Cα atoms to the model after or before a selected residue. After a Cα is inserted in the electron density map in the correct position, it is possible to substitute the Cα with the desired amino acid, which is automatically bound to the rest of the protein. The main chain of the missing protein region can also be constructed using the program database after a suitable piece of structure has been built.

4. Omit map. If a part of the structure does not match the map, this means that it is built incorrectly. Thus it is possible to use, as a major strategy for overcoming phase bias, the so-called "omit maps". In practice, the model region that has to be refitted is removed and the maps are recalculated after a few

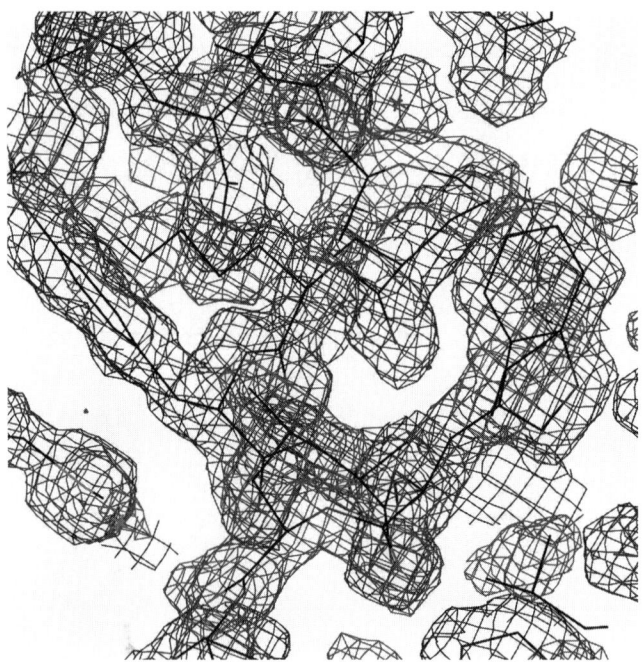

Fig. 3.5. Electronic density map contoured at 1.0 σ of Dps from *Thermosynechococcus elongatus* (*see* **Note 22**) calculated after many REFMAC5 refinement cycles. The final structure (thick lines) solved at 1.8 Å resolution is superimposed on the map.

refinement cycles. This method allows the phases calculated from the rest of the model to phase the area of interest with no bias from parts of the model left out.

5. Optimization. At this stage, large sections of the structure should be approximately fitting the electron density map (**Fig. 3.5**). The next step is the choice of the correct side-chain rotamers. This operation may be done by hand or by using real space refinement tools. Finally, water molecules and ions and/or ligands bound to the protein have to be identified and added to the model. For this purpose only the Fo-Fc map contoured at 3σ is used. The water molecules can be added either manually or automatically (*see* **Note 21**).

4. Notes

1. The intercepts of the planes with the cell edges must be fractions of the cell edge. Therefore, cell intercepts can be at 1/0 (= ∞), 1/1, 1/2, 1/3 ... 1/n. The conventional way of identifying these sets of planes is by using three integers that are the denominators of the intercepts along the three axes of the unit cell, hkl, called Miller indices. If a set of planes had intercepts at 1/2, 1/3, and 1/1, then the planes would be referred to as the (2 3 1) set of planes.

2. Another advantage of synchrotron radiation is its tunability, which allows the user to select radiation wavelengths higher or lower than 1.5418 Å (copper radiation). Collection of data at wavelengths below 1.5418Å results in a lower signal-to-noise ratio.

3. A crystal can be regarded as a three-dimensional grid and one can imagine that this will produce a three-dimensional x-ray diffraction pattern. As with electron microscope grids, the pattern is reciprocal to the crystal lattice. The planes that intersect the sphere in **Fig. 3.1** are layers in a three-dimensional lattice, called reciprocal lattice because the distances are related reciprocally to the unit cell dimensions. Each reciprocal lattice point corresponds to one diffracted reflection. The reciprocal lattice is an imaginary but extremely convenient concept to determine the direction of the diffracted beams. If the crystal rotates, the reciprocal lattice rotates with it. In an x-ray diffraction experiment the direction of the diffracted beams depends on two factors: the unit-cell distances in the crystal, from which the unit-cell distances in the reciprocal lattice are derived, and the x-ray wavelength. As indicated in **Fig. 3.1**, diffraction conditions are determined not only by the reciprocal lattice but also by

the radius of the sphere of the reflection or "Ewald sphere," of which the radius is $1/\lambda$.

4. The imaging plate detector is formed by a photosensitive plate, made of BaFBr:Eu. When hit by radiation, the plate produces a latent image that can be excited by a laser operating at 633 nm, which generates 390 nm radiation corresponding to the fluorescence transition of Europium. This radiation is collected in the photomultiplier and converted to an electric signal.

5. The CCD camera (charged coupled device) is another kind of area detector. The detector surface is constituted by voltage sensitive elements (pixels). They have a high dynamic range, combined with excellent spatial resolution, low noise, and high maximum count rate.

6. The resolution is defined as the minimum inter-planar spacing of the real lattice for the corresponding reciprocal lattice points (reflections) that are being measured. It is directly related to the optical definition, the minimum distance that two objects can be apart and still be seen as two separate objects. Thus, high resolution means low minimum spacing. Resolution is normally quoted in Ångstroms (Å).

7. An oscillation image (also called frame) is obtained by rotating a crystal continuously through 0.2–1.0° about a fixed axis, called ϕ axis, perpendicular to the incident x-ray beam.

8. The Patterson function is a Fourier summation with intensities as coefficients and without phase angles. It can be written as: $P(u,v,w) = \Sigma |F(hkl)|^2 \cos 2\pi (hu + kv + lw)$. Further, it can be demonstrated that the Patterson function can be alternatively written as the self-convolution of the electronic density: $P(u,v,w) = \int_r \rho(r)\rho(r + u)dr$.

9. Macromolecules in crystals are not static. Atoms vibrate around an equilibrium position and, as a consequence, the intensity of the diffracted beams are weakened. This phenomenon is expressed by the temperature factor $B = 8\pi^2 \times u^2$ where "u" is the mean square displacement of atoms around the atomic positions.

10. The maximum likelihood method involves determining the parameters of an assumed model that maximize the likelihood of the data. Thus, the most appropriate value for each variable (e.g., bond distances, angles, etc.) is that which maximizes the probability of observing the measured values.

11. Protein crystals are affected by lattice defects. Therefore, they are formed by different mosaic blocks with slightly different orientations. As an ideal single crystal has a mosaicism equal to 0 degrees, a good quality protein crystal should have a low mosaicism (0.2–0.5 degrees).

12. **Table 3.1** shows the output of the program Denzo after autoindexing. In this table, all the 14 possible Bravais lattices are listed from the highest symmetry (primitive cubic) to the lowest (primitive triclinic), allowing identification of the crystal lattice. After the lattice name, the table displays a percentage value that represents the amount of distortion that unit-cell parameters would suffer in order to fit the lattice. Next to this percentage, the "distorted-to-fit" unit-cell parameters are listed. Below these values, the undistorted unit-cell parameters are shown for comparison. The goal of the autoindexing procedure is to find the highest symmetry lattice which fits the data with minimal distortion. In the example shown in Table 3.1, the crystal lattice is primitive hexagonal, since 0.22% is an acceptable amount of distortion, especially given that the unit-cell parameters were refined from a single frame. The crystal lattice should be confirmed by the overall Denzo data reduction and Scalepack scaling procedure.

13. χ^2 is a parameter related to the ratio between intensity and its standard deviation σ for all measurements and its value should be around 1. The χ^2 is mathematically represented by the following equation:

$$\chi^2 = \frac{\sum_{ij}\left(\left|I_{ij}(hkl) - \langle I_j(hkl)\rangle\right|\right)^2}{\sigma i^2 \frac{N}{N-1}}$$

where hkl are the Miller indices and N indicates the number of observations.

14. R_{sym} is the parameter used to compare the intensity (I) of symmetry related reflections for n independent observations:

$$R_{sym} = \frac{\sum_{hkl}\sum_i \left|I_i(hkl) - \overline{I(hkl)}\right|}{\sum_{hkl}\sum_i I_i(hkl)}.$$

The index i indicates the experimental observations of a given reflection. $\overline{I(hkl)}$ is the average intensity for symmetry-related observations.

15. To avoid model bias often the model is transformed into a poli-Ala search probe. Only the coordinates of the polypeptide backbone and of Cβ atoms are conserved, whereas the side chain atoms are deleted.

16. The MolRep solutions represent the highest superposition peaks between the experimental Patterson function and the Patterson function calculated from the search probe, rotated, and translated in the real cell.

17. Correlation coefficient value (CC_f) lies between 0 and 1 and measures the agreement between the structure factors calculated from the rotated and translated model and the observed structure factors. The correlation coefficient is calculated by REFMAC5 using the following formula:

$$cc_f = \frac{\left[\sum_{hkl}(|F_{obs}||F_{calc}|) - (\langle|F_{obs}|\rangle \times \langle|F_{calc}|\rangle)\right]}{\left[\sum_{hkl}(F_{obs}^2 - \langle F_{obs}\rangle^2)\sum_{hkl}(F_{calc}^2 - \langle F_{calc}\rangle^2)\right]^{1/2}}$$

18. Figure of merit. The "figure of merit" m is: $m = \dfrac{|F(hkl)_{best}|}{|F(hkl)|}$

    where $F(hkl)_{best} = \dfrac{\sum_{\alpha} P(\alpha) F_{hkl}(\alpha)}{\sum_{\alpha} P(\alpha)}$,

    $P(\alpha)$ is the probability distribution for the phase angle α and F_{hkl}(best) represents the best value for the structure factors. The m value is between 0 and 1 and is a measure of the agreement between the structure factors calculated on the basis of the model and the observed structure factors. If the model is correct the figure of merit approaches 1.

19. Non-crystallographic symmetry (NCS) occurs when the asymmetric unit is formed by two or more identical subunits. The presence of this additional symmetry could help to improve the initial phases and obtain interpretable maps for model building using the so-called density modification techniques *(24)*.

20. Usually, the sequence region that contains the largest number of aromatic residues is chosen to start the search. The aromatic residues (especially tryptophan) contain a high number of electrons and display an electronic density shape that is easy to recognize (*see* **Fig. 3.4**).

21. All the mentioned programs (XFIT, O, QUANTA, etc.) are provided with functions that identify the maxima in the Fo-Fc map above a given threshold (usually 3σ is used) and place the water molecules at the maxima peaks.

22. DpsTe. DpsTe is a member of the Dps family of proteins (DNA binding proteins from starved cells). DpsTe has been isolated and purified from the cyanobacterium *Thermosynechococcus elongatus*. The structure has been solved by Molecular Replacement at 1.81 Å resolution and has been deposited in the Protein Data bank with the accession number 2C41.

References

1. Ducruix, A., Giegè, R. (1992) *Crystallization of Nucleic Acids and Proteins*, Oxford University Press, NY.
2. Hahn, T. (ed.) (2002) *International Table of Crystallography*, Kluwer Academic Publishers, Dordrecht.
3. McPherson, A. J. (1990) Current approach to macromolecular crystallization. *Eur J Biochem* 189, 1–23.
4. Matthews, B. W. (1968) Solvent content of protein crystals. *J Mol Biol* 33, 491–497.
5. Friedrich, W., Knipping P., Laue, M. (1981), *Structural Crystallography in Chemistry and Biology*, in (Glusker, J. P., ed.), Hutchinson & Ross, Stroudsburg, PA.
6. Bragg, W. L., Bragg, W. H. (1913) The structure of crystals as indicated by their diffraction of X-ray. *Proc Roy Soc London* 89, 248–277.
7. Chothia C., Lesk A. M. (1986) The relation between the divergence of sequence and structure in proteins. *EMBO J* 5, 823–826.
8. Berman, H. M., Westbrook, J., Feng, Z., et al. (2000) The Protein Data Bank. *Nucl Acids Res* 28, 235–242.
9. Crowther, R. A. (1972) *The Molecular Replacement Method*, in (Rossmann, M.G., ed.) Gordon & Breach, New York.
10. Hendrickson, W. A. (1985) Stereochemically restrained refinement of macromolecular structures. *Methods Enzymol* 115, 252–270.
11. Brunger, A. T., Adams, P. D., Rice, L. M. (1999) Annealing in crystallography: a powerful optimization tool. *Prog Biophys Mol Biol* 72, 135–155.
12. Rodgers, D. W., Rodgers D. W. (1994) Cryocrystallography. *Structure* 2, 1135–40.
13. Otwinoski, Z., Minor, W. (1997) Processing of X-ray diffraction data collected in oscillation mode. *Methods Enzymol* 276, 307–326.
14. CCP4 (Collaborative Computational Project, number 4) (1994) The CCP4 suite: programs for protein crystallography. *Acta Cryst.* D50, 760–763.
15. French, G.S., Wilson, K.S. (1978) On the treatment of negative intensity observations. *Acta Cryst* A34, 517–525.
16. Altshul S. F., Koonin, E. V. (1998) Iterated profile searches with PSI-BLAST: a tool for discovery in protein databases. *TIBS* 23, 444–447.
17. Navaza, G. (1994) AMORE: an automated package for Molecular Replacement. *Acta Crystallogr* A50, 157–163.
18. Vagin, A., Teplyakov, A. (1997) MOLREP: an automated program for Molecular Replacement. *J Appl Crystallogr* 30, 1022–1025.
19. Brünger, A. T., Adams, P. D., Clore, G. M., et al. (1998) Crystallography & NMR system: a new software suite for macromolecular structure determination. *Acta Crystallogr* D54, 905–921.
20. Murshudov, G. N., Vagin, A. A., Dodson, E. J. (1997) Refinement of macromolecular structures by the maximum-likelihood method. *Acta Cryst* D53, 240–255.
21. Emsley, P., Cowtan, K. (2004) Coot: model-building tools for molecular graphics. *Acta Cryst* D60, 2126–2132.
22. Jones, T. A., Zou, J. Y., Cowan, S. W., et al. (1991). Improved methods for building protein models in electron density maps and the location of errors in these models. *Acta Cryst* A47, 110–119.
23. McRee, D. E. (ed.) (1993) *Practical Protein Crystallography*. Academic Press, San Diego, CA.
24. Rossmann, M. G., Blow, D. M. (1962) The detection of subunits within the crystallographic asymmetric unit. *Acta Cryst* 15, 24–31.
25. Franceschini, S, Ilari A., Verzili D., et al. (2007) Molecular basis for the impaired function of the natural F112L sorcin mutant: X-ray crystal structure, calcium affinity, and interaction with annexin VII and the ryanodine receptor *Faseb J*, in press.

Chapter 4

Pre-Processing of Microarray Data and Analysis of Differential Expression

Steffen Durinck

Abstract

Microarrays have become a widely used technology in molecular biology research. One of their main uses is to measure gene expression. Compared to older expression measuring assays such as Northern blotting, analyzing gene expression data from microarrays is inherently more complex due to the massive amounts of data they produce. The analysis of microarray data requires biologists to collaborate with bioinformaticians or learn the basics of statistics and programming. Many software tools for microarray data analysis are available. Currently one of the most popular and freely available software tools is Bioconductor. This chapter uses Bioconductor to preprocess microarray data, detect differentially expressed genes, and annotate the gene lists of interest.

Key words: microarray; normalization; bioconductor; R; differential gene expression.

1. Introduction

1.1. Noisy Signals

Microarray data consist of noisy signal measurements. The real gene expression measure is masked by different sources of noise, such as labeling efficiency, print-tip effects, between-slide variation, and other factors (1, 2). Normalization of microarray data aims to correct the raw intensity data for these effects. After normalization, the data are ready for further analysis, such as determining differentially expressed genes and clustering. This chapter assumes that microarrays have been scanned and image processing has provided the intensity measurements for the channels involved. Methods for quality assessment, data normalization, detection of differential expression, and annotation of differentially expressed features are described.

1.2. Quality Assessment

Various methods exist to assess the quality of the microarray experiments and the need for normalization. Graphical representations of the array data can help to quickly identify bad arrays and the need for normalization.

Hybridization problems can be identified by plotting images of the foreground and background intensities for each channel for spotted arrays and of the PM and MM values of Affymetrix GeneChip™ chips. One usually does not expect to see any distinct pattern on these images, as intensity levels are expected to be spread randomly over the array. As such, any strikes or other patterns that show up indicate possible faults and contamination during hybridization. Depending on the severity of the fault, it may be possible to correct the problem during normalization.

For cDNA arrays, the effect of failing or suboptimal print tips can be visualized by using a boxplot of the intensities per print tip.

A general assessment of hybridization quality is to identify how many spots are above the background. If this number is too low, then the hybridization might have to be redone.

Between-array differences can be visualized by plotting boxplots of the raw intensities grouped per array. Failing arrays show up as outliers, and the need for between-array normalization is visualized by the boxplot pattern.

A second plot, useful for comparing different arrays, is a scatter plot of the intensities of every array against the corresponding intensities on every other array in the experiment. Alternatively, this can be represented by plotting a correlation heatmap (**Note 1**). Usually one has at least one repeat of each sample and one can expect the correlation between these hybridizations to be high and higher than when compared with a different hybridization. When the correlation between repeats is low this usually means a bad hybridization or in some cases a mix-up of samples. Hybridizations failing these quality assessments should be discarded and replaced by a new hybridization if possible or given a lower weight in the subsequent analysis.

1.3. Pre-Processing of Affymetrix Data

Affymetrix probe sets consist of probes, which individually measure a perfect match (PM) or a paired-mismatch (MM) signal. The PM probes measure the effective gene expression status and the MM probes estimate the amount of cross-hybridization and thus background noise. For each probe set there are usually 11 to 20 PM and MM probes. Each probe is 25 bp long. One can think of Affymetrix chip preprocessing as a three-step procedure, although steps are sometimes combined depending on the algorithm. These three steps are: estimation of the background signal, possible background correction normalization, and summarizing the separate normalized probe measurements into a single value per probe set.

The MAS5.0 normalization procedure is the normalization method developed by Affymetrix. It makes use of the MM measurements to compute an Ideal Mismatch value and then corrects the PM values with these Ideal Mismatch values. MAS5.0 then uses a one-step Tukey Biweight algorithm as summarization method *(3)*.

The first normalization method implementing a model to estimate the background corrected expression measure was the Model Based Expression Index (MBEI) (also called dChip algorithm) *(4)*. MBEI uses the observed PM and MM values to calculate a fitted value for the two properties and calculates the difference PM – MM.

The Robust Multi-array Average (RMA) *(5)* method does not use the MM values; the background is instead estimated by convoluting the signal and noise distributions from the PM values. After background correction, RMA performs a quantile normalization (**Note 2**) and then uses median polish (**Note 3**) as summarization method.

GCRMA *(6)* is a modified RMA procedure and includes the GC content of the probes in the background adjustment step.

Finally, the Variance Stabilizing Normalization (VSN) method combines background correction and normalization and can be used for two-color experiments as well *(7)*. VSN returns normalized data, which has an approximately constant variance independent of the spot intensity.

Normalized data are usually returned in a log2 scale, except for the VSN method, which returns data in a generalized log (*glog*) scale *(7)*.

Methods such as MAS5.0, MBEI, GCRMA, RMA, and VSN are used frequently as normalization methods, and depending on the method used, the final expression values will differ *(8–10)*. To date, none of these methods has been adopted as the standard best method and more research has to be done to investigate this.

1.4. Pre-Processing of Two-Color Data

Two-color microarray experiments are characterized by probes spotted on a microarray slide and the use of two labeled samples that are hybridized simultaneously on the same array and labeled with a different dye, usually Cy3 (green) and Cy5 (red). Image analysis software gives background and foreground intensity measurements for each channel. One can choose to subtract the background measurements from the foreground or not.

Print-tip lowess normalization is a robust local regression based normalization that accounts for intensity and spatial dependence in dye biases *(1)* and is frequently used to normalize cDNA microarray experiments. The lowess fit adjusts each feature with a different normalization value depending on its overall intensity. By generating a lowess fit for each print-tip separately one can correct for print-tip effects.

This type of normalization can be used under the assumption that only a small subset of genes is differentially expressed between the two samples. If this assumption does not hold, one can rely on spiked-in RNA to calculate the lowess fit and use that fit to adjust the data *(11)*. Other frequently used methods to normalize cDNA microarrays are qspline *(12)* and VSN *(7)* normalization.

A between-slide normalization, which corrects for a difference in scale, can be applied if the variance of the ratios differs a lot between the different slides. A boxplot of the ratios grouped per slide will reveal the necessity of such between-slide normalization.

ANOVA *(13)* is an alternative to the types of normalization methods described in the preceding and aims to estimate the size of different effects, including dye, gene, array, and sample. Changes in gene expression across the samples are estimated by the sample x gene interaction terms of the model *(13)*.

1.5. Data Filtering

Prior to detection of differentially expressed genes, a data-filtering step can be applied. It is recommended to eliminate all genes that are not expressed over all samples.

For cDNA microarrays, genes for which the foreground <f(background), where f can be for example background+2sd (background), can be chosen as threshold for expression. For Affymetrix chips the absent/present calls can be used. Genes that show a low variation over the whole dataset or are not expressed at all can be eliminated, thus reducing the number of genes that have to be tested for differential expression.

1.6. Detection of Differentially Expressed Genes

Different methods to investigate differential expression exist, and as with the different methods for normalization, no one method has become the standard. Some popular methods to detect differentially expressed genes are:

Fold change

T-test and ANOVA

SAM: Significance Analysis of Microarray data (implemented in the *siggenes* package) *(14)*

Linear models and empirical Bayes, implemented in the *limma* package *(15, 16)*

In this chapter, the linear models and empirical Bayes methods implemented in the *limma* package *(15, 16)* are used for detection of differential expression. With *limma* one first calculates a moderated t-statistic for differential expression for each gene by performing a linear model fit on the data. Second, an empirical Bayes step is applied that produces more stable estimates when the number of samples is small. This method for detection of differentially expressed genes is general and can be applied to both one- and two-color microarray data.

1.7. Annotation

Different annotation packages exist in Bioconductor. This chapter uses the *biomaRt* package *(17)* to annotate a list of differentially expressed features. Other annotation packages that can be used are *annaffy, annBuilder*, and *annotate (18)*. BioMart *(19)* is a generic data management system developed jointly by the European Bioinformatics Institute (EBI) and Cold Spring Harbor Laboratory (CSHL). Examples of BioMart databases are:

Ensembl: A software project that produces and maintains automatic annotation on selected eukaryotic genomes (http://www.ensembl.org)

GRAMENE: A data resource for comparative genome analysis in the grasses. (http://www.gramene.org)

Wormbase: A data resource for Caenorhabditis biology and genomics (http://www.wormbase.org)

HAPMART: A data resource containing the results of the HAPMAP project. (http://www.hapmap.org)

These databases provide annotation data and other biological information, which cover most of the microarray research needs.

The BioConductor package *biomaRt* enables fast real-time queries to these BioMart databases and their associated web services. Examples of information that can be retrieved starting with Affymetrix identifiers are gene names, chromosome locations, GO identifiers, and many others.

2. Materials

2.1. R Essentials

In order to use *biomaRt*, one needs to have a basic knowledge of R. Unlike many "easy-to-use" microarray analysis packages, which have GUI facilities, R is command line. This gives the advantage that users will have a better understanding of what they are doing and can become power-users who go beyond the limitations associated with GUIs, have access to the newest methods, and can gradually start to implement their own R methods. A disadvantage is that the learning curve for R is steep and it takes some endurance to become familiar with it. If the reader is already accustomed with R, he or she may wish to continue from **Section 2.3**.

2.1.1. Vectors

1. R is vector based and even if one assigns only one value to a variable, it will be handled as a vector.

2. To create a vector with length larger than 1 we use the *c()* function:

```
>vec=c(20,25,30,35,40)
```

3. To select the third element of this vector, type:

   ```
   vec[3]
   ```

2.1.2. Matrices

1. A microarray dataset can typically be represented as a matrix in which the rows represent genes and the columns represent different samples or conditions under which the gene expression is measured.

2. A matrix can be created using the *matrix()* function. The arguments *nrow* and *ncol* define the number of rows and columns in the matrix, and the argument *byrow* defines how the data should be put in the matrix, filling the rows first or the columns (by default this is columns first).

   ```
   >mat=matrix(c(30,40,50,60), nrow=2)
   >mat
        [,1] [,2]
   [1,]   30   50
   [2,]   40   60
   ```

3. To select the element of the first row and second column, type:

   ```
   >mat[1,2]
   [1] 50
   ```

2.1.3. Lists

1. A list is a collection of different objects such as matrices, vectors, and characters.

   One can create a list consisting of the vector and matrix created above and a character string by:

   ```
   >myList=list(vec, mat, "my first list")
   ```

2. To access the first element one can type:

   ```
   >myList[[1]]
   [1] 20 25 30 35 40
   ```

3. A second type of list is a named list. Here the elements of the list are named and these names can be used to access the separate elements by using a $ symbol between the name of the object and the name of the element in the list.

   ```
   >myNamedList=list(myVec=vec, myMat=mat, myText="Some text")
   >myNamedList$myText
   [1] "Some text"
   ```

2.1.4. data.frame

1. A data.frame is very similar to a list but has the restriction that the length of all the elements it contains should be the same. So in the preceding example, the matrix should have as many rows as the length of the vector.

2. This type of object is used very frequently in gene expression data. For example, one can think of putting a matrix with log ratios and a vector with gene annotations in a data.frame.

3. An example data.frame would be the matrix defined above combined with a vector of gene names.

```
>genes=c("BRCA2","CDH1")
>myFrame=data.frame(genes=genes,
exprs=mat)
```

4. As with the named list, different parts of the data.frame can be accessed using the $ symbol.

```
>myFrame$genes
[1] BRCA2 CDH1
Levels: BRCA2 CDH1
```

2.1.5. Help

1. R provides many ways to get help.

2. If one knows the function name but does not know how to use it, a question mark before a function name will display its help. For example, to display the help for plot, type:

```
>?plot
```

3. If one does not know the function name exactly you can try to find help by using the *help.search*( ) function

```
>help.search("plot")
```

2.2. R Packages

The core of R contains functions that are broadly applicable. R packages provide functions for specific applications. The R packaging system is a way to distribute functions that can be used to do a specific type of job, such as cDNA microarray normalization. Beside code, these packages contain help files, example data, and vignettes. Vignettes are a good way to explore how to use new functions. Most packages can be downloaded from a nearby CRAN mirror website (see http://www.r-project.org and follow link CRAN).

2.3. Bioconductor

Bioconductor is an open source and open development software project for the analysis and comprehension of genomic data *(20, 21)*. Bioconductor is implemented in R and makes use of the R packaging system to distribute its software. The project's home page can be found at http://www.bioconductor.org, and contains lots of information on packages, tutorials, and more.

2.4. Example datasets

The following Method section uses publicly available microarray datasets, which can be downloaded from ArrayExpress (http://www.ebi.ac.uk/arrayexpress) and the Bioconductor web site.

2.4.1. Affymetrix Experiments

1. Go to the ArrayExpress homepage.
2. Follow link "query database."
3. Query for experiment "E-MEXP-70."
4. Follow the FTP link and download the .cel files.

2.4.2. cDNA Experiment

1. The cDNA experiment used here comes as an example dataset with the *marray* package, downloadable from the Bioconductor web site.

3. Methods

3.1. Normalization of Affymetrix Data

3.1.1. Standard Functions

1. Affymetrix chips can be normalized using the *affy* and *gcrma* Bioconductor packages. The *affy* package provides data visualization methods and various normalization methods such as MAS5.0 and RMA *(22)*. The GCRMA normalization method is implemented in a separate package *gcrma* *(6)*. When applying GCRMA for the first time, a package containing the probe sequence information of the chip will be automatically downloaded; this will be done every time an array of a new chip design is normalized (**Note 4**).

2. After installing these packages, load them into your R session.

   ```
   >library(affy)
   >library(gcrma)
   ```

3. Read in the CEL files with the function ReadAffy and give the celfile.path as argument.

   ```
   >data=ReadAffy(celfile.path="D:/DATA/
   E-MEXP-70")
   ```

4. Check your slides by producing raw images. Consider, for example, the 3rd hybridization in the set under consideration (**Note 5**). These images allow one to assess the hybridization quality. Any strikes or bright zones indicate contamination with dust or other interfering materials.

   ```
   >image(data[,3])
   ```

   The result of this command is shown in **Fig. 4.1**.

5. Check for RNA degradation. Affymetrix probes in a probe set are designed to cover different parts of the transcript. By plotting the intensities according to location of these probes (5′ to 3′) in transcript sequence, one can see if the RNA was degraded or not. Degradation usually starts from the 5′ side of the transcript.

Pre-Processing of Microarray Data and Analysis of Differential Expression 97

Day0(3).CEL

Fig. 4.1. Image of raw intensities of an Affymetrix chip. This image was created using the 'image' function.

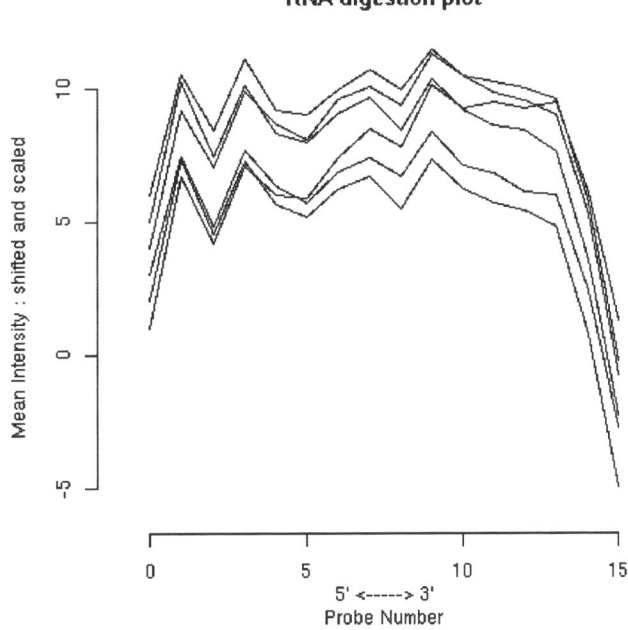

Fig. 4.2. RNA degradation plot created using the function 'plotAffyRNAdeg.'

```
>degrad=AffyRNAdeg(data)
>plotAffyRNAdeg(degrad)
```

The result of this command is shown in **Fig. 4.2**.

6. Normalize the data using GCRMA (**Note 6**).

```
>norm=gcrma(data)
Computing affinities.Done.
Adjusting for optical effect……Done.
Adjusting for non-specific binding……Done.
Normalizing
Calculating Expression
```

7. The object created after normalization is called an Expression Set (exprSet) and is the typical Bioconductor object to store normalized data of two and one color microarray data (**Note 7**).

```
>norm
Expression Set (exprSet) with
12625 genes
6 samples
phenoData object with 1 variables and 6 cases
varLabels
  sample: arbitrary numbering
```

8. Check the normalized data by plotting a boxplot.

    ```
    >boxplot(as.data.frame(exprs(norm)))
    ```

    The plot produced by this command is shown in **Fig. 4.3**.

9. Check correlation between samples by either pair-wise scatter plot or a heatmap of the correlation.

    ```
    >pairs(exprs(norm))
    ```

    The plot produced by the *pairs* command is shown in **Fig. 4.4**.

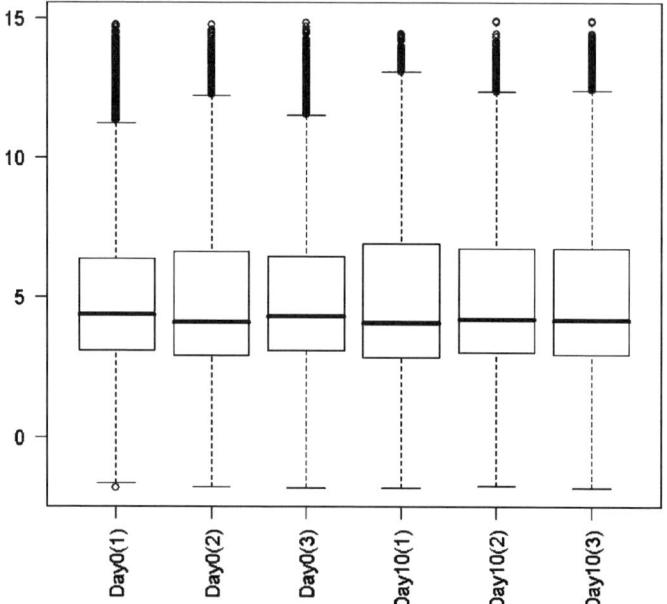

Fig. 4.3. Boxplot comparing the intensities grouped per array after normalization.

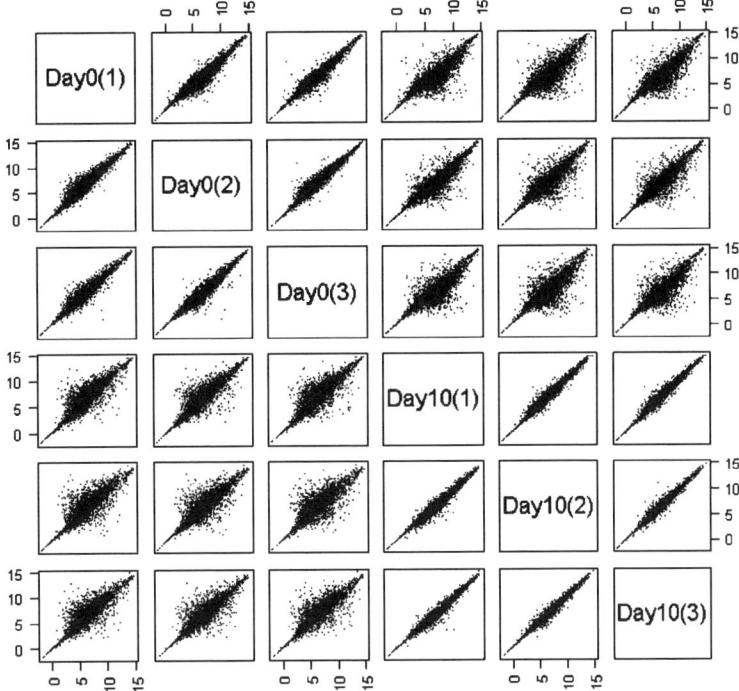

Fig. 4.4. Pairwise scatter plots of the intensities of each array against the corresponding intensities on all other arrays.

Note how the correlation between repeats is higher than between different samples.

10. Save the normalized data e.g. in a D:/DATA directory, make sure the file has the extension "RData" (**Note 8**).

    ```
    >save(norm, file="D:/DATA/emexp-70.RData")
    ```

11. The norm object is of the class "ExprSet" and the normalized expression data can be accessed using the *exprs* function, which returns the expression matrix.

3.1.2. Using Expresso

The function *expresso* is an alternative function that implements the steps described in the preceding.

1. Repeat the steps of 3.1.1. until Step 5

2. Now use the *expresso* command to normalize the data. In the following command, the VSN method is used without background correction as the normalization method and medianpolish is used as the summarization method.

    ```
    >norm=expresso(data, pmcorrect.method
    ="pmonly", bg.correct=FALSE, normalize.
    method = "vsn", summary. method =
    "medianpolish")
    ```

3.2. cDNA Microarray Data

1. There are many formats in which microarray data can be given to the data analyst. The following example involves importing .spot data.

   Load the *limma* and *marray* libraries.

   ```
   >library(limma)
   >library(marray)
   ```

2. In order to use a dataset contained in the *marray* package, one must first retrieve the directory where this dataset is installed.

   ```
   >dir=system.file("swirldata,"
   package="marray")
   ```

3. The read.maimages function can be used to import all .spot files (and some other file formats) that reside in this directory (**Note 9**).

   ```
   >raw=read.maimages(source="spot,"
   ext="spot", path=dir)
   ```

4. Now you should have a *RGList* object containing the raw intensity data. It is a names list containing the red and green foreground and background intensities and the filenames of the .spot files. The individual intensities can be accessed using a $ symbol as explained in the R essentials section.

   ```
   >raw
   ```

An object of class "RGList"
$R
```
         swirl.1      swirl.2     swirl.3     swirl.4
[1,]  19538.470   16138.720   2895.1600   14054.5400
[2,]  23619.820   17247.670   2976.6230   20112.2600
[3,]  21579.950   17317.150   2735.6190   12945.8500
[4,]   8905.143    6794.381    318.9524     524.0476
[5,]   8676.095    6043.542    780.6667     304.6190
8443 more rows ...
```

$G
```
         swirl.1      swirl.2     swirl.3     swirl.4
[1,]  22028.260   19278.770   2727.5600   19930.6500
[2,]  25613.200   21438.960   2787.0330   25426.5800
[3,]  22652.390   20386.470   2419.8810   16225.9500
[4,]   8929.286    6677.619    383.2381     786.9048
[5,]   8746.476    6576.292    901.0000     468.0476
8443 more rows ...
```

$Rb
```
         swirl.1      swirl.2     swirl.3     swirl.4
[1,]   174          136          82           48
[2,]   174          133          82           48
```

```
[3,] 174         133         76          48
[4,] 163         105         61          48
[5,] 140         105         61          49
8443 more rows …

$Gb
    swirl.1     swirl.2     swirl.3     swirl.4
[1,] 182         175         86          97
[2,] 171         183         86          85
[3,] 153         183         86          85
[4,] 153         142         71          87
[5,] 153         142         71          87
8443 more rows …

$targets
FileName
swirl.1 swirl.1
swirl.2 swirl.2
swirl.3 swirl.3
swirl.4 swirl.4
```

5. The gene names are stored in a separate .gal file. One can add the gene names to the raw object as follows:

   > raw$genes = readGAL (paste (dir, "fish. gal, "sep = "/"))

6. Now it is necessary to assign the array layout to the *RGList* object. This can be done using the function *getLayout*()

   > raw$printer = getLayout (raw$genes)

7. Once the layout is assigned, one can perform quality assessment by plotting, for example, the images of the different channels to detect hybridization errors. **Figure 4.5** shows a plot of the log2 intensities for the red background signal.

   > imageplot (log2 (raw$Rb[,1]), RG$printer, low = "white", high = "black")

8. An M versus A plot *(23)*, which is sometimes also called an R versus I plot, is a nice way to visualize the data and helps to see the effects of normalization and it is also a popular way to show differentially expressed genes (see the following). The function *plotMA*() draws such a plot, the argument *array* can be used to set which hybridization should be plotted.

   > plotMA (raw, array = 3)

The result of this plotting function is shown in **Fig. 4.6**. Note the typical banana shape of this plot of non-normalized two-color microarray data.

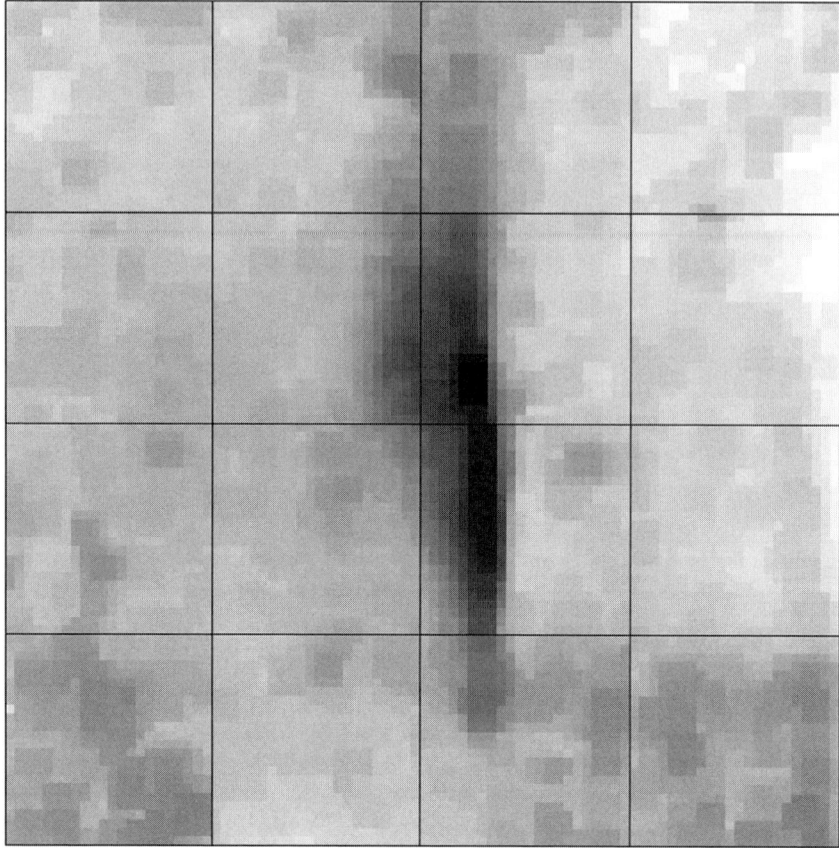

z-range 5.9 to 11.1 (saturation 5.9, 11.1)

Fig. 4.5. Image of the log2 intensities for the red background signal of cDNA microarray. The different zones in this plot represent different print-tips.

9. To assess the need for print tip specific lowess normalization, one can either plot boxplots where the intensities are grouped by print-tip or plot an MA plot for each print-tip separately using the *plotPrintTipLoess()* function.

   ```
   >plotPrintTipLoess(raw)
   ```

10. Now the raw microarray data can be normalized using the function *normalizeWithinArrays()* (**Note 10**). This function performs by default a print-tip specific lowess normalization on background subtracted red/green ratios. This normalization returns a *MAList* object.

    ```
    >norm=normalizeWithinArrays(raw)
    ```

11. To see the effect normalization has on the data one can plot again an M versus A plot.

    ```
    >plotMA(norm, array=3)
    ```

    The result of this plot is shown in **Fig. 4.7**. Notice that the banana shape visible in **Fig. 4.6** has disappeared.

Pre-Processing of Microarray Data and Analysis of Differential Expression 103

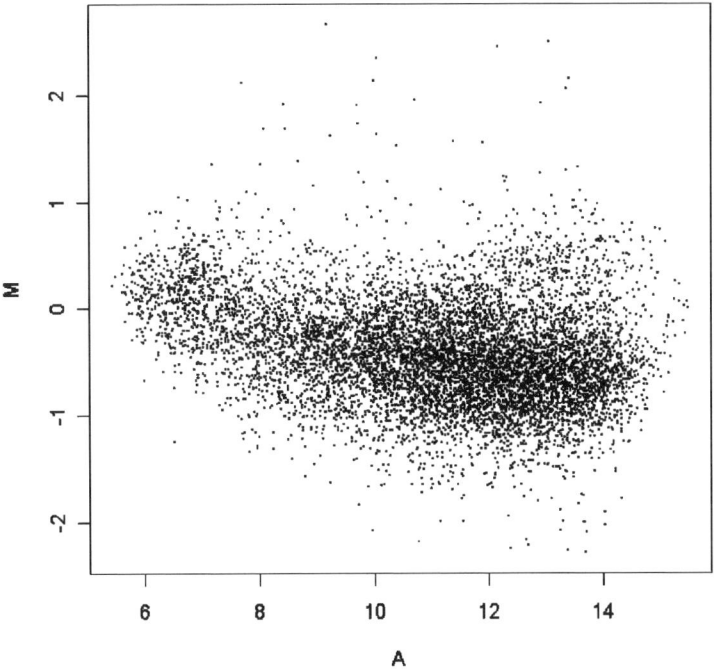

Fig. 4.6. M versus A plot of background corrected raw intensities.

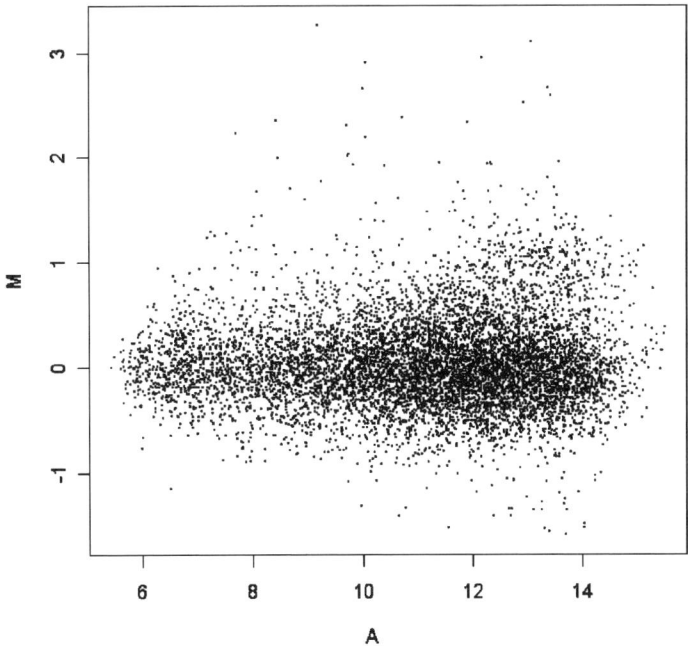

Fig. 4.7. M versus A plot of normalized data.

12. As with the Affymetrix normalization, a boxplot and pairwise plot can be used to further check quality and need for between-array normalization.
13. Use the function *normalizeBetweenArrays()* to do between-array normalization.

    > norm = normalizeBetweenArrays(norm)

14. Now the data are ready for further analysis.

3.3. Detection of Differential Expression Using limma

1. Load the *limma* package.

   > library(limma)

2. Create labels for the samples.

   > sample= c("day0","day0","day0","day10", "day10","day10")

3. Create the design matrix (**Note 11**).

   > design=model.matrix(~0+factor(sample))

4. Adjust the column names of the design matrix.

   > colnames(design) =c("day0","day10")
   > design
   day0 day10
   1 1   0
   2 1   0
   3 1   0
   4 0   1
   5 0   1
   6 0   1
   attr(,"assign")
   [1] 1 1
   attr(,"contrasts")
   attr(,"contrasts")$"factor(sample)"
   [1] "contr.treatment"

5. Fit the linear model.

   > fit=lmFit(norm, design=design)

6. Specify the contrast matrix to define which differences are of interest. In this simple example there is only one difference of interest and that is the difference between day0 and day10.

   > contrast.matrix=makeContrasts(day0-day10, levels=design)
   > contrast.matrix
   day0-day10
   day0 1
   day10 -1

7. Fit the contrast matrix.

    `>fit2=contrasts.fit(fit, contrast.matrix)`

8. Apply empirical Bayes on this fit

    `>eb=eBayes(fit2)`

9. Show the top 100 differentially expressed genes using false discovery rate correction for multiple testing.

    ```
    >top=topTable(eb, n=100, adjust.
    method="fdr")
    >top
          ID         M        A        t         P.Value    B
    7261  37192_at  -7.582453  5.073572 -56.48199 4.806783e-05 10.339575
    5808  35752_s_at -6.974551 6.422358 -49.68752 4.806783e-05 10.003077
    4059  34020_at   7.602192  5.399944  48.24657 4.806783e-05  9.918376
    892   1797_at   -5.355800  6.250378 -43.20631 6.505689e-05  9.575488
    10434 40335_at  -5.860488  5.388965 -41.81805 6.505689e-05  9.466281
    5683  35628_at  -3.714908  6.582599 -36.18743 1.212805e-04  8.940546
    5151  35101_at  -6.287120  5.722050 -35.32328 1.212805e-04  8.846069
    274   1252_at   -3.653588  4.055042 -33.78907 1.292874e-04  8.667675
    8130  38052_at  -8.984562  8.237705 -33.41587 1.292874e-04  8.622092
    7468  37397_at  -3.036533  7.160758 -30.83967 1.619917e-04  8.281642
    ```

10. A volcano plot, which plots the log odds score against the log fold change is a good way to visualize the results of the differential gene expression calculations. The argument *highlight* can be used to plot the names for a specified number of topmost differentially expressed features.

    `>volcanoplot(eb, highlight=4)`

    The volcano plot produced by this function is shown in **Fig. 4.8**.

3.4. Annotation

As described in the Introduction, the biomaRt package is used here to annotate the list of differentially expressed features. For further information on installation of biomaRt *see* **Note 12**.

The following example uses the list of upregulated genes found in **Section 3.3.1**.

1. Load the library.

    `>library(biomaRt)`

2. List the available BioMart databases using the function *listMarts*().

    ```
    >listMarts()
    name            version
    ensembl         ENSEMBL 42 GENE (SANGER)
    snp             ENSEMBL 42 VARIATION (SANGER)
    ```

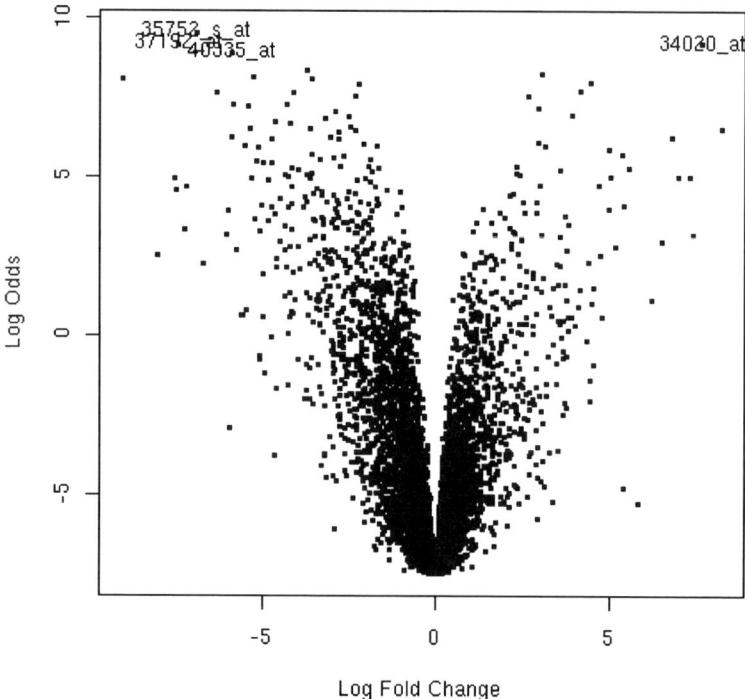

Fig. 4.8. Volcano plot, plotting the significance of differential expression versus log fold change.

```
vega         VEGA 21 (SANGER)
wormbase     WORMBASE (CSHL)
uniprot      UNIPROT PROTOTYPE (EBI)
msd          MSD PROTOTYPE (EBI)
dicty        DICTYBASE (NORTHWESTERN)
```

3. Select a BioMart database to use with the function *useMart*( ) and verify database connection.

    ```
    >ensembl=useMart("ensembl")
    ```

4. List the available datasets in the selected BioMart database.

    ```
    >listDatasets(ensembl)
         dataset                        version
      1  ptroglodytes_gene_ensembl      CHIMP1
      2  ggallus_gene_ensembl           WASHUC1
      3  rnorvegicus_gene_ensembl       RGSC3.4
      4  scerevisiae_gene_ensembl       SGD1
      5  tnigroviridis_gene_ensembl     TETRAODON7
      6  xtropicalis_gene_ensembl       JGI3
      7  frubripes_gene_ensembl         FUGU2
      8  cintestinalis_gene_ensembl     CINT1.95
    ```

```
 9  agambiae_gene_ensembl      MOZ2a
10  amellifera_gene_ensembl    AMEL2.0
11  btaurus_gene_ensembl       BDGP4
12  celegans_gene_ensembl      CEL140
13  mmusculus_gene_ensembl     NCBIM34
14  cfamiliaris_gene_ensembl   BROADD1
15  dmelanogaster_gene_ensembl BDGP4
16  drerio_gene_ensembl        ZFISH5
17  hsapiens_gene_ensembl      NCBI35
18  mdomestica_gene_ensembl    JGI3
```

5. Update the mart object by selecting a dataset using the *useDataset*() function.

   ```
   >ensembl=useDataset(dataset="hsapiens_
   gene_ensembl",mart=mart)
   
   Checking attributes and filters ... ok
   ```

6. In BioMart systems, attributes represent the annotation one wants to retrieve. The *listAttributes* function lists all the available attributes that can be retrieved from the selected dataset. One of them is 'hgnc_symbol,' which will be used later.

   ```
   >listAttributes(ensembl)
        name       description
   1    adf_embl   embl
   2    adf_go     go
   3    adf_omim   omim
   4    adf_pdb    pdb
   ```

7. In BioMart systems, filters make the application retrieve the features that pass these filters. The function *listFilters* shows all possible filters that can be used on this dataset. One of them is 'affy_hg_u95av2', which represents the Affymetrix identifiers of the example dataset.

   ```
   >listFilters(ensembl)
        name                description
   1    affy_hc_g110        Affy hc g 110 ID(s)
   2    affy_hg_focus       Affy hg focus ID(s)
   3    affy_hg_u133_plus_2 Affy hg u133 plus 2 ID(s)
   4    affy_hg_u133a       Affy hg u133a ID(s)
   ```

8. Get the gene symbols for the 100 differentially expressed features stored in the top object, which was created in **Section 3.3**, using 'hgnc_symbol' as attribute to retrieve and 'affy_hg_u95av2' as filter. The top list contains affymetrix identifiers from the hgu85av2 array that is to be annotated.

   ```
   >getBM(attributes=c("affy_hg_u95av2","hgnc_
   symbol,""chromosome_name"),filters="affy_
   ```

```
hg_u95av2",values=top$ID[1:10],
mart=ensembl)
```

affy_hg_u95av2	hgnc_symbol	chromosome_name
1252_at	REEP5	5
1797_at	CDKN2D	19
35752_s_at		3
35752_s_at	PROS1	3
37192_at	EPB49	8
37524_at	STK17B	2
38052_at	F13A1	6
39878_at	PCDH9	13
40335_at	CLCN4	X

9. Retrieving other information is similar to retrieving gene names. By using different attributes, one can retrieve information such as GO, INTERPRO protein domains, start and end positions on chromosomes, etc., starting from the Affymetrix identifiers or any other type of filter available.

4. Notes

1. A correlation heatmap plots the correlation of the intensities between all samples in a color-coded fashion. The image is displayed as a grid containing square fields. In a correlation heatmap the rows and columns of this grid represent the samples and the color of each field represents the correlation between the respective two samples. See the function *heatmap* for more details.

2. Quantile normalization aims to make the distribution of probe intensities for each array the same for a set of arrays *(8)*.

3. When applying the median polish algorithm to a matrix, one subtracts the row medians and then the column medians in an iterative manner. After convergence, the row and column medians are zero, and the remaining matrix is subtracted from the original data matrix.

4. The probe package contains the probe sequences used in GCRMA. The cdf package is needed for both GCRMA and RMA, as it contains the location and identifiers of the probes. When an array of a new design is processed, the cdf and probe packages of this array will be automatically downloaded and installed. If this fails, make sure you are connected to the Internet and have permission to install R packages.

5. To save an image or plot, the following three steps should be performed: determine the format of the picture (e.g. JPEG, PNG, postscript), and assign a name; do the plotting; close the plotting device to store the plot.

```
>png(file="myplot.png")
>plot(x,y) #The function used to make
the plot
>dev.off() # to close the plotting device
and save the plot
```

6. To normalize the data with the RMA method, use the following command:

   ```
   >eset=rma(data)
   ```

7. The different parts of an expression set object can be accessed using specific functions. Here is an example to access the intensity data.

   ```
   >int=exprs(eset)
   ```

8. To save the normalized data in a tab-delimited file, use the function *write.table*( ).

   ```
   >write.table(exprs(eset),    quotes=FALSE,
   sep='\t', file='D:/myExprsData.txt')
   ```

9. When the raw data are not in a format supported by the limma package, the *read.table*( ) function can be used to import it into R. The *skip* argument can be used to let the function know how many lines in the input file should be skipped before the real data row starts. Once the data are read, they need to be reformatted and put in a new *RGList* object which then can be used to normalize.

10. The *marray* package also provides good methods to normalize and visualize two-color experiments. Check out the functions *maNorm*( ), *maPlot*( ) and *maBoxplot*( ).

11. More advanced examples and details on the creation of design matrices can be found in the excellent *limma* users guide by G. Smyth.

12. Installation of biomaRt requires that the RCurl and XML packages are installed.

    RCurl can be found on the following URL: http://www.omegahat.org/RCurl

References

1. Yang, Y. H., Dudoit, S., Luu, P., et al. (2002) Normalization for cDNA microarray data: a robust composite method addressing single and multiple slide systematic variation. *Nucleic Acids Res* 30(4), e15.

2. Zakharin, S. O., Kim, K., Mehta, T., et al. (2005) Sources of variation in Affymetrix microarray experiments. *BMC Bioinformatics* 6, 214.

3. Affymetrix (2002) Statistical Algorithms Description Document http://www.affymetrix.com/support/technical/whitepapers/sadd_whitepaper.pdf

4. Li, C., Wong, W. H. (2001) Model-based analysis of oligonucleotide arrays: Expression index computation and outlier detection. *Proc Natl Acad Sci U S A* 98(1), 31–36.

5. Irizarry, R. A., Hobbs, B., Collin, F., et al. (2003) Exploration, normalization, and summaries of high-density oligonucleotide array probe level data. *Biostatistics* 4, 249–264.

6. Wu, Z., Irizarry, R., Gentleman, R., et al. (2004) A model based background adjustment for oligonucleotide expression arrays. *JAMA* 99(468), 909–917.

7. Huber, W., von Heydebreck, A., Sueltmann, H., et al. (2002) Variance stabilization applied to microarray data calibration and to the quantification of differential expression. *Bioinformatics* 18, S96–S104.

8. Bolstad, B. M., Irizarry, R. A., Astrand, M., et al. (2003) A comparison of normalization methods for high density oligonucleotide array data based on variance and bias. *Bioinformatics* 19(2), 185–193.

9. Cope, L., Irizarry, R, Jaffee, H., et al. (2004) A benchmark for Affymetrix GeneChip expression measures. *Bioinformatics* 20(3), 323–331.

10. Shedden, K., Chen, W., Kuick, R., et al. (2005) Comparison of seven methods for producing Affymetrix expression scores based on False Discovery Rates in disease profiling data. *BMC Bioinformatics* 6(1), 26.

11. Van de Peppel, J., Kemmeren, P., van Bakel, H., et al. (2003) Monitoring global messenger RNA changes in externally controlled microarray experiments. *EMBO Repts* 4(4), 387–393.

12. Workman, C., Jensen, L. J., Jarmer, H., et al. (2002) A new non-linear normailzation method for reducing variability in DNA microarray experiments. *Genome Biology* 3(9), research0048.

13. Kerr, K., Martin, M., Churchill, G. (2000) Analysis of Variance for gene expression microarray data. *J Comput Biol* 7, 819–837.

14. Tusher, V. G., Tibshirani, R., Chu, G. (2001) Significance analysis of microarrays applied to the ionizing radiation response. *Proc Natl Acad Sci U S A* 98(9), 5116–5121.

15. Smyth, G. K. (2004) Linear models and empirical Bayes methods for assessing differential expression in microarray experiments. *Stat Appl Gen Mol Biol* 3(1), Article 3.

16. Smyth, G. K., Michaus, J., Scott, H. (2005). The use of within-array replicate spots for assessing differential expression in microarray experiments. *Bioinformatics* 21(9), 2067–2075.

17. Durinck, S., Moreau, Y., Kasprzyk, A., et al. (2005). BioMart and Bioconductor: a powerful link between biological databases and microarray data analysis. *Bioinformatics* 21, 3439–3440.

18. Zhang, J., Carey, V., Gentleman, R. (2003) An extensible application for assembling annotation for genomic data. *Bioinformatics* 19(1), 155–156.

19. Kasprzyk, A., Keefe, D., Smedley, D., et al. (2004) EnsMart: a generic system for fast and flexible access to biological data. *Genome Res* 14(1), 160–169.

20. Gentleman, R. C., Carey, V. J., Bates, D. M., et al. (2004) Bioconductor: open software development for computational biology and bioinformatics. *Genome Biol* 5, R80.

21. Gentleman, R. C., Carey, V., Huber, W., et al. (2005) *Bioinformatics and Computational Biology Solutions Using R and Bioconductor*. Springer, NY.

22. Gautier, L., Cope L., Bolstad, B. M., et al. (2004) Affy: analysis of Affymetrix GeneChip data at the probe level. *Bioinformatics* 20(3), 307–315.

23. Dudoit, S., Yang, Y. H., Callow, M. J., et al. (2002) Statistical methods for identifying genes with differential expression in replicated cDNA microarray experiments. *Stat Sin* 12, 111–139.

Chapter 5

Developing an Ontology

Midori A. Harris

Abstract

In recent years, biological ontologies have emerged as a means of representing and organizing biological concepts, enabling biologists, bioinformaticians, and others to derive meaning from large datasets. This chapter provides an overview of formal principles and practical considerations of ontology construction and application. Ontology development concepts are illustrated using examples drawn from the Gene Ontology (GO) and other OBO ontologies.

Key words: ontology, database, annotation, vocabulary.

1. Introduction

1.1. Why Build an Ontology?

The ongoing accumulation of large-scale experimental and computational biological investigation is frequently noted, and has led to concomitant growth of biological databases, and the amount and complexity of their content. To take full advantage of this accumulated data, biologists and bioinformaticians need consistent, computable descriptions of biological entities and concepts represented in different databases. By providing a common representation of relevant topics, ontologies permit data and knowledge to be integrated, reused, and shared easily by researchers and computers. One example, describing the use of several different ontologies to integrate data for the laboratory mouse, is discussed in *(1)*. A number of ontologies are now emerging in the biomedical domain to address this need, and still others can be envisioned.

1.2. What Is an Ontology?

An ontology is a shared, common, backbone taxonomy of relevant entities, and the relationships between them within an application domain (*see* **Note 1**).

In other words, a particular subject area can be thought of as a domain, and ontologies represent domains by defining entities within the domain (and the terms used to refer to them) and the way in which the entities are related to each other. For example, the Gene Ontology (GO) describes three domains, molecular function, biological process, and cellular component (for further information on GO, *see (2–7)*); sample terms from the cellular component ontology are "plastid" and "chloroplast," and the latter is a subtype of the former (formally, the relationship is chloroplast *is_a* plastid). In addition to forming a computable representation of the underlying reality, an ontology thus provides a framework for communicating knowledge about a topic. Ontologies can take many forms, and can be represented in different formats, which support different levels of sophistication in computational analyses. For more detailed introductions to ontologies, especially as applied in biomedical domains, *see* references *(8–17)*.

2. Practical Matters

2.1. Online Resources

2.1.1. Ontology Editing Tools

A number of ontology development tools exist, of which the most commonly used for biomedical ontologies are Protégé and OBO-Edit. Both are open-source, cross-platform applications that provide graphical interfaces for ontology creation and maintenance (*see* **Note 2**).

2.1.1.1. Protégé

Protégé supports editors for two approaches to modeling ontologies: the Protégé-Frames editor is used to build and populate frame-based ontologies, and the Protégé-OWL editor allows users to build ontologies for the Semantic Web, using the W3C's Web Ontology Language (OWL; http://www.w3.org/2004/OWL/): available from http://protege.stanford.edu/download/download.html.

2.1.1.2. OBO-Edit

OBO-Edit (formerly DAG-Edit) is a Java application for viewing and editing OBO ontologies (see the following). OBO-Edit uses a graph-based interface that is very useful for the rapid generation of large ontologies focusing on relationships between relatively simple classes. Available from https://sourceforge.net/project/showfiles.php?group_id=36855

2.1.2. Ontology Collections The most comprehensive list available at present is the Open Biomedical Ontologies (OBO) collection (http://obofoundry.org). The OBO project not only collects ontologies, but also establishes guidelines for useful ontologies in biomedical domains (OBO inclusion criteria are shown in **Table 5.1**). Other useful ontology collections include the Protégé Ontologies Library (http://protege.cim3.net/cgi-bin/wiki.pl?ProtegeOntologiesLibrary) and the MGED Ontology Resources (http://mged.sourceforge.net/ontologies/OntologyResources.php).

2.2. Overview of Ontology Development Ontology building involves an initial phase of research and groundwork, followed by a (usually intensive) phase in which the ontology is first established and populated; finally, ongoing iterative refinement guided by usage should continue as long as the ontology is actively applied.

Table 5.1
Original OBO principles

For an ontology to be accepted as one of the Open Biomedical Ontologies, the following criteria must be met. These criteria are available online at http://obofoundry.org/crit.html

- The ontologies must be open and can be used by all without any constraint other than that their origin must be acknowledged and they cannot be altered and redistributed under the same name. The OBO ontologies are for sharing and are resources for the entire community. For this reason, they must be available to all without any constraint or license on their use or redistribution. However, it is proper that their original source is always credited and that after any external alterations, they must never be redistributed under the same name or with the same identifiers.

- The ontologies are in, or can be instantiated in, a common shared syntax. This may be either the OBO syntax, extensions of this syntax, or OWL.
  The reason for this is that the same tools can then be usefully applied. This facilitates shared software implementations. This criterion is not met in all of the ontologies currently listed, but we are working with the ontology developers to have them available in a common OBO syntax.

- The ontologies are orthogonal to other ontologies already lodged within OBO.
  The major reason for this principle is to allow two different ontologies, for example, anatomy and process, to be combined through additional relationships. These relationships could then be used to constrain when terms could be jointly applied to describe complementary (but distinguishable) perspectives on the same biological or medical entity.
  As a corollary to this, we would strive for community acceptance of a single ontology for one domain, rather than encouraging rivalry between ontologies.

- The ontologies share a unique identifier space.
  The source of concepts from any ontology can be immediately identified by the prefix of the identifier of each concept. It is, therefore, important that this prefix be unique.

- The ontologies include textual definitions of their terms.
  Many biological and medical terms may be ambiguous, so concepts should be defined so that their precise meaning within the context of a particular ontology is clear to a human reader.

2.2.1. Groundwork

2.2.1.1. Ontology Scope

The first step in building any ontology is to determine the scope: What subject area must it represent, and at what level of detail? Clearly specified subject matter is essential, along with a clear sense of what does and does not fall into the intended scope.

For example, GO covers three domains (molecular function, biological process, and cellular component) and defines them. Equally importantly, GO also explicitly documents some of the things that it does not include, such as gene products or features thereof, mutant or abnormal events, evolutionary relationships, and so forth *(2–4, 6, 18)*.

2.2.1.2. Evaluating Existing Ontologies

Once the scope has been decided upon, the next step is to find out whether any existing ontology covers the domain (e.g., see Ontology Collections); the obvious reason for this is to avoid duplicating other ontology development efforts. In most cases, developing a new ontology for a topic that is covered by an existing ontology not only wastes time and work (and therefore money), but also runs the risk that incompatible knowledge representations will result.

It is important, however, to evaluate existing ontologies and the efforts that support them, to ensure that an ontology project is viable and the ontology itself usable. The most important things to look for in an ontology are openness, active development, and active usage. Open access to an ontology ensures that it can be easily shared.

Ongoing development ensures that an ontology will be adapted to accommodate new information or changed understanding of the domain, essential for rapidly moving fields such as many biological research areas. Active usage feeds back to the development process, ensuring that future development maintains or improves the ontology's fitness for its purpose.

These and other considerations form the basis for the OBO Inclusion Criteria and the emerging OBO Foundry paper citation *(19)*, which offer additional guidance in creating ontologies and evaluating existing ontologies (*see* **Tables 5.1** and **5.2**).

2.2.1.3. Domain Knowledge

Should a new ontology be required, domain knowledge is indispensable in its development; the involvement of the community that best understands the domain is one of the most important elements of any successful ontology development project. For biological ontologies, of course, the relevant community consists of researchers in the area covered by a given ontology, as well as curators of any databases devoted to the subject area. The advantages of involving domain experts such as biologists in ontology development are twofold: first, computer scientists (and other non-experts) can be spared the effort of learning large amounts of domain-specific knowledge; and second, if community members feel involved in development, it helps ensure that the

Table 5.2
Additional principles for the OBO Foundry

The OBO Foundry is a more recent development that refines and extends the original OBO model, and includes a number of additional criteria:

- The ontology has a clearly specified and clearly delineated content.
- The ontology provider has procedures for identifying distinct successive versions.
- The ontology uses relations that are unambiguously defined following the pattern of definitions laid down in the OBO Relation Ontology.
- The ontology is well documented.
- The ontology has a plurality of independent users.

The OBO Foundry will also distinguish "Reference" and "Application" ontologies; the former are formally robust and application-neutral, whereas the latter are constructed for specific practical purposes. As the OBO Foundry matures, more inclusion criteria will be established. For further information on the OBO Foundry, see http://obofoundry.org/ and http://sourceforge.net/projects/obo

resulting ontology will be accepted and used within the biological community. In other words, people are more inclined to "buy into" resources developed within rather than outside their communities.

It is essential to consider how the ontology will be used: Many of the decisions that must be made about the structure and content of an ontology depend on how it will be applied, what sort of queries it must support, and so forth. The expected lifespan of the ontology is also an important consideration: especially in a rapidly changing domain (such as almost any area of biomedical research), an ontology must be updated to reflect accumulating knowledge and evolving usage. Ideally, ontology developers should commit to actively maintaining both the ontology itself and contacts with the user community, to respond to queries and incorporate growing and changing expertise.

Example 1: The Gene Ontology was originally designed, and continues to be used extensively, for the annotation of gene products in biological databases. It is often desirable to make annotations to very specific terms, to capture as much detail as is available in the literature, but also to be able to retrieve annotations to a less specific term using the relationships between GO terms (*see* **Note 3**). These features of the desired application (or "use case") not only ruled out a flat, unstructured vocabulary such as a key word list, but also inspired GO to adopt the so-called "true-path rule," whereby every path from specific to general to root terms must be biologically accurate. This requirement is consistent with sophisticated logical ontology representations that use transitive relations (see Relationships), and is effectively imposed on any formal ontology. Imposing this rule has thus had the additional effect of simplifying improvements in GO's underlying logical representation.

Example 2: GO term names tend to be a bit verbose, often with some repetition between parent and child term names. This allows each term to make sense on its own, which is necessary for GO because terms are often displayed outside the context of the rest of the ontology, e.g., in gene or protein records in databases. An example is the GO biological process term "propionate metabolic process, methylmalonyl pathway (GO:001968)": it is an *is_a* child of "propionate metabolic process (GO:0019541)", but is not called simply "methylmalonyl pathway."

2.2.2. Ontology Construction

The backbone of an ontology is formed of terms (also called classes, types, or, in formal contexts, universals), their definitions, and the relationships between terms. The actual work of developing an ontology therefore consists of identifying the appropriate classes, defining them, and creating relationships between them.

2.2.2.1. Classes

Classes represent types, not instances, of objects in the real world, and refer to things important to the domain; relationships organize these classes, usually hierarchically.

2.2.2.2. Definitions

In any controlled vocabulary, the precise meanings of terms must be clearly specified to ensure that each term is used consistently. Most biological ontology efforts start by using free text definitions that capture meaning, including the conditions necessary and sufficient to distinguish a term from any other, in language accessible to humans. Formal, machine-parsable definitions take relationships into account to generate a computable representation of necessary conditions.

2.2.2.3. Relationships

The key feature distinguishing ontologies from keyword lists is that ontologies capture relationships between terms as well as the terms themselves and their meanings. Relationships come in different types, of which *is_a* is the most important. Like classes, relationships must be clearly defined to ensure that they are used consistently. The *is_a* relationship indicates that one class is a subclass of another, as in the anthranilate pathway example cited in the preceding. The *is_a* relationship is transitive, meaning that if A *is_a* B and B *is_a* C, it can logically be inferred that A *is_a* C. Other relationship types used may be transitive or not, and include *part_of*, which is used in GO and many other OBO ontologies, and *develops_from*, which appears in anatomy ontologies and the OBO Cell Ontology *(20)*. A number of core relationships are defined in the OBO Relations Ontology *(21)*, and should be used wherever applicable.

2.2.2.4. Additional Information

Other useful features include term identifiers, synonyms, and cross-references (cross-references may be to database entries, entities from other ontologies or classification systems, literature, etc.). Every class in an ontology needs a unique identifier, which should not encode too much information; it is especially important not to encode position, i.e. information on relationships between terms, in an identifier, because to do so would mean that to move a term would change its ID. Synonyms and cross-references are optional but can be included to support searches and correlation with other information resources. See, for example, the GO term "ATP citrate synthase activity" (GO:0003878), which has a text synonym ("ATP-citrate (pro-S)-lyase activity") and cross-references to entries in the Enzyme Commission classification (EC:2.3.3.8) and MetaCyc (ATP-CITRATE-(PRO-S-)-LYASE-RXN).

2.2.2.5. Representation

In the early stages of ontology development, it is important to consider what representation will best support the intended use of the ontology. At the simple end of the scale, key word lists form controlled vocabularies, and may include text definitions of the term, but do not capture relationships between different terms.

The most basic vocabularies that can be regarded as ontologies are those that include relationships between terms as well as the terms themselves. Still more logically rigorous structural features include complete subsumption hierarchies (*is_a*, described above) and formal, computable term definitions that support reasoning over an ontology (see Reasoning, below). Regardless of what structure and representation an ontology uses at the outset of its development, each feature addition makes more powerful computing possible using the ontology, at a cost of increasing complexity, requiring more sophisticated maintenance tools. The most successful biomedical ontology projects such as GO, have started with simple representations, yet allowed sufficient flexibility to add logical rigor as the development project matures, to take advantage of the benefits of formal rigor (*see* below) *(18, 22–24)*.

Furthermore, these structures can be represented in a number of different formats. The World Wide Web Consortium (W3C) has established the Web Ontology Language (OWL) as a standard for ontologies used by the Semantic Web; OWL has three sub-languages, OWL-Lite, OWL-DL, and OWL-Full, of differing degrees of complexity and expressive potential (see http://www.w3.org/2004/OWL/ and *(25)*). The Open Biomedical Ontologies (OBO) format (see http://www.geneontology.org/GO.format.obo–1_2.shtml) provides a representation similar to that of OWL, but with some special features tailored to the needs of the biological community. Description logic systems (of which OWL-DL is an example), frame-based representations, Resource Description Framework (RDF; see http://www.w3.org/RDF/), and other (often much simpler and less expressive) representations have also been widely used; *see (13)* for an overview of ontology representations from a biological perspective.

2.2.3. Early Deployment

To maximize its utility, an ontology should be released to the public and applied early in the development process. Errors to be corrected, gaps to be filled, and other opportunities to improve the ontology become much more apparent when the ontology is put to use for its intended purpose, involving actual instances of data, and when these data are used to answer research questions.

Early adoption of an ontology, along with the use of important research data, also facilitates ongoing iterative development by making continued involvement in ontology maintenance advantageous and desirable to the wider community.

3. Formal Principles

3.1. Advantages of Adhering to Formal Principles

Formal ontologists have elucidated a number of rules governing the content and structure of "well-built" ontologies. As with the OBO Foundry inclusion criteria (with which they partly overlap), adherence to these rules requires development effort, but offers practical advantages, especially as an ontology project matures. The most obvious advantage is that if an ontology is consistent with formal rules, its developers can avail themselves of a growing body of shared software tools. Two other benefits are more particular to ontology development and use: ontology alignment and reasoning.

3.1.1. Alignment

"Ontology alignment" refers to explicit references made from classes in one ontology to classes in another ontology.

Many ontologies need to invoke terms (and the entities they represent) from domains covered by other ontologies. Although an ontology term may simply include the name of a class from another ontology, a formalized, identifier-based reference to an "external" ontology is highly preferable. An external ontology reference provides: a definition and relationships to other classes from that external ontology; the ability to track when the same class is referenced in yet a third ontology; and the ability to parallel any error corrections or other enhancements that the external ontology makes (keeping the ontologies "in sync"). In addition, using cross-references between ontologies limits the duplication of terms and helps manage what might otherwise be an "explosion" of combinatorial terms in any one ontology.

For example, the GO biological process ontology contains many terms describing the metabolism of specific compounds and differentiation of specific cell types. Work is under way to create explicit references to ChEBI (http://www.ebi.ac.uk/chebi/) for the former and the Cell Ontology for the latter, enabling these different ontologies to establish, maintain, and update mutually consistent representations *(24, 26)*.

3.1.2. Reasoning

Formally correct ontologies support reasoning: In reasoning over an ontology, software (the "reasoner") uses the network of classes and relationships explicitly specified in an ontology to check the ontology for errors, discover the logical implications of existing structures, and suggest new terms that might be added to the ontology. Reasoning is thus extremely useful for keeping an ontology internally consistent, in that it can spot certain kinds of inconsistencies, such as redundancies, missing terms, or missing relationships, and can make implied relationships explicit. Reasoning is only possible in ontologies that have complete *is_a* paths (see Relationship Types); this is a key reason why *is_a* completeness is important and desirable.

3.1.3. Fundamental Distinctions

Formal ontologies make three fundamental distinctions: *continuants* vs. *occurrents*; *dependents* vs. *independents*; and *types* vs. *instances*. Although the terminology is unfamiliar to biologists, these distinctions are readily illustrated in familiar terms, as well as being essential to ontological integrity.

Continuants have continuous existence in time. They can gain or lose parts, preserving their identity through change; they exist *in toto* whenever they exist at all. Occurrents are events that unfold in time, and never exist as wholes. For example, the complexes and structures represented in the GO cellular component ontology are continuants, whereas the events in the GO biological process ontology are occurrents.

Dependent entities require independent continuants as their bearers; for example, the molecular weight of a protein makes sense only if there is a protein.

All occurrents are dependent entities; they require independent entities to carry them out. For example, a biological process such as translation requires mRNA, ribosomes, and other molecules and complexes.

The type/instance distinction: Ontologies generally include types/classes, and not instances, as an example, the GO term "mitochondrion" is a class, and refers to mitochondria generally; any particular mitochondrion existing in any actual cell would be an instance of the class (*see* **Note 4**). Nevertheless, it is worthwhile to think about the instances, because they are the members (in reality) of the classes in the ontology, and considering instances helps developers figure out which classes are required and what the relationships between them should be. Instances are also highly relevant to the question of how the ontology will be used.

3.1.4. Relationship Types

Relationship types commonly used in biomedical ontologies include (in addition to *is_a*) *part_of* and *derives_from*; the latter is also known as *develops_from* and is used to capture lineage, as in anatomy ontologies and the Cell Type ontology (CL; http://obo.sourceforge.net/cgi-bin/detail.cgi?cell).

Additional relationships such as *has_participant*, which can relate events to the entities involved in them (see "continuants and occurrents" in the preceding), are also becoming more widely used in biological ontologies. The OBO relationship types ontology (OBO_REL) (21) provides a set of core relationships, with formal definitions, that can be used in any ontology; like any OBO ontology, OBO_REL can be extended with new types as the need arises. As with other ontological formalisms, using the OBO relationship types allows an ontology to take advantage of shared tools and to be usefully aligned with other ontologies.

Ideally, ontologies should be "*is_a* complete," i.e., every term should have an *is_a* parent, resulting in unbroken paths via *is_a* relationships from any term to the root (*see* **Note 5**). This completeness makes computational ontology use much easier; indeed, most tools for developing or using ontologies (including Protégé) require *is_a* completeness. Complete *is_a* paths are also achievable for a bio-ontology with a reasonable amount of effort, especially when starting from scratch. Retrofitting an existing ontology to add complete *is_a* paths is also feasible.

3.1.5. Additional Desirable Features

3.1.5.1. Univocity

Every term and relationship in an ontology should have a single meaning.

Examples: biology poses two types of challenge in achieving univocity. The first is simple: there may be many words or phrases used to refer to a single entity. This is easily resolved by the inclusion of synonyms; also, an ontology may include related but not exactly synonymous terms for query support, as in the GO term "protein polyubiquitination" (GO:0000209), for which both exact synonyms ("protein polyubiquitinylation" and "protein polyubiquitylation") and a related text string ("polyubiquitin") are noted.

A more difficult situation arises when a single word or phrase is used to mean different things; often different research communities have adopted terminology that is unambiguous within a field or for a species, only to "clash" with another community's usage when ontology development efforts attempt to unify the respective bodies of knowledge. One example in GO has been resolved fairly simply: the phrase "secretory vesicle" has been used with two meanings, which are captured in GO as two separate entities, "secretory granule" (GO:0030141) and "transport vesicle" (GO:0030133). The string "secretory vesicle" is included as a related string for each.

3.1.5.2. Positivity

Complements of classes are not themselves classes; terms in an ontology should be defined based on what they are (i.e., properties they possess), not on what they are not. For example, terms such as "non-mammal" or "non-membrane" do not designate genuine classes (*see* **Note 6**).

3.1.5.3. Objectivity

Which classes exist does not depend on our biological knowledge. Terms such as "unknown," "unclassified," or "unlocalized" thus do not designate biological natural kinds, and are not suitable ontology classes.

For example, GO contains terms representing G-protein coupled receptor (GPCR) activities, and at the time of writing these included the term "G-protein coupled receptor activity, unknown ligand (GO:0016526)." To improve objectivity, GO intends to make the "unknown ligand" term obsolete, and use the parent term "G-protein coupled receptor activity (GO:0004930)" to annotate gene products that are GPCRs whose ligands are unknown. Note that no actual information is lost by making this change.

3.1.5.4. Intelligibility of Definitions

The terms used in a definition should be simpler (more intelligible) than the term to be defined (*see* **Note 7**).

3.1.5.5. Basis in Reality

When building or maintaining an ontology, always think carefully about how classes relate to instances in reality. In GO, for example, the actual gene products that might be annotated with an ontology term provide guidance for the term name, relationships, definition, etc.

3.1.5.6. Single Inheritance

No class in a classification hierarchy should have more than one *is_a* parent on the immediate higher level. The rationale for this recommendation is that single inheritance will result if an ontology rigorously ensures that the *is_a* relationship is used univocally (*see* **Note 8**).

4. Conclusions

This chapter provides an overview of practical and formal aspects of ontology development. For any domain, the best ontological representation depends on what must be done with the ontology;

there is no single representation that will meet all foreseeable needs. Experience with bio-ontologies such as GO has shown that ontologies can and must continue to develop as their user communities grow and ontology use becomes more widespread. Many factors are now combining to speed and enhance bio-ontology development, such as the increasing availability of sophisticated tools, broader application of ontology alignment, and efforts to provide a shared set of core ontologies (OBO Foundry) and evaluate existing ontologies. The development and use of ontologies to support biomedical research can thus be expected to continue apace, providing more and better support for primary research and the dissemination of biomedical knowledge.

5. Notes

1. In philosophy, ontology is the study of what exists. The word "ontology" has also been adopted in the computer science community to refer to a conceptualization of a domain, although such conceptualizations might more accurately be called epistemologies. Further discussion of the various interpretations of "ontology" is beyond the scope of an introductory guide.

2. Protégé and OBO-Edit are both ontology editing tools, and have many similarities. The fundamental difference between them is that Protégé has a predominantly object-centered perspective, whereas OBO-Edit emphasizes a graph-structure-centered perspective. The two tools thus have similar capabilities but are optimized for different types of ontologies and different user requirements.

3. For additional information, see the documentation on "GO slim" sets at http://www.geneontology.org/GO.slims.shtml.

4. Some ontologies do include instances (such as the example ontology in the excellent tutorial by N. F. Noy and D. L. McGuinness; see http://protege.stanford.edu/publications/ontology_development/ontology101-noy-mcguinness.html), but in the biomedical domain ontologies that explicitly exclude instances are far more prevalent. For this reason inclusion of instances is not covered herein.

5. At present, several OBO ontologies, are not *is_a* complete, but adding complete *is_a* parentage is a development goal.

6. There are cases in which the language typically used in biological literature poses challenges for achieving positivity. For example, the GO terms "membrane-bound organelle (GO:0043227)"

and "non–membrane-bound organelle (GO:0043228)." The latter appears to violate the principle of positivity, but this is due to the limitations of the language used to express the concept—and represent the entity—of an organelle bounded by something other than a membrane. There is an essential distinction between "non–membrane-bound organelle" and "not a membrane-bound organelle": The former is still restricted to organelles, and excludes those surrounded by membranes, whereas the latter encompasses everything that is not a membrane-bound organelle.

7. In some cases, accurate but simple terminology is not readily available. The most notable examples are chemical names, which are often used in definitions of terms representing biochemical reactions or metabolic processes (GO molecular function and biological process, respectively), as well as being entity names in their own right (in ChEBI).

8. Of all the desired features of formal ontologies, single inheritance is the hardest to achieve in practice in an ontology for a biological domain. Because disentangling multiple parentage takes considerable effort, single inheritance is enforced in very few bio-ontologies currently in use.

Acknowledgments

The author thanks Tim Rayner and Jane Lomax for valuable comments on the manuscript. The chapter content makes extensive use of material made available from the ISMB 2005 tutorial on "Principles of Ontology Construction" prepared by Suzanna Lewis, Barry Smith, Michael Ashburner, Mark Musen, Rama Balakrishnan, and David Hill, and of a presentation by Barry Smith to the GO Consortium.

References

1. Blake, J. A., Bult, C. J. (2006) Beyond the data deluge: data integration and bio-ontologies. *J Biomed Inform* 39, 314–320.
2. The Gene Ontology Consortium (2000) Gene Ontology: tool for the unification of biology. *Nat Genet* 25, 25–29.
3. The Gene Ontology Consortium (2001) Creating the Gene Ontology resource: design and implementation. *Genome Res* 11, 1425–1433.
4. The Gene Ontology Consortium (2004) The Gene Ontology (GO) database and informatics resource. *Nucleic Acids Res* 32, D258–D261.
5. Blake, J. A., Harris, M. A. (2003) The Gene Ontology project: structured vocabularies for molecular biology and their application to genome and expression analysis, in (Baxevanis, A. D., Davison, D. B., Page, R. D. M., et al., eds.) *Current Protocols in Bioinformatics.* John Wiley & Sons, New York.
6. Harris, M. A., Lomax, J., Ireland, A., et al. (2005) The Gene Ontology project. In Subramaniam, S. (ed.) *Encyclopedia of*

Genetics, Genomics, Proteomics and Bioinformatics. John Wiley & Sons, New York.

7. The Gene Ontology Consortium (2006) The Gene Ontology (GO) project in 2006. *Nucleic Acids Res* 34, D322–D326.
8. Rojas, I., Ratsch, E., Saric, J., et al. (2004) Notes on the use of ontologies in the biochemical domain. *In Silico Biol* 4, 89–96.
9. Burgun, A. (2006) Desiderata for domain reference ontologies in biomedicine. *J Biomed Inform* 39, 307–313.
10. Yu, A. C. (2006) Methods in biomedical ontology. *J Biomed Inform* 39, 252–266.
11. Blake, J. (2004) Bio-ontologies—fast and furious. *Nat Biotechnol* 22, 773–774.
12. Bard, J. B. L., Rhee, S. Y. (2004) Ontologies in biology: design, applications and future challenges. *Nat Rev Genet* 5, 213–222.
13. Wroe, C., Stevens, R. (2006) Ontologies for molecular biology, in (Lengauer, T., ed.), *Bioinformatics: From Genomes to Therapies.* Wiley-VCH, Weinheim, Germany.
14. Gruber, T. R. (1993) A translation approach to portable ontology specifications. *Knowl Acq* 5, 199–220.
15. Jones, D. M., Paton, R. (1999) Toward principles for the representation of hierarchical knowledge in formal ontologies. *Data Knowl Eng* 31, 99–113.
16. Schulze-Kremer, S. (2002) Ontologies for molecular biology and bioinformatics. *In Silico Biol* 2, 179–193.
17. Stevens, R., Goble, C. A., Bechhofer, S. (2000) Ontology-based knowledge representation for bioinformatics. *Brief Bioinform* 1, 398–414.
18. Lewis, S. E. (2005) Gene Ontology: looking backwards and forwards. *Genome Biol* 6, 103.
19. Smith, B., Ashburtner, M., Rosse, C., et al. (2007) The OBO Foundry: coordinated evolution of ontologies to support biomedical data integration. *Nat Biotechnol* 25, 1251–1255.
20. Bard, J., Rhee, S. Y., Ashburner, M. (2005) An ontology for cell types. *Genome Biol* 6, R21.
21. Smith, B., Ceusters, W., Klagges, B., et al. (2005) Relations in biomedical ontologies. *Genome Biol* 6, R46.
22. Wroe, C. J., Stevens, R., Goble, C. A., et al. (2003) A methodology to migrate the gene ontology to a description logic environment using DAML+OIL. *Pac Symp Biocomput* 2003, 624–635.
23. Bada, M., Stevens, R., Goble, C., et al. (2004) A short study on the success of the Gene Ontology. *J Web Semantics* 1, 235–240.
24. Mungall, C. J. (2005) Obol: integrating language and meaning in bio-ontologies. *Comp Funct Genomics* 5, 509–520.
25. Lord, P., Stevens, R. D., Goble, C. A., et al. (2005) Description logics: OWL and DAML + OI, in (Subramaniam, S., ed.), *Encyclopedia of Genetics, Genomics, Proteomics and Bioinformatics.* John Wiley & Sons, New York.
26. Hill, D. P., Blake, J. A., Richardson, J. E., et al. (2002) Extension and integration of the Gene Ontology (GO): combining go vocabularies with external vocabularies. *Genome Res* 12, 1982–1991.

Chapter 6

Genome Annotation

Hideya Kawaji and Yoshihide Hayashizaki

Abstract

The dynamic structure and functions of genomes are being revealed simultaneously with the progress of genome analyses. Evidence indicating genome regional characteristics (genome annotations in a broad sense) provide the basis for further analyses. Target listing and screening can be effectively performed *in silico* using such data. This chapter describes steps to obtain publicly available genome annotations or construct new annotations based on your own analyses, as well as an overview of the types of available genome annotations and corresponding resources.

Keywords: genome annotation, the UCSC Genome Browser, Ensembl, the Generic Genome Browser, GFF format, database.

1. Introduction

Genome sequencing of human and model organisms has made possible genome analyses that reveal dynamic structures and functions of genomes, in addition to mere nucleotide sequences. Evidence indicating genome regional characteristics, which are genome annotations in a broad sense (*see* **Note 1**), provide the basis for further analyses as well as help us to understand the nature of organisms. All genome-wide analyses are in this sense potential resources for genome annotations.

At the time when genome sequencing itself was a challenging issue, genome annotation mainly referred to gene structure, that is, the boundaries of exons/introns and CDS (coding sequence)/UTR (untranslated regions) at protein-coding loci. Gene products and their loci were the major interest *(1)*. Genome-wide analyses in the "post-genome sequencing era" revealed additional aspects: a significant number of non-coding RNAs *(2)*, the fact that the majority of mammalian genomes are transcribed *(2)*, a large number of antisense transcriptions *(3)*, dynamic expression profiles under various conditions *(4)*, epigenetic modifications of nucleotides *(5)* and nucleoproteins *(6)* related to development and cancer, genome sequence conservation among various species *(7)*, and a large number of polymorphisms among individuals and haplotypes *(8)*. "Post-genome" annotations have accumulated at an increasing rate, especially for human and model organisms, and attract wide interest.

Genome annotations can be combined with other annotations derived from distinct methods or viewpoints based on genomic coordinates, even if they are less comprehensive. This kind of integration will result in a more complete view of genomes. Full utilization of annotations is essential not only for comprehensive "-omics" analysis, but also for systems-based analysis of specific biological phenomena. Such analyses require listing of possible targets, target screening and setting of priorities prior to analysis, and interpretation of the analysis results. Each step can be performed effectively by retrieving public annotations of interest, constructing in-house annotations, combining them, and extracting relevant attributes from these combined data.

This chapter explains how to use various genome annotations, including steps to make your own. Public annotations and resources are described in the Materials section, and the required steps for using genome annotations are described in the Methods.

2. Materials

A significant number of genome annotations are available and published already, and the number is increasing. This is a general overview of genome annotations and major resources to simplify your choice of annotations. Subsequently, data formats are described with which one should be familiar in order to understand and use annotation files.

2.1. Genome Annotation, an Overview

2.1.1. Transcription and Its Regulation

1. The complete structure of transcripts (or mRNA), in terms of exons and introns, can be derived from sequence alignments of full-length cDNA with the genome. ESTs (expressed sequence tags) can be sequenced at lesser cost, but contain only partial (5'- or 3'-end) structures. Short tag-based technologies, such as CAGE, 5'-SAGE, GIS, and GSC were developed for characterization of transcript boundaries with high throughput *(9)*. The first two techniques characterize 5'-ends of transcripts, and the other two characterize both ends using sequencing paired-end di-tags (called PETs).

2. Genome-wide transcription levels can be profiled by DNA microarrays. The number and properties of probes in an array limit the number of transcripts that can be profiled in a single experiment. However, high-density genomic oligonucleotide arrays enable us to profile entire genomes based on designed probes in certain intervals (10). Customized microarrays enable us to focus on regions of interest. High throughput sequencing of short tags also enables monitoring of transcription levels by counting sequenced tags, as well as characterization of transcript structures (9).

3. The binding of DNA to transcription factors (proteins participating in transcription and transcriptional regulation) can be identified by ChIP (chromatin immunoprecipitation). An antibody targets the protein of interest and its associated DNAs can be found by expression profiling techniques, for example, DNA microarrays (ChIP/chip) (11) and sequencing paired-end di-tags (ChIP/PETs).

4. Computational predictions, such as transcript structure and transcription factor binding sites (TFBSs), are also useful for finding possible targets for analysis.

2.1.2. Modifications and Structures of Nucleotides and Nucleosomes

1. Eukaryotic genomic DNA is coupled with histones to comprise nucleosomes, and chromatin on a higher level. Modifications and the status of these structures affect transcription regulation. The combination of these annotations with transcript structures and transcriptional activities will be useful for understanding of the detailed machinery of such modifications and structures in transcription regulation.

2. DNA methylation occurs in CpG dinucleotides, which are involved in epigenetic transcription regulation. It is called epigenetic because it is not encoded in the genome sequence itself. CpG methylation can be detected by bisulfite PCR. High throughput analyses based on methylation-sensitive restriction enzymes find differentially methylated DNA in several ways, such as two-dimensional electrophoretograms in Restriction Landmark Genomic Scanning (RLGS, a computational method to map each spot to the genome is called Vi-RLGS), sequencing in Methylation Sensitive Representational Difference Analysis (MS-RDA), and DNA microarrays (12).

3. Post-translational histone modifications, such as acetylation of histone H3 at K9 and methylation of histone H3 at K4, are also epigenetic modifications and are associated with transcriptional activities (6). A modified histone can be detected in ChIP analysis by using an antibody specific to modified histones.

4. Profiling nuclease sensitivity is an approach to reveal nucleosome structure. Both DNase-I and micrococcal nuclease

(MNase) digest nucleosome-free regions: The former tend to digest random sites and the latter digest linker sites. Of them, DNase-I hypersensitive sites have been widely used for identifying regulatory elements *(6)*.

2.1.3. Evolutionary Conservation and Variation

1. Genome sequence conservation among species provides significant information for genome analyses, because conservation reflects the action of natural selection. Coding regions in genomes are generally more conserved than non-coding regions. Cis-elements within promoters tend also to be conserved. Interestingly, promoter regions of non-coding transcripts (ncRNA) are also conserved, whereas ncRNA exons tend to be less conserved *(2)*. Conserved regions are possible candidates for functional elements in the genome, and comparisons between all types of genome annotations will be useful, especially for target screening or priority settings prior to analysis.

2. There are ultra-conserved regions in the human genome, where an almost complete identity is observed between orthologous regions of rat and mouse, and a high conservation with the corresponding sequence in chicken, dog, and fish *(13)*. The functions of these elements are still unclear. Clues to their function may be obtained by associating them with various genome annotations.

3. Genetic variations among individuals and groups have a great significance to our understanding and treatment of diseases. Single nucleotide polymorphisms (SNPs) and haplotypes (a particular combination of SNP alleles along a chromosome) *(8)* have proved to be a promising resource for analyses of genetically based human diseases.

2.1.4. Collaboration Efforts and Databases

1. International collaborations or projects have published many genome annotations, some of them focused on human (**Table 6.1**). For example, the ENCODE (ENCyclopedia Of DNA Elements) project focused on small regions (about 1%) of the human genome in a pilot phase, while evaluating various annotation methods. Selected methods will be expanded to the entire genome in future. The Genome network project/FANTOM (Functional Annotation Of Mammalian) focuses on mammalian genomes, especially human and mouse. The Human epigenome project focus on DNA methylation, and the international HapMap project focuses on variations among human individuals.

2. Databases storing genome annotations, including those maintained by the above collaborations, are listed in **Table 6.2**. Some of them maintain data derived from dozens of organisms: the UCSC Genome Browser Database *(14)*, Ensembl *(15)*, TIGR's

Table 6.1
Collaboration efforts for human genome annotations

Collaboration effort	URL
The ENCODE project	http://www.genome.gov/10005107
Genome network project/ FANTOM	http://genomenetwork.nig.ac.jp/ http://fantom.gsc.riken.jp/
Human epigenome project	http://www.epigenome.org/
International HapMap project	http://www.hapmap.org/

Table 6.2
Resources of genome annotations

Resources for wide range of organisms	URL
The UCSC Genome Browser Database	http://genome.ucsc.edu/
Ensembl	http://www.ensembl.org/
TIGR's genome project	http://www.tigr.org/db.shtml
MIPS	http://mips.gsf.de/

Resources for specific organisms	URL	Organism(s)
Genome network project/ FANTOM	http://genomenetwork.nig.ac.jp/ http://fantom.gsc.riken.jp/	Human, Mouse
MGD	http://www.informatics.jax.org/	Mouse
FlyBase	http://flybase.bio.indiana.edu/	*Drosophila*
WormBase	http://www.wormbase.org/	*Caenorhabditis elegans*
TAIR	http://www.arabidopsis.org/	*Arabidopsis thaliana*
GRAMENE	http://www.gramene.org/	Grass
RAP-DB	http://rapdb.lab.nig.ac.jp/	Rice
SGD	http://www.yeastgenome.org/	Yeast

genome project (http://www.tigr.org/db.shtml), and MIPS *(16)*. Others focus on specific organisms; for example, the Genome network project/FANTOM *(2)*, MGD *(17)*, FlyBase *(18)*, WormBase *(19)*, TAIR *(20)*, GRAMENE *(21)*, RAP-DB *(22)*, SGD *(23)*, and other databases (*see* **Note 2**).

2.2. Data Format

A standard web browser, such as Internet Explorer or Mozilla Firefox, is all that is needed to browse and acquire public annotations of interest. This section describes a format for genome annotation; General Feature Format (GFF). Although some variations of GFF have been proposed (**Note 3**), we mainly refer to GFF version 2 (http://www.sanger.ac.uk/Software/formats/GFF/), which is widely supported. All of these formats are easy to understand, parse, and process by lightweight programming languages, such as Perl, Python, or Ruby.

2.2.1. General Feature Format

1. This format consists of tab-delimited text with nine columns (**Fig. 6.1**). Although there are some variations of this format, the differences are only in the ninth column. The first column describes the sequence or chromosome name.
2. The second describes the source of this feature, or the method used for acquiring the sequence. Distinguishable data strings are required if you intend to compare the same type of annotations derived from distinct databases, experimental procedures, and computational methods.
3. The third describes the type of this feature, such as "exon," "CDS," and "gap." Using a standard nomenclature, such as DDBJ/EMBL/GenBank feature table (http://www3.ebi.ac.uk/Services/WebFeat/) and sequence ontology (http://song.sourceforge.net/), is recommended to distinguish features annotated for the same target.
4. The fourth describes the feature's start position.
5. The fifth describes the feature's end position.
6. The sixth describes a numerical value as an attribute to the feature. There is no recommendation for what kinds of values should be described, and "." (dot) is used for no value.

```
chr7   RefSeq     exon  86599192  86599411  0  -  .  accession "NM_007233";
chr7   RefSeq     exon  86615490  86615651  0  -  .  accession "NM_007233";
chr7   RefSeq     exon  86619010  86619482  0  -  .  accession "NM_007233";
chr7   DDBJ_EST   exon  86618772  86619496  0  -  .  accession "BG709171";
chr7   dbSNP      SNP   86619052  86619052  0  .  .  accession "rs5885579";
```

Fig. 6.1. GFF version 2 formatted annotations. A plain text formatted in GFF version 2, containing some genome annotations, where all blanks except for the spaces after "accession" are TAB characters. It contains annotations for a RefSeq transcript with three exons, an EST deposited in DDBJ, and a SNP deposited in dbSNP.

The ninth column described below can also be used for the feature's values, where multiple values can be described.

7. The seventh describes the feature's strand: "+" for forward and "−" for reverse strand. "." (dot) is used when the strand is not relevant, such as dinucleotide repeats and conserved regions.

8. The eighth column carries "0," "1," or "2," describing a phase of this feature relative to the codon. The value "0" means the first position of the feature corresponds to the first base of the codon, and "1" and "2" means for the second and third position of the codon, respectively. It is valid only for CDS features, and "." (dot) is used in other cases.

9. The ninth column describes multiple attributes, which consist of name and value pairs separated by " " (space), and multiple attributes separated by ";" (semicolon). You can use this column for additional information not described in the above columns.

3. Methods

A first step to use genome annotations is to browse and export annotations of interest from the publicly available databases. In addition to describing this, steps to construct and share annotations based on your own analysis are also described. This section focuses on three tools widely used to store and display genome annotations: the UCSC Genome Browser *(24)*, the Ensembl Genome Browser *(25)*, and the Generic Genome Browser (GBrowse) *(26)*. Some tools or software for further use are mentioned in **Notes 4–7**.

3.1. Browsing and Obtaining Public Annotations

1. Open a URL of a genome annotation database with a standard web browser.

2. Search for the element you are interested in. The UCSC browser has text boxes entitled "position or search term" or "position/search," GBrowse has "Landmark or region," and Ensembl has "Search" at the upper right of the page. If the entries found are directly hyperlinked to their main views, skip Step 3.

3. If you already know the genomic coordinates for your sequence, you can specify them directly. A concatenated string of chromosome number, start, and stop position (example: "chr7:127,471,196-127,495,720") works for the UCSC browser and GBrowse. Ensembl has separate interfaces for chromosome number and start and stop coordinates. It is important to consider the genome assembly to which your coordinates are referring.

4. A graphical view of genome annotations for the specified region is displayed after the above steps (**Figs. 6.2** to **6.4**).

5. Select annotations and type of views to be displayed. Although you can select views on the same page as the main browsing page, additional customization pages are available from the "configure" and "Set track options" buttons in the UCSC browser and GBrowse, respectively. You can reach an interface to select views from the menus at the top of "Detailed view" in Ensembl (**Figs. 6.2** to **6.4**).

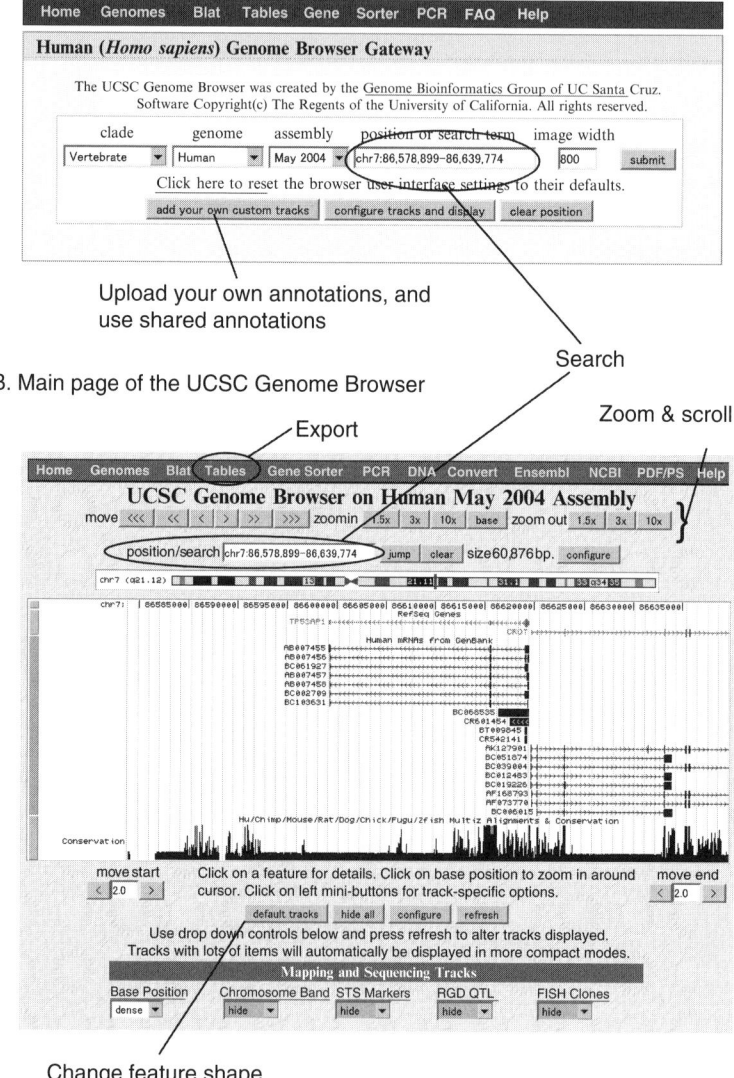

Fig. 6.2. The UCSC Genome Browser. Screenshots of the UCSC Genome Browser, displaying a locus of TP53 activated protein 1. (**A**) A gateway page. (**B**) A main graphical view of genome annotation.

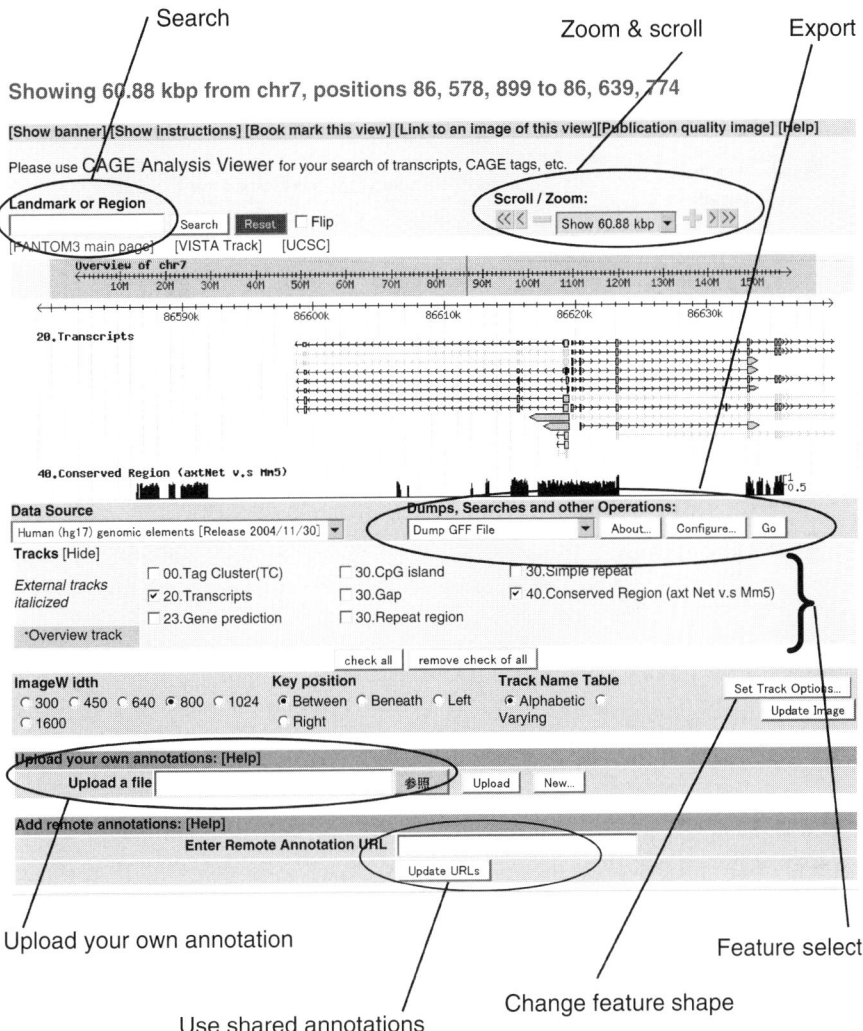

Fig. 6.3. The Generic Genome Browser. A screenshot of the Generic Genome Browser (GBrowse) used in FANTOM3, displaying the same region as Fig. 6.2.

6. Scroll up and down; zoom in and out of the displayed region. Intuitive interfaces are provided by the databases (**Figs. 6.2** to **6.4**).

7. Download the displayed annotations. For this purpose, the UCSC browser provides "Table browser," which is hyperlinked from the top of the main view. "Dump GFF file" plug-in in GBrowse and Export view in Ensembl are also available for this purpose (**Figs. 6.2** to **6.4**).

3.2. Make and Browse Your Own Annotations

1. Define or get genomic coordinates for your elements. If your aim is genome annotation based on your own transcript sequences with BLAST (27) as the alignment tool, a file of alignment results will be obtained by BLAST with –m9 option for a tab-delimited format (**Fig. 6.5A**). Note

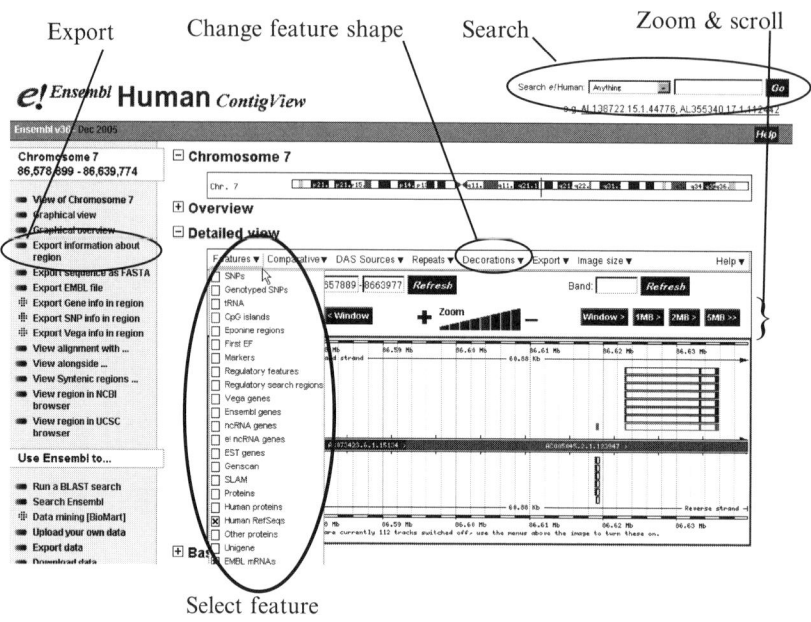

Fig. 6.4. The Ensembl Genome Browser. A screenshot of the Ensembl Genome Browser, displaying the same region as Figs. 6.2 and 6.3.

A. BLAST result

```
# BLASTN 2.2.13 [Nov-27-2005]
# Query: myseq_1
# Database: genome.fa
# Fields: Query id, Subject id, % identity, alignment length, mismatches, gap openings,
q. start, q. end, s. start, s. end, e-value, bit score
myseq_1  chr7  100.00  213  0  0  1    213   86619482  86619270  6e-117  422
myseq_1  chr7  100.00  195  0  0  210  404   86619202  86619008  3e-106  387
myseq_1  chr7  100.00  165  0  0  400  564   86615654  86615490  3e-88   327
myseq_1  chr7  100.00  146  0  0  565  710   86599411  86599266  6e-77   289
# BLASTN 2.2.13 [Nov-27-2005]
# Query: myseq_2
# Database: genome.fa
# Fields: Query id, Subject id, % identity, alignment length, mismatches, gap openings,
q. start, q. end, s. start, s. end, e-value, bit score
myseq_2  chr7  100.00  607  0  0  544  1150  86633414  86634020  0.0     1156
myseq_2  chr7  100.00  188  0  0  301  488   86633171  86633358  9e-102  373
myseq_2  chr7  100.00  117  0  0  166  282   86623014  86623130  2e-59   232
```

B. GFF formatted BLAST result

chr7	myseq	exon	86619270	86619482	.	-	.	myID myseq_1
chr7	myseq	exon	86619008	86619202	.	-	.	myID myseq_1
chr7	myseq	exon	86615490	86615654	.	-	.	myID myseq_1
chr7	myseq	exon	86599266	86599411	.	-	.	myID myseq_1
chr7	myseq	exon	86633414	86634020	.	+	.	myID myseq_2
chr7	myseq	exon	86633171	86633358	.	+	.	myID myseq_2
chr7	myseq	exon	86623014	86623130	.	+	.	myID myseq_2

Fig. 6.5. BLAST output and GFF format. (**A**) An output of BLAST with –m9 option, when two cDNA sequences are aligned with the human genome. (**B**) A GFF formatted data converted from the BLAST output.

that this is just a simple example for an explanation of these steps, and additional steps and/or other programs can be used for transcript structure annotations. If your aim is other types of annotation, you will have to prepare the required features and their genomic coordinates in different ways (*see* **Note 8**).

2. Make a genome annotation file from the coordinates. When the GFF format is used, source, type, and attributes columns should be included. In this example, "myseq" is used for its source to distinguish from public annotations, and "exon" is used for its type after checking the sequence ontology. A sample program written in Perl (**Fig. 6.6**) converts the previous BLAST result into GFF format (**Fig. 6.5B**).

3. Upload your own annotation to public genome browsers to get its graphical view. In this case, public databases and your own analyses have to use the same genome assembly to share the same coordinate system. You can use the "add your own custom tracks" button on the top page in the UCSC genome browser, the "Upload your own annotation" box in GBrowse, and the "URL based data" from the "DAS source" menu in Ensembl (**Fig. 6.2** to **6.4**) (*see* **Note 9**). An example of the image displayed by the UCSC browser is in **Fig. 6.7**.

4. In order to share your annotation, place it on your own ftp or http server. Others can also get the same graphical view by specifying its URL on genome browsers.

```perl
#!/usr/bin/env perl

while(<>){
  next if /^#/;                                    # Skip comments
  my @c = split(/\t/,$_);                          # Parse a feature
  my ($start,$end,$strand) = ($c[8],$c[9],"+");
  if ($start > $end) {                             # Correnct start, end, and strand
    $strand = "-";                                 #    when the feature is
    ($start,$end) = ($end,$start);                 #    on reverse strand
  }
                                                   # Print the feature's
  print "$c[1]\tmyseq\texon\t",                    #    chromosome, source, feature
        "$start\t$end\t.\t$strand\t.\t", #          start , end, score, strand, and phase
        "myID $c[0]\n"                   #          and attributes
}
```

Fig. 6.6. Sample code to convert BLAST output into GFF. A sample Perl code to convert a BLAST output (*see* **Fig. 6.5A**) into GFF (*see* **Fig. 6.5B**). It just parses columns in a line of its input, and prints the corresponding column in GFF.

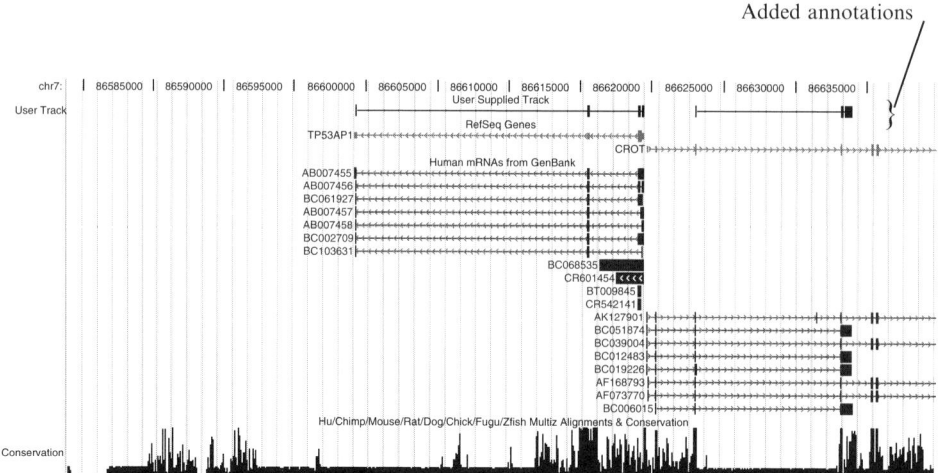

Fig. 6.7. Graphical display of uploaded annotations. An example of the graphical display in the UCSC Genome Browser when the annotations of the Fig. 6.5B are uploaded.

4. Notes

1. In a narrow sense, genome annotation means annotation of genes on genomes, based on supporting evidence, such as mRNA alignment with genome sequences, computational predictions of exons, and others.

2. Additional databases for genome annotations are not listed here, and the number is increasing. The annual database issue and the molecular biology database collection *(28)* published by Nucleic Acids Research will be helpful to find appropriate databases for your analysis.

3. Major variations of GFF are GFF version 1, 2 (http://www.sanger.ac.uk/Software/formats/GFF/), and 3 (http://song.sourceforge.net/gff3.shtml). The differences are only in the ninth column: version 1 requires just one string rather than a pair of name and value, and version 3 requires value name to be concatenated with its value by "=" (equal). Some keywords are reserved for specific use in GTF (Gene Transfer Format (http://genes.cs.wustl.edu/GTF2.html), a format based on GFF version 2 with some special attribute names.

4. Software used in the public genome annotation databases is available for setting up your own annotation database. Source codes of the UCSC Genome Browser *(24)* and Ensembl Genome Browser *(25)* are convenient for mirroring their annotations and setting up modified or additional version. The Generic Genome Browser

(GBrowse) *(26)* is convenient for setting up a genome annotation browser mainly based on your original annotations, as well as public ones.

5. The distributed annotation system (DAS, http://www.biodas.org/) also can be used to share your own annotations. Based on a client-server model, some genome annotation database systems, such as Ensembl and GBrowse, can also work as DAS clients, and use the DAS server to display annotations stored in the databases and DAS server.

6. The UCSC Table Browser *(29)* and BioMart *(30)* focus on exporting data with complex queries, or combinations of some conditions. Web interfaces to specify conditions are available by setting up those systems with your own annotations. These are convenient to retrieve annotations genome-wide, rather than just in the displayed region.

7. Stand-alone applications for browsing (and editing) genome annotations are also available, such as Integrated Genome Browser (IGB, http://genoviz.sourceforge.net/) and Apollo *(31)*. They are convenient if you need to avoid uploading your data for some reason.

8. Besides transcript sequences, experimental results based on DNA microarray and PCR can be mapped on genomes. Genomic coordinates of microarray probes will be derived from probe annotations or probe sequence alignments with genomes by general alignment tools such as BLAST *(27)*, FASTA *(32)*, SSAHA *(33)*, and BLAT *(34)*. More specific tools such as primersearch in EMBOSS *(35)* are convenient to identify genomic coordinates of PCR products.

9. There is an inconsistency in chromosome name between the UCSC Genome Browser Database and Ensembl. The former use a chromosome number with a string "chr," for example, "chr22," but the latter use a chromosome number such as "22." This is just a tiny difference, but you have to be careful because annotations will not be displayed if the wrong name is used to refer to a sequence.

Acknowledgment

The authors thank A. Karlsson and A. Sandelin for English editing. This study was supported by a Research Grant for the RIKEN Genome Exploration Research Project provided by the Ministry of Education, Culture, Sports, Science and Technology (MEXT), and a grant of the Genome Network Project from MEXT.

References

1. Lander, E. S., Linton, L. M., Birren, B., et al. (2001) Initial sequencing and analysis of the human genome. *Nature* 409, 860–921.
2. Carninci, P., Kasukawa, T., Katayama, S., et al. (2005) The transcriptional landscape of the mammalian genome. *Science* 309, 1559–1563.
3. Katayama, S., Tomaru, Y., Kasukawa, T., et al. (2005) Antisense transcription in the mammalian transcriptome. *Science* 309, 1564–1566.
4. Barrett, T., Suzek, T. O., Troup, D. B., et al. (2005) NCBI GEO: mining millions of expression profiles—database and tools. *Nucleic Acids Res* 33, D562–566.
5. Wilkins, J. F. (2005) Genomic imprinting and methylation: epigenetic canalization and conflict. *Trends Genet* 21, 356–365.
6. Huebert, D. J., Bernstein, B. E. (2005) Genomic views of chromatin. *Curr Opin Genet Dev* 15, 476–481.
7. Brudno, M., Poliakov, A., Salamov, A., et al. (2004) Automated whole-genome multiple alignment of rat, mouse, and human. *Genome Res* 14, 685–692.
8. Altshuler, D., Brooks, L. D., Chakravarti, A., et al. (2005) A haplotype map of the human genome, *Nature* 437, 1299–1320.
9. Harbers, M., Carninci, P. (2005) Tag-based approaches for transcriptome research and genome annotation. *Nat Methods* 2, 495–502.
10. Mockler, T. C., Chan, S., Sundaresan, A., et al. (2005) Applications of DNA tiling arrays for whole-genome analysis. *Genomics* 85, 1–15.
11. Cawley, S., Bekiranov, S., Ng, H. H., et al. (2004) Unbiased mapping of transcription factor binding sites along human chromosomes 21 and 22 points to widespread regulation of noncoding RNAs. *Cell* 116, 499–509.
12. Murrell, A., Rakyan, V. K., Beck, S. (2005) From genome to epigenome. *Hum Mol Genet* 14 Spec No 1, R3–R10.
13. Bejerano, G., Pheasant, M., Makunin, I., et al. (2004) Ultraconserved elements in the human genome. *Science* 304, 1321–1325.
14. Hinrichs, A. S., Karolchik, D., Baertsch, R., et al. (2006) The UCSC Genome Browser Database: update 2006. *Nucleic Acids Res* 34, D590–598.
15. Birney, E., Andrews, D., Caccamo, M., et al. (2006) Ensembl 2006. *Nucleic Acids Res* 34, D556–561.
16. Mewes, H. W., Frishman, D., Mayer, K. F., et al. (2006) MIPS: analysis and annotation of proteins from whole genomes in 2005. *Nucleic Acids Res* 34, D169–172.
17. Blake, J. A., Eppig, J. T., Bult, C. J., et al. (2006) The Mouse Genome Database (MGD): updates and enhancements. *Nucleic Acids Res* 34, D562–567.
18. Grumbling, G., Strelets, V. (2006) FlyBase: anatomical data, images and queries, *Nucleic Acids Res* 34, D484–488.
19. Schwarz, E. M., Antoshechkin, I., Bastiani, C., et al. (2006) WormBase: better software, richer content. *Nucleic Acids Res* 34, D475–478.
20. Rhee, S. Y., Beavis, W., Berardini, T. Z., et al. (2003) The Arabidopsis Information Resource (TAIR): a model organism database providing a centralized, curated gateway to Arabidopsis biology, research materials and community. *Nucleic Acids Res* 31, 224–228.
21. Jaiswal, P., Ni, J., Yap, I., et al. (2006) Gramene: a bird's eye view of cereal genomes. *Nucleic Acids Res* 34, D717–723.
22. Ohyanagi, H., Tanaka, T., Sakai, H., et al. (2006) The Rice Annotation Project Database (RAP-DB): hub for Oryza sativa ssp. japonica genome information. *Nucleic Acids Res* 34, D741–744.
23. Christie, K. R., Weng, S., Balakrishnan, R., et al. (2004) Saccharomyces Genome Database (SGD) provides tools to identify and analyze sequences from *Saccharomyces cerevisiae* and related sequences from other organisms. *Nucleic Acids Res* 32, D311–314.
24. Kent, W. J., Sugnet, C. W., Furey, T. S., et al. (2002) The human genome browser at UCSC. *Genome Res* 12, 996–1006.
25. Stalker, J., Gibbins, B., Meidl, P., et al. (2004) The Ensembl Web site: mechanics of a genome browser. *Genome Res* 14, 951–955.
26. Stein, L. D., Mungall, C., Shu, S., et al. (2002) The generic genome browser: a building block for a model organism system database. *Genome Res* 12, 1599–610.
27. McGinnis, S., Madden, T. L. (2004) BLAST: at the core of a powerful and diverse set of sequence analysis tools. *Nucleic Acids Res* 32, W20–25.
28. Galperin, M. Y. (2006) The Molecular Biology Database Collection: 2006 update. *Nucleic Acids Res* 34, D3–5.
29. Karolchik, D., Hinrichs, A. S., Furey, T. S., et al. (2004) The UCSC Table Browser

data retrieval tool. *Nucleic Acids Res* 32, D493–496.
30. Kasprzyk, A., Keefe, D., Smedley, D., et al. (2004) EnsMart: a generic system for fast and flexible access to biological data. *Genome Res* 14, 160–169.
31. Lewis, S. E., Searle, S. M., Harris, N., et al. (2002) Apollo: a sequence annotation editor. *Genome Biol* 3, RESEARCH0082.
32. Pearson, W. R., Lipman, D. J. (1988) Improved tools for biological sequence comparison. *Proc Natl Acad Sci U S A* 85, 2444–2448.
33. Ning, Z., Cox, A. J., Mullikin, J. C. (2001) SSAHA: a fast search method for large DNA databases. *Genome Res* 11, 1725–1729.
34. Kent, W. J. (2002) BLAT—the BLAST-like alignment tool. *Genome Res* 12, 656–664.
35. Rice, P., Longden, I., Bleasby, A. (2000) EMBOSS: the European Molecular Biology Open Software Suite. *Trends Genet* 16, 276–277.

Section II

Sequence Analysis

Chapter 7

Multiple Sequence Alignment

Walter Pirovano and Jaap Heringa

Abstract

Multiple sequence alignment (MSA) has assumed a key role in comparative structure and function analysis of biological sequences. It often leads to fundamental biological insight into sequence-structure-function relationships of nucleotide or protein sequence families. Significant advances have been achieved in this field, and many useful tools have been developed for constructing alignments. It should be stressed, however, that many complex biological and methodological issues are still open. This chapter first provides some background information and considerations associated with MSA techniques, concentrating on the alignment of protein sequences. Then, a practical overview of currently available methods and a description of their specific advantages and limitations are given, so that this chapter might constitute a helpful guide or starting point for researchers who aim to construct a reliable MSA.

Key words: multiple sequence alignment, progressive alignment, dynamic programming, phylogenetic tree, evolutionary scheme, amino acid exchange matrix, sequence profile, gap penalty.

1. Introduction

1.1. Definition and Implementation of an MSA

A multiple sequence alignment (MSA) involves three or more homologous nucleotide or amino acid sequences. An alignment of two sequences is normally referred to as a pairwise alignment. The alignment, whether multiple or pairwise, is obtained by inserting gaps into sequences such that the resulting sequences all have the same length L. Consequently, an alignment of N sequences can be arranged in a matrix of N rows and L columns, in a way that best represents the evolutionary relationships among the sequences.

Organizing sequence data in MSAs can be used to reveal conserved and variable sites within protein families. MSAs can provide essential information on their evolutionary and functional relationships. For this reason, MSAs have become an essential prerequisite for genomic analysis pipelines and many downstream computational modes of analysis of protein families such as homology modeling, secondary structure prediction, and phylogenetic reconstruction. They may further be used to derive profiles *(1)* or hidden Markov models *(2, 3)* that can be used to scour databases for distantly related members of the family. As the enormous increase of biological sequence data has led to the requirement of large-scale sequence comparison of evolutionarily divergent sets of sequences, the performance and quality of MSA techniques is now more important than ever.

1.2. Reliability and Evolutionary Hypothesis

The automatic generation of an accurate MSA is computationally a tough problem. If we consider the alignment or matching of two or more protein sequences as a series of hypotheses of positional homology, it would obviously be desirable to have *a priori* knowledge about the evolutionary (and structural) relationships between the sequences considered. Most multiple alignment methods attempt to infer and exploit a notion of such phylogenetic relationships, but they are limited in this regard by the lack of ancestral sequences. Naturally, only observed taxonomic units (OTUs), i.e., present-day sequences, are available. Moreover, when evolutionary distances between the sequences are large, adding to the complexity of the relationships among the homologous sequences, the consistency of the resulting MSA becomes more uncertain (*see* **Note 1**).

When two sequences are compared it is important to consider the evolutionary changes (or sequence edits) that have occurred for the one sequence to be transformed into the second. This is generally done by determining the minimum number of mutations that may have occurred during the evolution of the two sequences. For this purpose several amino acid exchange matrices, such as the PAM *(4)* and BLOSUM *(5)* series, have been developed, which estimate evolutionary likelihoods of mutations and conservations of amino acids. The central problem of assembling an MSA is that a compromise must be found between the evolutionarily most likely pairwise alignments between the sequences, and the embedding of these alignments in a final MSA, where changes relative to the pairwise alignments are normally needed to globally optimize the evolutionary model and produce a consistent multiple alignment.

1.3. Dynamic Programming

Pairwise alignment can be performed by the dynamic programming (DP) algorithm *(6)*. A two-dimensional matrix is constructed based on the lengths of the sequences to be aligned, in which each possible alignment is represented by a unique path through the matrix. Using a specific scoring scheme, which defines scores for residue matches, mismatches, and gaps, each position of the matrix is filled. The DP algorithm guarantees that, given a specific scoring scheme, the optimal alignment will be found. Although dynamic programming is an efficient way of aligning sequences, applying the technique to more than two sequences quickly becomes computationally unfeasible. This is due to the fact that the number of comparisons to be made increases exponentially with the number of sequences. Carrillo and Lipman *(7)* and more recently Stoye et al. *(8)* proposed heuristics to reduce the computational requirements of multidimensional dynamic programming techniques. Nonetheless, computation times required remain prohibitive for all but the smallest sequence sets.

1.4. The Progressive Alignment Protocol

An important breakthrough in multiple sequence alignment has been the introduction of the progressive alignment protocol *(9)*. The basic idea behind this protocol is the construction of an approximate phylogenetic tree for the query sequences and repeated use of the aforementioned pairwise alignment algorithm. The tree is usually constructed using the scores of all-against-all pairwise alignments across the query sequence set. Then the alignment is build up by progressively adding sequences in the order specified by the tree (**Fig. 7.1**), which is therefore referred to as the *guide tree*. In this way, phylogenetic information is incorporated to guide the alignment process, such that sequences and blocks of sequences become aligned successively to produce a final MSA. Fortunately, as the pairwise DP algorithm is only repeated a limited number of times, typically on the order of the square of the number of sequences or less, the progressive protocol allows the effective multiple alignment of large numbers of sequences.

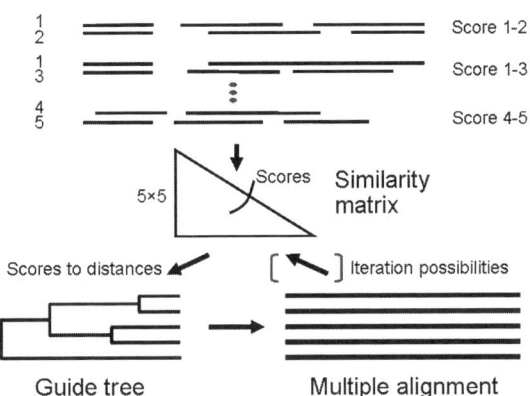

Fig. 7.1. Schematic representation of the progressive alignment protocol. A similarity (distance) matrix, which contains scores from all pairwise alignments, is used to construct a guide tree. The final alignment is built up progressively following the order of the guide tree. The black arrow between brackets indicates possible iterative cycles.

However, the obtained accuracy of the final MSA suffers from the so-called greediness of the progressive alignment protocol; that is, alignment errors cannot be repaired anymore and will be propagated into following alignment steps ("Once a gap, always a gap"). In fact, it is only later during the alignment progression that more information from other sequences (e.g., through profile representation) *(1)* becomes employed in the alignment steps.

1.5. Alignment Iteration

Triggered by the main pitfall of the progressive alignment scenario, some methods try to alleviate the greediness of this strategy by implementing an iterative alignment procedure. Pioneered by Hogeweg and Hesper *(10)*, iterative techniques try to enhance the alignment quality by gleaning increased information from repeated alignment procedures, such that earlier alignments are "corrected" *(10, 11)*. In this scenario, a previously generated MSA is used for improvement of parameter settings, so that the initial guide tree and consequently the alignment can be optimized. Apart from the guide tree, the alignment procedure itself can also be adapted based on observed features of a preceding MSA. The iterative procedure is terminated whenever a preset maximum number of iterations or convergence is reached. However, depending on the target function of an iterative procedure, it does not always reach convergence, so that a final MSA often depends on the number of iterations set by the user. The alignment scoring function used during progressive alignment can be different from the target function of the iteration process, so a decision has to be made whether the last alignment (with the maximal iterative target function value) or the highest scoring alignment encountered during iteration will be taken as the final result upon reaching convergence or termination of the iterations by the user.

Currently, a number of alternative methods are able to produce high-quality alignments. These are discussed in Section 3, as well as the options and solutions they offer, also with respect to the considerations outlined in the preceding.

2. Materials

2.1. Selection of Sequences

Since sequence alignment techniques are based upon a model of divergent evolution, the input of a multiple alignment algorithm should be a set of homologous sequences. Sequences can be retrieved directly from protein sequence databases, but usually a set is created by employing a homology searching technique for a provided query sequence. Widely used programs such as BLAST *(12)* or FASTA *(13)* employ carefully crafted heuristics to perform a rapid search over sequence databases and recover putative homologues.

Selected sequences should preferably be orthologous but in practice it is often difficult to ensure that this is the case. It is important to stress that MSA routines will also be capable of producing alignments of unrelated sequences that can appear to have some realistic patterns, but these will be biologically meaningless ("garbage in, garbage out"). For example, it is possible that some columns appear to be well conserved, although in reality no homology exists. Such misinterpretation could well have dramatic consequences for conclusions and further analysis modes. Although the development of P- and E-values to estimate the statistical significance of putative homologues found by homology searching techniques limits the chance of false positives, it is entirely possible that essentially non-homologous sequences enter the alignment set, which might confuse the alignment method used.

2.2. Unequal Sequence Lengths: Global and Local Alignment

Query sequence sets comprise sequences with unequal length. The extent of such length differences requires a decision whether a *global* or *local* alignment should be performed. A *global* alignment strategy *(6)* aligns sequences over their entire length. However, many biological sequences are modular and contain shuffled domains *(14)*, which can render a global alignment of two complete sequences meaningless (*see* **Note 2**). Moreover, global alignment can also lead to incorrect alignment when large insertions of gaps are needed, for example, to match two domains A and B in a two-domain protein against the corresponding domains in a three-domain structure ACB. In general, the global alignment strategy is appropriate for sequences of high to medium sequence similarity. At lower sequence identities, the global alignment technique can still be useful provided there is confidence that the sequence set is largely colinear without shuffled sequence motifs or insertions of domains. Whenever such confidence is not present, the *local* alignment technique *(15)* should be attempted. This technique selects and aligns the most conserved region in either of the sequences and discards the remaining sequence fragments. In cases of medium to low sequence similarity, local alignment is generally the most appropriate approach with which to start the analysis. Techniques have also been developed to align remaining sequence fragments iteratively using the local alignment technique (e.g., *(16)*).

2.3. Type of Alignment

A number of different alignment problems have been identified in the literature. For example, the BAliBASE MSA benchmark database *(17)* groups these in five basic categories that contain sequence sets comprising the following features:

1. *Equidistant sequences.* Pairwise evolutionary distances between the sequences are approximately the same.
2. *Orphan sequences.* One or more family members of the sequence set are evolutionarily distant from all the others (which can be considered equidistant).

3. *Subfamilies.* Sequences are distributed over two or more divergent subfamilies.

4. *Extensions.* Alignments contain large N- and/or C-terminal gaps.

5. *Insertions.* Alignments have large internal gap insertions.

The preceding classification of alignment problems opens up the possibility of developing different alignment techniques that are optimal for each individual type of problem. Other cases that are challenging for alignment engines include repeats, where different repeat types and copy numbers often lead to incorrect alignment (*see* **Note 3**), and transmembrane segments, where different hydrophobicity patterns confuse the alignment (*see* **Note 4**). However, one would then need *a priori* knowledge about the alignment problem at hand (*see* **Note 5**), which can be difficult to obtain. A suggestion for investigators is to make a first (quick) multiple alignment using general parameter settings. Often, after this first round, it becomes clear in which problem category the chosen sequence set falls, so that for further alignment parameters can be set accordingly. Remember that alignments always can be manually adjusted by using one of the available alignment editors (*see* **Note 6**).

3. Methods

This section highlights a selection of the most accurate MSA methods to date (**Table 7.1**). Each of these follows one or both of two main approaches to address the greediness of the progressive MSA protocol (see the preceding): the first is trying to avoid early match errors by using increased information for aligning pairwise sequences; the second is reconsidering alignment results and improving upon these using iterative strategies.

3.1. PRALINE

PRALINE is an online MSA toolkit for protein sequences. It includes a web server offering a wide range of options to optimize the alignment of input sequences, such as global or local pre-processing, predicted secondary structure information, and iteration strategies (**Fig. 7.2**).

1. *Pre-profile processing options.* Pre-profile processing is an optimization technique used to minimize the incorporation of erroneous information during progressive alignment. The difference between this strategy and the standard global strategy is that the sequences to be aligned are represented by pre-profiles instead of single sequences. Three different options are available: *(1)* global pre-processing

Table 7.1
Web sites of multiple alignment programs mentioned in this chapter

Name	Web site
PRALINE	http://ibi.vu.nl/programs/pralinewww/
MUSCLE	http://www.drive5.com/muscle/
T-Coffee and 3D-Coffee	http://igs-server.cnrs-mrs.fr/Tcoffee/tcoffee_cgi/index.cgi
MAFFT	http://align.bmr.kyushu-u.ac.jp/mafft/online/server/
ProbCons	http://probcons.stanford.edu/
SPEM & SPEM-3D	http://sparks.informatics.iupui.edu/Softwares-Services_files/spem_3d.htm

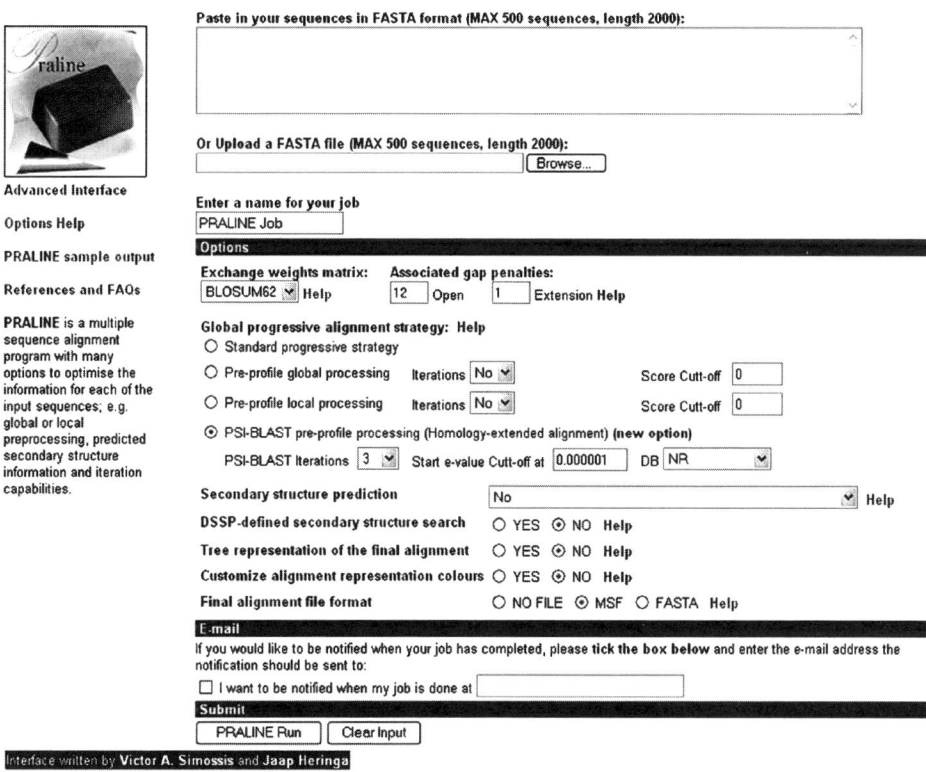

Fig. 7.2. The PRALINE standard web interface. Protein sequences can be pasted in the upper box in FASTA format or directly uploaded from a file. In addition to using default settings, various alignment strategies can be selected (*see* **Section 3.1**) as well as the desired number of iterations or preprocessing cut-off scores.

(18,19), *(2)* local pre-processing *(19)*, and *(3)* PSI-Praline *(20)*. The first two options attempt to maximize the information from each sequence. For each sequence, a pre-profile is built containing information from other sequences in the query set. Under global pre-processing, other sequences can be selected according to a preset minimal pairwise alignment score with the main sequence within each pre-profile. Under local pre-processing, segments of other sequences in the query set are selected based on local alignment scores. The PSI-Praline pre-profile processing strategy employs the PSI-BLAST homology search engine *(21)* to enrich the information of each of the pre-profiles. Based on a user-specified E-value, PSI-BLAST selects sequence fragments from a large non-redundant sequence database, building more consistent and useful pre-profiles for the alignment. The alignment quality of the PSI-Praline strategy is among the highest in the field *(20)*, but the technique is relatively slow as a PSI-BLAST run needs to be conducted for every sequence in the input set.

2. *DSSP or predicted secondary structure information.* PRALINE currently allows the incorporation of DSSP-defined secondary structure information *(22)* to guide the alignment. If no DSSP is available, a choice of seven secondary structure prediction methods is provided to determine the putative secondary structure of those sequences that do not have a PDB structure. In addition, two different consensus strategies are also included, both relying on the prediction methods PSIPRED *(23)*, PROFsec *(24)*, and YASPIN *(25)*.

3. *Iteration.* For the above global and local pre-processing strategies, iterative optimization is possible. Iteration is based on the consistency of a preceding multiple alignment, in which consistency is defined as the agreement between matched amino acids in the multiple alignment and those in corresponding pairwise alignments. These consistency scores are then fed as weights to a next round of dynamic programming. During iteration, therefore, consistent multiple alignment positions tend to be maintained, whereas inconsistent segments are more likely to become re-aligned. Iterations are terminated upon reaching convergence or limit cycle (i.e., a number of cyclically recurring multiple alignments), whereas the user can also specify a maximum number of iterations.

3.2. MUSCLE

MUSCLE *(26, 27)* is multiple alignment software for both nucleotide and protein sequences. It includes an online server, but the user can also choose to download the program and run it locally. The web server performs calculations using pre-defined default parameters, albeit the program provides a large number of options. MUSCLE is a very fast algorithm, which should be particularly considered when aligning large datasets. Basically, the progressive alignment protocol is sped up due to a clever pairwise sequence comparison that avoids the slow DP technique for the construction of the so-called guide tree. Because of the computational efficiency gained, MUSCLE by default employs iterative refinement procedures that have been shown to produce high-quality multiple alignments.

1. *Iteration.* The full iteration procedure used by MUSCLE consists of three steps, although only the last can be considered truly iterative.

    a. In the first step sequences are clustered according to the number of *k-mers* (contiguous segment of length *k*) that they share using a compressed amino acid alphabet (28). From this the guide tree is calculated using UPGMA, after which the sequences are progressively aligned following the tree order.

    b. During the next step the obtained MSA is used to construct a new tree by applying the Kimura distance correction. This step is executed at least twice and can be repeated a number of times until a new tree does not achieve any improvements anymore. As a measure to estimate improvement, the number of internal nodes for which the branching order has changed is taken. If this number remains constant or increases, the iteration procedure terminates and a last progressive alignment is built for this step.

    c. Finally, the third step involves refinement of the alignment using the now fixed tree-topology. Edges from the tree are deleted in order of decreasing distance from the root. For each subdivision of the tree, the two corresponding profiles are aligned (tree-dependent refinement step). If a resulting alignment has a higher score than the previously retained alignment, the new alignment is taken. Iteration terminates if after traversing all tree edges no new alignment is produced or the user-defined number of iterations has been reached.

2. *Large datasets.* As outlined, one of the most important advantages of MUSCLE is that it is very fast and therefore allows handling large datasets in reasonable time. A good compromise between time and accuracy can be made by the user who can decide for all stages and actions whether to include them or not. As an additional option, the user can also define a time range in which the program will select the best solution so far. Another possibility to speed up the program during pairwise k-mer alignment is provided by allowing the user to switch off extending the k-words by dynamic programming (see the preceding). A final option, called "anchor optimization," is designed to reduce computations

during tree-dependent refinement by dividing a given alignment in vertical blocks and aligning the associated profiles separately.

3.3. T-Coffee

The T-Coffee program *(29)* can also handle both DNA and protein sequences. It includes a web server (following the default settings) as well as an option to download the program. The algorithm derives its sensitivity from combining both local and global alignment techniques. Additionally, transitivity is exploited using triplet alignment information including each possible third sequence. A pairwise alignment is created using a protocol named *matrix extension* that includes the following steps:

1. *Combining local and global alignment.* For each pairwise alignment, the match scores obtained from local and global alignments are summed, where for every matched residue pair the identity score of the associated (global or local) alignment is taken. For each sequence pair, the 10 highest scoring local alignments are compiled using Lalign *(30)* and a global alignment is calculated using ClustalW *(31)*.

2. *Transitivity.* For each third sequence C relative to a considered sequence pair A and B, the alignments A-C and C-B together constitute an alignment A-B. For each matched residue x in A and y in B, the minimum of the score of the match between residue x in A with residue z in C (alignment A-C) and that of residue z in C with y in B (alignment C-B) is taken; identity scores of associated alignments are taken as in the preceding step and all scores from the direct alignment as well as through all third sequences are summed.

3. For each sequence pair, dynamic programming is performed over the thus extended matrices. Owing to the fact that the signal captured in the extended scores is generally more consistent than noise, the scores are generally salient such that gap penalties can be set to zero.

   From the extended alignment scores a guide tree is calculated using the Neighbor-Joining technique, and sequences are progressively aligned following the dynamic programming protocol. The combined use of local alignment, global alignment, and transitivity effectively alleviates error propagation during progressive alignment. However, the program is constrained by computational demands when aligning larger sets. As a consequence, the T-Coffee web server constrains the allowed number of input sequences to 50. T-Coffee permits the following further features:

4. *Integrating tertiary structures with 3D-Coffee.* A variant of the described protocol, 3D-Coffee *(32)* allows the inclusion of tertiary structures associated with one or more of the input sequences for guiding the alignment based upon the

principle that "Structure is more conserved than sequence." If a partial sequence of a structure is given, the program will only take the corresponding structural fragment into account. The 3D-Coffee web server incorporates two default pairwise structural alignment methods: SAP *(33)* and FUGUE *(34)*. The first method is a structure superposition package, which is useful if more than one structure is included. The latter is a threading technique that can improve the multiple alignment process when local structural fragments are available. The advanced interface of the program allows the user to select alternative structural alignment methods.

5. *Accelerating the analyses.* Speed limitations of the T-Coffee program can be partially reduced by running a less demanding version. As an alternative, sequences can be divided into subgroups and aligned separately. To assist in this scenario, the program offers an option to compile a final alignment of these previously aligned subgroups.

6. *Consensus MSA.* A recent extension is the method M-Coffee *(35)*, which uses the T-Coffee protocol to combine the outputs of other MSA methods into a single consensus MSA.

3.4. MAFFT

The multiple sequence alignment package MAFFT *(36, 37)* is suited for DNA and protein sequences. MAFFT includes a script and a web server that both incorporate several alignment strategies. An alternative solution is proposed for the construction of the guide tree, which usually requires most computing time in a progressive alignment routine. Instead of performing all-against-all pairwise alignments, Fast Fourier Transformation (FFT) is used to rapidly detect homologous segments. The amino acids are represented by volume and polarity values, yielding high FFT peaks in a pairwise comparison whenever homologous segments are identified. The segments thus identified are then merged into a final alignment by dynamic programming. Additional iterative refinement processes, in which the scoring system is quickly optimized at each cycle, yield high accuracy of the alignments.

1. *Fast alignment strategies.* Two options are provided for large sequence sets: FFT-NS-1 and FFT-NS-2, both of which follow a strictly progressive protocol. FFT-NS-1 generates a quick and dirty guide tree and compiles a corresponding MSA. If FFT-NS-2 is invoked, it takes the alignment obtained by FFT-NS-1 but now calculates a more reliable guide tree, which is used to compile another MSA.

2. *Iterative strategies.* The user can choose from several iterative approaches. The FFT-NS-i method attempts to further refine the alignment obtained by FFT-NS-2 by re-aligning subgroups until the maximum weighted sum of pairs (WSP) score *(38)* is reached. Two more recently included iterative

refinement options (MAFFT version 5.66) incorporate local pairwise alignment information into the objective function (sum of the WSP scores). These are L-INS-i and E-INS-i, which use standard affine and generalized affine gap costs *(39,40)* for scoring the pairwise comparisons, respectively.

3. *Alignment extension.* Another tool included in the MAFFT alignment package is mafftE. This option enhances the original dimension of the input set by including other homologous sequences, retrieved from the SwissProt database with BLAST *(12)*. Preferences for the exact number of additional sequences and the e-value can be specified by the user.

3.5. ProbCons

ProbCons *(41)* is a recently developed progressive alignment algorithm for protein sequences. The software can be downloaded but sequences can also be submitted to the ProbCons web server. The method follows the T-Coffee approach in spirit, but implements some of the steps differently. For example, the method uses an alternative scoring system for pairs of aligned sequences. The method starts by using a pair-HMM and expectation maximization (EM) to calculate a posterior probability for each possible residue match within a pairwise comparison. Next, for each pairwise sequence comparison, the alignment that maximizes the "expected accuracy" is determined *(42)*. In a similar way to the T-Coffee algorithm, information of pairwise alignments is then extended by considering consistency with all possible third "intermediate" sequences. For each pairwise sequence comparison, this leads to a so-called "probabilistic consistency" that is calculated for each aligned residue pair using matrix multiplication. These changed probabilities for matching residue pairs are then used to determine the final pairwise alignment by dynamic programming. Upon construction of a guide tree, a progressive protocol is followed to build the final alignment.

ProbCons allows a few variations of the protocol that the user can decide to adopt:

1. *Consistency replication.* The program allows the user to repeat the probabilistic consistency transformation step, by recalculating all posterior probability matrices. The default setting includes two replications, which can be increased to a maximum of 5.

2. *Iterative refinement.* The program also includes an additional iterative refinement procedure for further improving alignment accuracy. This is based on repeated random subdivision of the alignment in two blocks of sequences and realignment of the associated profiles. The default number of replications is set to 100, but can be changed from 0 to 1000 iterations (for the web server one can select 0, 100, or 500).

3. *Pre-training*. Parameters for the pair-HMM are estimated using unsupervised expectation maximization (EM). Emission probabilities, which reflect substitution scores from the BLOSUM-62 matrix *(5)*, are fixed, whereas gap penalties (transition probabilities) can be trained on the whole set of sequences. The user can specify the number of rounds of EM to be applied on the set of sequences being aligned. The default number of iterations should be followed, unless there is a clear need to optimize gap penalties when considering a particular dataset.

3.6. SPEM

The SPEM-protocol *(43)*, designed for protein MSA, is a recent arrival in the field. Both a SPEM server and downloadable software are available. Two online SPEM protocols are available: SPEM (normal) and SPEM-3D. Each follows a standard routine so that the user cannot change many options. The 3D-variant SPEM-3D, which allows the inclusion of information from tertiary structure, can only be used through the Web. The SPEM approach focuses on the construction of proper pairwise alignments, which constitute the input for the progressive algorithm. To optimize pairwise alignment, the method follows the PRALINE approach (see the preceding) in that it combines information coming from sequence pre-profiles (constructed *a priori* with homology searches performed by PSI-BLAST) *(21)*, and knowledge about predicted and known secondary structures. However, the latter knowledge is exploited in the dynamic programming algorithm by applying secondary structure dependent gap penalty values, whereas PRALINE in addition uses secondary structure-specific residue exchange matrices. The pairwise alignments are further refined by a consistency-based scoring function that is modelled after the T-Coffee scenario (see the preceding) based on integrating information coming from comparisons with all possible third sequences.

Next, a guide tree is calculated based on sequence identities and followed to determine the progressive multiple alignment path, leading to a final MSA based on the refined pairwise alignments. The web servers for SPEM and SPEM-3D can handle up to 200 sequences, whereas for the 3D version maximally 100 additional structures can be included.

4. Notes

1. *Distant sequences*. Although high throughput alignment techniques are now able to make very accurate MSAs, alignment incompatibilities can arise under divergent evolution. In practice, it has been shown that the accuracy of all

alignment methods decreases dramatically whenever a considered sequence shares <30% sequence identity *(44)*. Given this limitation, it is advisable to compile a number of MSAs using different amino acid substitution matrices. Among these, the PAM *(4)* and BLOSUM *(5)* series of substitution matrices are the most widely used (especially BLOSUM62). It is helpful to know that higher PAM numbers and low BLOSUM numbers (e.g., PAM250 or BLOSUM45) correspond to exchange matrices that have been designed for the alignment of increasingly divergent sequences, respectively, whereas matrices with lower PAM and higher BLOSUM numbers are suitable for more closely related sequence sets. Furthermore, it is crucial to attempt different gap penalty values, as these can greatly affect the alignment quality. Gap penalties are an essential part of protein sequence alignment when using dynamic programming. The higher the gap penalties, the stricter the insertion of gaps into the alignment and consequently the fewer gaps inserted. Gap regions in an MSA often correspond to loop regions in the associated tertiary structures, which are preferentially altered by divergent evolution. Therefore, it can be useful to lower the gap penalty values for more divergent sequence sets, although care should be taken not to deviate too much from the recommended settings. Excessive gap penalty values will enforce a gap-less alignment, whereas low gap penalties will lead to alignments with very many gaps, allowing (near) identical amino acids to be matched. In both cases the resulting alignment will be biologically inaccurate. The way in which gap penalties affect the alignment also depends on the residue exchange matrix used. Although recommended combinations of exchange matrices and gap penalties have been described in the literature and most methods include default matrices and gap penalty settings, there is no formal theory yet as to how gap penalties should be chosen given a particular residue exchange matrix. Therefore, gap penalties are set empirically: for example, penalties of 11 and 1 are recommended for BLOSUM62, whereas the suggested values for PAM250 are 10 and 1.

2. *Multi-domain proteins (Dialign, T-Coffee)*: Multi-domain proteins can be a particular challenge for multiple alignment methods. Whenever there has been an evolutionary change in the domain order of the query protein sequences, or if some domains have been inserted or deleted across the sequences, this leads to serious problems for global alignment engines. Global methods are not able to deal with permuted domain orders and normally exploit gap penalty regimes that make it difficult to insert long gaps corresponding to the length of one or more protein domains. For the alignment of

multi-domain protein sequences, it is advisable to resort to a local multiple alignment method. Alternatively, the T-Coffee *(29)* and Dialign *(46, 47)* methods might provide a meaningful alignment of multi-domain proteins, as they are (partly) based on the local alignment technique.

3. *Repeats*: The occurrence of repeats in many sequences can seriously compromise the accuracy of MSA methods, mostly because the techniques are not able to deal with different repeat copy numbers. Recently, an MSA strategy has become available that keeps track of various repeat types *(45)*. The method requires the specification of the individual repeats, which can be obtained by running one of the available repeat detection algorithms, after which a repeat-aware MSA is produced. Although the alignment result can be markedly improved by this method, it is sensitive to the accuracy of the repeats information provided.

4. *TM regions*: A special class of proteins is comprised of membrane-associated proteins. The regions within such proteins that are inserted in the cell membrane display a profoundly changed hydrophobicity pattern as compared with soluble proteins. Because the scoring schemes (e.g., PAM *(4)* or BLOSUM *(5)*) normally used in MSA techniques are derived using sequences of soluble proteins, the alignment methods are in principle not suitable to align membrane-bound protein regions. This means that great care should be taken when using general MSA methods. Fortunately, trans-membrane (TM) regions can be reliably recognized using state-of-the-art prediction techniques such as TMHMM or Phobius *(48, ref)*. Therefore, it can be advisable to mark the putative TM regions across the query sequences, and if their mutual correspondence would be clear, to align the blocks of intervening sequence fragments separately.

5. *Preconceived knowledge*: In many cases, there is already some preconceived knowledge about the final alignment. For instance, consider a protein family containing a disulfide bridge between two specific cysteine residues. Given the structural importance of a disulfide bond, constituent Cys residues are generally conserved, so that it is important that the final MSA matches such Cys residues correctly. However, depending on conservation patterns and overall evolutionary distances of the sequences, it can well happen that the alignment engine needs special guidance for matching the Cys residues correctly. Currently none of the approaches has a built-in tool to mark particular positions and assign specific parameters for their consistency, although the library structure of the T-Coffee method allows the specification of weights for matching individual amino acids across the

input sequences. However, exploiting this possibility can be rather cumbersome. The following suggestions are therefore offered for (partially) resolving this type of problem:

   a. *Chopping alignments.* Instead of aligning whole sequences, one can decide to chop the alignment in different parts. For example, this could be done if the sequences have some known domains for which the sequence boundaries are known. An added advantage in such cases is that no undesirable overlaps will occur between these pre-marked regions if aligned separately. Finally, the whole alignment can be built by concatenating the aligned blocks. It should be stressed that each of the separate alignment operations is likely to follow a different evolutionary scenario, as for example the guide tree or the additionally homologous background sequences in the PSI-PRALINE protocol can well be different in each case. It is entirely possible, however, that these different scenarios reflect true evolutionary differences, such as for instance unequal rates of evolution of the constituent domains.

   b. *Altering amino acid exchange weights.* Multiple alignment programs make use of amino acid substitution matrices in order to score alignments. Therefore, it is possible to change individual amino acid exchange values in a substitution matrix. Referring to the disulfide example mentioned in the preceding, one could decide to up-weight the substitution score for a cysteine self-conservation. As a result, the alignment will obtain a higher score when cysteines are matched, and as a consequence the method will attempt to create an alignment where this is the case. However, some protein families have a number of known pairs of Cys residues that form disulfide bonds, where mixing up of the Cys residues involved in different disulfide bridges might happen in that Cys residues involved in different disulfide bonds become aligned at a given single position. To avoid such incorrect matches in the alignment, some programs (e.g., PRALINE) allow the addition of a few extra amino acid designators in the amino acid exchange matrix that can be used to identify Cys residue pairs in a given bond (e.g., J, O, or U). The exchange scores involving these "alternative" Cys residues should be identical to those for the original Cys, except for the cross-scores between the alternative letters for Cys that should be given low (or extreme negative) values to avoid cross alignment. It must be stressed that such alterations are heuristics that can violate the evolutionary model underlying a given residue exchange matrix.

6. *Alignment editors*: A number of multiple alignment editors are available for editing automatically generated alignments,

which often can be improved manually. Posterior manual adjustments can be helpful, especially if structural or functional knowledge of the sequence set is at hand. The following editing tools are available:

a. *Jalview* (www.jalview.org) *(50)* is a protein multiple sequence alignment editor written in Java. In addition to a number of editing options, it also provides a wide scale of sequence analysis tools, such as sequence conservation, UPGMA, and NJ *(51)* tree calculation, and removal of redundant sequences. Color schemes can also be customized according to amino acid physicochemical properties, similarity to consensus sequence, hydrophobicity, or secondary structure.

b. *SeaView* (http://pbil.univ-lyon1.fr/software/seaview.html) *(52)* is a graphical editor suited for Mac, Windows, Unix, and Linux. The program includes a dot-plot routine for pairwise sequence comparison *(53)* or the ClustalW *(31)* multiple alignment program to locally improve the alignment and can also perform phylogenetic analyses. Again, color schemes can be customized.

c. *STRAP* (http://www.charite.de/bioinf/strap/) *(54)* is an interactively extendable and scriptable editor program, able to manipulate large protein alignments. The software is written in Java and is compatible with all operating systems. Among the many extra features provided are: enhanced alignment of low-similarity sequences by integrating 3D-structure information, determination of regular expression motifs, and transmembrane and secondary structure predictions.

d. *CINEMA* (http://umber.sbs.man.ac.uk/dbbrowser/CINEMA2.1/) *(55)* is a Java interactive tool for editing either nucleotide or amino acid sequences. The flexible editor permits color scheme changes and motif selection. Hydrophobicity patterns can also be viewed. Furthermore, there is an option to load prepared alignments from the PRINTS fingerprint database *(56)*.

References

1. Gribskov, M., McLachlan, A. D., Eisenberg, D. (1987) Profile analysis: detection of distantly related proteins. *Proc Natl Acad Sci U S A* 84, 4355–4358.
2. Haussler, D., Krogh, A., Mian, I. S., et al. (1993) Protein modeling using hidden Markov models: analysis of globins, in *Proceedings of the Hawaii International Conference on System Sciences.* Los Alamitos, CA: IEEE Computer Society Press.
3. Bucher, P., Karplus, K., Moeri, N., et al. (1996) A flexible motif search technique based on generalized profiles. *Comput Chem* 20, 3–23.
4. Dayhoff, M. O., Schwart, R. M., Orcutt, B. C. (1978) A model of evolutionary change in proteins, in (Dayhoff, M., ed.), *Atlas of Protein Sequence and Structure.* National Biomedical Research Foundation, Washington, DC.

5. Henikoff, S., Henikoff, J. G. (1992) Amino acid substitution matrices from protein blocks. *Proc Natl Acad Sci U S A* 89, 10915–10919.
6. Needleman, S. B., Wunsch, C. D. (1970) A general method applicable to the search for similarities in the amino acid sequence of two proteins. *J Mol Biol* 48, 443–453.
7. Carillo, H., Lipman, D. J. (1988) The multiple sequence alignment problem in biology. *SIAM J Appl Math* 48, 1073–1082.
8. Stoye, J., Moulton, V., Dress, A. W. (1997) DCA: an efficient implementation of the divide-and-conquer approach to simultaneous multiple sequence alignment. *Comput Appl Biosci* 13, 625–626.
9. Feng, D. F., Doolittle, R. F. (1987) Progressive sequence alignment as a prerequisite to correct phylogenetic trees. *J Mol Evol* 25, 351–360.
10. Hogeweg, P., Hesper, B. (1984) The alignment of sets of sequences and the construction of phyletic trees: an integrated method. *J Mol Evol* 20, 175–186.
11. Gotoh, O. (1996) Significant improvement in accuracy of multiple protein sequence alignments by iterative refinement as assessed by reference to structural alignments. *J Mol Biol* 264, 823–838.
12. Altschul, S. F., Gish, W., Miller, W., et al. (1990) Basic local alignment search tool. *J Mol Biol* 215, 403–410.
13. Pearson, W. R. (1990) Rapid and sensitive sequence comparison with FASTP and FASTA. *Methods Enzymol* 183, 63–98.
14. Heringa, J., Taylor, W. R. (1997) Three-dimensional domain duplication, swapping and stealing. *Curr Opin Struct Biol* 7, 416–421.
15. Smith, T. F., Waterman, M. S. (1981) Identification of common molecular subsequences. *J Mol Biol* 147, 195–197.
16. Waterman, M. S., Eggert, M. (1987) A new algorithm for best subsequence alignments with application to tRNA-rRNA comparisons. *J Mol Biol* 197, 723–728.
17. Thompson, J. D., Plewniak, F., Poch, O. (1999) BAliBASE: a benchmark alignment database for the evaluation of multiple alignment programs. *Bioinformatics* 15, 87–88.
18. Heringa, J. (1999) Two strategies for sequence comparison: profile-preprocessed and secondary structure-induced multiple alignment. *Comput Chem* 23, 341–364.
19. Heringa, J. (2002) Local weighting schemes for protein multiple sequence alignment. *Comput Chem* 26, 459–477.
20. Simossis, V. A., Heringa, J. (2005) PRALINE: a multiple sequence alignment toolbox that integrates homology-extended and secondary structure information. *Nucleic Acids Res* 33, W289–294.
21. Altschul, S. F., Madden, T. L., Schaffer, A. A., et al. (1997) Gapped BLAST and PSI-BLAST: a new generation of protein database search programs. *Nucleic Acids Res* 25, 3389–3402.
22. Kabsch, W., Sander, C. (1983) Dictionary of protein secondary structure: pattern recognition of hydrogen-bonded and geometrical features. *Biopolymers* 22, 2577–2637.
23. Jones, D. T. (1999) Protein secondary structure prediction based on position-specific scoring matrices. *J Mol Biol* 292, 195–202.
24. Rost, B., Sander, C. (1993) Prediction of protein secondary structure at better than 70% accuracy. *J Mol Biol* 232, 584–599.
25. Lin, K., Simossis, V. A., Taylor, W. R et al. (2005) A simple and fast secondary structure prediction method using hidden neural networks. *Bioinformatics* 21, 152–159.
26. Edgar, R. C. (2004) MUSCLE: a multiple sequence alignment method with reduced time and space complexity. *BMC Bioinformatics* 5, 113.
27. Edgar, R. C. (2004) MUSCLE: multiple sequence alignment with high accuracy and high throughput. *Nucleic Acids Res* 32, 1792–1797.
28. Edgar, R. C. (2004) Local homology recognition and distance measures in linear time using compressed amino acid alphabets. *Nucleic Acids Res* 32, 380–385.
29. Notredame, C., Higgins, D. G., Heringa, J. (2000) T-Coffee: A novel method for fast and accurate multiple sequence alignment. *J Mol Biol* 302, 205–217.
30. Huang, X., Miller, W. (1991) A time-efficient, linear-space local similarity algorithm. *Adv Appl Math* 12, 337–357.
31. Thompson, J. D., Higgins, D. G., Gibson, T. J. (1994) CLUSTAL W: improving the sensitivity of progressive multiple sequence alignment through sequence weighting, position-specific gap penalties and weight matrix choice. *Nucleic Acids Res* 22, 4673–4680.
32. O'Sullivan, O., Suhre, K., Abergel, C., et al. (2004) 3DCoffee: combining protein sequences and structures within multiple sequence alignments. *J Mol Biol* 340, 385–395.
33. Taylor, W. R., Orengo, C. A. (1989) Protein structure alignment. *J Mol Biol* 208, 1–22.

34. Shi, J., Blundell, T. L., Mizuguchi, K. (2001) FUGUE: sequence-structure homology recognition using environment-specific substitution tables and structure-dependent gap penalties. *J Mol Biol* 310, 243–257.
35. Wallace, I. M., O'Sullivan, O., Higgins, D. G., et al. (2006) M-Coffee: combining multiple sequence alignment methods with T-Coffee. *Nucleic Acids Res* 34, 1692–1699.
36. Katoh, K., Misawa, K., Kuma, K., et al. (2002) MAFFT: a novel method for rapid multiple sequence alignment based on fast Fourier transform. *Nucleic Acids Res* 30, 3059–3066.
37. Katoh, K., Kuma, K., Toh, H., et al. (2005) MAFFT version 5: improvement in accuracy of multiple sequence alignment. *Nucleic Acids Res* 33, 511–518.
38. Gotoh, O. (1995) A weighting system and algorithm for aligning many phylogenetically related sequences. *Comput Appl Biosci* 11, 543–551.
39. Altschul, S. F. (1998) Generalized affine gap costs for protein sequence alignment. *Proteins* 32, 88–96.
40. Zachariah, M. A., Crooks, G. E., Holbrook, S. R., et al. (2005) A generalized affine gap model significantly improves protein sequence alignment accuracy. *Proteins* 58, 329–338.
41. Do, C. B., Mahabhashyam, M. S., Brudno, M., et al. (2005) ProbCons: Probabilistic consistency-based multiple sequence alignment. *Genome Res* 15, 330–340.
42. Holmes, I., Durbin, R. (1998) Dynamic programming alignment accuracy. *J Comput Biol* 5, 493–504.
43. Zhou, H., Zhou, Y. (2005) SPEM: improving multiple sequence alignment with sequence profiles and predicted secondary structures. *Bioinformatics* 21, 3615–3621.
44. Rost, B. (1999) Twilight zone of protein sequence alignments. *Protein Eng* 12, 85–94.
45. Sammeth, M., Heringa, J. (2006) Global multiple-sequence alignment with repeats. *Prot Struct Funct Bioinf* 64, 263–274.
46. Morgenstern, B., Dress, A., Werner, T. (1996) Multiple DNA and protein sequence alignment based on segment-to-segment comparison. *Proc Natl Acad Sci U S A* 93, 12098–12103.
47. Morgenstern, B. (2004) DIALIGN: multiple DNA and protein sequence alignment at BiBiServ. *Nucleic Acids Res* 32, W33–36.
48. Krogh, A., Larsson, B., von Heijne, G., et al. (2001) Predicting transmembrane protein topology with a hidden Markov model: application to complete genomes. *J Mol Biol* 305, 567–580.
49. Kall, L., Krogh, A., Sonnhammer, E. L. (2004) A combined transmembrane topology and signal peptide prediction method. *J Mol Biol* 338, 1027–1036.
50. Clamp, M., Cuff, J., Searle, S. M., et al. (2004) The Jalview Java alignment editor. *Bioinformatics* 20, 426–427.
51. Saitou, N., Nei, M. (1987) The neighbor-joining method: a new method for reconstructing phylogenetic trees. *Mol Biol Evol* 4, 406–425.
52. Galtier, N., Gouy, M., Gautier, C. (1996) SEAVIEW and PHYLO_WIN: two graphic tools for sequence alignment and molecular phylogeny. *Comput Appl Biosci* 12, 543–548.
53. Li, W.-H., Graur, D. (1991) *Fundamentals of Molecular Evolution*. Sinauer, Sunderland, MA.
54. Gille, C., Frommel, C. (2001) STRAP: editor for STRuctural Alignments of Proteins. *Bioinformatics* 17, 377–378.
55. Parry-Smith, D. J., Payne, A. W., Michie, A. D., et al. (1998) CINEMA–a novel colour INteractive editor for multiple alignments. *Gene* 221, GC57–63.
56. Attwood, T. K., Beck, M. E., Bleasby, A. J., et al. (1997) Novel developments with the PRINTS protein fingerprint database. *Nucleic Acids Res* 25, 212–217.

Chapter 8

Finding Genes in Genome Sequence

Alice Carolyn McHardy

Abstract

Gene-finding is concerned with the identification of stretches of DNA in a genomic sequence that encode biologically active products, such as proteins or functional non-coding RNAs. This is usually the first step in the analysis of any novel piece of genomic sequence, which makes it a very important issue, as all downstream analyses depend on the results. This chapter focuses on the biological basis, computational approaches, and corresponding programs that are available for the automated identification of protein-coding genes. For prokaryotic and eukaryotic genomes, as well as the novel, multi-species sequence data originating from environmental community studies, the state of the art in automated gene finding is described.

Key words: Gene prediction, genomic sequence, protein-coding sequences, prokaryotic, eukaryotic, environmental sequence samples.

1. Introduction

The coding regions of a genome contain the instructions to build functional proteins. This fact gives rise to several characteristic features that can be used universally for their identification. First, if complementary DNA (cDNA), expressed sequence tags (EST), or protein sequences are already known for an organism, these can be used to determine the location of the corresponding genes in the genomic sequence. Although this seems fairly straightforward, the complex gene structure, along with the low sequence quality of EST reads, makes this a non-trivial task for eukaryotic organisms. Second, natural selection for the encoded protein product to remain functional restricts the rate of mutation in coding sequences compared with non-functional genomic DNA. Thus, many protein-coding genes can be identified based on statistically significant sequence similarities that they exhibit toward evolutionarily related proteins from other organisms. Selection is also evident in genome sequence comparisons between closely related species, in which stretches of more highly conserved sequence correspond to the functionally active regions in the genome. Here, protein-coding genes in particular can be identified by their characteristic 3-periodic pattern of conservation. Due to the degeneracy of the genetic code, the third codon position is freer to change than the other positions without changing the encoded protein product. Approaches that use information about expressed sequences or sequence conservation are called "extrinsic," as they require additional knowledge besides the genomic sequence of the organism being analyzed. Then, there is the "intrinsic" approach to gene identification, which is based on the evaluation of characteristic differences between coding and non-coding genomic sequence. In particular, there are characteristic differences in the distribution of short DNA oligomers between the two. One biological reason for these differences is the influence of "translational selection" on the usage of synonymous codons and codon combinations in protein-coding genes. Synonymous codons that encode the same amino acid are not used with equal frequencies in coding sequences. Instead, there is a genome-wide preference in many organisms for the codons that are read by the more frequent tRNAs during the translation process. This effect is especially pronounced for highly expressed genes *(1–4)*. There is also evidence that codon combinations that are prone to initiate frame shift errors during the translation process tend to be avoided *(5)*. Another influence is evident in GC-rich genomes, in which the genome-wide tendencies toward high GC usage establish themselves in the form of a 3-periodic skew toward high GC-content in the generally more flexible third codon position (*see* **Note 1**). In the early 1990s, Fickett systematically evaluated the suitability of a large variety of intrinsic coding measures to discriminate between the coding and non-coding sequence that had by then been proposed, and found that simply counting DNA oligomers is the most effective *(6)*.

2. Methods

2.1. Gene Finding in Prokaryotes

Finding genes in genome sequences is a simpler problem for prokaryotes than it is for eukaryotic organisms. First, the prokaryotic gene structure is less complex: A protein-coding gene corresponds to a single Open Reading Frame (ORF) in the genome sequence, which begins with a start and ends with a stop codon.

The translation start site is defined by a ribosome binding site that is typically located about 3–16 bp upstream of the start codon. Second, more than 90% of prokaryotic genome sequence is coding, as opposed to higher organisms, in which vast stretches of non-coding DNA exist.

For a prokaryotic gene-finding program, the task is to: *(1)* discriminate between the coding and non-coding ORFs in the genome, as there are usually many more ORFs than protein-coding genes in the sequence; and *(2)* identify the correct start codon for those genes, as these can also appear internally in a gene sequence, or even upstream of the actual translation start site. The protein-coding ORFs (ORFs that are transcribed and translated *in vivo*) are commonly referred to as coding sequences (CDSs).

Many programs make use of both extrinsic and intrinsic sources of information to predict protein-coding genes, as they complement each other nicely. External evidence of evolutionary conservation provides strong evidence for the presence of a biologically active gene, but genes without (known) homologs cannot be detected this way. By contrast, intrinsic methods do not need external knowledge; however, an accurate model of oligonucleotide differences between coding and non-coding sequences is required.

A wide variety of techniques is used to solve the gene finding problem (**Table 8.1**). These include probabilistic methods such as Hidden Markov Models (HMMs) *(7–9)* or Interpolated Context Models *(10)*, and machine learning techniques for supervised or unsupervised classification, such as Support Vector Machines (SVMs) *(11)* and Self-Organizing Maps (SOMs) *(12)*. Most gene finders apply their classification models locally to individual sequence fragments (using a sliding window approach) or ORFs in the sequence. In the "global" classification approach implemented with the HMM architecture of GeneMark.hmm, a maximum likelihood parse is derived for the complete genomic sequence, to either coding or non-coding states. An interesting property of this HMM technique is that the search for the optimal model and creation of the final prediction occur simultaneously. The final model that results from the optimization procedure in the training phase uses the hereby found maximum likelihood sequence parse, which is also the most likely assignment of genes to the sequence.

Another issue of importance to gene prediction relates to the fact that many prokaryotic genomes exhibit a considerable portion of sequence that is "atypical" in terms of sequence composition compared with other parts. This, in combination with other properties of such regions, usually indicates a foreign origin and acquisition of the respective region by lateral gene transfer *(13)*. To enable the accurate detection of genes based on sequence composition in such regions, some gene-finding

Table 8.1.
Publicly available prokaryotic gene-finding software

Program	I[a]	E[b]	Comments	URL
BDGF (19)	−	+	Classifies based on universal CDS-specific usage of short amino acid "seqlets"	http://cbcsrv.watson.ibm.com/Tgi.html
Critica (20)	+	+		http://www.ttaxus.com/index.php?pagename=Software
EasyGene (8)	+	+	Uses HMMs. Model training based on BLAST-derived "reliable" genes.	http://www.cbs.dtu.dk/services/EasyGene
GeneMark.hmm/S (7, 9)	+	−	Uses HMMs	http://Opal.biology.gatech.edu/GeneMark
Gismo (11)	+	+	Uses SVMs. Model training based on 'reliable' genes found with PFAM protein domain HMMs.	http://www.cebitec.uni-bielefeld.de/groups/brf/software/gismo See **Fig. 8.2** for a schematic version of the program output.
Orpheus (21)	+	+		http://pedant.gsf.de/orpheus
Reganor (22)	+	+	Utilizes Glimmer and Critica	http://www.cebitec.uni-bielefeld.de/groups/brf/software/reganor/cgi-bin/reganor_upload.cgi
Yacop (23)	+	+	Utilizes Glimmer, Critica, and Orpheus	http://gobics.de/tech/yacop.php
ZCurve (24)	+	−	Uses the "Z-transform" of DNA as information source for classification	http://tubic.tju.edu.cn/Zcurve_B
RescueNet (12)	+	+	Unsupervised discovery of multiple gene classes using a SOM. No exact start/stop prediction.	http://bioinf.nuigalway.ie/RescueNet

[a]Uses intrinsic evidence.
[b]Uses extrinsic evidence.

programs provide a state for genes with atypical sequence properties within the HMM architecture, or by the use of techniques such as the SVM, classifiers can be created that are able to also optimally distinguish between "atypical" CDSs and non-coding ORFs based on the input features. With even more generalized approaches, the unsupervised discovery of characteristic CDS classes in the input data set initiates the gene-finding procedure (12, 14).

The following sequence of steps is often employed by a prokaryotic gene finder (**Fig. 8.1**):

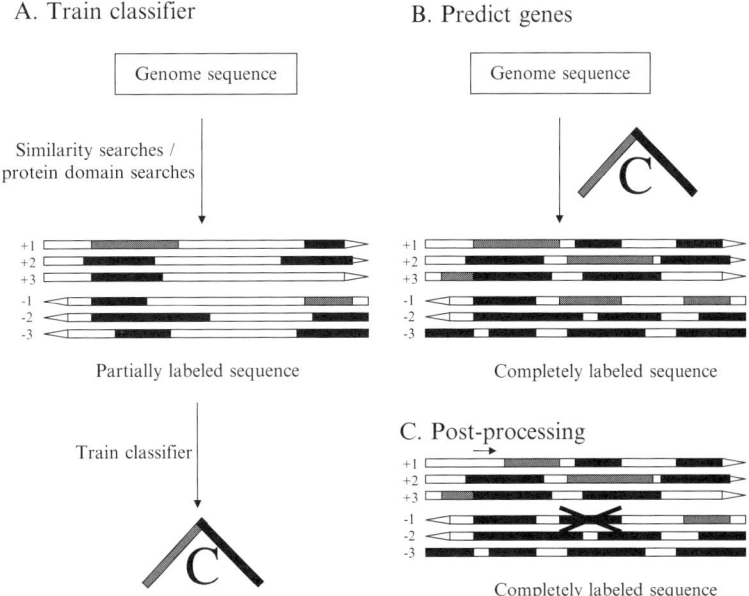

Fig. 8.1. Overview of a sequence of steps employed by a prokaryotic gene finder. (**A**) The sequence is initially searched for regions that exhibit significant conservation on amino acid level relative to other protein-coding regions or show motifs of protein domains. By extension of such regions to a start and stop codon, a partial labeling of the genome sequence into coding regions *(light gray)* and non-coding ORFs, which significantly overlap with such coding sequences in another frame *(dark gray)*, can be obtained. The labeled parts can be used as training sequences to derive vectors of intrinsic sequence features for the training of a binary classifier. (**B**) The classifier is applied to classify all ORFs above a certain length in the sequence as either CDS or nORF. In the post-processing phase C, start positions of the predicted CDSs are reassigned with the help of translation start site models, and conflicts between neighboring predictions are resolved.

1. In the initial phase, an intrinsic classifier for the discrimination between nORFs and CDSs is learnt. This is often initiated with similarity searches of the genomic sequences against protein or DNA sequence databases or a search for motifs of protein domains. The hereby generated information can be used for a partial labeling of the sequence into nORF and CDSs and also to generate reliable training data for the classifier. ORFs supported by external sequence similarities can be used as training material for the protein-coding genes, and ORFs that significantly overlap with such regions (that are located in their "shadow") as training material for the non-coding ORF class.

2. In the prediction phase, the classifier is applied to identify the protein-coding genes of the sequence.

3. In the post-processing phase, start positions of predicted CDSs are relocated to the most likely translation start site

```
##gff-version 2
##date 2006-03-21

##source-version Parser::GFF 1.2
36521  CDS  381    82     -0.262462    .  -1  rbs "381"   ; reliable "0"
36521  CDS  1353   1967   1.00021      3  .   rbs "1353"  ; evalue "5.3e-18" ; description "Molybdopterin oxidoreductase Fe4S4 do" ; reliable "1"
36521  CDS  1980   4457   1.06453      3  .   rbs "1980"  ; evalue "6e-22"   ; description "Molybdopterin oxidoreductase" ; reliable "1"
36521  CDS  4454   5446   0.420936     .  2   rbs "4454"  ; evalue "2.8e-05" ; description "4Fe-4S binding domain" ; reliable "1"
36521  CDS  5424   6110   0.105856     3  .   rbs "5424"  ; comment "overlapping" ; evalue "0.0041" ; description "Cytochrome b561 family" ; reliable "1"
36521  CDS  5519   7069   1.38546.     2  .   rbs "5519"  ; comment "overlapping" ; evalue "1.7e-39" ; description "Protein involved in formate dehydrogen" ; reliable "1"
36521  CDS  7318   6782   -0.536073    .  -2  rbs "7318"  ; discard "1"    ; reliable "0"
36521  CDS  7074   8474   1.0716       3  .   rbs "7074"  ; comment "overlapping" ; evalue "3.7e-255" ; description "L-seryl-tRNA selenium transferase" ; reliable "1"
36521  CDS  8302   9255   1.08006.     1  .   rbs "8302"  ; comment "overlapping" ; evalue "4.5e-10"  ; description "Sell repeat" ; reliable "1"
36521  CDS  9252   11252  1.00039      3  .   rbs "9252"  ; evalue "4.4e-30" ; description "Elongation factor Tu GTP binding doma" ; reliable "1"
36521  CDS  11453  11776  -0.0957853   .  2   rbs "11453" ; reliable "0"
36521  CDS  11506  12063  -0.338037    .  1   rbs "11506" ; reliable "0"
36521  CDS  12026  11763  -0.596191    .  -3  rbs "12026" ; discard "1"  ; reliable "0"
36521  CDS  12546  12259  0.233941     .  -1  rbs "12546" ; reliable "1"
36521  CDS  12754  13161  -0.579761    .  1   rbs "12754" ; reliable "0"
36521  CDS  13227  13700  -0.00217621  .  3   rbs "13227" ; reliable "0"
36521  CDS  13718  14068  -0.564218    .  2   rbs "13718" ; reliable "0"
36521  CDS  15094  14450  1.0002       .  -2  rbs "15094" ; evalue "0.0034" ; description "Response regulator receiver domain" ; reliable "1"
36521  CDS  15753  15358  -0.525293    .  -1  rbs "15753" ; reliable "0"
36521  CDS  15727  16209  0.247978     .  1   rbs "15727" ; reliable "1"
36521  CDS  16418  16251  -0.466092    .  -3  rbs "16418" ; reliable "0"
```

Fig. 8.2. Output of the program GISMO (see **Table 8.1**). The output is in GFF format. For each prediction, the contig name, start and stop position, SVM score, and reading frame are given. The position of a ribosome binding site (RBS) and a confidence assignment ("1" for high confidence, "0" for low confidence) is also given. If a prediction has a protein domain match in PFAM, additionally the e-value of that hit and the description of the PFAM-entry are reported. Predictions that were discarded in the post-processing phase (removal of overlapping low-confidence predictions) are labeled as discard ("1").

using translation start site models. Such models usually incorporate information about the ribosome binding site, and are applied to search for the start codon with the strongest upstream signal for any given gene. Programs that have been specifically designed for this task are also available (15–18). The above-mentioned HMM-based gene finders deviate from this procedure, as they incorporate the ribosome binding site signal directly into the HMM model, and identify the optimal start sites simultaneously with the overall optimal sequence parse. Conflicts of overlapping predictions that could not be solved by relocation of start sites are resolved by the removal of the "weaker" prediction, whereby this call can be based on intrinsic and also the extrinsic information generated in phase A.

2.2. Gene Finding in Environmental Sequence Samples

Recently, the application of genome sequencing techniques to DNA samples obtained directly from microbial communities inhabiting a certain environment has spawned the novel field of environmental or community genomics, or metagenomics *(25)*. The field promises to deliver insights into numerous issues that cannot be addressed by the sequencing of individual, lab-cultivated organisms. By estimates, <1% of all microorganisms can be grown in pure culture with standard techniques. As a result of this, our current knowledge of prokaryote biology and also the sequenced genomes exhibits a strong bias toward four phyla, which contain many cultivable organisms, out of an estimated total of 53 or more existing prokaryotic phyla (*26, 27*). By bypassing the need for pure culture, environmental genomics allows the discovery of novel organisms and the analysis of sequences that could not have been obtained otherwise. These studies also increase understanding of the processes and interactions that shape the metabolism, structure, function, and evolution of a microbial community seen as a whole.

The sequences of an environmental sample sometimes can represent a very large number of organisms, each of which is represented in proportion to its abundance in the habitat. Sufficient sequence is typically sampled to allow the reconstruction of nearly complete genomes for the most abundant organisms in a sample. In addition, numerous short fragments and unassembled singleton reads are generated from the many less abundant organisms. The new data type creates challenges at all levels for the bioinformatics tools that have been established in genome sequence analysis *(28)*, including the available prokaryotic gene-finding programs. The short sequences that are created frequently contain truncated and frame-shifted genes, which many of the standard gene finders have not been designed to identify. Recently, a program specifically designed for this application has become available that allows identification of the homology-supported genes, including truncated and frame-shifted versions, with good specificity *(29)*. Second, the sequence properties of both coding and non-coding sequences can vary considerably between genomes *(30)*, which might cause difficulties in creating an appropriate intrinsic model (*see* **Note 2**). A systematic comparison and evaluation of different feasible approaches is still missing at this point, but procedures employing a variable number of CDS models might fare best. Such models could be derived by the initial unsupervised discovery of distinct CDS classes (12, 14) in the data. Alternatively, an amino-acid composition-based approach such as the Bio-Dictionary gene finder (BDGF) *(19)* might perform well. The BDGF identifies CDSs based on a universal model of amino acid composition (using short amino acid "seqlets") and does not require the construction of organism-specific intrinsic models to find genes without homologs.

2.3. Gene Finding in Eukaryotes

Compared with gene identification in prokaryotic organisms, in which automated prediction methods have reached high levels of accuracy (*see* **Note 3**), eukaryotic gene prediction is currently still a highly challenging problem for a number of reasons. First, only a small fraction of eukaryotic genome sequences correspond to protein-encoding exons, which are embedded in vast amounts of non-coding sequence. Second, the gene structure is complex. The Open Reading Frame encoding the final protein product can be located discontinuously in two or more sometimes very short exonic regions, which are separated from each other by intronic sequence. The junctions of exon-intron boundaries are characterized by splice sites on the initial transcript, which guide intron removal in fabrication of the ripe mRNA transcript at the spliceosome. Additional signal sequences, such as an adenylation signal, are found in the proximity of the transcript end and determine a cleavage site corresponding to the end of the ripe transcript to which a poly-A tail is added for stability. The issue is further complicated by the fact that genes can have alternative splice and polyadenylation sites, as well as alternative translation and transcription initiation sites. Third, due to the massive sequencing requirements, additional eukaryotic genomes that can be used to study sequence conservation are becoming available more slowly than their prokaryotic counterparts.

The complex organization of eukaryotic genes makes determination of the correct gene structure the most difficult problem in eukaryotic gene prediction. Signals of functional sites are very informative; for instance, splice site signals are the best means to locate exon-intron boundaries. Methods that are designed to identify these or other functional signals from promoter or polyadenylation (polyA) sites are generally referred to as "signal sensors." Methods that classify genomic sequence into coding or non-coding content are called "content sensors." Content sensors can utilize all of the above-mentioned sources of information for gene identification: extrinsic information such as EST/cDNA or protein sequences, comparisons to genomic sequences from other organisms, or intrinsic information about coding sequence-specific sequence properties. Extrinsic evidence, such as the sequences of known proteins and EST/cDNA libraries, tends to produce reliable predictions, but it is biased toward genes that are highly and ubiquitously expressed. The programs for *"de novo"* gene identification are also categorized as intrinsic "one genome" approaches, and "dual" or "multi-genome" approaches, that utilize comparative information from other genomes *(31)*. Many gene finders use HMMs for the task of "decoding" the unknown gene structure from the sequence, which allows the combination of content and signal sensor modules into a single, coherent probabilistic model with biological meaning. Typically, these models include states for exons, introns, splice sites, polyadenylation signals, start codons, stop codons, and intergenic regions, as well as an additional model for single exon genes. See (31, 32) for recent reviews of the programs and techniques used in the field.

2.3.1. The Ensembl Gene Prediction Pipeline

As Ensembl is a very widely used resource, a brief description of the Ensembl gene prediction pipeline *(33)* is included at this point. The Ensembl pipeline leans strongly toward producing a specific prediction with few false-positives rather than a sensitive prediction with few false-negatives. Every gene that is produced is supported by direct extrinsic evidence, such as experimental EST/cDNA data, known protein sequences from the organism's proteome, or known proteins of related organisms. The complete procedure involves a wide variety of programs (**Fig. 8.3**). Repetitive elements are initially identified and masked, to remove them from the analyzed input.

1. The sequences of known proteins from the organism are mapped to a location on the genomic sequence. Local alignment programs are used for a prior reduction of the sequence search space, and the HMM-based GeneWise program *(34)* for the final sequence alignment.
2. An *ab initio* predictor, such as Genscan *(35)*, is run, and for predictions confirmed by the presence of homologs, those homologs are aligned to the genome using GeneWise, as before.
3. Simultaneously to Step 1, Exonerate *(36)* is used to align known cDNA sequences to the genome sequence.
4. The candidates that were found with this procedure are merged to create consensus transcripts with 3′ UTRs, coding sequence and 5′ UTRs.
5. Redundant transcripts are merged, and genes are identified, which by Ensembl's definition correspond to sets of transcripts with overlapping exons.

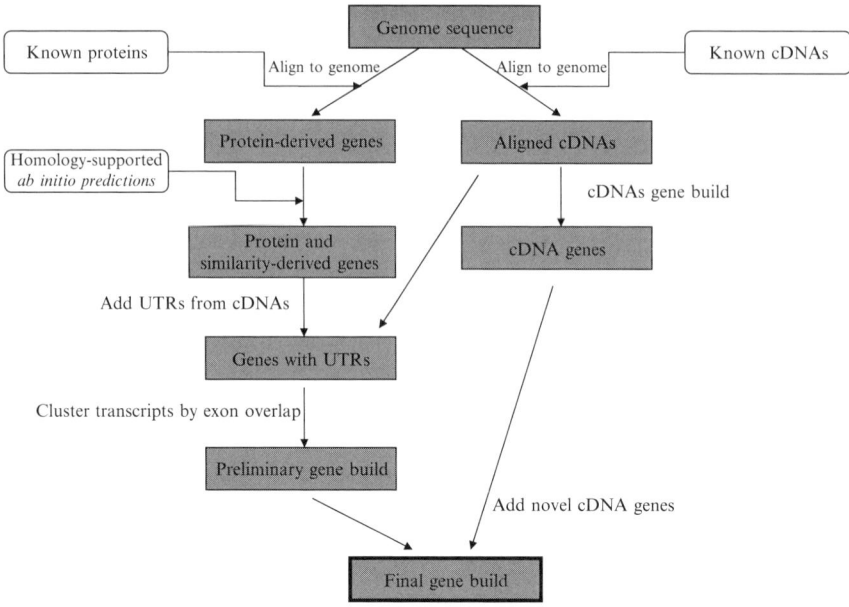

Fig. 8.3. Overview of the Ensembl pipeline for eukaryotic gene prediction.

6. Novel cDNA genes that do not match with any exons of the protein-coding genes identified so far are added to create the final gene set.

It is important to note that for the genomes of different organisms, this procedure is applied with slight variations, depending on the available resources.

2.3.2. De Novo Prediction in Novel Genomes

Traditionally, the models of eukaryotic *de novo* gene-finding programs are trained in a supervised manner, in order to learn the optimal, species-specific parameters for splice site, intron length, and content sensor models. Their complexity requires large and reliable training sets, which can best be derived by mapping EST or cDNA sequence information to the genome. In cases in which sufficient information is not available, models that have been created for other species can be used, although this can deliver suboptimal results. The programs SNAP *(37)* and Genemark. hmm ES-3.0 *(38)* employ two different techniques to circumvent this problem, and allow high-quality gene prediction for genomes in which little reliable annotation is available. SNAP utilizes an iterative "bootstrap" procedure, in which the results initially derived by application of models from other organisms are used in an iterative manner for model training and prediction refinement. Genemark.hmm derives its model in a completely unsupervised manner from the unannotated genomic sequence, starting with broadly set, generic model parameters, which are iteratively refined based on the prediction/training set obtained from the raw genomic sequence.

2.3.3. Accuracy Assessment and Refining Annotation

The increasing availability of high-coverage genomic sequence for groups of related eukaryotic organisms has resulted in significant advances in *de novo* gene-finding accuracy. For compact eukaryotic genomes, dual genome approaches achieve up to 67% accuracy in the *de novo* prediction of complete gene structures. Experimental follow-up experiments led to the biological verification of to up to 75% of the novel predictions selected for testing (*39, 40*). Performance is still markedly lower for the mammalian genomes, however, due to their larger fractions of non-coding sequence and pseudogenes.

As long stretches of high-quality, experimentally verified genomic DNA are still scarce for eukaryotes, it is often impossible to unequivocally decide whether novel predictions correspond to false-positives or are, in fact, real novel genes that are missing from the current annotations.

For the human genome, the ENCODE (ENCyclopedia of DNA Elements) project was launched in 2004, which ultimately aims to identify all functional elements in the human genome. In the current pilot phase, the goal is to produce an accurate, experimentally verified annotation for 30 megabases (1%) of the human genome, and delineate a high throughput strategy suitable for application to the complete genome sequence. As part of this effort, the EGASP (ENCODE Genome Annotation Assessment Project) 2005 was organized *(41)*, which brought together more than 20 teams working on computational gene prediction. Evaluation of the different programs on the high-quality annotation map showed, not surprisingly, that the extrinsic programs that utilize very similar information as the human annotators perform best. Of the *de novo* prediction programs, those including comparative genomic analyses came in second, and lastly, the programs predicting solely based on intrinsic genomic evidence. The comparison with existing annotation and the subsequent experimental evaluation of several hundred *de novo* predictions showed that only very few human genes are not detected by computational means, but that even for the best programs, the accuracy of exactly predicting the genomic structure was only about 50% *(42)*. Of the experimentally validated *de novo* predictions, only few (3.2%) turned out to correspond to real, previously unknown exons. Another insight gained in this project was that alternative splicing seems to occur in most human genes, which marks one more tough challenge eukaryotic computational gene finders will have to face in order to create highly accurate, automated predictions.

3. Notes

1. The periodic skew toward high GC-content in the third codon position of coding sequences in GC-rich genomes can be visualized with "frame-plots," in which the phase-specific GC content of the sequence (averaged over a sliding sequence window) is visualized in three different curves (**Fig. 8.4**). Such plots are commonly used by annotators for start site annotation and are part of many genome annotation packages.

2. Currently, genes in metagenome samples in some cases are identified based on sequence similarities only, which leaves genes that do not exhibit similarities to known genes (or to other genes found in the sample) undiscovered. An alternative procedure is to create a universal intrinsic model of prokaryotic sequence composition from known genomic sequences. This prevents a bias of the model toward the dominating organisms of a metagenome community, but will still be biased toward phyla that have been characterized so far.

3. The number of completely sequenced prokaryotic organisms has grown exponentially within the last 5 years. The existence of such a large and diverse data set allows a thorough assessment of gene-finding accuracy across the wide variety of sequenced organisms. For several programs, evaluations on more than 100 genomes sequences have been undertaken. Prediction of CDSs for prokaryotic genomes generally has become very accurate. The gene finder Reganor *(22)*, which incorporates evidence from the programs Glimmer and

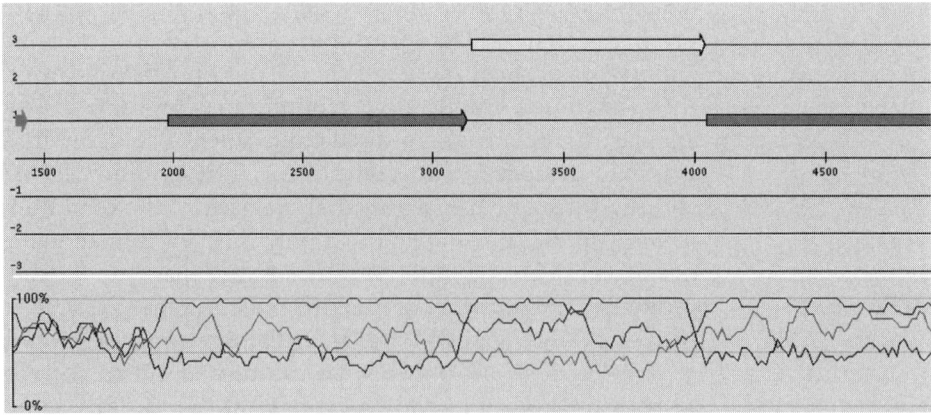

Fig. 8.4. The location of genes in GC-rich genomes is indicated by the frame-specific GC content. A plot of the frame-specific GC content was computed for a sliding window of size 26 bps that was moved with a step size of 5 across the sequence. The GC content is plotted in the lower panel for the three frames: frame 3, frame 2, and frame 1. The upper panel shows the location of annotated protein-coding genes in the genome *(arrows)*.

Critica, reproduces bacterial annotations with 95% specificity, and 98% sensitivity in identification of "certain" genes that are supported by external evidence. The SVM-based gene finder GISMO *(11)* and the HMM-based EasyGene *(43)* reach sensitivity levels of up to 99%, while also displaying high levels of specificity. Automated gene prediction has become so accurate that a considerable fraction of the additional "false-positive" predictions according to the manually compiled annotations in fact seem to represent real, but not annotated genes *(44)* (*see* **Note 4**). Programs for start site identification, such as TICO *(17)* and GS-Finder *(15)*, have also been found to perform with >90% accuracy. However, due to the currently limited numbers of CDSs with experimentally verified N-termini it is too early to generalize this observation. A challenge that remains to be addressed by prokaryotic gene finders is the correct identification of "short" genes with fewer than 100 codons. Evidence indicates that the automatic identification as well as the current annotation for such short genes need to be improved *(22, 43, 45)*.

4. Annotations are not perfect, and not supported by experimental evidence in every case, so accuracy estimates obtained with this standard of truth can be questionable. Accordingly, claims that additional predictions might in fact correspond to real, but currently missing genes, could be correct, just as possibly genes that were predicted in accordance to the annotation might be false-positive in both cases. Thus, accuracy estimates should generally be based on further evidence that supports or refutes the predictions. A strong indicator of a valid prediction for instance is its location in a cluster of genes with homologs in a similar arrangement in related genomes *(46)*.

Acknowledgments

The author thanks Lutz Krause, Alan Grossfield, and Augustine Tsai for their comments.

References

1. Dong, H., Nilsson, L., Kurland, C. G. (1996) Co-variation of tRNA abundance and codon usage in *Escherichia coli* at different growth rates. *J Mol Biol* 260, 649–663.
2. Ikemura, T. (1981) Correlation between the abundance of *Escherichia coli* transfer RNAs and the occurrence of the respective codons in its protein genes: a proposal for a synonymous codon choice that is optimal for the *E. coli* translational system. *J Mol Biol* 151, 389–409.
3. Sharp, P. M., Bailes, E., Grocock, R. J., et al. (2005) Variation in the strength of selected codon usage bias among bacteria. *Nucleic Acids Res* 33, 1141–1153.
4. Rocha, E. P. (2004) Codon usage bias from tRNA's point of view: redundancy,

specialization, and efficient decoding for translation optimization. *Genome Res* 14, 2279–2286.

5. Hooper, S. D., Berg, O. G. (2000) Gradients in nucleotide and codon usage along Escherichia coli genes. *Nucleic Acids Res* 28, 3517–3523.

6. Fickett, J. W., Tung, C. S. (1992) Assessment of protein coding measures. *Nucleic Acids Res* 20, 6441–6450.

7. Besemer, J., Lomsadze, A., Borodovsky, M. (2001) GeneMarkS: a self-training method for prediction of gene starts in microbial genomes. Implications for finding sequence motifs in regulatory regions. *Nucleic Acids Res* 29, 2607–2618.

8. Larsen, T. S., Krogh, A. (2003) EasyGene—a prokaryotic gene finder that ranks ORFs by statistical significance. *BMC Bioinformatics* 4, 21.

9. Lukashin, A. V., Borodovsky, M. (1998) GeneMark.hmm: new solutions for gene finding. *Nucleic Acids Res* 26, 1107–1115.

10. Delcher, A. L., Harmon, D., Kasif, S., et al. (1999) Improved microbial gene identification with GLIMMER. *Nucleic Acids Res* 27, 4636–4641.

11. Krause, L., McHardy, A. C., Nattkemper, T. W., et al. (2007) GISMO—gene identification using a support vector machine for ORF classification. *Nucleic Acids Res* 35, 540–549.

12. Mahony, S., McInerney, J. O., Smith, T. J., et al. (2004) Gene prediction using the Self-Organizing Map: automatic generation of multiple gene models. *BMC Bioinformatics* 5, 23.

13. Ochman, H., Lawrence, J. G., and Groisman, E. A. (2000) Lateral gene transfer and the nature of bacterial innovation. *Nature* 405, 299–304.

14. Hayes, W. S., Borodovsky, M. (1998) How to interpret an anonymous bacterial genome: machine learning approach to gene identification. *Genome Res* 8, 1154–1171.

15. Ou, H. Y., Guo, F. B., Zhang, C. T. (2004) GS-Finder: a program to find bacterial gene start sites with a self-training method. *Int J Biochem Cell Biol* 36, 535–544.

16. Suzek, B. E., Ermolaeva, M. D., Schreiber, M., et al. (2001) A probabilistic method for identifying start codons in bacterial genomes. *Bioinformatics* 17, 1123–1130.

17. Tech, M., Pfeifer, N., Morgenstern, B., et al. (2005) TICO: a tool for improving predictions of prokaryotic translation initiation sites. *Bioinformatics* 21, 3568–3569.

18. Zhu, H. Q., Hu, G. Q., Ouyang, Z. Q., et al. (2004) Accuracy improvement for identifying translation initiation sites in microbial genomes. *Bioinformatics* 20, 3308–3317.

19. Shibuya, T., Rigoutsos, I. (2002) Dictionary-driven prokaryotic gene finding. *Nucleic Acids Res* 30, 2710–2725.

20. Badger, J. H., Olsen, G. J. (1999) CRITICA: coding region identification tool invoking comparative analysis. *Mol Biol Evol* 16, 512–524.

21. Frishman, D., Mironov, A., Mewes, H. W., et al. (1998) Combining diverse evidence for gene recognition in completely sequenced bacterial genomes. *Nucleic Acids Res* 26, 2941–2947.

22. McHardy, A. C., Goesmann, A., Puhler, A., et al. (2004) Development of joint application strategies for two microbial gene finders. *Bioinformatics* 20, 1622–1631.

23. Tech, M., Merkl, R. (2003) YACOP: Enhanced gene prediction obtained by a combination of existing methods. *In Silico Biol* 3, 441–451.

24. Guo, F. B., Ou, H. Y., Zhang, C. T. (2003) ZCURVE: a new system for recognizing protein-coding genes in bacterial and archaeal genomes. *Nucleic Acids Res* 31, 1780–1789.

25. Venter, J. C., Remington, K., Heidelberg, J. F., et al. (2004) Environmental genome shotgun sequencing of the Sargasso Sea. *Science* 304, 66–74.

26. Hugenholtz, P. (2002) Exploring prokaryotic diversity in the genomic era. *Genome Biol* 3, REVIEWS0003.

27. Rappe, M. S., Giovannoni, S. J. (2003) The uncultured microbial majority. *Annu Rev Microbiol* 57, 369–394.

28. Chen, K., Pachter, L. (2005) Bioinformatics for whole-genome shotgun sequencing of microbial communities. *PLoS Comput Biol* 1, 106–112.

29. Krause, L., Diaz, N. N., Bartels, D., et al. (2006) Finding novel genes in bacterial communities isolated from the environment. *Bioinformatics* 22, e281–289.

30. Sandberg, R., Branden, C. I., Ernberg, I., et al. (2003) Quantifying the species-specificity in genomic signatures, synonymous codon choice, amino acid usage and G+C content. *Gene* 311, 35–42.

31. Brent, M. R., Guigo, R. (2004) Recent advances in gene structure prediction. *Curr Opin Struct Biol* 14, 264–272.

32. Mathe, C., Sagot, M. F., Schiex, T., et al. (2002) Current methods of gene prediction,

33. Curwen, V., Eyras, E., Andrews, T. D., et al. (2004) The Ensembl automatic gene annotation system. *Genome Res* 14, 942–950.
34. Birney, E., Clamp, M., Durbin, R. (2004) GeneWise and Genomewise. *Genome Res* 14, 988–995.
35. Burge, C., Karlin, S. (1997) Prediction of complete gene structures in human genomic DNA. *J Mol Biol* 268, 78–94.
36. Slater, G. S., Birney, E. (2005) Automated generation of heuristics for biological sequence comparison. *BMC Bioinformatics* 6, 31.
37. Korf, I. (2004) Gene finding in novel genomes. *BMC Bioinformatics* 5, 59.
38. Lomsadze, A., Ter-Hovhannisyan, V., Chernoff, Y. O., et al. (2005) Gene identification in novel eukaryotic genomes by self-training algorithm. *Nucleic Acids Res* 33, 6494–6506.
39. Tenney, A. E., Brown, R. H., Vaske, C., et al. (2004) Gene prediction and verification in a compact genome with numerous small introns. *Genome Res* 14, 2330–2335.
40. Wei, C., Lamesch, P., Arumugam, M., et al. (2005) Closing in on the C. elegans ORFeome by cloning TWINSCAN predictions. *Genome Res* 15, 577–582.
41. Guigo, R., Reese, M. G. (2005) EGASP: collaboration through competition to find human genes. *Nat Methods* 2, 575–577.
42. Guigo, R., Flicek, P., Abril, J. F., et al. (2006) EGASP: the human ENCODE Genome Annotation Assessment Project. *Genome Biol* 7 Suppl 1, S2 1–31.
43. Nielsen, P., Krogh, A. (2005) Large-scale prokaryotic gene prediction and comparison to genome annotation. *Bioinformatics* 21, 4322–4329.
44. Linke, B., McHardy, A. C., Krause, L., et al. (2006) REGANOR: A gene prediction server for prokaryotic genomes and a database of high quality gene predictions for prokaryotes. *Appl Bioinformatics* 5, 193–198.
45. Skovgaard, M., Jensen, L. J., Brunak, S., et al. (2001) On the total number of genes and their length distribution in complete microbial genomes. *Trends Genet* 17, 425–428.
46. Osterman, A., Overbeek, R. (2003) Missing genes in metabolic pathways: a comparative genomics approach. *Curr Opin Chem Biol* 7, 238–251.

their strengths and weaknesses. *Nucleic Acids Res* 30, 4103–4117.

Chapter 9

Bioinformatics Detection of Alternative Splicing

Namshin Kim and Christopher Lee

Abstract

In recent years, genome-wide detection of alternative splicing based on Expressed Sequence Tag (EST) sequence alignments with mRNA and genomic sequences has dramatically expanded our understanding of the role of alternative splicing in functional regulation. This chapter reviews the data, methodology, and technical challenges of these genome-wide analyses of alternative splicing, and briefly surveys some of the uses to which such alternative splicing databases have been put. For example, with proper alternative splicing database schema design, it is possible to query genome-wide for alternative splicing patterns that are specific to particular tissues, disease states (e.g., cancer), gender, or developmental stages. EST alignments can be used to estimate exon inclusion or exclusion level of alternatively spliced exons and evolutionary changes for various species can be inferred from exon inclusion level. Such databases can also help automate design of probes for RT-PCR and microarrays, enabling high throughput experimental measurement of alternative splicing.

Key words: alternative splicing, genome annotation, alternative donor, alternative acceptor, exon skipping, intron retention, tissue-specific, cancer-specific, exon inclusion, exon exclusion, RT-PCR, microarray.

1. Introduction

One area in which genomics and bioinformatics have made a dramatic impact is the field of alternative splicing. Since its discovery in 1978 *(1)*, alternative splicing was widely considered to be a relatively uncommon form of regulation affecting perhaps 5–15% of genes. Recently, thanks to the vast amounts of public data produced by EST, cDNA, and genome sequencing projects, and fast alignment tools, genome-wide studies on alternative splicing have become feasible. Surprisingly, genome-wide studies on alternative splicing have consistently demonstrated that 40–70% of human genes are alternatively spliced *(2–24)*. Now the major challenge is to make sense of the functional impact of this huge new catalog of alternative splicing, by making these data easy for biologists to incorporate into their research. This effort has several components. Alternative splicing plays an important role in increasing protein diversity and function. Many splice variants with missing domains or motifs can be related to various diseases; therefore, alternatively spliced genes can be therapeutic targets for drug development *(25–27)*. This chapter reviews the data, methodology, and technical challenges of these genome-wide analyses of alternative splicing, and briefly surveys some of the major applications that such alternative splicing databases have been designed to address.

1.1. Types of Alternative Transcript Diversity

First, it is important to understand the diverse types of alternative transcript variation, since these frame the technical challenges that must be considered. Alternative splicing patterns can be classified into alternative donor site (alternative 5′ splicing), alternative acceptor site (alternative 3′ splicing), exon skipping, alternative initiation site, alternative polyadenylation, and intron retention (**Fig. 9.1**). Technically, alternative initiation and alternative polyadenylation are due not to alternative *splicing* per se, but instead to usage of a different transcriptional start site (alternative initiation) or of a different polyadenylation site. This has immediate technical implications. In general, alternative splicing patterns such as alternative donor, alternative acceptor, and exon skipping sites are easier to detect and interpret, because of the extra technical challenges of assigning confidence to the other

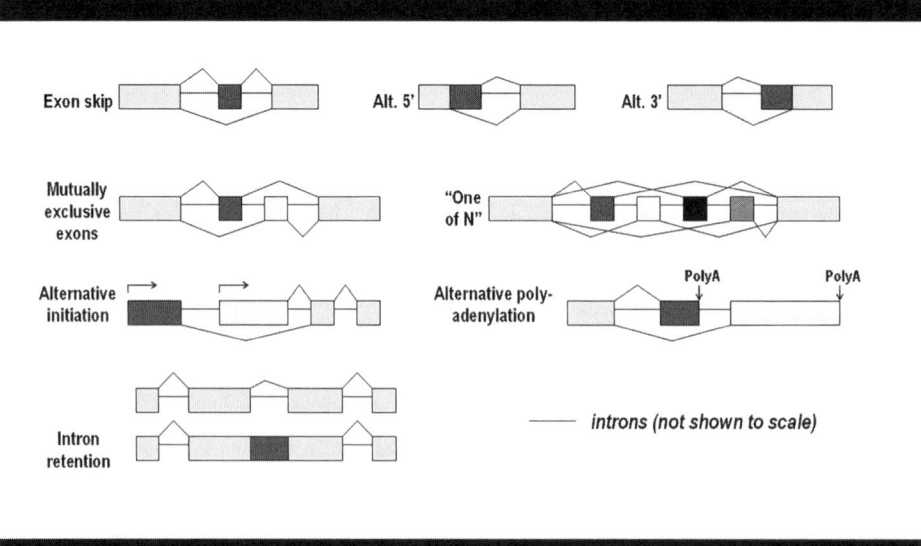

Fig. 9.1. Types of alternative transcripts. Gray and black boxes represent alternative exons. White boxes represent constitutive exons.

events, as examined in detail in the following. Intron retention is generally the most problematic type, because it is hard to distinguish from experimental artefacts (*see* **Note 1**).

1.2. How to Detect Alternative Splicing Patterns

All attempts to identify alternative splicing start by comparing EST sequences with mRNA and genomic sequences *(4, 6, 8–15, 17, 18, 20–24, 28–42)*. First, the detection method should report all differences between transcript fragments that could be real alternative transcript variation, while excluding the many types of artefacts that are possible (*see* **Table 9.1**; **Notes 1** and **2**). For example, one could focus on pairs of observed splices that share one splice site but differ at the other splice site (*see* **Fig. 9.1**). This simple but highly specific pattern detects most possible AS events; by contrast, genomic contamination and other artefacts do not produce this pattern. In addition to the transcript sequences themselves, genomic sequence also plays a critical role. First, it provides the essential foundation for detecting and validating splicing events in the transcript fragments. Since much of the information for splice sites is intronic, the genomic sequence is vital for checking candidate splices (which should correspond to introns). For this reason, comparing only the mRNA and EST sequences *(20, 31, 35, 36, 41)* yields more limited predictive power, because it does not utilize information in the intronic part of the genome. mRNA splicing is a carefully controlled process. Since more than 95% of intron sequences match the GT…AG consensus of U1/U2 splice sites, even simple genomic sequence checks can be broadly useful. Now that many genome assemblies are available in public, it is possible to exploit the information hidden in the intronic part of the genome. Second, genomic sequence provides the best foundation for determining which EST fragments are "from the same gene" (i.e., EST clustering), and distinguishing ESTs from similar genes (paralogs), which otherwise could cause incorrect predictions of alternative splicing. Thanks to the development of fast sequence alignment tools such as BLAT *(43)* and GMAP *(44)*, it is feasible to map all mRNA and EST sequences against genome assembly within a few days, even using a single CPU.

2. Materials

2.1. Types and Amount of Available Public Data

It is important to understand the available public resources for transcript sequencing data. EST sequences are small pieces of DNA sequences, usually 200–500 bp, generated by shotgun sequencing of one or both strands of an expressed gene. EST sequencing can be produced rapidly and inexpensively, and has been the primary source of raw data for discovering new forms of transcript diversity. To date, nearly 36 million EST sequences have

Table 9.1
Artefacts and issues in high throughput alternative transcript analysis

Experimental factors	Type of errors caused: (+) false-positive, (−) false-negative
EST coverage limitations, bias	(−). Only a few genomes have good EST coverage. Even for those, coverage across divergent tissues is largely incomplete.
RT/PCR artefacts	(+) in methods that do not screen for fully valid splice sites (which requires genomic mapping, intronic sequence)
Library artefacts	(+) in methods with weaker detection criteria (especially intron retention)
Genomic coverage, assembly errors	(−) in methods that map ESTs on the genome. Incomplete genome sequence (especially for plants) limits AS discovery.
Chimeric ESTs	(+) in methods that simply compare ESTs
Genomic contamination	(+) in methods that do not screen for *pairs* of mutually exclusive splices
EST orientation error, uncertainty	(+/−) in methods that don't correct misreported orientation, or don't distinguish overlapping genes on opposite strands
Sequencing error	(+/−). Single pass EST sequencing error can be very high locally (e.g. >10% at the ends). Need chromatograms.
EST fragmentation	Where ESTs end cannot be treated as significant.

Bioinformatics factors

Mapping ESTs to the genome	(−) in methods that map genomic location for each EST
Alternative initiation sites?	(+/−). Distinguishing alternative transcription start sites from alternative splicing is currently imperfect.
Alternative poly-A sites?	(+/−). Alternative poly-A site prediction is perhaps harder than just detecting mutually exclusive splice events.
"Liberal" vs. "conservative" isoform prediction	(+/−). Some approaches predict all possible combinations of alternative splicing events in a gene, whereas others predict the minimum set of isoforms capable of explaining the observations.

Issue	Comments
Non-standard splice sites?	(+) in methods that do not fully check splice sites; (−) in methods that restrict to standard GT-AG splice sites
Alignment degeneracy	(+/−). Alignment of ESTs to genomic sequence can be degenerate around splice sites. New methods incorporate splicing into sequence alignment.
Pairwise vs. true multiple alignment	(−). For ESTs (high sequencing error rate; random fragmentation), pairwise alignment can yield inconsistent results.
Lack of stable identifiers	(+/−). As experimental data are updated, AS databases should keep the same ID for a specific AS event. Such stability is largely lacking.
Paralogous genes	(+) in all current methods, but mostly in those that do not map genomic location or do not check all possible locations
Rigorous measures of evidence	(+/−). New methods measure the strength of experimental evidence for a specific splice form probabilistically.
Arbitrary cutoff thresholds	(+/−) in methods that use cutoffs (e.g., "99% identity")
Alignment size limitations	(−) in methods that can not align >10^2, >10^3 sequences
Biological interpretation factors	Comments
Spliceosome errors	Bioinformatics criteria for predicting whether AS events are functional exist, but should be validated experimentally in each case.
What's truly functional?	Just because a splice form is real (i.e., present in the cell) does not mean it's biologically functional. Conversely, even an mRNA isoform that makes a truncated, inactive protein might be a biologically valid form of functional regulation.
Defining the coding region	Predicting ORF in novel genes; splicing may change ORF.
Predicting impact in protein	Motif, signal, domain prediction, and functional effects
Predicting impact in UTR	Knowledge about effects of mRNA stability, localization, and other possible UTR sequences is incomplete.
Integrated query schema	AS databases should integrate splicing, gene annotation, and EST library data to enable query of tissue-specific, disease-specific AS, etc.
Assessing and correcting for bias	Our genome-wide view of function is under construction. Until then, we have unknown selection bias.

Adapted and rewritten from Modrek et al. (172). Note that we have listed items in each category loosely in descending order of importance.

been deposited in the public dbEST database *(45)*. Furthermore, since EST sequences have linked information about their source sample (commonly including tissue, cancer, gender, and developmental stage as well as organism information), many efforts have been made to discover biologically important patterns from those data.

Since ESTs are short fragments of mRNA transcripts, it has also been necessary to sequence full-length cDNAs each representing a single sub-cloned mRNA. In theory, if only one transcript for a given gene existed, fragment assembly methods could easily predict that full-length isoform from EST fragments. However, in the presence of alternative splicing, the assembly problem becomes much more challenging and in fact has no guaranteed solution; bioinformatics can only provide predictions. Thus there is no true substitute for experimental sequencing of full-length cDNA sequences. A number of programs and databases can provide predicted transcript sequences, e.g., RefSeq *(46–50)*, GenScan *(51)*, Twinscan *(52)*, ECgene *(14, 15)*, Fgenesh++ *(53)*, GenomeScan *(54)*, and others. However, gene prediction methods have suboptimal false-positive rates and (just as importantly) false-negative rates for prediction of alternatively spliced transcripts, so existing alternative splicing databases generally do not use such predicted sequences.

It is worth noting that there are relatively little data available concerning alternatively spliced protein products. The Swiss-Prot database *(55)* includes alternative splicing information for a small fraction of its protein records, 14,158 out of 217,551 (UniProtKB/Swiss-Prot Release 49.6 of 02-May-2006), via the VAR-SPLIC feature annotation *(56)*. Often it is unclear whether the SwissProt alternative splicing information is based on direct experimental identification of the *protein* isoform (i.e., direct detection of the precise protein form, as opposed to inferring the existence of this form from mRNA sequence). It has long been hoped that mass spectrometry would yield large-scale identification of alternative protein isoforms, but this has not yet happened.

Another important class of data is EST clustering *(13–15, 57–68)*. The main uncertainty about any EST is what gene it is from, and the initial focus of EST research was on finding novel genes. Based on similarity and genomic location, EST sequences are grouped into clusters to identify those that originated from the same gene. For reasons of space, EST clustering is not covered in detail in this chapter. In general, EST databases include clustering information. Among the major EST databases (UniGene *(57, 67, 68)*, TIGR Gene Indices *(61, 62, 64–66)*, STACKdb *(60, 63)*, and ECgene *(13–15)*), TIGR and ECgene provide transcript assembly but UniGene does not. UniGene clusters for human have the fastest update cycle and are updated every 2–3 months. Other databases have about a 1-year update cycle. The faster EST data increase, the shorter the average update cycle. For examples of alternative splicing detection, this chapter refers to UniGene as a primary database.

2.2. Completeness and Updating of Public Data

There are nearly 36 million EST sequences available from dbEST, of which the largest individual component is human (7.7 million EST sequences). After human, mouse is the next largest (4.7 million EST sequences), followed by rice, cattle, and frog (each more than 1 million EST sequences). It is worth emphasizing that the main limitation on alternative transcript discovery is the amount of EST and mRNA sequence coverage, both across all *regions* of a given gene and across all *tissues* and developmental/ activation states in which that gene is expressed. Even for human EST data, there are relatively few genes with high levels of EST coverage (e.g., 10× coverage) in each one of the individual tissues in which the gene is expressed. Thus it is likely that even human and mouse alternative splicing databases are far from complete. However, it is unclear that EST sequencing is an efficient method for obtaining a complete picture of alternative splicing across diverse tissues. Microarray approaches appear likely to take over this role in the future. For other organisms, the data are even more incomplete. Their alternative splicing data probably only scratch the surface, and thus represent a new frontier for alternative splicing discovery.

It is more difficult to obtain full-length cDNA sequences, and in general it appears that many genes lack full-length mRNA sequences. For example, of the 95,887 total UniGene clusters for human (Release #190), 69,242 clusters do not have a single mRNA sequence. (It should be emphasized that many of these clusters may represent microRNAs or other transcription units of unknown function.) To increase the number of mRNA sequences, several HTC sequencing projects have been launched *(69–74)*. GenBank *(75–80)* has established an HTC division for such data, and these HTC sequences will be moved to the appropriate taxonomic division after annotation.

Another essential category of public data is genomic sequence. In April 2003, the Human Genome Project announced the completion of the DNA reference sequence of *Homo sapiens (81–85)*. The latest human genome assembly version was released recently, NCBI build 36.1 (UCSC hg18). For analyses of other organisms, it is important to distinguish different stages of genome assembly sequences. In the draft stage, the assembly consists of small contigs not assembled into chromosomes. Some mammalian genome assemblies are released as these small contigs (termed as "scaffold" sequences). In a final stage, these contigs will be assembled further to make chromosome sequences. There are 41 published complete eukaryotic genomes and 607 eukaryotic ongoing genome projects as of today *(86–88)*. Major genome assemblies are available at NCBI *(89)* (http://www.ncbi.nlm.nih.gov/), UCSC Genome Bioinformatics *(90–92)* (http://genome.ucsc.edu/), and Ensembl *(93–101)* (http://www.ensembl.org/).

2.2.1. Downloading NCBI, UniGene Database

The UniGene database is available at NCBI. UniGene distributes clusters for almost 70 species, and the number of species is still growing. With each new release, the older versions are no longer kept and comparing results between versions can be challenging. Thus, it is best to use the latest version. Each UniGene database contains files with the following extensions: ".info" for statistics, ".data.gz" for cluster results and annotation, ".lib.info.gz" for EST library information, ".seq.all.gz" for all sequences, ".seq.uniq.gz" for representative sequences for each cluster, and ".retired.lst.gz" for retired sequences information of the previous version.

2.2.2. Downloading Genome Assemblies

Genome assembly sequences are available at NCBI, UCSC Genome Bioinformatics, and Ensembl. These three genome centers use different assembly names based on their own methods. UCSC Genome Bioinformatics reports another chromosome named "random." These "random" chromosomes files contain clones that are not yet finished or cannot be placed with certainty at a specific place on the chromosome or haplotypes that differ from the main assembly. In general, two archives are distributed: soft- and hard-masked. Repeat sequences are represented in lower case in soft-masked flat files and by Ns in hard-masked flat files. Usually, if you are working with nucleotides, soft-masked fasta sequences are recommended and hard-masked fasta sequences for research related to proteins.

2.2.3. Downloading Multiple Alignments

One of the best ways to study orthologous genes is to use multiple alignments of genome assemblies. UCSC Genome Bioinformatics distributes multiple alignments as MAF (multiple alignment formats) files *(102)*. Currently, 17way MAF is available at UCSC Genome Bioinformatics; that is, multiple alignments of 17 genome assemblies.

2.2.4. Parsing and Uploading into Relational Databases

The best way to store all data files is to use a relational database, such as MySQL. Most genome browsers use MySQL database to retrieve data. Data files are often distributed as MySQL-ready format. Database schemas should be carefully designed to avoid inefficient queries.

2.3. Downloading and Compiling Bioinformatics Tools

This section briefly summarizes the availability of the software described in the Methods section. BLAST *(103–108)* is a database search engine for sequence similarities and is available at NCBI, including pre-compiled binaries or source package (NCBI Toolkit). BLAT *(43)* is a sequence alignment tool using a hashing and indexing algorithm and is available from the web site of Jim Kent (http://

www.soe.ucsc.edu/~kent/). Most of the alignments shown in the UCSC genome browser are generated by BLAT. SPA *(109)* uses a probabilistic algorithm for spliced alignment. It was developed using the BLAT library and is available at http://www.biozentrum.unibas.ch/personal/nimwegen/cgi-bin/spa.cgi. SPA considers a consensus of GT...AG splice site. GMAP *(44)* is a recently developed fastest sequence alignment tool and is available at http://www.gene.com/share/gmap. GMAP considers a consensus of GT...AG splice sites. SIM4 *(110)* is based on dynamic programming and is available at http://globin.cse.psu.edu/html/docs/sim4.html. SIM4 considers a consensus of GT...AG splice sites. Calculation speed for BLAST and SIM4 is somewhat slow. In order to align all mRNA and EST sequences against a genome, SPA or GMAP would be primary choices because calculation speed is fast and splice site consensus is well detected. RepeatMasker is a program that screens DNA sequences for interspersed repeats and low complexity DNA sequences *(111)*. RepeatMasker is available at http://www.repeatmasker.org and additional packages such as Cross_Match *(112)*, WUBlast *(113)* and Repbase *(114, 115)* are needed. CLUSTALW *(116, 117)* is a multiple alignment tool and is available at http://www.ebi.ac.uk/clustalw/.

EMBOSS *(118, 119)* is a free open source analysis package specially developed for the needs of the molecular biology community. EMBOSS is available at http://emboss.sourceforge.net. Most of the data files are distributed as flat files, that is, simple text files. There are several modules for parsing flat files. For example, biopython is a collection of python modules for computational molecular biology and available at http://www.biopython.org *(120, 121)*.

3. Methods

Alternative splicing detection methods can be broken down into three distinct stages. First, transcript and genomic sequence data are pre-processed to eliminate possible problems. Second, transcript sequences are aligned to each other and to genomic sequence to obtain experimental "gene models." Third, alternative splicing and other transcript variation is detected and analyzed from these alignments. This section considers each of these phases separately.

3.1. Pre-Processing

Unfortunately, EST data contain a wide variety of experimental artefacts that can lead to incorrect prediction of alternative splicing (**Table 9.1**). **Note 2** discusses these artefacts at length. To reduce the number of such artefacts, several types of pre-processing are commonly performed. First, repetitive sequences must be masked to prevent artefactual alignment of sequences from different genes. Since repeat sequences are ubiquitous in mammalian genomes, and can also occur in EST and mRNA sequences, they should be masked using programs such as RepeatMasker. For example, in human UniGene data, almost 1 million EST sequences are discarded during the clustering step because they contain repeat sequences or other contaminating sequences, or are too short. Furthermore, even in the absence of repeat sequences, some EST sequences can align well to multiple locations in the genome. Such sequences are also considered ambiguous (putative "repeats"), and are removed from the analysis. PolyA/T tails pose a similar masking challenge. Since polyA/T sequences may not be found in the genome (they are added during mRNA processing), their presence can artefactually depress the apparent alignment quality of an EST vs. genomic sequence. For this reason, polyA/T tails are removed by some methods using EMBOSS, TRIMEST. It is important to remember that putative poly-A sequences may be genuine genomic sequence, so trimming them may itself introduce ambiguity or cause artefacts.

3.2. Aligning Transcript Sequences to Genome Assembly

Alignment of transcript sequences to the genome sequence is important for reducing incorrect mixing of paralogous sequences, and for validating putative splices (*see* **Note 2**). Several methods are commonly used to align all EST and mRNA sequences against complete genome sequences. Previously, researchers used a combination of BLAST and SIM4 *(12, 33)*. However, BLAST can take several months on human ESTs vs. the human genome (on a single CPU). BLAT is a faster alignment tool than BLAST, but gives fragmented alignments when alignment quality is imperfect (e.g., due to EST sequencing error and fragmentation). To get better results, researchers have combined BLAT and SIM4 *(13–15)*, because SIM4 considers GT…AG canonical consensus splice sites; gaps (splices) in the alignment should normally match this consensus. To simplify this process, recently GMAP and SPA have been introduced. Both alignment tools consider GT…AG consensus splice sites, and also run faster. SPA was developed on the foundation of the BLAT library; GMAP shows superior speed in alignment. They can produce slightly different alignments. Since ESTs can have high rates of sequencing error (up to 10%) within local regions, and are also randomly fragmented, pairwise alignment (i.e., aligning each EST individually against the genomic sequence) can give suboptimal results compared with multiple sequence alignment (i.e., all ESTs for a gene simultaneously against the genomic sequence). Multiple alignment methods that can deal with the branching structure of alternative splicing (e.g., Partial Order Alignment, POA) *(122–126)* can improve AS detection sensitivity and reduce alignment artefacts.

3.3. Alignment Filtering

Most alignment tools give several possible alignments for a given EST sequence, either to different genomic locations, or suboptimal alignments to the same genomic locus. AS detection methods generally screen out ambiguous cases (multiple, indistinguishable hits), and choose only the best hit as measured by percent identity and alignment coverage. If there are many high-scoring "best hits," these alignments should be dealt with carefully, because such sequences may come from repeat sequences. Strictly for the purposes of detecting alternative splicing, all unspliced alignments can be removed at this step. Most ESTs with unspliced alignments are from the 3′UTR, but some may also represent experimental artefacts such as genomic contamination (*see* **Note 2**). Additionally, the total aligned region should be larger than 100bp after repeat masking, with a minimum of 96% identity and 50% coverage to reduce artefacts (*see* **Note 3** for more details). More than 95% of intron sequences have GT…AG consensus, therefore most alternative splicing patterns can be found without losing information even if we use only the GT…AG canonical splice site. The genome alignments can be further corrected by sequence alignment tools such as SIM4, GMAP and SPA. These alignment tools consider GT…AG consensus to generate valid alignments. As an alternative to such arbitrary cutoffs, van Nimwegen and co-workers have developed a rigorous probabilistic method for measuring the quality of EST evidence for alternative splicing *(109)*.

3.4. Detection and Classification of Candidate Alternative Splicing

Procedures to detect alternative splicing patterns from genome alignments can be subdivided into two major steps: identification of individual alternative splicing events and construction of alternative "gene models" incorporating all these data. An individual alternative splicing event consists of a pair of mutually exclusive splicing events. Two splices are mutually exclusive if their genomic intervals overlap. (These two splices could not co-occur in the same transcript.) It should be noted that this "mutually exclusive splicing" criterion excludes many artefacts, such as genomic contamination. This simple criterion can be made more or less exact. At one extreme, any pair of overlapping splices is reported; at the other extreme one can require that the pair of splices must share one splice site but differ at the other splice site. This narrow definition works not only for alternative-5′ and alternative-3′ splicing, but also for most other cases of alternative exon usage and even alternative initiation and polyadenylation (which typically also cause a change in splicing). Exons can be represented as alignment blocks in mRNA-genome and EST-genome map. Different types of alternative splicing can be distinguished by refining the basic rule for finding mutually exclusive splices into subtypes based upon the specific relation of the splices to each other and to overlapping exons, following **Fig. 9.1**. For example, if an exonic region is located in the middle of an intronic region, it is a skipped exon. On the other hand, if an intronic region is located in the middle of an exonic region, it is a retained intron. Thus, one can compare all exon coordinates with all intron coordinates in order to detect exon skipping and intron retention.

Construction of gene models involves integration of all the evidence, beginning from splices and exons and ultimately producing full-length isoforms. Whereas early methods relied on describing gene models purely in terms of genomic intervals (e.g., start and stop coordinates for an exon or intron), Heber and co-workers demonstrated that graph representations ("splicing graphs") have many advantages for dealing with the branched structure of multiple isoform models *(32)*. However, it is important to stress that there is no guaranteed optimal solution for prediction of alternatively spliced isoforms from raw fragment data (e.g., ESTs or microarray probe data). Bioinformatics methods fall broadly into two classes: "liberal" methods (e.g., Heber et al.) that generate *all possible* isoforms (by producing all possible combinations of the observed alternative splicing events); and more "conservative" methods based on maximum likelihood, that seek to predict the *minimal set of isoforms* capable of explaining the observed experimental data *(127)*. This is a complex topic; for a detailed review see the chapter by Xing and Lee in this volume.

3.5. Alternative Splicing Database Schema and User Analysis

Careful design of the alternative splicing database schema is essential to enable powerful "data mining" of these biological data. Several principles should be emphasized. First, the database should fully integrate biological data (e.g., gene annotation information) with the alternative splicing results. Second, the schema design should link bioinformatics results (e.g., detection of splices and exons) with the underlying experimental evidence for these results, as connected but separate tables. **Figure 9.2** illustrates an example alternative splicing database that follows this "Bayesian" schema design, pairing each bioinformatics *interpretation* table (e.g., the "splice" table) with an experimental *observation* table (e.g., the "splice_obs" table) that makes it possible to trace in detail the strength and sources of the experimental evidence for each result. Collectively, these design principles make it possible to query the database for tissue-specific alternative splicing (see the following), and other interesting biological questions. In the future, new types of alternative splicing can be analyzed from the same database, by querying data such as the intron coordinates stored in the "splice" table. Pairwise comparison of intron coordinates or exon-intron coordinates can be performed by SQL queries using this database schema.

As one example application of this schema, one can mine tissue-specific alternative splicing patterns from the database. Since the database connects each splice to individual ESTs and thus to

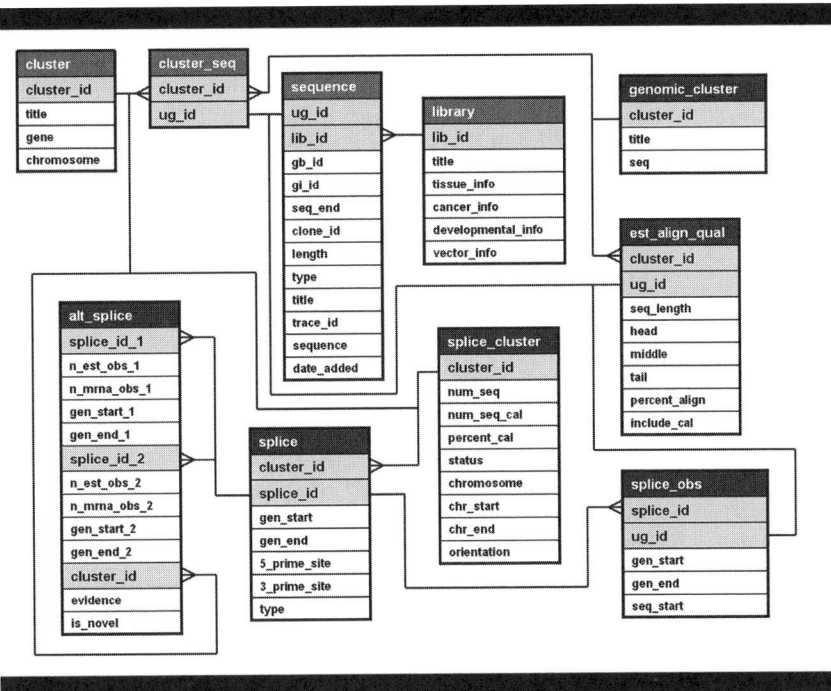

Fig. 9.2. Database scheme for UniGene & ASAP. Upper left four tables represents UniGene database and others ASAP database. Modified and adapted from ASAP database *(17)*.

library source information (e.g., tissue type), we can directly mine tissue-specific, cancer-specific, or developmental state-specific alternative splicing simply by running queries on the database. These queries count the number of ESTs for a given splice subdivided by tissue (or tumor vs. normal, etc.), and perform statistical tests to identify statistically significant differences. Standard tests such as Fisher's Exact Test are available in software packages such as R (http://www.r-project.org/). Such queries have been used to identify hundreds of alternative splicing events that are regulated in a tissue-specific manner *(38)*; these results have been experimentally validated. For example, this analysis predicted a novel kidney-specific isoform of *WNK1* that disrupts its kinase domain; subsequent experimental studies have shown that this form is expressed specifically in distal tubules and is thought to be involved in PHA type II hypertension *(128, 129)*.

Similarly, such a database schema can be used to automate design of probe sequences for primer sequences for RT-PCR detection of alternative splicing, or microarray detection of alternative splicing *(16, 27, 130–137)*. Many biologists are interested in verifying bioinformatics predictions of alternative splicing using RT-PCR (Reverse Transcription Polymerase Chain Reaction). RT-PCR is a modification of PCR for amplifying data containing

RNA by converting it into DNA, then amplifying it. PRIMER3 *(138)* can be used to design adequate primers. New microarray designs include both exon and splice junction probes that can detect alternative splicing, and typically include systematic coverage of oligonucleotides for all exons in a gene, both constitutive exons and alternative exons. However, in addition to such "all-exon" probes, research has shown that splice junction probes greatly improve sensitivity for detecting alternative splicing. Inclusion junction probes give sensitive detection of inclusion of alternatively spliced exons, and exclusion junction probes give sensitive detection of exon skipping isoforms. These junction probes have been shown to yield good measurements of the level of alternative splicing isoforms. These experiments can be performed using cell lines from various tissues or cancer/normal cells depending on the purpose of the experiments. However, this is a complex topic in its own right; see *(16, 27, 130–136)* for further details.

Comparative genomics data (multiple genome alignments) are another important source of information for biologists studying alternative splicing. One good way to tap this information is via online genome browsers. For example, the UCSC Genome Browser *(90–92, 139)* includes many third party tracks and multiple alignments for various species. DNA fragments including alternative exon with color coding can be viewed. Also, multiple alignments from various species, currently 17 species, can give useful information about gene evolution or conservation between species.

4. Notes

A wide variety of experimental and bioinformatics issues can cause false-positive and -negative errors (**Table 9.1**). Existing alternative splicing databases have been designed to minimize such artefacts. This section reviews types of artefacts and the technical issues for catching and reducing them.

1. More challenging types of transcript variation. Recently there has been growing interest in detecting additional forms of transcript variation besides alternative splicing, such as alternative initiation and polyadenylation. These are somewhat more challenging to predict and to distinguish definitively from true alternative splicing which can produce somewhat similar patterns. In order to find alternative initiation and termination events, we need transcript assemblies that have full-length cDNA evidences because both events can not be found using only "short" EST sequences. Nevertheless, several studies have been reported based on transcript assemblies of mRNA and EST alignments due to the lack of full-length

cDNA. These results may contain false-positive results if the transcript assemblies are comprised of EST evidences only. Before suitable 3′ UTR databases were developed, detection of alternative polyadenylation events was focused on alignments of EST against mRNA sequences *(141–143)*. However, evidences from sequence itself are not enough to find alternative polyadenylation events because those polyA sequences may be part of genomic sequences or sequence errors. Legendre et al. developed the ERPIN program to enhance the accuracy of polyA site detection using both upstream and downstream elements of cleavage sites, which distinguish true polyadenylation sites from randomly occurring AAUAAA hexamers *(144)*. By combined use of UTR databases *(145–149)* and genomic alignments, genome-wide studies on alternative polyadenylation have been performed *(150, 151)*.

Another frontier is the investigation of intron retention in plants. In the past, many AS databases have ignored intron retention because such EST patterns can arise as experimental artefacts, e.g., by incomplete splicing of mRNA, or genomic contamination. However, recent researches on plant alternative splicing show that more than half of alternative splicing patterns are intron retention *(24, 152)*. Also, researches on human intron retention have been performed *(153–155)*. More experimental studies will be required to validate the possible biological functions of such intron retention events.

2. Removal of artefacts and contaminated libraries. EST data contain significant experimental artefacts that must be screened out during genome-wide analyses of alternative transcript variation. Genomic contamination, incomplete mRNA processing, and library preparation artefacts can produce the appearance of "alternative transcripts," particularly intron retention or skipping. In general, methods have sought to screen these out by setting conservative criteria. For example, rather than simply reporting every pair of transcripts that differ from each other, it is common to restrict the analysis to validated splices, and specifically to pairs of validated splices that share one splice site but differ at the other splice site. This very specific requirement excludes the above types of artefacts while detecting normal types of alternative splicing, although it may cause false negatives for certain alternative transcript events. In some cases it is possible to directly identify specific EST libraries that are contaminated by artefacts *(140)*.

3. Fused genes. Chromosome translocation and gene fusion are frequent events in various species and are often the cause of many types of tumor. The most famous example is the

fusion protein BCR-ABL *(156–160)*, the target protein of the drug *gleevec* treating chronic myeloid leukemia (CML). CML is associated in most cases with a chromosomal translocation between chromosome 9 and 22 that creates the Philadelphia chromosome. Besides chromosomal translocation *(161–163)*, two adjacent independent genes may be co-transcribed and the intergenic region spliced out so that the resulting fused transcript possesses exons from both genes *(164–166)*. These two kinds of gene fusion can cause trans-splicing and co-transcription with intergenic splicing. Several studies seeking gene fusion events have been performed *(161, 162, 167–171)*. These events can interfere with detection of alternative splicing. For example, BCR-ABL fused sequences can be aligned in two genomic locations from BCR and ABL genes. We should find fused genes after aligning sequences against the genome.

Acknowledgments

This work was funded by a Dreyfus Foundation Teacher-Scholar Award to C.J.L., and by the National Institutes of Health through the NIH Roadmap for Medical Research, Grant U54 RR021813 entitled Center for Computational Biology (CCB). Information on the National Centers for Biomedical Computing can be obtained from http://nihroadmap.nih.gov/bioinformatics.

References

1. Gilbert, W. (1978) *Nature* 271, 501.
2. Black, D. L. (2003) *Annu Rev Biochem* 72, 291–336.
3. Black, D. L., Grabowski, P. J. (2003) *Prog Mol Subcell Biol* 31, 187–216.
4. Brett, D., Pospisil, H., Valcarcel, J., et al. (2002) *Nat Genet* 30, 29–30.
5. Burke, J., Wang, H., Hide, W., et al. (1998) *Genome Res* 8, 276–290.
6. Croft, L., Schandorff, S., Clark, F., et al. (2000) *Nat Genet* 24, 340–341.
7. Grabowski, P. J., Black, D. L. (2001) *Prog Neurobiol* 65, 289–308.
8. Graveley, B. R. (2001) *Trends Genet* 17, 100–107.
9. Gupta, S., Zink, D., Korn, B., et al. (2004) *Bioinformatics* 20, 2579–2585.
10. Huang, H. D., Horng, J. T., Lee, C. C., et al. (2003) *Genome Biol* 4, R29.
11. Huang, Y. H., Chen, Y. T., Lai, J. J., et al. (2002) *Nucleic Acids Res* 30, 186–190.
12. Kan, Z., Rouchka, E. C., Gish, W. R., et al. (2001) *Genome Res* 11, 889–900.
13. Kim, N., Shin, S., Lee, S. (2004) *Nucleic Acids Res* 32, W181–186.
14. Kim, N., Shin, S., Lee, S. (2005) *Genome Res* 15, 566–576.
15. Kim, P., Kim, N., Lee, Y., et al. (2005) *Nucleic Acids Res* 33, D75–79.
16. Kochiwa, H., Suzuki, R., Washio, T., et al. (2002) *Genome Res* 12, 1286–1293.
17. Lee, C., Atanelov, L., Modrek, B., et al. (2003) *Nucleic Acids Res* 31, 101–105.
18. Lee, C., Wang, Q. (2005) *Brief Bioinform* 6, 23–33.
19. Maniatis, T., Tasic, B. (2002) *Nature* 418, 236–243.

20. Mironov, A. A., Fickett, J. W., Gelfand, M. S. (1999) *Genome Res* 9, 1288–1293.
21. Modrek, B., Lee, C. J. (2003) *Nat Genet* 34, 177–180.
22. Modrek, B., Resch, A., Grasso, C., et al. (2001) *Nucleic Acids Res* 29, 2850–2859.
23. Thanaraj, T. A., Stamm, S., Clark, F., et al. (2004) *Nucleic Acids Res* 32 Database issue, D64–69.
24. Wang, B. B., Brendel, V. (2006) *Proc Natl Acad Sci U S A* 103, 7175–7180.
25. Caceres, J. F., Kornblihtt, A. R. (2002) *Trends Genet* 18, 186–193.
26. Levanon, E. Y., Sorek, R. (2003) *TARGETS* 2, 109–114.
27. Li, C., Kato, M., Shiue, L., et al. (2006) *Cancer Res* 66, 1990–1999.
28. Bailey, L. C., Jr., Searls, D. B., Overton, G. C. (1998) *Genome Res* 8, 362–376.
29. Boue, S., Vingron, M., Kriventseva, E., et al. (2002) *Bioinformatics* 18 Suppl 2, S65–S73.
30. Gelfand, M. S., Dubchak, I., Dralyuk, I., et al. (1999) *Nucleic Acids Res* 27, 301–302.
31. Gopalan, V., Tan, T. W., Lee, B. T., et al. (2004) *Nucleic Acids Res* 32 Database issue, D59–63.
32. Heber, S., Alekseyev, M., Sze, S. H., et al. (2002) *Bioinformatics* 18 Suppl 1, S181–188.
33. Kan, Z., States, D., Gish, W. (2002) *Genome Res* 12, 1837–1845.
34. Kawamoto, S., Yoshii, J., Mizuno, K., et al. (2000) *Genome Res* 10, 1817–1827.
35. Krause, A., Haas, S. A., Coward, E., et al. (2002) *Nucleic Acids Res* 30, 299–300.
36. Pospisil, H., Herrmann, A., Bortfeldt, R. H., et al. (2004) *Nucleic Acids Res* 32 Database issue, D70–74.
37. Xie, H., Zhu, W. Y., Wasserman, A., et al. (2002) *Genomics* 80, 326–330.
38. Xu, Q., Modrek, B., Lee, C. (2002) *Nucleic Acids Res* 30, 3754–3766.
39. Yeo, G., Holste, D., Kreiman, G., et al. (2004) *Genome Biol* 5, R74.
40. Zavolan, M., Kondo, S., Schonbach, C., et al. (2003) *Genome Res* 13, 1290–300.
41. Zavolan, M., van Nimwegen, E., Gaasterland, T. (2002) *Genome Res* 12, 1377–1385.
42. Nagasaki, H., Arita, M., Nishizawa, T., et al. (2006) *Bioinformatics* 22, 1211–1216.
43. Kent, W. J. (2002) *Genome Res* 12, 656–664.
44. Wu, T. D., Watanabe, C. K. (2005) *Bioinformatics* 21, 1859–1875.
45. Boguski, M. S., Lowe, T. M., Tolstoshev, C. M. (1993) *Nat Genet* 4, 332–333.
46. Maglott, D. R., Katz, K. S., Sicotte, H., et al. (2000) *Nucleic Acids Res* 28, 126–128.
47. Pruitt, K. D., Katz, K. S., Sicotte, H., et al. (2000) *Trends Genet* 16, 44–47.
48. Pruitt, K. D., Maglott, D. R. (2001) *Nucleic Acids Res* 29, 137–140.
49. Pruitt, K. D., Tatusova, T., Maglott, D. R. (2005) *Nucleic Acids Res* 33, D501–504.
50. Pruitt, K. D., Tatusova, T., Maglott, D. R. (2003) *Nucleic Acids Res* 31, 34–37.
51. Burge, C., Karlin, S. (1997) *J Mol Biol* 268, 78–94.
52. Korf, I., Flicek, P., Duan, D., et al. (2001) *Bioinformatics* 17 Suppl 1, S140–148.
53. Salamov, A. A., Solovyev, V. V. (2000) *Genome Res* 10, 516–522.
54. Yeh, R. F., Lim, L. P., Burge, C. B. (2001) *Genome Res* 11, 803–816.
55. Shomer, B. (1997) *Comput Appl Biosci* 13, 545–547.
56. Kersey, P., Hermjakob, H., Apweiler, R. (2000) *Bioinformatics* 16, 1048–1049.
57. Boguski, M. S., Schuler, G. D. (1995) *Nat Genet* 10, 369–371.
58. Burke, J., Davison, D., Hide, W. (1999) *Genome Res* 9, 1135–1142.
59. Cariaso, M., Folta, P., Wagner, M., et al. (1999) *Bioinformatics* 15, 965–973.
60. Christoffels, A., van Gelder, A., Greyling, G., et al. (2001) *Nucleic Acids Res* 29, 234–238.
61. Liang, F., Holt, I., Pertea, G., et al. (2000) *Nat Genet* 25, 239–240.
62. Liang, F., Holt, I., Pertea, G., et al. (2000) *Nucleic Acids Res* 28, 3657–3665.
63. Miller, R. T., Christoffels, A. G., Gopalakrishnan, C., et al. (1999) *Genome Res* 9, 1143–1155.
64. Pertea, G., Huang, X., Liang, F., et al. (2003) *Bioinformatics* 19, 651–652.
65. Quackenbush, J., Cho, J., Lee, D., et al. (2001) *Nucleic Acids Res* 29, 159–164.
66. Quackenbush, J., Liang, F., Holt, I., et al. (2000) *Nucleic Acids Res* 28, 141–145.
67. Schuler, G. D. (1997) *J Mol Med* 75, 694–698.
68. Schuler, G. D., Boguski, M. S., Stewart, E. A., et al. (1996) *Science* 274, 540–546.
69. Bono, H., Kasukawa, T., Furuno, M., et al. (2002) *Nucleic Acids Res* 30, 116–118.
70. Carter, M. G., Piao, Y., Dudekula, D. B., et al. (2003) *C R Biol* 326, 931–940.
71. Gerhard, D. S., Wagner, L., Feingold, E. A., et al. (2004) *Genome Res* 14, 2121–2127.

72. Hayashizaki, Y. (2003) *C R Biol* 326, 923–929.
73. Lamesch, P., Milstein, S., Hao, T., et al. (2004) *Genome Res* 14, 2064–2069.
74. Sasaki, D., Kawai, J. (2004) *Tanpakushitsu Kakusan Koso* 49, 2627–2634.
75. Benson, D. A., Karsch-Mizrachi, I., Lipman, D. J., et al. (2000) *Nucleic Acids Res* 28, 15–18.
76. Benson, D. A., Karsch-Mizrachi, I., Lipman, D. J., et al. (2002) *Nucleic Acids Res* 30, 17–20.
77. Benson, D. A., Karsch-Mizrachi, I., Lipman, D. J., et al. (2003) *Nucleic Acids Res* 31, 23–27.
78. Benson, D. A., Karsch-Mizrachi, I., Lipman, D. J., et al. (2006) *Nucleic Acids Res* 34, D16–20.
79. Benson, D. A., Karsch-Mizrachi, I., Lipman, D. J., et al. (2004) *Nucleic Acids Res* 32 Database issue, D23–26.
80. Burks, C., Fickett, J. W., Goad, W. B., et al. (1985) *Comput Appl Biosci* 1, 225–233.
81. Arnold, J., Hilton, N. (2003) *Nature* 422, 821–822.
82. Carroll, S. B. (2003) *Nature* 422, 849–857.
83. Collins, F. S., Green, E. D., Guttmacher, A. E., et al. (2003) *Nature* 422, 835–847.
84. Collins, F. S., Morgan, M., Patrinos, A. (2003) *Science* 300, 286–290.
85. Frazier, M. E., Johnson, G. M., Thomassen, D. G., et al. (2003) *Science* 300, 290–293.
86. Bernal, A., Ear, U., Kyrpides, N. (2001) *Nucleic Acids Res* 29, 126–127.
87. Kyrpides, N. C. (1999) *Bioinformatics* 15, 773–774.
88. Liolios, K., Tavernarakis, N., Hugenholtz, P., et al. (2006) *Nucleic Acids Res* 34, D332–334.
89. Jenuth, J. P. (2000) *Methods Mol Biol* 132, 301–312.
90. Hinrichs, A. S., Karolchik, D., Baertsch, R., et al. (2006) *Nucleic Acids Res* 34, D590–598.
91. Karolchik, D., Baertsch, R., Diekhans, M., et al. (2003) *Nucleic Acids Res* 31, 51–54.
92. Kent, W. J., Sugnet, C. W., Furey, T. S., et al. (2002) *Genome Res* 12, 996–1006.
93. Birney, E. (2003) *Cold Spring Harb Symp Quant Biol* 68, 213–215.
94. Birney, E., Andrews, D., Bevan, P., et al. (2004) *Nucleic Acids Res* 32, D468–470.
95. Birney, E., Andrews, D., Caccamo, M., et al. (2006) *Nucleic Acids Res* 34, D556–561.
96. Birney, E., Andrews, T. D., Bevan, P., et al. (2004) *Genome Res* 14, 925–928.
97. Butler, D. (2000) *Nature* 406, 333.
98. Clamp, M., Andrews, D., Barker, D., J., et al. (2003) *Nucleic Acids Res* 31, 38–42.
99. Hammond, M. P., Birney, E. (2004) *Trends Genet* 20, 268–272.
100. Hubbard, T., Andrews, D., Caccamo, M., et al. (2005) *Nucleic Acids Res* 33, D447–453.
101. Hubbard, T., Barker, D., Birney, E., et al. (2002) *Nucleic Acids Res* 30, 38–41.
102. Blanchette, M., Kent, W. J., Riemer, C., et al. (2004) *Genome Res* 14, 708–715.
103. Altschul, S. F., Gish, W., Miller, W., et al. (1990) *J Mol Biol* 215, 403–410.
104. Altschul, S. F., Madden, T. L., Schaffer, A. A., et al. (1997) *Nucleic Acids Res* 25, 3389–3402.
105. Gish, W., States, D. J. (1993) *Nat Genet* 3, 266–272.
106. Madden, T. L., Tatusov, R. L., Zhang, J. (1996) *Methods Enzymol* 266, 131–141.
107. Zhang, J., Madden, T. L. (1997) *Genome Res* 7, 649–656.
108. Zhang, Z., Schwartz, S., Wagner, L., et al. (2000) *J Comput Biol* 7, 203–214.
109. van Nimwegen, E., Paul, N., Sheridan, A., et al. (2006) *PLoS Genetics* 2, 587–605.
110. Florea, L., Hartzell, G., Zhang, Z., et al. (1998) *Genome Res* 8, 967–974.
111. Bedell, J. A., Korf, I., Gish, W. (2000) *Bioinformatics* 16, 1040–1041.
112. Lee, W. H., Vega, V. B. (2004) *Bioinformatics* 20, 2863–2864.
113. Lopez, R., Silventoinen, V., Robinson, S., et al. (2003) *Nucleic Acids Res* 31, 3795–3798.
114. Jurka, J. (2000) *Trends Genet* 16, 418–420.
115. Jurka, J., Kapitonov, V. V., Pavlicek, A., et al. (2005) *Cytogenet Genome Res* 110, 462–467.
116. Aiyar, A. (2000) *Methods Mol Biol* 132, 221–241.
117. Thompson, J. D., Higgins, D. G., Gibson, T. J. (1994) *Nucleic Acids Res* 22, 4673–4680.
118. Olson, S. A. (2002) *Brief Bioinform* 3, 87–91.
119. Rice, P., Longden, I., Bleasby, A. (2000) *Trends Genet* 16, 276–277.
120. Hamelryck, T., Manderick, B. (2003) *Bioinformatics* 19, 2308–2310.
121. Mangalam, H. (2002) *Brief Bioinform* 3, 296–302.
122. Grasso, C., Lee, C. (2004) *Bioinformatics* 20, 1546–1556.
123. Grasso, C., Modrek, B., Xing, Y., et al. (2004) *Pac Symp Biocomput* 29–41.

124. Grasso, C., Quist, M., Ke, K., et al. (2003) *Bioinformatics* 19, 1446–1448.
125. Lee, C. (2003) *Bioinformatics* 19, 999–1008.
126. Lee, C., Grasso, C., Sharlow, M. F. (2002) *Bioinformatics* 18, 452–464.
127. Xing, Y., Resch, A., Lee, C. (2004) *Genome Res* 14, 426–441.
128. Hollenberg, N. K. (2002) *Curr Hypertens Rep* 4, 267.
129. Wilson, F. H., Disse-Nicodeme, S., Choate, K. A., et al. (2001) *Science* 293, 1107–1112.
130. Huang, X., Li, J., Lu, L., et al. (2005) *J Androl* 26, 189–196.
131. Le, K., Mitsouras, K., Roy, M., et al. (2004) *Nucleic Acids Res* 32, e180.
132. Pan, Q., Saltzman, A. L., Kim, Y. K., et al. (2006) *Genes Dev* 20, 153–158.
133. Pan, Q., Shai, O., Misquitta, C., et al. (2004) *Mol Cell* 16, 929–941.
134. Shai, O., Morris, Q. D., Blencowe, B. J., et al. (2006) *Bioinformatics* 22, 606–613.
135. Wang, H., Hubbell, E., Hu, J. S., et al. (2003) *Bioinformatics* 19 Suppl 1, i315–322.
136. Zheng, C. L., Kwon, Y. S., Li, H. R., et al. (2005) *Rna* 11, 1767–1776.
137. Kim, N., Lim, D., Lee, S., et al. (2005) *Nucleic Acids Res* 33, W681–685.
138. Rozen, S., Skaletsky, H. (2000) *Methods Mol Biol* 132, 365–386.
139. Karolchik, D., Hinrichs, A. S., Furey, T. S., et al. (2004) *Nucleic Acids Res* 32 Database issue, D493–496.
140. Sorek, R., Safer, H. M. (2003) *Nucleic Acids Res* 31, 1067–1074.
141. Edwalds-Gilbert, G., Veraldi, K. L., Milcarek, C. (1997) *Nucleic Acids Res* 25, 2547–2561.
142. Gautheret, D., Poirot, O., Lopez, F., et al. (1998) *Genome Res* 8, 524–530.
143. Pauws, E., van Kampen, A. H., van de Graaf, S. A., et al. (2001) *Nucleic Acids Res* 29, 1690–1694.
144. Legendre, M., Gautheret, D. (2003) *BMC Genomics* 4, 7.
145. Mignone, F., Grillo, G., Licciulli, F., et al. (2005) *Nucleic Acids Res* 33, D141–146.
146. Pesole, G., Liuni, S., Grillo, G., et al. (1999) *Nucleic Acids Res* 27, 188–191.
147. Pesole, G., Liuni, S., Grillo, G., et al. (2000) *Nucleic Acids Res* 28, 193–196.
148. Pesole, G., Liuni, S., Grillo, G., et al. (2002) *Nucleic Acids Res* 30, 335–340.
149. Pesole, G., Liuni, S., Grillo, G., et al. (1998) *Nucleic Acids Res* 26, 192–195.
150. Beaudoing, E., Freier, S., Wyatt, J. R., et al. (2000) *Genome Res* 10, 1001–1010.
151. Beaudoing, E., Gautheret, D. (2001) *Genome Res* 11, 1520–1526.
152. Ner-Gaon, H., Halachmi, R., Savaldi-Goldstein, S., et al. (2004) *Plant J* 39, 877–885.
153. Galante, P. A., Sakabe, N. J., Kirschbaum-Slager, N., et al. (2004) *Rna* 10, 757–765.
154. Hiller, M., Huse, K., Platzer, M., et al. (2005) *Nucleic Acids Res* 33, 5611–5621.
155. Michael, I. P., Kurlender, L., Memari, N., et al. (2005) *Clin Chem* 51, 506–515.
156. Mauro, M. J., Druker, B. J. (2001) *Curr Oncol Rep* 3, 223–227.
157. Mauro, M. J., Druker, B. J. (2001) *Oncologist* 6, 233–238.
158. Mauro, M. J., O'Dwyer, M., Heinrich, M. C., et al. (2002) *J Clin Oncol* 20, 325–334.
159. Mauro, M. J., O'Dwyer, M. E., Druker, B. J. (2001) *Cancer Chemother Pharmacol* 48 Suppl 1, S77–78.
160. Rowley, J. D. (2001) *Nat Rev Cancer* 1, 245–250.
161. Finta, C., Warner, S. C., Zaphiropoulos, P. G. (2002) *Histol Histopathol* 17, 677–682.
162. Finta, C., Zaphiropoulos, P. G. (2002) *J Biol Chem* 277, 5882–5890.
163. Flouriot, G., Brand, H., Seraphin, B., et al. (2002) *J Biol Chem* 277, 26244–26251.
164. Communi, D., Suarez-Huerta, N., Dussossoy, D., et al. (2001) *J Biol Chem* 276, 16561–16566.
165. Magrangeas, F., Pitiot, G., Dubois, S., et al. (1998) *J Biol Chem* 273, 16005–16010.
166. Roux, M., Leveziel, H., Amarger, V. (2006) *BMC Genomics* 7, 71.
167. Dandekar, T., Sibbald, P. R. (1990) *Nucleic Acids Res* 18, 4719–4725.
168. Kim, N., Kim, P., Nam, S., et al. (2006) *Nucleic Acids Res* 34, D21–24.
169. Kim, N., Shin, S., Cho, K.-H., et al. (2004) *Genomics & Informatics* 2, 61–66.
170. Romani, A., Guerra, E., Trerotola, M., et al. (2003) *Nucleic Acids Res* 31, e17.
171. Shao, X., Shepelev, V., Fedorov, A. (2006) *Bioinformatics* 22, 692–698.
172. Modrek, B., Lee, C. (2002) *Nat Genet* 30, 13–19.

Chapter 10

Reconstruction of Full-Length Isoforms from Splice Graphs

Yi Xing and Christopher Lee

Abstract

Most alternative splicing events in human and other eukaryotic genomes are detected using sequence fragments produced by high throughput genomic technologies, such as EST sequencing and oligonucleotide microarrays. Reconstructing full-length transcript isoforms from such sequence fragments is a major interest and challenge for computational analyses of pre-mRNA alternative splicing. This chapter describes a general graph-based approach for computational inference of full-length isoforms.

Key words: alternative splicing, splice graph, ESTs, microarray, dynamic programming, sequence assembly.

1. Introduction

Researchers often observe a new splice variant and ask, "What does this splice variant do to the protein product of this gene? What is the functional consequence of this alternative splicing event?" Answers to such questions, however, are not easy to get. This is because most novel splice variants are detected using sequence fragments (see details in the previous chapter). On the other hand, to elucidate the functional impact of alternative splicing on proteins, we need to know the full-length sequences of the resulting protein isoforms. This is commonly known as the "isoform problem" in analyses of alternative splicing (1–3). This section describes a general graph-based approach for inferring full-length isoforms of novel splice variants.

2. The Isoform Problem

The vast majority of alternative splicing events in humans and other eukaryotic species are discovered by high throughput genomic methodologies, such as EST sequencing and splicing-sensitive oligonucleotide microarrays *(4)*. These genomics approaches do not detect full-length mRNA transcripts. Instead, they target a specific region of a gene. For example, ESTs are widely used for discoveries of novel splice variants. ESTs are shotgun fragments of full-length mRNA sequences. We can use ESTs to infer local information about the gene structure (e.g., whether one exon is included or skipped from the transcript), but we cannot directly infer full-length isoform sequences from ESTs. In fact, over 80% of splice variants in the human transcriptome are detected using EST data *(5)*, with no corresponding full-length isoforms immediately available. The increasingly popular splicing-sensitive microarrays, on which each probe detects signals from a specific exon or an exon-exon junction, generates even more fragmented information than ESTs. In addition, many genes have multiple alternatively spliced regions. Multiple alternative splicing events of a single gene can be combined in a complex manner, which further complicates the inference of full-length isoforms from individual alternative splicing events.

The isoform problem can be formulated as a sequence assembly problem. Given all the sequence observations for a gene, including full-length (e.g., cDNAs) and fragmentary (e.g., ESTs and microarray probe intensities) sequences, the goal is to infer the most likely set of full-length isoforms that explain the observed data. Specifically, we need to assemble multiple consensus sequences from a mixture of fragmentary sequences, corresponding to multiple full-length isoforms of a gene.

3. Splice Graph and Maximum Likelihood Reconstruction of Full-Length Isoforms

Over the years, several groups have developed computational methods for constructing full-length isoforms (reviewed in *(2)*). In a key study published in 2002, Heber and colleagues introduced a new approach of representing gene structure and alternative splicing, which is often referred to as the "splice graph" *(3)*. This section describes how to use splice graphs to infer full-length isoforms from sequence fragments.

3.1. Splice Graph Representation of Gene Structure and Alternative Splicing

Conventionally, a gene structure is represented as a linear string of exons (**Fig. 10.1A**), ordered according to the positions of exons from 5′ to 3′. This simple representation, however, is insufficient for the analyses of alternative splicing and the inference of full-length isoforms. By definition, alternative splicing introduces branches to the gene structure, disrupting the validity of a *single* linear order of all exons.

In a pioneering study, Heber and colleagues introduced the concept of "splice graph" *(3)*, which is a directed acyclic graph representation of gene structure. In the splice graph, each exon is represented as a node, and each splice junction is represented as a directed edge between two nodes (i.e., exons) (**Fig. 10.1B**). Different types of alternative splicing events, such as exon skipping, alternative 5′/3′ splicing, and intron retention, can be easily represented using splice graphs. **Fig. 10.2** shows the splice graph of a multi-exon human gene *TCN1*. The observed exon skipping event of exon 2 is represented as a directed edge from node 1 to node 3. Similarly, the observed exon skipping event of exon 5/6 is represented as a directed edge from node 4 to node 7. One EST from unspliced genomic DNA is represented as a single isolated node of the splice graph (node 8). Under such a representation, the isoform problem becomes a graph traversal problem. Multiple traversals of the splice graph correspond to multiple isoforms of a gene. Furthermore, the splice graph can be weighted. The edge weight reflects the strength of experimental evidence for a particular splice junction. For EST data, this can be the number of ESTs on which two exons are connected by a splice junction. For microarray data, this can be the signal intensity of a particular exon junction probe.

To construct a splice graph, we start from sequence-based detection of exon-intron structure and alternative splicing (which is described in detail in the previous chapter). We treat each exon as a node in the splice graph. Alternative donor/acceptor splicing can produce two exon forms with a common splice site at one end, and different splice sites at the other end. We treat these two exon forms as different nodes in the splice graph. Next, we go through each expressed sequence to obtain edge information of the splice graph. We connect two nodes with a directed edge if the two exons are linked by a splice junction in the expressed sequences. The edge weight is set as N if the connection between two nodes is observed in N expressed sequences. In the end, we obtain a directed acyclic graph (DAG). This graph represents all splicing events of a gene and their numbers of occurrences in the sequence data.

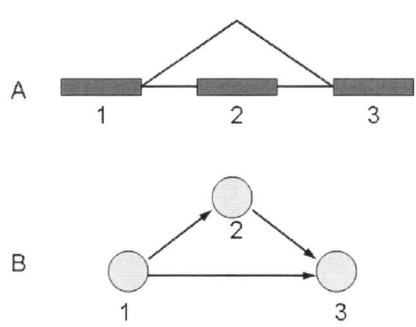

Fig. 10.1. Splice graph representation of gene structure and alternative splicing. (**A**) The exon-intron structure of a three-exon gene. The middle exon is alternatively spliced. (**B**) The splice graph representation of the gene structure. Alternative splicing of the second exon is represented by a directed edge from node 1 to node 3.

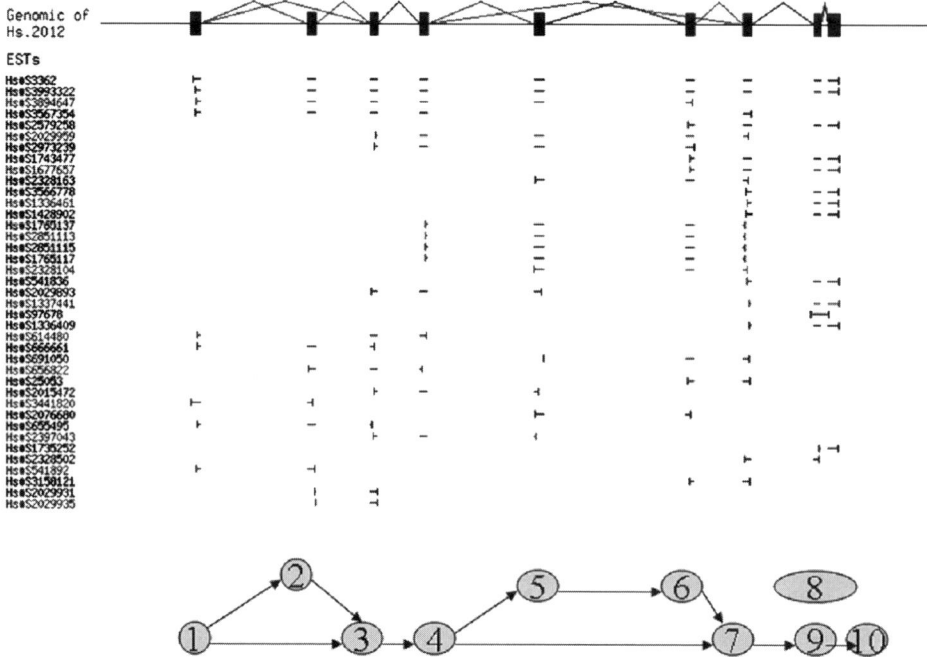

Fig. 10.2. Splice graph representation of gene structure of Hs.2012 (TCN1). The top figure shows the gene structure and EST-genome alignment of TCN1. The bottom figure shows the splice graph of TCN1.

3.2. Splice Graph Traversal Using Heaviest Bundling Algorithm

Once a splice graph is constructed, we can apply graph algorithms to obtain traversals of the splice graph. There are different ways to do this. Heber and colleagues enumerated all the possible traversals of the splice graph *(3, 6)*. A few other studies used a set of rules or dynamic programming algorithms to obtain the minimal set of traversals that were most likely given the sequence observations *(5, 7–10)*. The second strategy is more favorable, because in genes with multiple alternative splicing events, exons are not randomly joined to create exon combinations with no evidence in the observed sequence data *(2)*. A more recent method uses the Expectation-Maximization algorithm to compute the probability of each possible traversal of the splice graph *(11)*.

Figure 10.3A illustrates an isoform reconstruction pipeline using a splice graph traversal algorithm *Heaviest Bundling (12)*. *Heaviest Bundling* is a dynamic programming algorithm that searches for the splice graph traversal with the maximum overall edge weight (**Fig. 10.3B**). Details of the *Heaviest Bundling* algorithm were described in *(12)*. For each gene, the isoform reconstruction is an iterative process. Each round of the iteration consists of two steps: *templating*, which up-weights the longest expressed sequence that is not explained by any constructed isoforms; and *Heaviest Bundling*, which uses the *HB* algorithm to

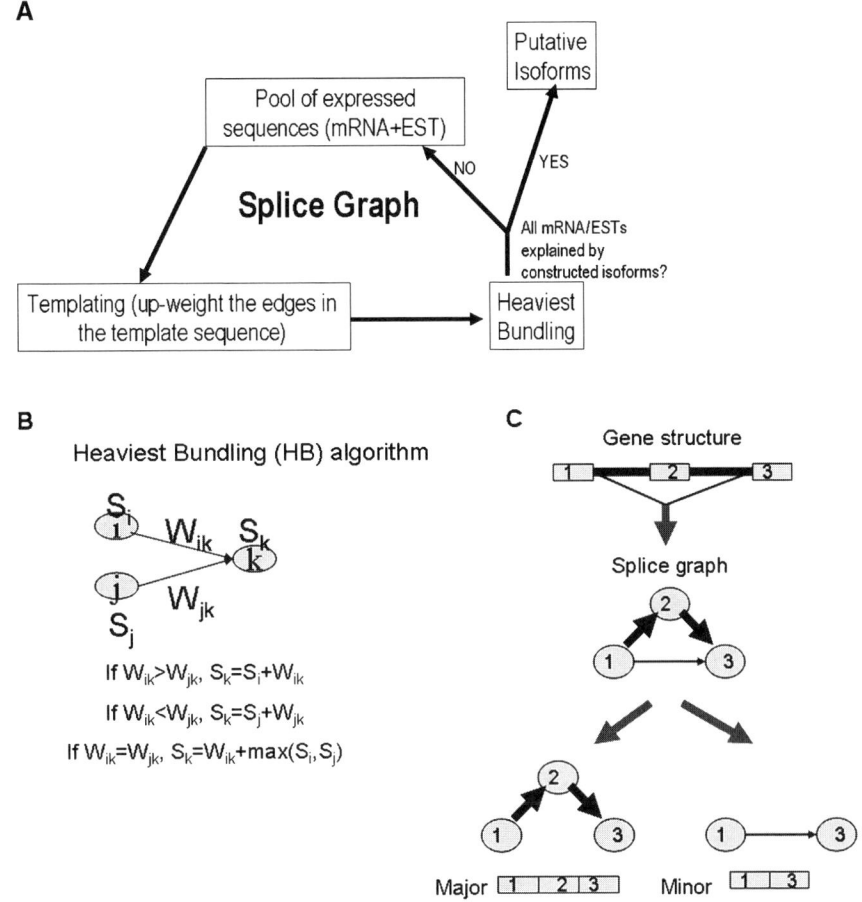

Fig. 10.3. Heaviest bundling and isoform reconstruction. (**A**) An overview of the isoform reconstruction pipeline. (**B**) Heaviest bundling algorithm. (**C**) Heaviest bundling and isoform reconstruction for a hypothetical three-exon gene.

search for the mostly likely traversal of the graph, based on its current edge weight values. The iteration is terminated once the constructed isoforms can explain all the sequence observations. **Figure 10.3C** shows the splice graph and isoform reconstruction for a hypothetical 3′-exon gene. The exon-inclusion form has a higher edge weight and is recognized as the "major" isoform. The exon-skipping form has a lower edge weight and is recognized as a "minor" isoform.

3.3. Filtering of Putative Isoforms

Once the putative isoforms are constructed, it is important to assess whether a putative isoform might result from artifacts in the EST data. Multiple rules can be used to filter putative isoforms, by checking: *(1)* whether the translated protein sequence is too short (e.g., <50 aa); *(2)* whether the similarity of the translated protein sequence to the major protein isoform is too low (e.g., <50%); *(3)* whether the transcript has no stop codon at all; (4) whether the transcript contains a premature stop codon and is a likely target of the mRNA nonsense-mediated decay pathway *(13)*. Such filters are heuristic; nevertheless, they can eliminate a lot of noise and artefacts in the EST sequence data. After a set of high-confidence protein isoforms are obtained, we can apply many sequence analysis tools to assess how alternative splicing modifies the protein product (for two examples, see *(14, 15)*).

4. Web Resources and Computational Tools for Full-Length Transcript/Protein Isoforms

A few labs have constructed online databases with computationally inferred full-length transcript and protein isoforms. A list of such online resources is provided in **Table 10.1**. In addition, there are several new isoform construction algorithms *(10, 11, 16)*, although no database has been built using these new methods.

Table 10.1
Web resources for full-length isoforms

Resources	Descriptions/URLs
Alternative Splicing Gallery (ASG) *(6)*	22127 splice graphs of human genes. Exhaustive enumerations of splice graphs produced 1.2 million putative transcripts. http://statgen.ncsu.edu/asg/
ASAP Database *(17)*	Alternative splicing, transcript and protein isoforms of human and mouse genes. The most likely isoforms are constructed for each gene, followed by heuristic filtering of putative isoforms. http://www.bioinformatics.ucla.edu/ASAP/
ECgene *(8)*	A genome browser that combines results from EST clustering and splice-graph based transcript assembly. ECgene has alternative splicing data for human, mouse and rat genomes. http://genome.ewha.ac.kr/ECgene/
DEDB (*Drosophila melanogaster* Exon Database) *(9)*	An exon database for *Drosophila melanogaster*. Gene structure is represented in a splice-graph format. http://proline.bic.nus.edu.sg/dedb/
ESTGenes *(7)*	Alternative splicing and isoform annotation integrated with the Ensembl genome browser. A minimal set of compatible isoforms is derived for each gene. http://www.ensembl.org/index.html

Acknowledgments

This work was funded by a Dreyfus Foundation Teacher-Scholar Award to C.J.L., and by the National Institutes of Health through the NIH Roadmap for Medical Research, Grant U54 RR021813 entitled Center for Computational Biology (CCB). Information on the National Centers for Biomedical Computing can be obtained from http://nihroadmap.nih.gov/bioinformatics.

References

1. Boue, S., Vingron, M., Kriventseva, E., et al. (2002) Theoretical analysis of alternative splice forms using computational methods. *Bioinformatics* 18 Suppl 2, S65–S73.
2. Lee, C., Wang, Q. (2005) Bioinformatics analysis of alternative splicing. *Brief Bioinform* 6, 23–33.
3. Heber, S., Alekseyev, M., Sze, S. H., et al. (2002) Splicing graphs and EST assembly problem. *Bioinformatics* 18 Suppl. 1, S181–188.
4. Blencowe, B. J. (2006) Alternative splicing: new insights from global analyses. *Cell* 126, 37–47.
5. Xing, Y., Resch, A., Lee, C. (2004) The Multiassembly Problem: reconstructing multiple transcript isoforms from EST fragment mixtures. *Genome Res* 14, 426–441.
6. Leipzig, J., Pevzner, P., Heber, S. (2004) The Alternative Splicing Gallery (ASG): bridging the gap between genome and transcriptome. *Nucleic Acids Res* 32, 3977–3983.
7. Eyras, E., Caccamo, M., Curwen, V., et al. (2004) ESTGenes: alternative splicing from ESTs in Ensembl. *Genome Res* 14, 976–987.
8. Kim, P., Kim, N., Lee, Y., et al. (2005) ECgene: genome annotation for alternative splicing. *Nucleic Acids Res* 33, D75–79.
9. Lee, B. T., Tan, T. W., and Ranganathan, S. (2004) DEDB: a database of Drosophila melanogaster exons in splicing graph form. *BMC Bioinformatics* 5, 189.
10. Florea, L., Di Francesco, V., Miller, J., et al. (2005) Gene and alternative splicing annotation with AIR. *Genome Res* 15, 54–66.
11. Xing, Y., Yu, T., Wu, Y. N., et al. (2006) An expectation-maximization algorithm for probabilistic reconstructions of full-length isoforms from splice graphs. *Nucleic Acids Res* 34, 3150–3160.
12. Lee, C. (2003) Generating consensus sequences from partial order multiple sequence alignment graphs. *Bioinformatics* 19, 999–1008.
13. Maquat, L. E. (2004) Nonsense-mediated mRNA decay: splicing, translation and mRNP dynamics. *Nat Rev Mol Cell Biol* 5, 89–99.
14. Resch, A., Xing, Y., Modrek, B., et al. (2004) Assessing the impact of alternative splicing on domain interactions in the human proteome. *J Proteome Res* 3, 76–83.
15. Xing, Y., Xu, Q., Lee, C. (2003) Widespread production of novel soluble protein isoforms by alternative splicing removal of transmembrane anchoring domains. *FEBS Lett* 555, 572–578.
16. Malde, K., Coward, E., Jonassen, I. (2005) A graph based algorithm for generating EST consensus sequences. *Bioinformatics* 21, 1371–1375.
17. Lee, C., Atanelov, L., Modrek, B., et al. (2003) ASAP: The Alternative Splicing Annotation Project. *Nucleic Acids Res* 31, 101–105.

Chapter 11

Sequence Segmentation

Jonathan M. Keith

Abstract

Whole-genome comparisons among mammalian and other eukaryotic organisms have revealed that they contain large quantities of conserved non–protein-coding sequence. Although some of the functions of this non-coding DNA have been identified, there remains a large quantity of conserved genomic sequence that is of no known function. Moreover, the task of delineating the conserved sequences is non-trivial, particularly when some sequences are conserved in only a small number of lineages. Sequence segmentation is a statistical technique for identifying putative functional elements in genomes based on atypical sequence characteristics, such as conservation levels relative to other genomes, GC content, SNP frequency, and potentially many others. The publicly available program changept and associated programs use Bayesian multiple change-point analysis to delineate classes of genomic segments with similar characteristics, potentially representing new classes of non-coding RNAs (contact web site: http://silmaril.math.sci.qut.edu.au/~keith/).

Key words: Comparative genomics, non-coding RNAs, conservation, segmentation, change-points, sliding window analysis, Markov chain Monte Carlo, Bayesian modeling.

1. Introduction

1.1. Sliding Window Analysis

A common practice in genomic and comparative genomic studies is to analyze whole-genome profiles via sliding window analysis—a form of moving average or Loess analysis. Some prominent examples include: sliding window analyses of G+C content, gene density, and repeat density in the human genome *(1, 2)*; G+C content and repeat distribution in the mouse genome, and human versus mouse conservation *(3)*; and human versus chimpanzee conservation *(4)*. Some packages that can perform this kind of analysis include ConSite *(5)* and rVista *(6)*. The purpose of this type of analysis is to identify parts of the genome or genomes under study that are atypical in terms of some property, whether it be GC content, conservation level, gene frequency, transposon frequency, or a combination of several properties. Atypical regions can be interesting for a variety of reasons, but one is that they may contain functional genomic components. Sophisticated methods for detecting protein-coding genes are available (*see* **Chapter 8**), and indeed most protein-coding genes have been identified in the human genome and other sequenced genomes. However, it is thought that most large, eukaryotic genomes contain numerous unidentified functional, non–protein-coding components. For example, it is estimated that the human genome contains at least twice as much conserved (and hence functional) non–protein-coding sequence as protein-coding sequence *(3, 7–9)*. Methods for detecting these elements are much less well-developed. Sliding window analysis may provide clues that will help to identify such elements.

Figure 11.1 displays an example of a sliding window analysis for an alignment of a part of the genomes of two fruit fly (*Drosophila*) species *D. melanogaster* and *D. simulans*. The two profiles shown were constructed as follows. First, a pairwise alignment of these two species was downloaded from the UCSC web site (http://genome.ucsc.edu/), encompassing genomic coordinates 1580575 to 1584910 of chromosome arm 3R of *D. melanogaster*. For each window of length 11 bases in this region of *D. melanogaster*, the proportion of bases that are aligned to a matching base was calculated and this value was assigned to the position at the centre of the window. This process

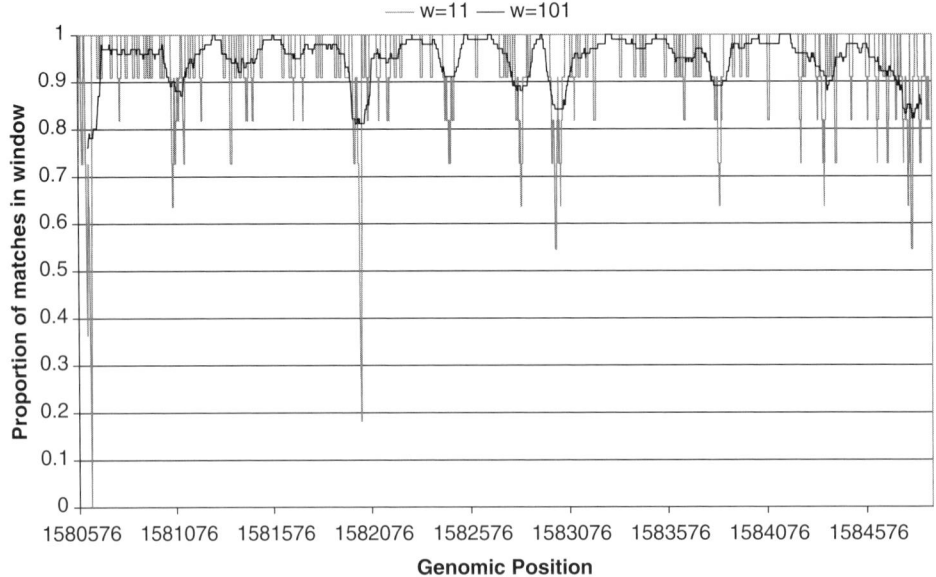

Fig. 11.1. Profiles of the proportion of bases at which matches occur in an alignment of the fruit fly species *D. melanogaster* and *D. simulans*, for sliding window sizes of 11 bases *(gray trace)* and 101 bases *(black trace)*.

was repeated for a window size of 101 bases. Note that, for both window sizes, there are several positions where the proportion of matches seems to be higher than the surroundings, and these may correspond to conserved features. However, the two profiles are quite different. The profile produced with a smaller window size is noisier, but also exhibits sharper changes, and in particular results in much deeper troughs at some positions. As shown in **Section 3.6**, some of these change-points correspond to exon-intron boundaries.

1.2. Change-Point Analysis

Although a useful, quick, and easy analysis for identifying regions of high conservation or other irregularity, sliding window analysis has a number of problems. In particular, it is usually not clear what the size of the sliding window should be. Large window sizes tend to "blur" sharp changes in the property of interest, as **Fig. 11.1** illustrates, because each value in a sliding window profile is actually an average over a region of length equal to the window size. Using smaller windows can reduce this problem, but this decreases the signal-to-noise ratio.

An alternative analysis, more difficult but potentially more sensitive, is *sequence segmentation*, also called *multiple change-point analysis*. In this approach, one attempts to identify the locations of *change-points* at which sharp changes in a particular property of interest occur. The best known of these approaches is the phylo-HMM based method PhastCons developed at UCSC *(9, 10)* specifically for analyzing multi-species conservation. However, numerous other approaches have been developed (for some examples, see *(11–21)*), including a number of Bayesian approaches *(22–31)*.

This chapter describes some of the practical issues involved in implementing the Bayesian approach of Keith and co-workers *(25, 26, 32)*, encoded as the C program changept. An attractive feature of this approach is that it includes a classification of segments into groups that share similar properties, such as conservation level and G+C content. Moreover, it estimates, for each genomic position, the probability that that genomic position belongs to each class. The ability to estimate probabilities in this way derives from the Bayesian modeling framework, but changept is orders of magnitude faster than alternative Bayesian genomic segmentation methods, and unlike them it can feasibly be applied to whole eukaryotic genomes.

The approach can be used to segment according to various properties, but this chapter focuses specifically on its use to analyze conservation between two species (*see* **Note 1**). To illustrate the method, data and results obtained for the closely related fruit fly species *D. melanogaster* and *D. simulans* are used throughout.

A feature of changept is that it does not merely generate a single segmentation, optimized according to some scoring function.

Rather, it generates multiple segmentations, sampled from a posterior distribution over the space of all possible segmentations. This feature is typical of Bayesian methodologies that rely on Markov chain Monte Carlo (MCMC) simulation. Although a full description of the Bayesian model and sampling algorithm is beyond the scope of this chapter, a few brief explanations are necessary. Further details of the model and MCMC sampler are found in papers by Keith and co-workers *(25, 26)* and in a submitted paper *(32)*.

1.3. A Model for Sequence Segmentation

Each segmentation output by changept consists of the number k of change-points (segment boundaries) and their positions in the input sequence. The conservation level (that is, the probability of a match) is assumed to be drawn from a mixture of beta distributions. The number of components g in the mixture model must be specified by the user, but we discuss in the following a method for deciding the appropriate number. The mixture proportions $\pi = (\pi_1, \ldots, \pi_g)$ and the parameters of the beta distributions (collected into a single vector α) are also sampled from a posterior distribution and are output along with each segmentation.

1.4. MCMC Simulation

Two features of MCMC algorithms that need to be explained here are the *burn-in* phase and *sub-sampling*. MCMC methods involve a Markov chain for which the limiting distribution is the distribution from which one wishes to sample, in this case a posterior distribution over the space of segmentations. However, the chain approaches this distribution asymptotically, and thus the elements generated early in the chain are not typical and need to be discarded. This early phase of sampling is known as burn-in, and we discuss in the following ways to determine at what point to end the burn-in phase. Even after burn-in, it is an inefficient (and often infeasible) use of disk space to record all of the segmentations generated by the algorithm. The algorithm therefore asks the user to specify a *sampling block length*. The sequence of segmentations will be divided into blocks of this length and only one element in each block will be recorded for future processing. Ways of deciding the sampling block length are discussed in the following.

1.5. Future Developments

The changept program has recently been generalized to allow for segmentation on the basis of multiple data types simultaneously (*see* **Note 2**). Although this work is still at an early stage, preliminary results suggest that the three classes of genomic region identified in *D. melanogaster* based on conservation level alone (see the following) can be divided into a large number of sub-classes. Some of these sub-classes may represent collections of functional non-coding RNAs. Thus this work is at an exciting stage, and may lead to the discovery of whole new classes of non-protein-coding RNAs.

The changept program should be regarded as a work in progress for the sensitive detection and classification of genomic elements. Software will be made available from the contact web site as it is developed and updated.

2. Systems, Data, and Databases

The illustrative analyses described below were performed on a cluster of approximately 120 PCs, each with 800 MHz Celeron CPUs and 128 Mb of memory. However, the programs can also be run on a single desktop or laptop. Source code for some of the programs described in the following is available from the contact web site. This code has been successfully compiled and run under both Unix, Windows, and Mac OS X operating systems with 32-bit processors. Some of the programs described in this chapter are only available as Windows executables, without source code, also at the contact web site.

The raw data used in the illustrative analyses consisted of a pairwise whole genome alignment of the fruit fly species *D. melanogaster* to the species *D. simulans*. These species diverged approximately 3–5 Mya *(33, 34)*. The alignments were downloaded from the UCSC Genome Browser (http://genome.ucsc.edu/) in axt format (see http://genome.ucsc.edu/goldenPath/help/axt.html for a description of this format).

3. Methods

3.1. Generating the Binary Files

A small extract of the pairwise alignment of *D. melanogaster* and *D. simulans* in axt format is shown in **Fig. 11.2A**. The main segmentation algorithm changept takes as input, not axt files, but rather a text file containing a binary sequence representing matches and mismatches, such as that shown in **Fig. 11.2B**. The conversion of axt format to changept format is straightforward and consists of the following steps:

1. Data cleaning
2. Generation of binary sequences
3. Pooling

The first step involves discarding sequences for which the alignments are suspect, either because they involve a high proportion of indels and mismatches, because the alignment blocks are very short, or because they are likely to have atypical characteristics and would thus be better analyzed separately. For the illustrative *Drosophila* analyses, this step consisted of discarding

(A)

```
0 chr2L 2 133 chrX 86 212 + 8034
gacaatgcacgacagaggaagcagaacagatatttagattgcctctcattttctctcccatattatagggagaaatatgatcgcgtatgcgagagtagtgccaacatattgtgctctttgatttttttggcaa
gacaacgcacgacaaggagagcaagagagatagtcagattgcctctaattttctctcccatattaccaaga-----atgatcgcttatgcgagaggagtcccaacatattgtcctcttcgaattttttggcaa

1 chr2L 134 253 chr3R 133 256 - 4104
cccaaaatggtggcggatgaaCGAGATG-----ATAATATA-TTCAAGTTGCCGCTAATCAGAAATAAATTCATTGCAACGTTAAATACAGCACAATATATGATCGCGTATGCGAGAGTAGTGCCA
ccccatagtgCAGTGGTTAACCGATATGTGTGCATACTACAATTCAAATTGCCGCTAATCAGAAAGAAATTTATCGCAACGTTAAATGCAGCACAAAATAGG--CGCTCAAAATGGGGTTGAATTA
```

(B)

```
1111101111111100100111100101111101011111111111011111111111111111000011IIIII111
111110111111111101110111111111111101111101101111111111#11101010010010110101011 10
111IIIIII1110110II1111101111111111111111101111101101111111111111011111111011101II
1110010000010110100001#1111100111111101II11111111111111111111101111111011111001
```

Fig. 11.2. The top few lines of: (**A**) a .axt file for the UCSC pairwise alignment of *D. melanogaster* and D. simulans, and (**B**) the input file to changept. The .axt file gives the block number (starting from zero) and genomic coordinates of the aligned sequences, followed by the sequences themselves on separate lines. Columns of the alignment with matching characters are converted to "1"s, mismatches are converted to "0"s, and columns containing an indel are converted to "I"s. Block boundaries are marked with a "#".

all but the aligned 2L, 2R, 3L, 3R, and X chromosome arms of *D. melanogaster* (*see* **Note 3**).

The second step involves substituting single labels for the columns of the alignment. A match becomes a "1," a mismatch becomes a "0" and an indel becomes an "I." Note that the indels are ignored by the main program changept. The boundaries of alignment blocks are marked with a # symbol. The hashes are treated as change-points with fixed positions. For the *Drosophila* analysis, binary sequences were individually generated for each of the 2L, 2R, 3L, 3R, and X chromosome arms. In the third step, binary sequences are concatenated in various combinations for the purpose of evaluating whether they are better modeled separately or together. For the *Drosophila* analysis, the 2L, 2R, 3L, and 3R sequences were concatenated to generate a binary sequence labeled E, and this was concatenated with the X sequence to generate a binary sequence labeled EX.

Two programs genbiomultialign.exe and genchangepts.exe (available at the contact web site) convert .axt files to the required binary format. These programs are currently only available as Windows executables. **Figure 11.3** displays an example of an MSDOS script used to run these programs. The script first defines the directories where various input and output files are located. The input files are the .axt files and the file chromlens.csv. This latter file contains the chromosome lengths and is available at the contact web site. The program genbiomultialign outputs the file dm2vsdroSim1.algn, which is used by other programs in this suite, including genchangepts. The latter program outputs the main input files for changept. It also produces files with the extension .map, which specify the genomic coordinates of each alignment block. This is required later to map profiles back to genomic coordinates for display purposes. Finally, genchangepts also produces files with an extension .sym25, which are used by other programs in this suite.

3.2. Running changept

Once the binary sequences have been generated and the code compiled, the segmentation algorithm is ready to run. Typing changept at the command line produces the following message:

```
    To correctly use this program, you must have
the following parameters:
-i   input_file_name (<99 char)
-sf  initial segmentation file (optional, <99 char)
-o   output_file_name (<99 char)
-n   num_samples (integer, >=1)
-b   num_burn (optional, integer, <num_iterations,
     default 0)
-s   sampling_block_size (optional, integer, >=1,
     default 1)
```

```
set experiment=exp1
set dm2vsds1axts=d:\goldenPath\dm2\vsDroSim1
set basedir=d:\bioexplorer
set bin=%basedir%\bin
set aligns=%basedir%\alignments
set resources=%basedir%\resources
set results=%basedir%\results\dm2centric
set expdir=%results%\%experiment%
set logs=%expdir%\logs

set LogLev=-f3 -F%logs%\gendm2vsdroSim1.log -S3

%bin%\genbiomultialign.exe %LogLev% -m0 -x -rdm2 -RdroSim1 -c%resources%\chromle
ns.csv -i%dm2vsds1axts%\*.axt -o%aligns%\dm2droSim1.algn -t"dm2vsdroSim1" -d"UCS
C dm2 vs droSim1"

call :genchpts dm2 DroSim1 dm2droSim1.algn dm2vsdroSim1_2L  chr2L
call :genchpts dm2 DroSim1 dm2droSim1.algn dm2vsdroSim1_2R  chr2R
call :genchpts dm2 DroSim1 dm2droSim1.algn dm2vsdroSim1_3L  chr3L
call :genchpts dm2 DroSim1 dm2droSim1.algn dm2vsdroSim1_3R  chr3R
call :genchpts dm2 DroSim1 dm2droSim1.algn dm2vsdroSim1_X   chrX

goto :completed

:genchpts
SETLOCAL

set refspecies=%1
set relspecies=%2
set alignfile=%3
set outprfx=%4
set chrom=%5

%bin%\genchangepts.exe %LogLev% -i%aligns%\%alignfile% -o%expdir%\%outprfx%_
chgpt.txt -r%refspecies% -R%relspecies% -c"%chrom%" -m0
%bin%\genchangepts.exe %LogLev% -i%aligns%\%alignfile% -o%expdir%\%outprfx%_
sym25.txt -r%refspecies% -R%relspecies% -c"%chrom%" -m1

ENDLOCAL
goto :EOF

:completed
echo All Processed
```

Fig. 11.3. Script used to execute the programs genbiomultialign and genchangepts, to produce input files for changept and other programs.

```
-a   alphabet_size (optional, integer, 2-10,
     default 2)
-p   prop_changepts_init (optional, double, 0<p<1,
     default 0.01)
-pa  hyperparameter for phi (optional, double, pa>0,
     default 1.0)
-pb  hyperparameter for phi (optional, double, pb>0,
     default 1.0)
-ng  number of groups (optional, integer, 0<ng<=10,
     default 1)
-nc  number of chains (optional, integer, 1<=nc<=10,
     default 1)
```

```
-hp  heating parameter for AP (optional, double,
     hp>=0, default 1.0)
-r   random number seed (optional, integer,
     default=time)
-pf  samples_per_file (optional, integer, >=1,
     default num_samples)
-nf  number_of_output_files (optional, integer,
     >=1, default num_samples/samples_per_file)
```

The switches introduce the following terms:

-i: The name of the text file containing the binary sequence to be segmented. As always, care should be taken to use file names that clearly identify the data. At the very least, the file name should indicate the aligned species and the chromosome.

-sf: The name of a text file containing a segmentation output by a previous changept run. If the file contains multiple segmentations, all but the last one will be ignored. This option is useful for restarting the Markov chain where a previous run ended.

-o: The name of the file to which segmentations will be output. Again care should be taken to clearly identify the source of the segmentations. At the very least, the output file name should include the input file name and the number of groups used (see -ng in the following).

-n: The number of segmentations to be sampled. These will be output to output_file_name. A sample size of 1000 is typical.

-b: The number of sampling blocks to discard before sampling begins. This option is useful for preventing changept from producing output during burn-in. However, since the user typically cannot predict how many sampling blocks will be required to achieve burn-in, it should not be assumed that burn-in has occurred once sampling begins. The user may prefer to retain the default "-b 0" so that no initial sampling blocks are discarded. However, if disk space is an issue, the user may wish to discard 100 or more initial segmentation blocks.

-s: The number of updates performed in each sampling block. A useful heuristic is to set this value to about one tenth of the length of the input sequence. Alternatively, if the user knows roughly how many change-points will be identified (and it is a good idea to perform a preliminary run to determine this), this parameter can be set to about ten times this number.

-a: The number of labels used in the input sequence. This parameter allows for non-binary input sequences. Up to 10 labels are permitted. (*See* **Note 1**.)

-p: This parameter controls the randomly chosen initial segmentation. Its value is irrelevant if the '-sf' switch is used. Otherwise the initial segmentation will be generated by throwing a uniform random number between 0 and 1 for each sequence position at which a change-point is not fixed, making that position a change-point if the random number is less than the value of this parameter. The algorithm is not sensitive to the value of this parameter, and the default value seems to work well in all cases so far.

-pa, -pb: The parameters of a beta distribution specifying the prior probability distribution for a parameter ϕ, which is the probability that any given sequence position is a change-point. More specifically, ϕ is the probability that a sequence position at which there is not a fixed change-point (i.e., an alignment block boundary) is a change-point. These values are useful if there is prior information about how many non-fixed change-points are present in the sequence. Otherwise, the default parameters specify a uniform (and hence uninformative) prior.

-ng: The number of groups into which to classify segments.

-nc, -hp: These parameters are used to implement a Metropolis-coupled Markov chain Monte Carlo algorithm *(35)*. They are currently experimental and may be removed from future versions.

-pf: The number of segmentations to output to each file. This switch is useful if the user wishes to break the output file up into several smaller files. This is sometimes necessary, as the output files can be very large and may exceed the maximum length of text files expected by some operating systems. Moreover, should one wish to restart the Markov chain where a previous run finished (using the –sf switch), the algorithm can more rapidly find the final segmentation in a smaller file. Note that the output files are differentiated by the addition of extensions ".1", ".2" and so on.

-nf: The number of output files to generate. This will only affect the output if one specifies fewer output files than are necessary given the values of the –n and –pf switches. In that case, the earlier output files will be overwritten. This may be useful if the user wants to monitor convergence while changept is running, and thus sets the sample size very large, with the intention of stopping the algorithm once the files generated during burn-in are overwritten.

For the *Drosophila* analysis of the EX file, using a three-group model, the following command was executed:

```
changept -i dm2vsdroSim1_EX_chgpt.txt -o
dm2vsdroSim1_EX_chgpt.txt.3grps -b 200 -n 1000
-s 30000000 -ng 3 -pf 100
```

3.3. Assessing Convergence

As always in MCMC, it is important to ensure that the chain has converged and that samples obtained during burn-in are discarded. Numerous methods for assessing convergence have been developed, but a common and successful method is to visually inspect graphs of time series for various parameters and statistics of interest. To facilitate this, changept outputs a log file (the file name of which is the output file name with the extension ".log" appended). An excerpt of the top of a log file for a three-group model is shown in **Fig. 11.4**. The first few lines contain the command used to execute the run, the seed used by the random number generator (based on the current time), the length of the input sequence and the number of fixed change-points. After the heading "Beginning MCMC" each line corresponds to a sampled segmentation in the main output file. Each line contains:

1. The sample number
2. The number of change-points
3. The mixture proportions of each group (three columns in **Fig. 11.4**)
4. The parameters of the beta distribution for each group, in pairs (six columns—three pairs—in **Fig. 11.4**)
5. The log-likelihood

---

changept -i dm2vsdroSim1_EX_chgpt.txt -o dm2vsdroSim1_EX_chgpt.txt.3grps -n 2000 -s 10000000 -ng 3 -pf 100
Random number seed=1176272106
Reading sequence.
Sequence length=104551330.
numfixed=29046
Beginning MCMC.
1. 547665 0.228599 0.047880 0.723521 1.255652 23.554376 326.997638 443.251581 1.028164 13.318207 -21261316.897484
2. 837554 0.565539 0.028305 0.406156 1.610463 36.877827 678.826174 913.076209 1.054476 10.865174 -21221275.319483
3. 921873 0.586653 0.026666 0.386682 1.642488 39.649917 620.570893 835.114015 1.129389 11.510267 -21213618.884855
4. 1040743 0.535950 0.023767 0.440283 1.805758 51.334139 860.390289 1152.575155 1.300227 13.694245 -21209539.357014
5. 1130941 0.525753 0.024949 0.449298 1.857796 57.372117 522.588598 702.253217 1.364315 14.486255 -21201871.683409

Fig. 11.4. Top lines of the log file output by changept for the three-group EX run. The header contains the command executed, the random number seed, length of the binary input sequence, and the number of fixed change-points (block boundaries). The next five lines (wrapped for this display) contain key parameters for the first five segmentations output. Each line shows the number of change-points, mixture proportions π (3 values), beta parameters α (3 pairs of values) and the log-likelihood.

It is highly recommended that time-series be plotted for the log-likelihood, the number of change points, the mixture proportions, and the means of the beta distributions for each group (given by $\alpha/(\alpha + \beta)$, where α and β are the second and first values in each pair of beta parameters, respectively) and the sum of the beta parameters for each group. It is important that *all* parameters tested reach a stable regime before the burn-in phase is judged complete.

To facilitate graphing of these time series, a Microsoft Excel file named "changept_templates.xls" may be downloaded from the contact web site. The content of log files below the heading "Beginning MCMC" can be simply cut and pasted into the appropriate worksheet of this Excel file, depending on the number of groups. The above-mentioned time-series will then be displayed automatically, to the right of the data.

An example of a graph of a time-series for the mixture proportions is shown in **Fig. 11.5**. Note that the chain appears to have converged prior to the beginning of sampling (note that 200 initial sampling blocks were discarded in this run) since the mixture proportions vary only slightly around stable means for the entire run. The same is true of the other parameters graphed using the Excel template. **Note 4** describes another technique for assessing convergence.

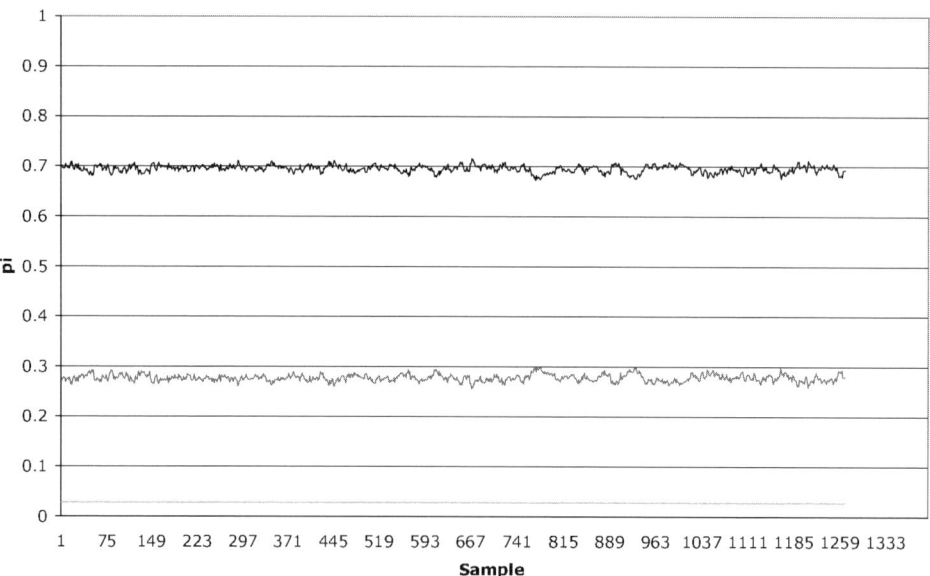

Fig. 11.5. The mixture proportions (π) for each of the three groups in the analysis of the *D. melanogaster* versus *D. simulans* EX data. These values represent the approximate proportion of segments (not bases) in each group. The proportions are stable from the very first sample, indicating that convergence probably occurred during the discarded iterations specified using the –b option.

3.4. Determining Degree of Pooling and Number of Groups

In principle, the number of groups in the mixture model is a parameter for which posterior probabilities can be calculated. Similarly, the algorithm itself should determine whether all chromosomes should be analyzed separately or whether some should be pooled. However, changept does not currently have this capability. At present, it is recommended that independent runs of changept with different numbers of groups and patterns of pooling be performed. These can then be assessed using the model comparison technique described in the following.

For the *Drosophila* analysis, three different pooling scenarios were explored:

1. 2L, 2R, 3L, 3R, X each analyzed independently
2. 2L, 2R, 3L, 3R pooled (i.e., concatenated to form a sequence labeled 'E'), with X analyzed separately
3. 2L, 2R, 3L, 3R, X pooled (i.e., concatenated to form a sequence labeled 'EX')

The changept program was run for each of these scenarios and with the number of groups ranging from 2 to 6. Independent runs were executed on different processors in the computing cluster. Since the first pooling scenario treats each chromosome arm independently, it required a total of 25 independent runs (one for each chromosome arm and number of groups). Similarly, the second pooling scenario required 10 runs and the third required 5.

The following score function can be used to compare the various models. It is an information criterion similar to the Akaike Information Criterion. For each pooling scenario and number of groups, we compute the value:

$$2\bar{k} - 2\overline{\ln L}$$

where $\bar{k}$ is the average number of change-points in the sample generated by that run and $\overline{\ln L}$ is the average log-likelihood. Note that for the first pooling scenario, the average number of change-points is the sum of the averages for the individual chromosomes, and similarly for the average log-likelihood and for the second pooling scenario. The preferred model is the one with the lowest value of the information criterion.

For the *Drosophila* analysis, it was found that the third pooling scenario—complete pooling—produced the lowest score, regardless of the number of groups. The scores of the EX runs for different numbers of groups are shown in **Table 11.1**. In this example, the best scoring model was the three-group model, although the six-group model also had a low score. It should be noted that the three and six-group models have similar distributions of conservation levels (see following section) the only differences being the addition of two groups with a low proportion of segments and the splitting of the most rapidly evolving group into two closely spaced sub-groups.

Table 11.1
The Complexity, Fit, and Information scores obtained using various numbers of groups for the *D. melanogaster* versus *D. simulans* EX data

# Groups	Complexity	Fit	Information
2	1421883.2	−21199193.8	45242153.9
3	1389290.2	−21191548.4	45161677.1
4	1389821.5	−21191266.7	45162176.3
5	1390087.7	−21191181.7	45162538.6
6	1390763.6	−21190944.2	45163415.5

3.5. Plotting the Distribution of Conservation Levels

As mentioned, a mixture of beta distributions is fitted to the distribution of conservation levels. The fitted distribution can be plotted for any sample using the mixture proportions and beta parameters obtained in that sample. The density is given by:

$$p(x \mid \pi, \alpha) = \sum_{i=1}^{k} \pi_i B(x \mid \alpha^{(i)})$$

where $B(x \mid \alpha^{(i)})$ is a beta density given by:

$$B(x \mid \alpha^{(i)}) = \frac{\Gamma(\alpha_0^{(i)} + \alpha_1^{(i)})}{\Gamma(\alpha_0^{(i)})\Gamma(\alpha_1^{(i)})} x^{\alpha_1^{(i)}-1}(1-x)^{\alpha_0^{(i)}-1}$$

and $\alpha^{(i)} = (\alpha_0^{(i)}, \alpha_1^{(i)})$.

The Excel template pages mentioned above and available from the contact web site contain calculation of these densities at intervals of 0.001 in the interval [0,1] and a graph of the beta density for each mixture component normalized so that the area under each density is the mixture proportion. The sum of the components is also shown. To display the graph for any given sample, execute the following steps.

1. Paste the log file output so that the upper left corner is the A1 cell of the appropriate worksheet (based on number of mixture components).

2. Locate the graph of the mixture density (top centre of the six graphs displayed to the right of the calculations).

3. Enter the row number containing the desired sample in the cell at the top left of the graph.

An example for the three-group *Drosophila* results is shown in **Fig. 11.6**.

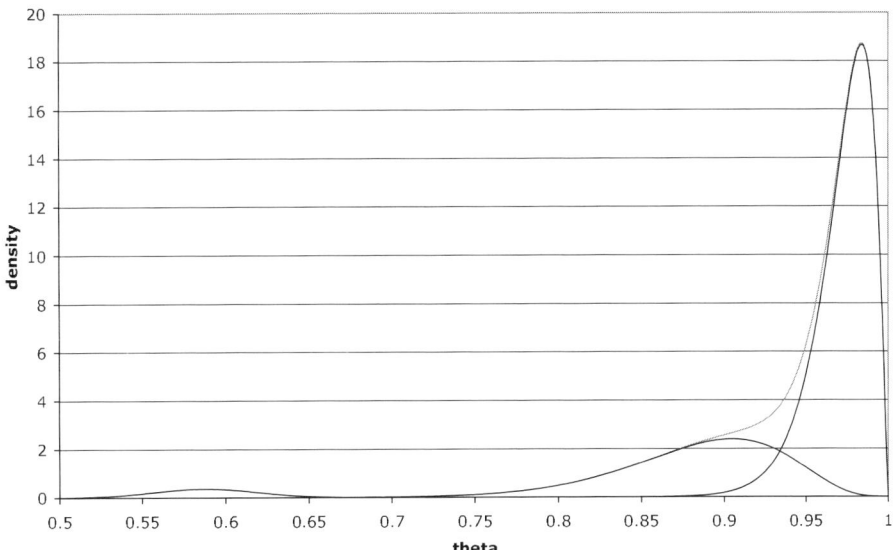

Fig. 11.6. Beta distributions for the three classes of conservation levels observed in the EX alignment of *D. melanogaster* to *D. simulans*, normalized in their mixture proportions. The weighted sum of the three beta distributions is also shown (in gray). The values of π and α are those obtained for the final sample of a run (sample 1267).

3.6. Generating and Viewing Profiles

The program changept outputs segmentations sampled from the posterior distribution in one or more text files, with user specified file names as discussed. The sampled segmentations can be used to compute various probabilities. In particular, they can be used to estimate, for each position in the input sequence, the probability that that position belongs to a given group. The result is a *profile* for that group that can be plotted across the genome. **Figure 11.7** shows a plot of the profile for the most slowly evolving group in the *Drosophila* analysis. Note that there are several well-defined slowly evolving regions, and that these correspond to exons of a known protein-coding gene.

The program readcp, which can be downloaded from the contact web site, takes a segmentation file as input and generates a profile for the desired group. Typing readcp at the command line produces the following message:

```
To correctly use this program, you must have
the following parameters:
-i   sequence_file_name
-c   cut_point_file_name1...(up to 150 file names)
-b   num_burn (optional, default 0)
-s   num_skip (optional, default 0)
-a   alphabet_size (optional, default 2)
-ng  num_groups (optional, default 2)
-pg  list of profiled group numbers (optional,
     default ng-1)
-notheta suppress theta profile (optional)
-nop suppress p profile (optional)
```

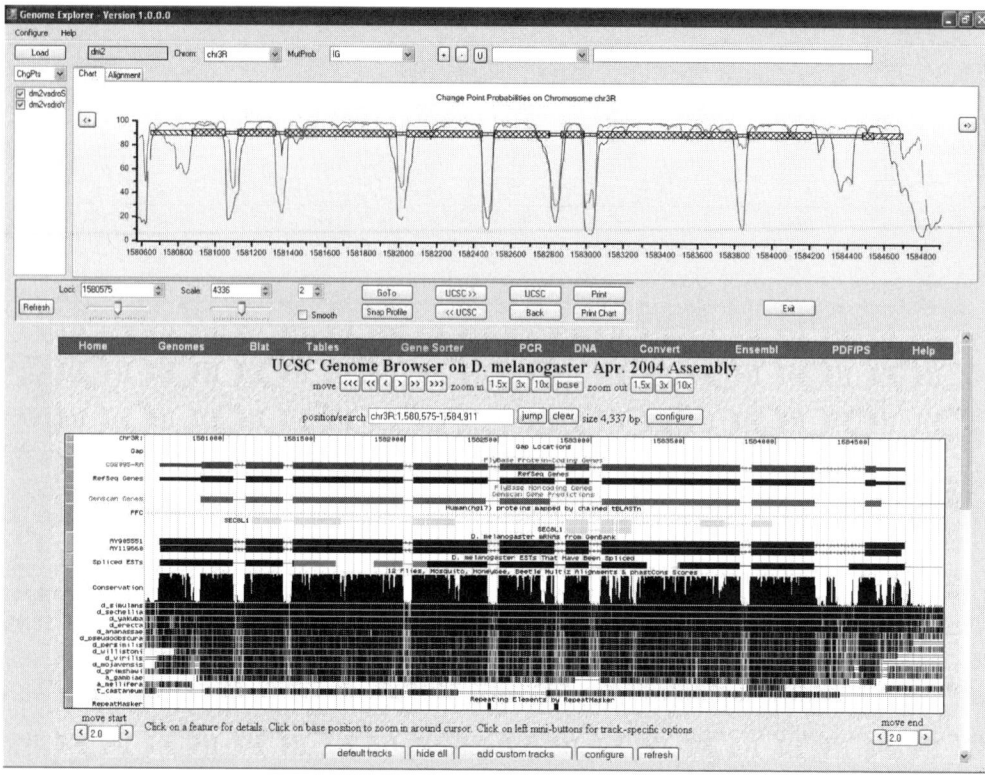

Fig. 11.7. Profiles obtained using changept for the *D. melanogaster* versus *D. simulans* EX analysis and for a similar analysis based on an alignment of *D. melanogaster* versus *D. yakuba*. The top pane shows the two profiles obtained. The *D. simulans* profile has shallower troughs than the *D. yakuba* profile for the same region of *D. melanogaster*. Note that the profiles are for the same region of the 3R chromosome arm as that shown in **Fig. 11.1**. The bottom pane shows this same region of the *D. melanogaster* genome, displayed with the UCSC genome browser. The conservation profile displayed in this pane is based on a multiple alignment of 15 insect species. The changept profiles, each based on only pairwise alignments, are nevertheless able to clearly delineate exons, with less noise.

The switches introduce the following terms:

-i: The file name of the binary input sequence. This must be the same as the binary sequence supplied as input to changept. Note that, although changept is equipped to handle sequences with up to 10 distinct labels, readcp can currently only process binary files. This is likely to change in the near future.

-c: names of up to 150 segmentation files generated by changept. These will be processed to generate the profile. The ability to process multiple files is useful if the user wants to combine the results of multiple independent runs or if the output of changept was broken up into smaller files using the –pf switch.

-b: The value of the segmentation index to be regarded as the last burn-in sample. (The "segmentation index" referred to here is merely the ordinal number of the segmentation output by changept; numbering starts at one.) Segmentations with this index or lower will not be processed. Note that if the output of multiple independent changept runs is combined, so that multiple segmentations share the same index, all segmentations with index less than or equal to this input parameter are regarded as part of the burn-in phase of their respective runs, and are discarded.

-s: The number of segmentations to skip between those processed to produce a profile. For example, if the options "-b 100 –s 1" are used, readcp will only process segmentations with the indices 101, 103, 105 and so on, skipping one segmentation between those processed. This option is useful if the user wishes to reduce the dependence of adjacent segmentations (but *see* **Note 5**).

-a: The number of labels used in the input sequence. Currently, readcp can only handle binary sequences, so the default should always be used.

-ng: The number of groups used to generate the segmentation file or files. This must be the same as the number of groups specified at the command line for changept to generate the segmentation files.

-pg: A list of the group indices to use when constructing the profile. Note that numbering of groups starts at zero and is ordered according to increasing mean of the fitted beta distributions for each group. The default value—the number of groups minus one—thus results in a profile for the most slowly evolving group. If more than one group index is listed, the resulting profile specifies, for each sequence position, the probability that that position belongs to any one of the listed groups.

-notheta: This option suppresses the production of the theta profile (see the following).

-nop: This option suppresses the production of the usual profile, and instead produces only a theta profile (see the following).

In addition to group profiles, readcp can also produce a *theta profile*. This provides, for each position in the input sequence, the average conservation level for that position, regardless of group index.

The profile shown in **Fig. 11.7** was obtained by executing the command:

```
readcp -i dm2vsdroSim1_EX_chgpt.txt -c dm2vsdro
Yak1_EX_chgpt.3grps1...dm2vsdroYak1_chgpt.3grps.10
-ng 3 -pg2
```

Group profiles and theta profiles can both be viewed using the purpose-designed genome browser bioexplorer available at the contact web site. This browser displays .dps files, which are generated using the program genchpts2dpts based on a profile generated by readcp, and on the .map and .sym25 files generated by genchangepts (*see* **Section 3.1**). An example script for executing genchpts2dpts is available from the contact web site.

3.7. Testing for Enrichment and Depletion of Genomic Regions

The groups identified via the above analysis may have functional significance. As a first step toward identifying the functional significance of each group, it is useful to investigate where the high-probability regions for a given group are located relative to known protein-coding genes. This can be done using the region file generated by genchpts2dpts and the program getdist, available at the contact web site. Typing getdist at the command line results in the following message:

```
Command line input should consist of:
  profile_name - name of the file containing
real values in the interval [0,1]
  type_name - name of the file containing the
types for each value in the profile.
  numtypes - number of types (2<=numtypes<=7).
  codetype - 0 for bit codes, 1 for integer indices.
The output file will be the profile name with
.hist appended.
```

Here the profile name is the name of the text file containing the profile for the group that we want to test for enrichment or depletion of coding and other sequences. The type name is the name of the region file produced by genchpts2dpts. The value numtypes is the number of different types identified in the type file. There are seven region types in a region file, representing coding sequence (CDS), 3' UTR regions, 5' UTR regions, introns, regions within 1000 bp upstream of known genes, regions within 1000 bp downstream of known genes, and intergenic sequence. The "known genes" here are RefSeq genes. The code type specifies how the type file represents a type. Region files are represented as bit codes. Bit codes allow genomic positions to belong to more than one type. Each bit of the binary representation for the code is one if the position is of the type corresponding to that bit, and zero otherwise.

This program was invoked to process the profile for the most slowly evolving group of the *Drosophila* analysis with the following command:

```
getdist dm2vsdroSim1_EX_chgpt.3grps.p2 dm2vs-
droSIm1_EX_region.txt 7 0
```

The output of getdist is a file with the extension ".hist" containing a column of values for each region type. The value in row i and column j is the number of sequence positions that are contained in the region of type j and have a profile value of $0.001i$, rounded to three decimal places. Thus, by dividing each value in the table by the column sum, and by accumulating column sums down each column, one can obtain the cumulative distribution of profile values for each region. **Figure 11.8** shows a plot of the cumulative distributions for the seven regions of the *Drosophila melanogaster* genome for the most slowly evolving group. For example, the graph shows that only about 13% of known coding sequence in *D. melanogaster* has a profile value of <0.5 for the most slowly evolving group. Thus, the most slowly evolving group is enriched in coding sequence relative to, say, intergenic sequence, for which about 25% has a slowly evolving profile less than 0.5. Similarly, the slowly evolving group is enriched in 3'UTR and 5' UTR sequence.

3.8. Generating Segments

For the purpose of further investigation of the properties of each group, it can be useful to obtain a list of delineated segments contained in each group. A simple way to do this is to identify all genomic segments for which the profile value for that group is higher than some threshold. The most natural threshold value is 0.5, since sequence positions with a profile value higher than 0.5 are more likely than not to belong to the group in question. However, more or less stringent thresholds can also be employed.

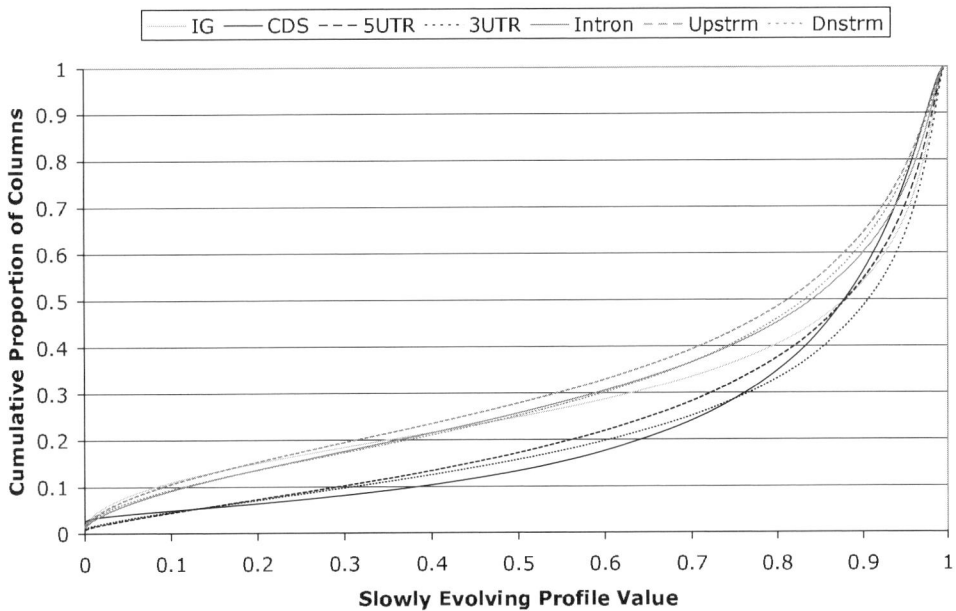

Fig. 11.8. Cumulative proportion of columns in the *D. melanogaster* versus *D. simulans* alignment having a profile value less than or equal to the values shown on the horizontal axis, for each of seven genomic fractions. The profile value here is the probability that the column belongs to the most slowly evolving group in the three-group analysis.

The simple program outputsegments, available from the contact web site, can be used to threshold profiles in this way. The program is called from the command line and must have five input parameters. These are:

1. The name of the .map file for the original input sequence
2. The name of the .sym file for the original input sequence
3. The name of the file containing the profile to be thresholded
4. The name of the output file to which segments will be written
5. The threshold value to use

Segment files for each of the four groups identified in this analysis were obtained using a threshold of 0.5 and are available for download from the contact web site for further investigation.

3.9. Attribution of Function

Once segments have been delineated for each group, the next step is to determine whether any of the groups corresponds to a well-defined class of functional elements, and determine its function. It is tempting in the *Drosophila* example to assume that the most rapidly evolving group corresponds to elements under positive selection, the most slowly evolving group corresponds to conserved sequence, and the intermediate group is neutrally evolving. However, preliminary investigations indicate that the inclusion of GC content and other data types in the analysis enables these groups to be further subdivided (*see* **Note 2**). Thus it may be that these subgroups correspond to classes of functional elements, and that all three of the groups we have identified contain one or more such classes of functional elements. Therefore, it is prudent to wait until the integration of multiple data types into the analysis is complete. Once this has been done, the author intends to attribute function to the groups by a variety of means, including testing for enrichment of GO terms in the segments of each group and their neighborhoods.

4. Notes

1. The algorithm has also been used to analyze G+C content in human chromosome 1 *(25)* and can be used for many other purposes. The changept program takes a binary sequence as input, and the distribution of many properties throughout the genome could be represented in this way. For example, one could segment according to the frequency of a specific motif by representing the genome as a binary sequence in which a "1" represents the first base of an instance of the motif and all other positions are "0." In fact, the program also accepts input that is not binary, and can handle sequences composed of up

to 10 labels: "0" to "9." The number of labels in the input sequence must be specified at the command line using the "-a *alphabet_size*" option, where *alphabet_size* is an integer in the range 2–10. The default value is "-a 2." An application that uses four labels is discussed in **Note 2**. Modifications of the algorithm for continuous data types may be made available at the contact web site in future.

2. A simple way to analyze two different data types simultaneously is the following. Suppose that one wants to segment based on conservation between two species and on the GC content of one of the species. Then one can construct a four-character sequence based on a pairwise alignment of the two species, using the following code for each column of the alignment.

    0: Mis-match in alignment, A or T in reference sequence

    1: Mis-match in alignment, G or C in reference sequence

    2: Match in alignment, A or T in reference sequence

    3: Match in alignment, G or C in reference sequence

    The author is currently investigating using this technique, as well as other more sophisticated approaches, to perform segmentation based on multiple data types. Preliminary results indicate that the groups observed using pairwise alignments alone can be resolved into numerous sub-groups by including GC content in the analysis.

3. Alignments are also available for other parts of the *D. melanogaster* genome. There are alignments for chromosome 4, for heterochromatic parts of chromosomes 2, 3, 4, and X, for the heterochromatic Y chromosome, for the mitochondrial chromosome and for unplaced reads (chrU). Most of these were omitted from the analyses because they are enriched in poor quality alignments. The mitochondrial chromosome was omitted because its distribution of conservation levels is expected to be quite different from the nuclear chromosomes, and thus should be treated in a separate analysis.

    Parts of the *D. melanogaster* genome that were aligned to atypical parts of the *D. simulans* genome (e.g., unplaced reads) were included in the analyses described here. These constitute such a small part of the total that they are unlikely to have a noticeable influence on the results. Nevertheless, it is also possible to eliminate such sequences.

4. Another useful way to assess convergence is to plot time-series for two parameters of interest simultaneously as an X-Y plot. For example, one can plot the means of the beta distributions versus the mixture proportions for each group. Such plots show clearly the trends during burn-in and the random-walk that occurs after burn-in.

5. It is standard practice among Bayesian practitioners to use all of the available segmentations produced by MCMC when estimating an average. No bias is introduced by using dependent samples, although the *effective sample size* is reduced. Moreover, it is an inefficient use of disk space to store segmentations that are not processed. Thus it is generally preferable to use the default value of "-s 0" in readcp. The effective sample size can be increased by specifying a large sampling block size using the –s switch in changept. In the absence of information about the sampling block size needed to achieve effective independence, a size one-tenth the length of the input sequence, or alternatively about 10 times the number of change-points, is recommended for binary sequences.

Acknowledgments

The author thanks Peter Adams for assistance in running simulations; Stuart Stephen for assisting in the development of much of the code; Benjamin Goursaud and Rachel Crehange for assisting in the generalization of the code for multiple data types; and John Mattick, Kerrie Mengersen, Chris Ponting, and Mark Borodovski for helpful discussions. This work was partially funded by Australian Research Council (ARC) Discovery Grants DP0452412 and DP0556631 and a National Health and Medical Research Council (NHMRC) grant entitled "Statistical methods and algorithms for analysis of high-throughput genetics and genomics platforms" (389892).

References

1. Lander, E. S., Linton, L. M., Birren, B., et al. (2001) Initial sequencing and analysis of the human genome. *Nature* 409, 860–921.
2. Venter, J. C., Adams, M. D., Myers, E. W., et al. (2001) The sequence of the human genome. *Science* 291, 1304–1351.
3. Waterston, R. H., Lindblad-Toh, K., Birney, E., et al. (2002) Initial sequencing and comparative analysis of the mouse genome. *Nature* 420, 520–562.
4. Mikkelsen, T. S., Hillier, L. W., Eichler, E. E., et al. (2005) Initial sequence of the chimpanzee genome and comparison with the human genome. *Nature* 437, 69–87.
5. Sandelin, A., Wasserman, W. W., Lenhard, B. (2004) ConSite: web-based prediction of regulatory elements using cross-species comparison. *Nucleic Acids Res* 32, W249–W52.
6. Loots, G. G., Ovcharenko, I., Pachter, L., et al. (2002) rVista for comparative sequence-based discovery of functional transcription factor binding sites. *Genome Res* 12, 832–839.
7. Cooper, G. M., Stone, E. A., Asimenos, G., et al. (2005) Distribution and intensity of constraint in mammalian genomic sequence. *Genome Res* 15, 901–913.
8. Gibbs, R. A., Weinstock, G. M., Metzker, M. L., et al. (2004) Genome sequence of the Brown Norway Rat yields insights into mammalian evolution. *Nature* 428, 493–521.
9. Siepel, A. C., Bejerano, G., Pedersen, J. S., et al. (2005) Evolutionarily conserved

elements in vertebrate, insect, worm, and yeast genomes. *Genome Res* 15, 1034–1050.

10. Siepel, A. C., Haussler, D. (2004) Combining phylogenetic and hidden Markov models in biosequence analysis. *J Com Biol* 11, 413–428.

11. Bernaola-Galvan, P., Grosse, I., Carpena, P., et al. (2000) Finding borders between coding and non-coding regions by an entropic segmentation method. *Phys Rev Letts* 85, 1342–1345.

12. Bernaola-Galvan, P., Roman-Roldan, R., Oliver, J. (1996) Compositional segmentation and long-range fractal correlations in DNA sequences. *Phys Rev E* 53, 5181–5189.

13. Braun, J. V., Braun, R. K., Muller, H.-G. (2000) Multiple changepoint fitting via quasilikelihood, with application to DNA sequence segmentation. *Biometrika* 87, 301–314.

14. Braun, J. V., Muller, H.-G. (1998) Statistical methods for DNA sequence segmentation. *Stat Sci* 13, 142–162.

15. Gionis, A., Mannila, H. (2003) Finding recurrent sources in sequences. In *Proceedings of the Seventh Annual International Conference on Research in Computational Molecular Biology*, 123–130.

16. Li, W. (2001) DNA segmentation as a model selection process. In *Proceedings of the Fifth Annual International Conference on Research in Computational Molecular Biology*, 204–210.

17. Li, W., Bernaola-Galvan, P., Haghighi, F., et al. (2002) Applications of recursive segmentation to the analysis of DNA sequences. *Comput Chem* 26, 491–510.

18. Oliver, J. L., Bernaola-Galvan, P., Carpena, P., et al. (2001) Isochore chromosome maps of eukaryotic genomes. *Gene* 276, 47–56.

19. Oliver, J. L., Carpena, P., Roman-Roldan, R., et al. (2002) Isochore chromosome maps of the human genome. *Gene* 300, 117–127.

20. Oliver, J. L., Roman-Roldan, R., Perez, J., et al. (1999) SEGMENT: identifying compositional domains in DNA sequences. *Bioinformatics* 15, 974–979.

21. Szpankowski, W., Ren, W., Szpankowski, L. (2005) An optimal DNA segmentation based on the MDL principle. *Int J Bioinformat Res Appl* 1, 3–17.

22. Boys, R. J., Henderson, D. A. (2002) On determining the order of Markov dependence of an observed process governed by a hidden Markov model. *Sci Prog* 10, 241–251.

23. Boys, R. J., Henderson, D. A. (2004) A Bayesian approach to DNA sequence segmentation. *Biometrics* 60, 573–588.

24. Boys, R. J., Henderson, D. A., Wilkinson, D. J. (2000) Depicting homogenous segments in DNA sequences by using hidden Markov models. *Appl Stat* 49, 269–285.

25. Keith, J. M. (2006) Segmenting eukaryotic genomes with the generalized Gibbs sampler. *J Comput Biol* 13, 1369–1383.

26. Keith, J. M., Kroese, D. P., Bryant, D. (2004) A Generalized Markov Sampler. *Methodol Comput Appl Prob* 6, 29–53.

27. Minin, V. N., Dorman, K. S., Fang, F., et al. (2005) Dual multiple change-point model leads to more accurate recombination detection. *Bioinformatics* 21, 3034–3042.

28. Husmeier, D., Wright, F. (2002) A Bayesian approach to discriminate between alternative DNA sequence segmentations. *Bioinformatics* 18, 226–234.

29. Liu, J. S., Lawrence, C. E. (1999) Bayesian inference on biopolymer models. *Bioinformatics* 15, 38–52.

30. Ramensky, V. E., Makeev, V. J., Toytberg, M. A., et al. (2000) DNA segmentation through the Bayesian approach. *J Comput Biol* 7, 215–231.

31. Salmenkivi, M., Kere, J., Mannila, H. (2002) Genome segmentation using piecewise constant intensity models and reversible jump MCMC. *Bioinformatics* 18, S211–S218.

32. Keith, J. M., Adams, P., Stephen, S., et al. Delineating slowly and rapidly evolving fractions of the *Drosophila* genome, submitted.

33. Russo, C. A. M., Takezaki, N., Nei, M. (1995) Molecular phylogeny and divergence times of Drosophilid species. *Mol Biol Evol* 12, 391–404.

34. Tamura, K., Subramanian, S., Kumar, S. (2004) Temporal patterns of fruit fly (*Drosophila*) evolution revealed by mutation clocks. *Mol Biol Evol* 21, 36–44.

35. Geyer, C. J. (1991) Markov chain Monte Carlo maximum likelihood, in (Keramidas, E. M., ed.), *Computing Science and Statistics: Proceedings of the 23rd Symposium on the Interface*, pp. 156–163. Interface Foundation, Fairfax Station, VA.

Chapter 12

Discovering Sequence Motifs

Timothy L. Bailey

Abstract

Sequence motif discovery algorithms are an important part of the computational biologist's toolkit. The purpose of motif discovery is to discover patterns in biopolymer (nucleotide or protein) sequences in order to better understand the structure and function of the molecules the sequences represent. This chapter provides an overview of the use of sequence motif discovery in biology and a general guide to the use of motif discovery algorithms. The chapter discusses the types of biological features that DNA and protein motifs can represent and their usefulness. It also defines what sequence motifs are, how they are represented, and general techniques for discovering them. The primary focus is on one aspect of motif discovery: discovering motifs in a set of unaligned DNA or protein sequences. Also presented are steps useful for checking the biological validity and investigating the function of sequence motifs using methods such as motif scanning—searching for matches to motifs in a given sequence or a database of sequences. A discussion of some limitations of motif discovery concludes the chapter.

Key words: Motif discovery, sequence motif, sequence pattern, protein domain, multiple alignment, position-specific scoring matrix, PSSM, position-specific weight matrix, PWM, transcription factor binding site, transcription factor, promoter, protein features.

1. Sequence Motifs and Biological Features

Biological sequence motifs are short, usually fixed-length, sequence patterns. Many features of DNA, RNA, and protein molecules can be well approximated by motifs. For example, sequence motifs can represent transcription factor binding sites (TFBSs), splice junctions, and binding domains in DNA, RNA, and protein molecules, respectively. Consequently, discovering sequence motifs can lead to a better understanding of transcriptional regulation, mRNA splicing and the formation of protein complexes.

Regulatory elements in DNA are among the most important biological features that are represented by sequence motifs. The DNA footprint of the binding sites for a transcription factor (TF) is often well described by a sequence motif. These TFBS motifs specify the order and nucleotide preference at each position in the binding sites for a particular TF. Discovering TFBS motifs and relating them to the TFs that bind to them is a key challenge in constructing a model of the regulatory network of the cell *(1, 2)*. Motif discovery algorithms have been used to identify many candidate TFBS motifs that were later validated by experimental methods.

Protein motifs can represent, among other things, the active sites of enzymes. They can also identify protein regions involved in determining protein structure and stability. The PROSITE, BLOCKS, and PRINTS databases *(3–5)* contain hundreds of protein motifs corresponding to enzyme active sites, binding sites, and protein family signatures. Motifs can also be used to identify features that confer particular chemical characteristics (e.g., thermal stability) on proteins *(6)*. Protein sequence motifs can also be used to classify proteins into families *(5)*.

The importance of motif discovery is born out by the growth in motif databases such as TRANSFAC, JASPAR, SCPD, DBTBS, and RegulonDB *(7–11)* for DNA motifs and PROSITE, BLOCKS, and PRINTS *(3–5)* for protein motifs. However, far more motifs remain to be discovered. For example, TFBS motifs are known for only about 500 vertebrate transcription factors TFs, but it is estimated that there are about 2,000 TFs in mammalian genomes alone *(6, 12)*.

Fixed-length motifs cannot represent all interesting patterns in biopolymer sequences. For instance, they are obviously not ideal for representing variable-length protein domains. For representing long, variable-length patterns, profiles *(13)* or HMMs *(14, 15)* are more appropriate. However, the dividing line between motifs and other sequence patterns (e.g., HMMs and profiles) is fuzzy, and is often erased completely in the literature. Some of the motif discovery algorithms discussed in the following sections, for example, do allow a single, variable-length "spacer," thus violating (slightly) our definition of motifs as being of fixed length. However, this chapter does not consider patterns that allow free insertions and deletions, even though these are sometimes referred to as motifs in the literature.

2. Representing Sequence Motifs

Biological sequence motifs are usually represented either as regular expressions (REs) or position weight matrices (PWMs). These two ways of describing motifs have different strengths and weaknesses when it comes to expressive power, ease of discovery, and usefulness for scanning. Motif discovery algorithms exist that output their results in each of these types of motif representation. Some motif discovery algorithms do not output a description of the motif at all, but, rather, output a list of the "sites" (occurrences) of the motif in the input sequences. Any set of sites can easily be converted to a regular expression or to a PWM.

Regular expressions are a way to describe a sequence pattern by defining exactly what sequences of letters constitute a match. The simplest regular expression is just a string of letters. For example, "T-A-T-A-A-T" is a DNA regular expression that matches only one sequence: "TATAAT". (This chapter follows the PROSITE convention of separating the positions in an RE by a hyphen ("-") to distinguish them from sequences.) To allow more than one sequence to match an RE, extra letters (ambiguity codes) are added to the four-letter DNA sequence alphabet. For example, the IUPAC (16) code defines "W = A or T", so the RE "T-A-T-A-W-T" matches both "TATATT" and "TATAAT". For the 20-letter protein alphabet, ambiguity codes would be unwieldy, so sets of letters (enclosed in square brackets) may be included in an RE. Any of the letters within the square brackets is considered a match. As an added convenience, PROSITE protein motif REs allow a list of letters in curly braces, and any letter *except* the enclosed letters matches at that position. For example, the PROSITE N-glycosylation site motif is "N-{P}-[ST]-{P}". This RE matches any sequence starting with "N", followed by anything but "P", followed by an "S" or a "T", ending with anything but "P". As noted, some motif discovery programs allow for a variable-length spacer separating the two, fixed-length ends of the motif. This is particularly applicable to dyad motifs in DNA *(17, 18)*. The RE "T-A-C-N(2,4)-G-T-A" describes such a motif, in which "N" is the IUPAC "match anything" ambiguity code. The entry "-N(2,4)-" in the RE matches any DNA sequence of length from two to four, so sequences matching this RE have lengths from eight to ten, and begin and end with "TAC" and "GTA", respectively.

Whereas REs define the set of letters that may match at each position in the motif, PWMs define the *probability* of each letter in the alphabet occurring at that position. A PWM is an n by w matrix, where n is the number of letters in the sequence alphabet (four for DNA, 20 for protein), and w is the number of positions in the motif. The entry in row a, column i in the PWM, designated $P_{a,i}$, is the probability of letter a occurring at position i in the motif. Mathematically, PWMs specify the parameters of a position-specific multinomial sequence model that assumes each position in the motif is statistically independent of the others. A PWM defines a probability for every possible sequence of the correct width (w). The positional independence assumption implies that the probability of a sequence is just the product of the corresponding entries in the PWM. For example, the probability of the sequence "TATAAT" according to a PWM (with six columns) is:

$$Pr(\text{``TATAAT''}) = P_{T,1} \cdot P_{A,2} \cdot P_{T,3} \cdot P_{A,4} \cdot P_{A,5} \cdot P_{T,6}.$$

As with REs, it is possible to extend the concept of PWMs to allow for variable-length spacers, but this is not commonly done by existing motif discovery algorithms.

For the purposes of motif scanning, many motif discovery algorithms also output a position-specific scoring matrix (PSSM), which is often confusingly referred to as a PWM. The entries in a PSSM are usually defined as:

$$S_{a,j} = \log_2 \frac{P_{a,j}}{f_a}, \qquad [1]$$

where f_a is the overall probability of letter a in the sequences to be scanned for occurrences of the motif. The PSSM score for a sequence is given by *summing* the appropriate entries in the PSSM, so the PSSM score of the sequence "TATAAT" is:

$$S(\text{``TATAAT''}) = S_{T,1} + S_{A,2} + S_{T,3} + S_{A,4} + S_{A,5} + S_{T,6}.$$

PSSM scores are more sensitive for scanning than probabilities because they take the "background" probability of different letters into account. This increases the match score for uncommon letters and decreases the score for common letters, thus reducing the rate of false-positives caused by non-uniform distribution of letters in sequences.

Underlying both REs and PWMs are the actual occurrences (sites) of the motif in the input sequences. The relationship among the motif sites, an RE and a PWM is illustrated in **Fig. 12.1**, which shows the JASPAR "*broad-complex 1*" motif. The nine motif sites from which this motif was constructed are shown aligned with each other at the top of the figure. The corresponding RE motif (using the IUPAC DNA ambiguity codes) is shown beneath the alignment. Below that, the counts of each letter in the corresponding alignment columns are shown. Below those, the corresponding PWM entries are shown. They were computed by normalizing each column in the counts matrix so that it sums to one. Beneath the PWM, the "LOGO" representation *(19)* for the motif is shown, where the height of each letter corresponds to its contribution to the motif's information content *(2)*.

Any alignment of motif sites can be converted into either an RE or PWM motif in the manner illustrated in **Fig. 12.1**. Usually a small amount (called a "pseudocount") is added to the counts in the position-specific count matrix before the PWM is

					Aligned Sites								
site1	C	T	A	A	T	T	G	G	C	A	A	A	T	G
site2	A	T	A	A	T	A	A	A	C	A	A	A	A	C
site3	G	A	C	A	T	A	G	A	C	A	A	G	A	C
site4	G	T	C	T	T	T	C	A	C	A	A	A	T	A
site5	G	T	G	A	A	A	G	A	C	A	A	G	T	T
site6	A	T	A	A	T	A	A	A	C	A	A	A	A	T
site7	G	T	T	G	A	A	A	A	C	A	A	T	A	G
site8	T	C	A	A	A	T	A	T	C	A	A	A	T	C
site9	A	G	A	A	T	A	G	A	A	A	G	G	T	A

Regular Expression (RE)

N -- N -- N -- D -- W -- W -- V -- D -- M -- A -- R -- D -- W -- N

Position-specific Count Matrix (PSCM)

A [	3	1	5	7	3	6	4	7	1	9	8	5	4	2]
C [	1	1	2	0	0	0	1	0	8	0	0	0	0	3]
G [	4	1	1	1	0	0	4	1	0	0	1	3	0	2]
T [	1	6	1	1	6	3	0	1	0	0	0	1	5	2]

Position-specific Weight Matrix (PWM)

A	[0.33	0.11	0.56	0.78	0.33	0.67	0.44	0.78	0.11	1.00	0.89	0.56	0.44	0.22]
C	[0.11	0.11	0.22	0.00	0.00	0.00	0.11	0.00	0.89	0.00	0.00	0.00	0.00	0.33]
G	[0.44	0.11	0.11	0.11	0.00	0.00	0.44	0.11	0.00	0.00	0.11	0.33	0.00	0.22]
T	[0.11	0.67	0.11	0.11	0.67	0.33	0.00	0.11	0.00	0.00	0.00	0.11	0.56	0.22]

LOGO

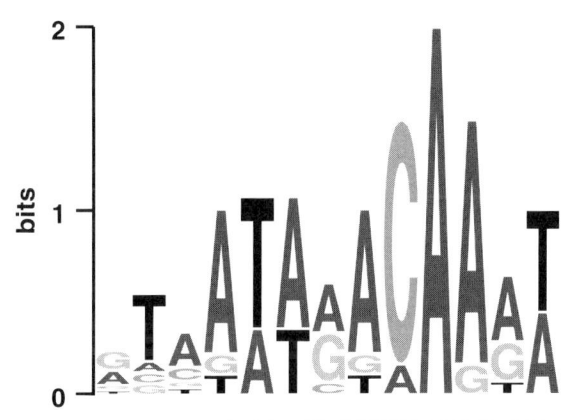

Fig. 12.1. Converting an alignment of sites into an RE and a PWM. The alignment of DNA sites is shown at the top. The RE (using the IUPAC ambiguity codes) is shown aligned below the sites. The corresponding counts of each letter in each alignment column—the position-specific count matrix (PSCM)—are shown in the next box. The PWM is shown below that. The last box shows the information content "LOGO" for the motif.

created by normalization in order to avoid probabilities of zero being assigned to letters that were not observed. This is sensible because, based on only a fraction of the actual sites, one cannot be certain that a particular letter *never* occurs in a real site.

Both PWMs and regular expressions are used by motif discovery algorithms because each has advantages. The main advantages of regular expressions are that they are easy for humans to visualize and for computers to search for. It is also easier to compute the statistical significance of a motif defined as a regular expression *(17, 20)*. On the other hand, PWMs allow for a more nuanced description of motifs than regular expressions, because each letter can "match" a particular motif position to varying degrees, rather than simply matching or not matching. This makes PWM motifs (converted to PSSMs using [1]) more suitable for motif scanning than REs in most applications. When used to model binding sites in nucleotide molecules, there is evidence that PWMs capture some of the statistical mechanics of protein-nucleotide binding *(21–23)*. An extension of PWMs, called hidden Markov models (HMMs), has also been shown to be an invaluable way to represent protein domains (e.g., the PFAM database of protein domains) *(24)*. The main disadvantage of PWMs for motif discovery is that they are far more difficult for computer algorithms to search for. This is true precisely because PWMs are so much more expressive than REs.

3. General Techniques for Motif Discovery

Many approaches have been tried for *de novo* motif discovery. In general, they fall into four broad classes. The predominant approach can be called the "focused" approach: assemble a small set of sequences and search for over-represented patterns in the sequences relative to a background model. Numerous examples of available algorithms that use this approach are given in **Table 12.3**. A related approach can be called the "focused discriminative" approach: Assemble two sets of sequences and look for patterns relatively over-represented in one of the input sets *(25, 26)*. The "phylogenetic" approach uses sequence conservation information about the sequences in a single input set *(27–30)*. The "whole-genome" approach looks for over-represented, conserved patterns in multiple alignments of the genomes of two or more species *(31, 32)*. This chapter does not describe the "whole-genome" approach in any detail.

A sequence motif describes a pattern that recurs in biopolymer sequences. To be interesting to biologists, the pattern should correspond to some functional or structural feature that the underlying molecules have in common. None of the computational techniques for motif discovery listed in the preceding can guarantee to find only biologically relevant motifs. The most that can generally be said about a computationally discovered motif is that it is statistically significant, given underlying assumptions about the sequences in which it occurs.

The predominant approach to sequence motif discovery is the focused approach, which searches for novel motifs in a set of unaligned DNA or protein sequences suspected to contain a common motif. The next section discusses how the sequences can be selected. RE-based motif discovery algorithms for the focused approach search the space of all possible regular expressions either exhaustively or heuristically (incompletely). Their objective is usually to identify the REs whose matches are most over-represented in the input sequences (relative to a background sequence model, randomly generated background sequences, or a set of negative control sequences). PWM-based motif discovery algorithms search the space of PWMs for motifs that maximize an objective function that is usually equal to (or related to) log likelihood ratio (LLR) of the PWM:

$$LLR(PWM) = \sum_{j=1}^{w} \sum_{a \in A} P_{a,j} \log_2 \frac{P_{a,j}}{f_a}, \qquad [2]$$

where the $P_{a,j}$ are estimated from the predicted motif sites as illustrated in **Fig. 12.1**. The appropriateness of this objective function is justified by both Bayesian decision theory *(33)*, and, in the case of TFBSs, by binding energy considerations *(21, 23)*. When the background frequency model is uniform, LLR is equivalent to "information content".

4. Discovering Motifs in Sets of Unaligned Sequences

This section describes the steps necessary for successfully discovering motifs using the "focused" approach. Each motif discovery application is different, but most have the following steps in common:

1. Assemble: Select the target sequences.
2. Clean: Mask or remove "noise."
3. Discover: Run a motif discovery algorithm.
4. Evaluate: Investigate the validity and function of the motifs.

In the first step, you assemble a "dataset" of DNA or protein sequences that you believe may contain an unknown motif encoding functional, structural, or evolutionary information. Next, if appropriate, you mask or remove confounding sequence regions such as low-complexity regions and known repeat elements. You then run a motif discovery algorithm using your set of sequences and with parameter settings appropriate to your application. The next step is intended to weed out motifs that are likely to be chance artefacts rather than motifs corresponding to functional or structural features, and to try to glean more information about them. This step can involve determining if a discovered motif is similar to a known motif, or if its occurrences are conserved in orthologous genes. Each of these steps is described in more detail in the following sections.

4.1. Assemble: Select the Target Sequences

The most important step in motif discovery is to assemble a set of sequences that is likely to contain multiple occurrences of one or more motifs (*see* **Note 2**). For motif discovery algorithms to successfully discover motifs, it is important that the sequence set be as "enriched" as possible in the motifs. Obviously, if the sequences consist entirely of motif occurrences for a single motif, the problem of motif discovery is trivial (*see* **Fig. 12.1**). In practice, the guiding idea behind assembling a sequence set is to come as close as possible to such a set. To achieve this, all available background knowledge should be applied in order to:

- Include as many sequences as possible that contain the motifs.
- Keep the sequences as short as possible.
- Remove sequences that are unlikely to contain any motifs.

How you assemble your input sequence set depends, of course, on what type of motifs you are looking for and where you expect them to occur. In most applications, there are two basic steps:

1. Clustering
2. Extraction

First, you cluster genes (or other types of sequences) based on information about co-expression, co-binding, function, environment, or orthology to select ones likely to have a common motif. Second, you extract the relevant (portions of) sequences from an appropriate sequence database.

As an example, to discover regulatory elements in DNA, you might select upstream regions of genes that show co-expression in a microarray experiment *(34)*. Co-expression can be determined by clustering of expression profiles. Alternatively, you could use the sequences that bound to a TF in a ChIP-chip experiment *(1, 35)*. A third possibility is to use information on co-expressed promoters from CAGE tag experiments *(36, 37)*. To these sequence sets you might also add orthologous sequences from related organisms, the assumption being that the regulatory elements have been conserved in them.

To discover protein functional or structural sequence motifs, you could select proteins belonging to a given protein family based on sequence similarity, structure, annotation, or other means *(24, 38, 39)*. You might further refine the selection to only include proteins from organisms with a particular feature, such as the ability to live in extreme environments *(40)*. Another protein motif discovery application uses information from protein–protein interaction experiments. You can assemble a set of proteins that bind to a common host protein, in order to discover sequence motifs for the interacting domains.

Most algorithms require sets of sequences in FASTA format. Proteins are usually easily extracted directly from the available sequence databases. Genomic DNA is more problematic, since annotation of genes, promoters, transcriptional start sites, introns, exons, and other important features is not always reliable. Several web servers available to aid you in extracting the relevant sequences for discovering regulatory elements in genomic DNA are shown in **Table 12.1**.

4.2. Clean: Mask or Remove "Noise"

Many genomic "phenomena" can masquerade as motifs and fool motif discovery algorithms (*see* **Note 3**). Things such as low-complexity DNA, low-complexity protein regions, tandem repeats, SINES, and ALUs all contain repetitive patterns that are problematic for existing motif-finding algorithms. It is therefore advisable to filter out these features from the sequences in the input set. This is done by running one or more of the programs described in **Table 12.2** on your set of sequences. Typically, the programs replace regions containing genomic "noise" with the ambiguity code for "match anything" in the appropriate sequence alphabet. This usually means "N" for DNA sequences and "X" for protein. Most motif discovery algorithms will not find motifs containing large numbers of these ambiguity codes, so they are effectively made invisible by this replacement process.

Table 12.2 lists some of the programs available to help you mask or remove confounding regions from your input sequence set. The DUST program *(41)* can be used to filter out low-complexity DNA. The XNU program *(42)* filters low-complexity (short period repeat) amino acid sequences. An alternative program for filtering out low-complexity protein

Table 12.1
Web servers for extracting upstream regions and other types of genomic sequence

Web server name	Function
RSA tools	Retrieve upstream regions for a large number of organisms. http://rsat.ulb.ac.be/rsat/
PromoSer	Retrieve human, rat, and mouse upstream regions, including alternative promoters. http://biowulf.bu.edu/zlab/PromoSer
UCSC genome browser *(74)*	View and extract genomic sequences and alignments of multiple genomes. http://genome.ucsc.edu

Table 12.2
Programs for filtering "noise" in DNA and protein sequences

Program name	Function
DUST	Filter low-complexity DNA. http://blast.wustl.edu/pub/dust
XNU	Filter low-complexity protein. http://blast.wustl.edu/pub/xnu
SEG	Filter low-complexity protein. http://blast.wustl.edu/pub/seg
RepeatMasker	Filter interspersed DNA repeats and low-complexity sequence. http://www.repeatmasker.org/cgi-bin/WEBRepeatMasker
Tandem Repeats Finder	Identify the positions of DNA tandem repeats. http://tandem.bu.edu/trf/trf.html

sequences is the SEG program *(43)*. Interspersed DNA repeats and low-complexity DNA sequence can both be filtered using the RepeatMasker program *(44)*. A web server is available for RepeatMasker, whereas at the time of this writing it was necessary to download, compile, and install the DUST, XNU, and SEG programs on your own computer. Tandem repeats can be identified in DNA using the "Tandem Repeats Finder" program. It has a web server that allows you to upload your sequence set (in FASTA format) for analysis. Of course, you should be aware that functional motifs can sometimes occur in the types of regions filtered by these programs, so caution is advised. It is important to study the documentation available with the programs to be sure you know what types of sequence they mask or identify. If you suspect that they may be masking regions containing your motifs of interest, you can always try running motif discovery algorithms on both the original and cleaned sets of sequences, and compare the results.

4.3. Discover: Run a Motif Discovery Algorithm

Many motif discovery algorithms are currently available. Most require installation of software on your computer. **Table 12.3** lists a variety of algorithms that have web servers in which you can upload your sequences directly, thus avoiding the need to install any new software. The table groups the algorithms according to whether they search for motifs expressed as REs or PWMs. Some of the algorithms are general purpose and can discover motifs in either DNA or protein sequences (MEME *(45)* and Gibbs *(46)*).

Table 12.3
Some motif discovery algorithms with web servers

PWM-Based algorithms

MEME	DNA or protein motifs using EM. http://meme.nbcr.net
Gibbs	DNA or protein motifs using Gibbs sampling. http://bayesweb.wadsworth.org/gibbs/gibbs.html
AlignACE	DNA motifs using Gibbs sampling. http://atlas.med.harvard.edu
CompareProspector	DNA motifs in eukaryotes using "biased" Gibbs sampling; requires multiple alignment. http://seqmotifs.stanford.edu
BioProspector	DNA motifs in prokaryotes and lower eukaryotes using Gibbs sampling. http://seqmotifs.stanford.edu
MDscan	DNA motifs; specialized for ChIP-chip probes. http://seqmotifs.stanford.edu

RE-Based algorithms

BlockMaker	Protein motifs. http://blocks.fhcrc.org/blocks/make_blocks.html
RSA tools	DNA motifs using RE-based or Gibbs sampler-based algorithms http://rsat.ulb.ac.be/rsat/
Weeder	DNA motifs using RE-based algorithm. http://www.pesolelab.it
YMF	DNA motifs using RE-based algorithm. http://wingless.cs.washington.edu/YMF

Combination algorithms

TAMO	Yeast, mouse, human; input as gene names or probe names, fetches upstream regions for you. http://fraenkel.mit.edu/webtamo

Some algorithms are specialized only for DNA (AlignACE *(47)*, BioProspector *(30)*, MDscan *(48)*, RSA Tools *(17, 49)*, Weeder *(50)*, and YMF *(51)*). CompareProspector *(52)* is specialized for DNA sequences and requires that you input your sequence set and conservation levels for each sequence position derived from a multiple alignment. BlockMaker *(53)* finds motifs only in protein sequences. The TAMO algorithm *(54)* runs multiple motif discovery algorithms (MEME, AlignACE, and MDscan) and combines the results.

Many excellent algorithms are not included in **Table 12.3** because they did not appear to have a (working) web server at the time of this writing. Motif discovery algorithms require a great deal of computational power, so most authors have elected to distribute their algorithms rather than provide a web server. Other motif discovery algorithms include ANN-Spec *(26)*, Consensus *(55)*, GLAM *(56)*, Improbizer *(57)*, MITRA *(58)*, MotifSampler *(59)*, Phyme *(27)*, QuickScore *(60)*, and SeSiMCMC *(61)*.

Different classes of algorithms (RE- and PWM-based) have different strengths and weaknesses, so it is often helpful to run one or more motif discovery algorithms of each type on your sequence set. Doing this can increase the chances of finding subtle motifs. Also, the confidence in a given motif is increased when it is found by multiple algorithms, especially if the algorithms belong to different classes (*see* **Note 4**).

Some motif discovery algorithms (e.g., CompareProspector) can take direct advantage of conservation information in multiple alignments of orthologous sequence regions. This has been shown to improve the detection of TFBSs because they tend to be over-represented in sequence regions of high conservation *(62, 63)*. To find subtle motifs, it can also be useful to run each motif discovery algorithm with various settings of the relevant parameters. What the relevant parameters are depends on the particular problem at hand and the motif discovery algorithm you are using. You should read the documentation for the algorithm you are using for hints about what non-default parameter settings may be appropriate for different applications. In general, important parameters to vary include the limits on the width of the motif, the model used to model background (or "negative" sequences), the number of sites expected (or required) in each sequence, and the number of motifs to be reported (if the algorithm can detect multiple motifs).

4.4. Evaluate: Investigate the Validity and Function of the Motifs

One of the most difficult tasks in motif discovery is deciding which, if any, of the discovered motifs is "real." Three complementary approaches can aid you in this. First, you can attempt to determine whether a given motif is statistically significant. Second, you can investigate whether the function of the motif is already known or can be inferred. Third, you can look for corroborating evidence for the motif. Each of these approaches is discussed in the following.

Most motif discovery algorithms report motifs regardless of whether they are likely to be statistical artefacts. In other words, they "discover" motifs even in randomly generated (or randomly selected) sequences. This is sometimes referred to as the "GIGO" rule: garbage-in, garbage-out. This, however, is not necessarily a bad thing; many truly functional DNA motifs are not statistically significant in the context of the kinds of sequence sets that can be assembled using clustered data from co-expression, ChIP-chip, CAGE, or other current technologies. So, it is important that motif discovery algorithms be able to detect these types of motifs even if they lie beneath the level of statistical significance that we might like. Measures of the statistical significance of a motif above the 0.05 significance level are still useful because they can be used to prioritize motifs for further validation.

Some motif discovery algorithms report an estimate of the statistical significance of the motifs they report. For example, MEME *(45)*, Consensus *(55)*, and GLAM *(56)* report the *E*-value of the motif: the probability of a motif of equal or greater information content occurring in a sequence set consisting of shuffled versions of each sequence. Motifs with very small (<0.05) *E*-values are statistically significant according to the given definition of random (shuffled sequences). The reported *E*-values are known to be conservative (too large), so motifs with *E*-values <0.05 may still be significant. Gibbs *(46)* uses a different statistical test (Wilcoxon signed-rank test) to determine motif significance. The relative merits of these two methods of assessing motif significance have not been studied.

Sometimes it is advisable to estimate motif significance empirically *(64)*. Many motif discovery algorithms do not make any attempt to report the statistical significance of the motifs they discover relative to the number of possible motifs that might have appeared in a randomly selected or generated sequence-set, so empirical estimation is the only available approach. Another reason to evaluate the significance of motifs empirically is that the motif significance estimates given by algorithms such as those named in the previous paragraph tend to be conservative, causing some biologically significant motifs to appear to be artefacts (*see* **Note 5**).

Empirical significance testing is very computationally expensive and therefore should generally be done using motif discovery algorithms installed on your local computer. Empirical significance testing is done by running the motif discovery algorithm hundreds of times on random sets of sequences of the same type and length, and with the same input parameters to the program, as were used in finding the motifs you are interested in evaluating. The motif scores for all the motifs found in the random runs are plotted as a histogram—the empirical score distribution. The significance of your real motifs' scores can be estimated by seeing where they lie on the histogram. The motif score can be either the information content score or the objective function score of the particular motif discovery method—usually some measure of over-representation. How you select (or generate) the random sequence sets depends on your application. For example, if your real sequences are selected upstream regions of genes from a single organism, a reasonable random model would be to use randomly chosen upstream regions from the same organism.

Whether or not you choose to determine their statistical significance, you will probably want to determine as much as possible about the function of your motifs (*see* **Note 6**). To do this, you can use your motifs to search databases of motifs and motif families, and you can use your motifs individually and in groups to search databases of sequences for single matches and local clusters of matches. DNA motifs can be searched against known vertebrate TF motifs in JASPAR. The JASPAR database also contains motifs that represent the binding affinities of whole families of TFs. If your motif matches one of these family motifs, it may be the TFBS motif of a TF in that structural family. You can search your protein motif against the BLOCKS or PRINTS *(5)* database using the LAMA program *(65)* to identify if it corresponds to a known functional domain. These databases are summarized in **Table 12.4**.

You will also want to see if your motif occurs in sequences other than those in the sequence set in which it was discovered. This is done by scanning a database of sequences using your motif (or motifs) as the query. This can help validate the motif(s) and shed light on its (their) function. If the novel occurrences have a positional bias relative to some sequence landmark (e.g., the transcriptional start site), then this can be corroborating evidence that the motif may be functional *(47)*. In bacteria, real TFBSs are more likely to occur relatively close to the gene for their TF, so proximity to the TF can increase confidence in TFBSs predicted by motif scanning *(2)*. Similarly, when the occurrences of two or more motifs cluster together in several sequences, it may be evidence that the motifs are functionally

Table 12.4
Some searchable motif databases with web servers

Database	Description
JASPAR	Searchable database of vertebrate TF motifs and TF-family motifs. http://jaspar.genegre.net/
BLOCKS PRINTS	Databases of protein signatures. http://blocks.fhcrc.org/blocks-bin/LAMA_search.sh

related. (Care must be taken that the clustering of co-occurrences is not simply due to sequence homology.) The functions of the sequences in which novel motif occurrences are detected can also provide a hint to the motif's function. Scanning with multiple motifs can shed light on the interaction/co-occurrence of protein domains and on cis-regulatory modules (CRMs) in DNA.

Numerous programs are available to assist you in determining the location, co-occurrence and correlation with functional annotation of your motifs in other sequences. The MAST program *(66)* allows you to search a selection of sequence databases with one or more unordered protein or DNA motifs. The PATSER program *(55)* allows you to search sequences that you upload for occurrences of your DNA motif. Several tools are available for searching for cis-regulatory modules that include your TFBS motifs. They include MCAST *(67)*, Comet *(68)* and Cluster-buster *(69)*. To determine if the genomic positions of the matches to your motif or motifs are correlated with functional annotation in the GO (Genome Ontology) database *(70)*, you can use GONOME *(71)*. If the genomic positions are strongly correlated with a particular type of gene, this can shed light on the function of your motif. Some tools for motif scanning that are available for direct use via web servers are listed in **Table 12.5**.

Table 12.5
Some web servers for scanning sequences for occurrences of motifs

Program	Description
MAST	Search one or more motifs against a sequence database; provides a large number of sequence databases or allows you to upload a set of sequences. http://meme.nbcr.net
PATSER	Search a motif against sequences you upload. http://rsat.ulb.ac.be/rsat/patser_form.cgi
Comet, Clusterbuster	Search for cis-regulatory modules. http://zlab.bu.edu/zlab/gene.shtml
GONOME	Find correlations between occurrences of your motif and genome annotation in the GO database. http://gonome.imb.uq.edu.au/index.html

An important way to validate DNA motifs is to look at the conservation of the motif occurrences in both the original sequences and in sequences you scan as described in the previous paragraph. It has been shown that TFBSs exhibit higher conservation than the surrounding sequence in both yeast and mammals *(31, 32)*. Motifs whose sites (as determined by the motif discovery algorithm) and occurrences (as determined by scanning) show preferential conservation are less likely to be statistical artefacts. Databases such as the UCSC genome browser (*see* **Table 12.1**) can be consulted to determine the conservation of motif sites and occurrences.

5. Limitations of Motif Discovery

Awareness of the limitations of motif discovery can guide you to more success in the use of the approaches outlined in this chapter. Some limitations have to do with the difficulty of discovering weak motifs in the face of noise. Spurious motifs are another source of difficulty. Another limitation is caused by the difficulty in determining which sequences to include in the input sequence set (*see* **Note 1**).

You can often think of motif discovery as a "needle-in-a-haystack" problem where the motif is the "needle" and the sequences in which it is embedded is the "haystack." Because motif discovery algorithms depend on the relative over-representation of a motif in the input set of sequences, a motif is "weak" if it is not significantly over-represented in the input sequences relative to what is expected by chance (or relative to a negative set of sequences) *(72)*.

Over-representation is a function of several factors, including:
- The number of occurrences of the motif in the sequences
- How similar all the occurrences are to each other
- The length of the input sequences

The more occurrences of the motif the sequences contain, the easier they will be to discover. Therefore, adding sequences to the input set that have a high probability of containing a motif will increase the likelihood of discovering it. Conversely, it can be helpful to reduce the number of sequences by removing ones unlikely to contain motif occurrences. Many DNA motifs (e.g., TFBSs) tend to have low levels of similarity among occurrences, so it is especially important to limit sequence length and the number of "noise" sequences (ones not containing occurrences) in the input sequence set. Over-representation depends inversely on the length of the sequences, so it is always good to limit the length of the input sequences as much as possible. Current motif discovery algorithms perform poorly at discovering TFBS when the sequences are longer than 1,000 bp.

Spurious motifs are motifs caused by non-functional, repetitive elements such as SINES, ALUs, and by skewed sequence composition in regions such as CpG islands. Such regions will contain patterns that are easily detected by motif discovery algorithms and may obscure real motifs. To help avoid this, you can pre-filter the sequences using the methods described in **Section 4.2**. In some cases, pre-filtering is not an option because the motifs of interest may lie in the regions that would be removed by filtering. For example, DNA regulatory elements often occur in or near CpG islands. In such cases, manual inspection using the methods of the previous section is necessary to remove spurious motifs. Using an organism-specific (or genomic-region–specific) random model is possible with some motif discovery algorithms, and may help to reduce the number of spurious motifs.

It is also important to be aware of the reliability of the methodologies used in selecting the input sequences for motif discovery. For example, sequences selected based on microarray expression data may miss many TFs because their level of expression is too low for modern methods to detect reliably *(2)*. ChIP-on-chip has become a popular procedure for studying genome-wide protein-DNA interactions and transcriptional regulation, but it can only map the probable protein-DNA interaction loci within 1-2Kbp resolution. Even if the input sequences all contain a TFBS motif, many TFBS motifs will not be detected in such long sequences using current motif discovery algorithms *(73)*. Another difficulty in discovering regulatory elements in DNA is that they can lie very far from the genes they regulate in eukaryotes, making sequence selection difficulty.

6. Notes

1. Be aware of the limitations of the motif discovery algorithms you use. For example, do not input an entire genome to most motif discovery algorithms—they are not designed for that and will just waste a lot of computer time without finding anything.

2. Use all available background information to select the sequences in which you will discover motifs. Include as many sequences as possible that contain the motifs. Keep the sequences as short as possible. Remove sequences that are unlikely to contain any motifs.

3. Prepare the input sequences carefully by masking or removing repetitive features that are not of interest to you such as ALUs, sines, and low-complexity regions. Filtering programs such as DUST, XNU, SEG, and RepeatMasker can help you do this.

4. Try more than one motif discovery algorithm on your data. They have different strengths and one program will often detect a motif missed by other programs.

5. Evaluate the statistical significance of your motifs. Remember that most motif discovery algorithms report motifs in any dataset, even though they may not be statistically significant. Even if the algorithm estimates the significance of the motifs it finds, these estimates tend to be very conservative, making it easy to reject biologically important motifs. So you should re-run the motif discovery algorithm on many sets of sequences that you select to be similar to your "real" sequences, but that you do not expect to be enriched in any particular motif. Compare the scores of your "real" motifs with those of motifs found in the "random" sequences to determine if they are statistically unusual.

6. Compare the motifs you discover to known motifs contained in appropriate motif databases such as those in **Table 12.4**.

References

1. Blais, A., Dynlacht, B. D. (2005) Constructing transcriptional regulatory networks. *Genes Dev* 19, 1499–1511.
2. Tan, K., McCue, L. A., Stormo, G. D. (2005) Making connections between novel transcription factors and their DNA motifs. *Genome Res* 15, 312–320.
3. Hulo, N., Bairoch, A., Bulliard, V., et al. (2006) The PROSITE database. *Nucleic Acids Res* 34, D227–D230.
4. Henikoff, J. G., Greene, E. A., Pietrokovski, S., et al. (2000) Increased coverage of protein families with the Blocks Database servers. *Nucleic Acids Res* 28, 228–230.
5. Attwood, T. K., Bradley, P., Flower, D. R., et al. (2003) PRINTS and its automatic supplement, prePRINTS. *Nucleic Acids Res* 31, 400–402.
6. La, D., Livesay, D. R. (2005) Predicting functional sites with an automated algorithm suitable for heterogeneous datasets. *BMC Bioinformatics* 6, 116.
7. Matys, V., Kel-Margoulis, O. V., Fricke, E., et al. (2006) TRANSFAC and its module TRANSCompel: transcriptional gene regulation in eukaryotes. *Nucleic Acids Res* 34, D108–D110.
8. Sandelin, A., Alkema, W., Engstrom, P., et al. (2004) JASPAR: an open-access database for eukaryotic transcription factor binding profiles. *Nucleic Acids Res* 32, D91–D94.
9. Zhu, J., Zhang, M. Q. (1999) SCPD: a promoter database of the yeast Saccharomyces cerevisiae. *Bioinformatics* 15, 607–611.
10. Makita, Y., Nakao, M., Ogasawara, N., et al. (2004) DBTBS: database of transcriptional regulation in Bacillus subtilis and its contribution to comparative genomics. *Nucleic Acids Res* 32, D75–D77.
11. Salgado, H., Gama-Castro, S., Peralta-Gil, M., et al. (2006) RegulonDB (version 5.0): *Escherichia coli* K-12 transcriptional regulatory network, operon organization, and growth conditions. *Nucleic Acids Res* 34(Database issue), D394–397.
12. Waterston, R. H., Lindblad-Toh, K., Birney, E., et al. (2002) Initial sequencing and comparative analysis of the mouse genome. *Nature* 420, 520–562.

13. Gribskov, M., Veretnik, S. (1996) Identification of sequence pattern with profile analysis. *Methods Enzymol* 266, 198–212.
14. Eddy, S. R. (1998) Profile hidden Markov models. *Bioinformatics* 14, 755–763.
15. Krogh, A., Brown, M., Mian, I. S., et al. (1994) Hidden Markov models in computational biology. Applications to protein modeling. *J Mol Biol* 235, 1501–1531.
16. IUPAC-IUB Commission on Biochemical Nomenclature (1970) Abbreviations and symbols for nucleic acids, polynucleotides and their constituents. recommendations 1970. *Eur J Biochem* 15, 203–208.
17. van Helden, J., Andre, B., Collado-Vides, J. (1998) Extracting regulatory sites from the upstream region of yeast genes by computational analysis of oligonucleotide frequencies. *J Mol Biol* 281, 827–842.
18. van Helden, J., Rios, A. F., Collado-Vides, J. (2000) Discovering regulatory elements in non-coding sequences by analysis of spaced dyads. *Nucleic Acids Res* 28, 1808–1818.
19. Schneider, T. D., Stephens, R. M. (1990) Sequence logos: a new way to display consensus sequences. *Nucleic Acids Res* 18, 6097–6100.
20. Reinert, G., Schbath, S., Waterman, M. S. (2000) Probabilistic and statistical properties of words: an overview. *J Comput Biol* 7, 1–46.
21. Schneider, T. D., Stormo, G. D., Gold, L., et al. (1986) Information content of binding sites on nucleotide sequences. *J Mol Biol* 188, 415–431.
22. Berg, O. G., von Hippel, P. H. (1987) Selection of DNA binding sites by regulatory proteins. Statistical-mechanical theory and application to operators and promoters. *J Mol Biol* 193, 723–750.
23. Berg, O. G., von Hippel, P. H. (1988) Selection of DNA binding sites by regulatory proteins. II. The binding specificity of cyclic AMP receptor protein to recognition sites. *J Mol Biol* 200, 709–723.
24. Finn, R. D., Mistry, J., Schuster-Bockler, B., et al. (2006) Pfam: clans, web tools and services. *Nucleic Acids Res* 34, D247–D251.
25. Sinha, S. (2003) Discriminative motifs. *J Comput Biol* 10, 599–615.
26. Workman, C. T., Stormo, G. D. (2000) ANN-Spec: a method for discovering transcription factor binding sites with improved specificity. *Pac Symp Biocomput*, 467–478.
27. Sinha, S., Blanchette, M., Tompa, M. (2004) PhyME: a probabilistic algorithm for finding motifs in sets of orthologous sequences. *BMC Bioinformatics* 5, 170.
28. Moses, A. M., Chiang, D. Y., Eisen, M. B. (2004) Phylogenetic motif detection by expectation-maximization on evolutionary mixtures. *Pac Symp Biocomput* 324–335.
29. Siddharthan, R., Siggia, E. D., van Nimwegen, E. (2005) PhyloGibbs: a Gibbs sampling motif finder that incorporates phylogeny. *PLoS Comput Biol* 1, e67.
30. Liu, X., Brutlag, D. L., Liu, J. S. (2001) BioProspector: discovering conserved DNA motifs in upstream regulatory regions of co-expressed genes. *Pac Symp Biocomput*, 127–138.
31. Xie, X., Lu, J., Kulbokas, E. J., et al. (2005) Systematic discovery of regulatory motifs in human promoters and 3 UTRs by comparison of several mammals. *Nature* 434, 338-345.
32. Kellis, M., Patterson, N., Birren, B., et al. (2004) Methods in comparative genomics: genome correspondence, gene identification and regulatory motif discovery. *J Comput Biol* 11, 319–355.
33. Duda, R. O., Hart, P. E. (1973) *Pattern Classification and Scene Analysis.* John Wiley & Sons, New York.
34. Seki, M., Narusaka, M., Abe, H., et al. (2001) Monitoring the expression pattern of 1300 Arabidopsis genes under drought and cold stresses by using a full-length cDNA microarray. *Plant Cell* 13, 61–72.
35. Harbison, C. T., Gordon, D. B., Lee, T. I., et al. (2004) Transcriptional regulatory code of a eukaryotic genome. *Nature* 431, 99–104.
36. Kawaji, H., Kasukawa, T., Fukuda, S., et al. (2006) CAGE Basic/Analysis Databases: the CAGE resource for comprehensive promoter analysis. *Nucleic Acids Res* 34, D632–D636.
37. Kodzius, R., Matsumura, Y., Kasukawa, T., et al. (2004) Absolute expression values for mouse transcripts: re-annotation of the READ expression database by the use of CAGE and EST sequence tags. *FEBS Lett* 559, 22–26.
38. Tatusov, R. L., Fedorova, N. D., Jackson, J. D., et al. (2003) The COG database: an updated version includes eukaryotes. *BMC Bioinformatics* 4, 41.
39. Andreeva, A., Howorth, D., Brenner, S. E., et al. (2004) SCOP database in 2004: refinements integrate structure and sequence family data. *Nucleic Acids Res* 32, D226–D229.

40. La, D., Silver, M., Edgar, R. C., Livesay, D. R. (2003) Using motif-based methods in multiple genome analyses: a case study comparing orthologous mesophilic and thermophilic proteins. *Biochemistry* 42, 8988–8998.

41. Tatusov, R. L., Lipman, D. J. Dust, in the NCBI/Toolkit available at http://blast.wustl.edu/pub/dust/.

42. Claverie, J.-M., States, D. J. (1993) Information enhancement methods for large scale sequence analysis. *Comput Chem* 17, 191–201.

43. Wootton, J. C., Federhen, S. (1996) Analysis of compositionally biased regions in sequence databases. *Methods Enzymol* 266, 554–571.

44. Smit, A., Hubley, R., Green, P. Repeatmasker, available at http://www.repeatmasker.org.

45. Bailey, T. L., Elkan, C. (1994) Fitting a mixture model by expectation maximization to discover motifs in biopolymers. *Proc Int Conf Intell Syst Mol Biol* 2, 28–36.

46. Thompson, W., Rouchka, E. C., Lawrence, C. E. (2003) Gibbs Recursive Sampler: finding transcription factor binding sites. *Nucleic Acids Res* 31, 3580–3585.

47. Roth, F. P., Hughes, J. D., Estep, P. W., et al. (1998) Finding DNA regulatory motifs within unaligned non-coding sequences clustered by whole-genome mRNA quantitation. *Nat Biotechnol* 16, 939–945.

48. Liu, X. S., Brutlag, D. L., Liu, J. S. (2002) An algorithm for finding protein-DNA binding sites with applications to chromatin immunoprecipitation microarray experiments. *Nat Biotechnol* 20, 835–839.

49. van Helden, J., Andre, B., Collado-Vides, J. (2000) A web site for the computational analysis of yeast regulatory sequences. *Yeast* 16, 177–187.

50. Pavesi, G., Mereghetti, P., Mauri, G., et al. (2004) Weeder Web: discovery of transcription factor binding sites in a set of sequences from co-regulated genes. *Nucleic Acids Res* 32, W199–W203.

51. Sinha, S., Tompa, M. (2003) YMF: A program for discovery of novel transcription factor binding sites by statistical overrepresentation. *Nucleic Acids Res* 31, 3586–3588.

52. Liu, Y., Liu, X. S., Wei, L., Altman, R. B., et al. (2004) Eukaryotic regulatory element conservation analysis and identification using comparative genomics. *Genome Res* 14, 451–458.

53. Henikoff, S., Henikoff, J. G., Alford, W. J., et al. (1995) Automated construction and graphical presentation of protein blocks from unaligned sequences. *Gene* 163, GC17–GC26.

54. Gordon, D. B., Nekludova, L., McCallum, S., et al. (2005) TAMO: a flexible, object-oriented framework for analyzing transcriptional regulation using DNA-sequence motifs. *Bioinformatics* 21, 3164–3165.

55. Hertz, G. Z., Stormo, G. D. (1999) Identifying DNA and protein patterns with statistically significant alignments of multiple sequences. *Bioinformatics* 15, 563–577.

56. Frith, M. C., Hansen, U., Spouge, J. L., et al. (2004) Finding functional sequence elements by multiple local alignment. *Nucleic Acids Res* 32, 189–200.

57. Ao, W., Gaudet, J., Kent, W. J., et al. (2004) Environmentally induced foregut remodeling by PHA4/FoxA and DAF-12/NHR. *Science* 305, 1742–1746.

58. Eskin, E., Pevzner, P. A. (2002) Finding composite regulatory patterns in DNA sequences. *Bioinformatics* 18, S354–S363.

59. Thijs, G., Marchal, K., Lescot, M., et al. (2002) A Gibbs sampling method to detect overrepresented motifs in the upstream regions of coexpressed genes. *J Comput Biol* 9, 447–464.

60. Regnier, M., Denise, A. (2004) Rare events and conditional events on random strings. *Discrete Math Theor Comput Sci* 6, 191–214.

61. Favorov, A. V., Gelfand, M. S., Gerasimova, A. V., et al. (2005) A Gibbs sampler for identification of symmetrically structured, spaced DNA motifs with improved estimation of the signal length. *Bioinformatics* 21, 2240–2245.

62. Tagle, D. A., Koop, B. F., Goodman, M., et al. (1988) Embryonic epsilon and gamma globin genes of a prosimian primate *(Galago crassi caudatus)*. Nucleotide and amino acid sequences, developmental regulation and phylogenetic footprints. *J Mol Biol* 203, 439–455.

63. Duret, L., Bucher, P. (1997) Searching for regulatory elements in human non-coding sequences. *Curr Opin Struct Biol* 7, 399–406.

64. Macisaac, K. D., Gordon, D. B., Nekludova, L., et al. (2006) A hypothesis-based approach for identifying the binding specificity of regulatory proteins from chromatin immunoprecipitation data. *Bioinformatics* 22, 423–429.

65. Pietrokovski, S. (1996) Searching databases of conserved sequence regions by aligning protein multiple-alignments. *Nucleic Acids Res* 24, 3836–3845.
66. Bailey, T. L., Gribskov, M. (1998) Combining evidence using p-values: application to sequence homology searches. *Bioinformatics* 14, 48–54.
67. Bailey, T. L., Noble, W. S. (2003) Searching for statistically significant regulatory modules. *Bioinformatics* 19, II16–II25.
68. Frith, M. C., Spouge, J. L., Hansen, U., et al. (2002) Statistical significance of clusters of motifs represented by position specific scoring matrices in nucleotide sequences. *Nucleic Acids Res* 30, 3214–3224.
69. Frith, M. C., Li, M. C., Weng, Z. (2003) Cluster-Buster: finding dense clusters of motifs in DNA sequences. *Nucleic Acids Res* 31, 3666–3668.
70. Ashburner, M., Ball, C. A., Blake, J. A., et al. (2000) Gene ontology: tool for the unification of biology. *Nat Genet* 25, 25–29.
71. Stanley, S., Bailey, T., Mattick, J. (2006) GONOME: measuring correlations between gene ontology terms and genomic positions. *BMC Bioinformatics* 7, 94.
72. Keich, U., Pevzner, P. A. (2002) Subtle motifs: defining the limits of motif finding algorithms. *Bioinformatics* 18, 1382–1390.
73. Tompa, M., Li, N., Bailey, T. L., et al. (2005) Assessing computational tools for the discovery of transcription factor binding sites. *Nat Biotechnol* 23, 137–144.
74. Kent, W. J., Sugnet, C. W., Furey, T. S., et al. (2002) The human genome browser at UCSC. *Genome Res* 12, 996–1006.

Section III

Phylogenetics and Evolution

Chapter 13

Modeling Sequence Evolution

Pietro Liò and Martin Bishop

Abstract

DNA and amino acid sequences contain information about both the phylogenetic relationships among species and the evolutionary processes that caused the sequences to divergence. Mathematical and statistical methods try to detect this information to determine how and why DNA and protein molecules work the way they do. This chapter describes some of the models of evolution of biological sequences most widely used. It first focuses on single nucleotide/amino acid replacement rate models. Then it discusses the modelling of evolution at gene and protein module levels. The chapter concludes with speculations about the future use of molecular evolution studies using genomic and proteomic data.

Key words: Models of evolution, DNA mutations, amino acid substitution, Markov property, human genome evolution, ALU.

1. Introduction

Molecular evolutionary studies offer an effective method for using genomic information to investigate many biomedical phenomena. DNA and amino acid sequences may contain both information about the phylogenetic relationships between species and information about the evolutionary processes that have caused the sequences to diverge. The analysis of phylogenetic relationships enables us to recognize and even exploit the statistical dependencies among sequence data that are due to common ancestry. The study of phylogenetic relationships among species has moved from being a mainly descriptive and speculative discipline to being a valuable source of information in a variety of biological fields, particularly in biotechnology. The process of phylogeny reconstruction requires four steps. The first step comprises sequence selection and alignment to determine site-by-site DNA or amino acid differences. The second step is to build a mathematical model describing the evolution in time of the sequences. A model can be built empirically using properties calculated through comparisons of observed sequences or parametrically using chemical and biological properties of DNA and amino acids. Such models permit estimation of the genetic distance between two homologous sequences, measured by the expected number of nucleotide substitutions per site that have occurred on the evolutionary lineages between them and their most recent common ancestor. Such distances may be represented as branch lengths in a phylogenetic tree; the extant sequences form the tips of the tree, whereas the ancestral sequences form the internal nodes and are generally not known. Note that alignment can also be considered as part of the evolutionary model. Although successful attempts to combine sequence alignment and phylogeny do exist *(1)*, current methods for phylogeny reconstruction require a known sequence alignment and neglect alignment uncertainty. Alignment columns with gaps are either removed from the analysis or are treated in an *ad hoc* fashion. As a result, evolutionary information from insertions and deletions is typically ignored during phylogeny reconstruction. The third step involves applying an appropriate statistical method to find the tree topology and branch lengths that best describe the phylogenetic relationships of the sequences. One of the most important method for phylogeny reconstruction is that of maximum likelihood *(2)*. The fourth step consists of the interpretation of results. **Figure 13.1** shows that most of the human genome is populated by DNA repeats of different length, number, and degree of dispersion. Long repeats in few copies are usually orthologous genes, which may contain hidden repeats in the form of runs of amino acids, and retroviruses inserted in the genome. For example, the human genome contains more than 50 chemokine receptor genes that have high sequence similarity *(3)* and almost 1,000 olfactory receptor genes and pseudogenes *(4)*. Protein-coding sequences (specifically exons) comprise <1.5% of the human genome. Short repetitive DNA sequences may be categorized into highly and moderately repetitive. The first is formed by tandemly clustered DNA of variable length motifs (5–100 bp) and is present in large islands of up to 100 Mb. The second can be either short islands of tandemly repeated microsatellites/minisatellites ("CA repeats," tri- and tetra-nucleotide repeats) or mobile genetic elements. Mobile elements include DNA transposons, short and long interspersed elements (SINEs and LINEs), and processed pseudogenes *(5, 6)*. This chapter describes the most widely used models of evolution for the different elements of a genome.

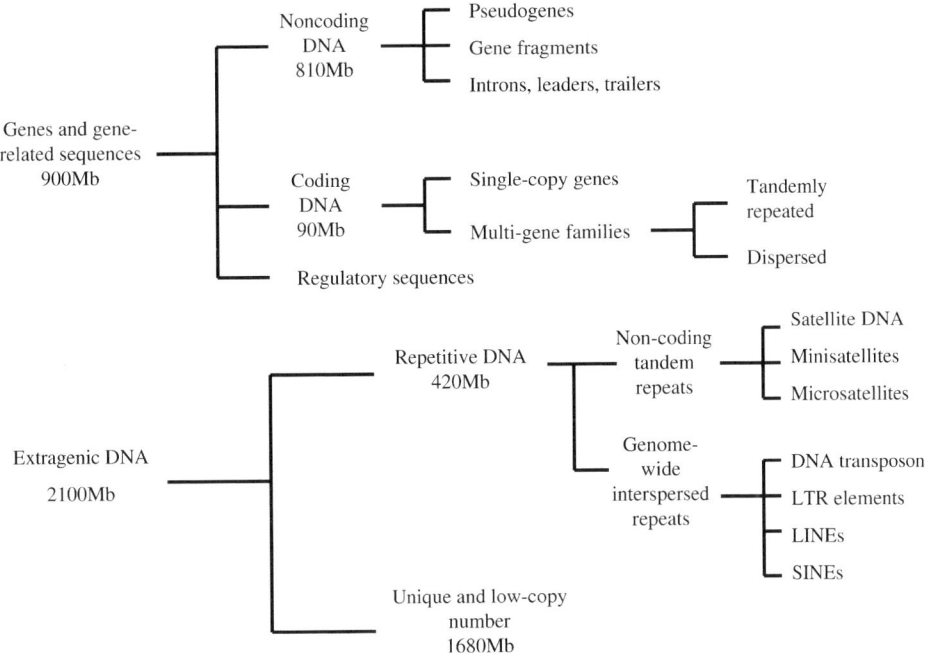

Fig. 13.1. Schematic description of the repetitious organization of the human genome.

2. Markov Models of DNA Evolution

Consider a stochastic model for DNA or amino acid sequence evolution. For the sake of brevity, we for the most part refer to DNA models, since the modifications for amino acid models are straightforward. We assume independence of evolution at different sites so that the probability of a set of sequences for some tree is the product of the probabilities for each of the sites in the sequences. At any single site, the model works with probabilities $P_{ij}(t)$ that base i will have changed to base j after a time t. The subscripts i and j take the values 1, 2, 3, 4 representing bases A, C, G, T, or 1, 2,..., 20 representing amino acids. The state space of the Markov chain is S_{DNA} = {A, C, G, T} for DNA sequences and $S_{protein}$ = {aa_1,...,aa_{20}} for amino acid sequences. A random variable, $X(t) \in S$, describes the substitution process of a sequence position. In general, a variable X follows a Markov process if:

$$\mathbf{P}(X(t_n) = j | X(t_1) = i_1.....X(t_{n-1}) = i_{n-1}) = \mathbf{P}(X(t_n) = j | X(t_{n-1}) = i_{n-1})$$

for all $j, i_1,....,i_n$. The Markov assumption asserts that $\mathbf{P}(X(t + s) = j | X(s) = i)$ is independent of $s \geq 0$. Assume that if a base mutates, it changes to a type i with a constant probability π_i. The simplest model considers a constant rate μ of mutation per unit time (generation). The probability of no mutations at a site after t generations is $(1 - \mu)^t$. The probability p that a mutation has occurred is thus: $p = 1 - (1-\mu)^t \approx 1 - e^{-\mu t}$. The probability of a change from base i to base j after time t can therefore be written as:

$$\mathbf{P}_{ii}(t) = (1-p) + p\pi_j, \; i = j$$
$$\mathbf{P}_{ij}(t) = p\pi_j, \; j \neq i$$

Note that probabilities involve mutation rate and time only through their product μt, which represents the expected number of substitutions along the branches of the tree.

A Markov process can have three important properties: homogeneity, reversibility, and stationarity. Homogeneity means that the rate matrix is independent of time, i.e., that the pattern of nucleotide distribution remains the same in different parts of the tree. This is not strictly true for DNA sequences because of the dependence of mutation on local sequence context. A homogeneous process has an equilibrium distribution, which is also the limiting distribution when time approaches infinity, i.e., $\lim_{t \to \infty} \mathbf{P}_{ij}(t) = \pi_j$ (and $\lim_{t \to 0^+} \mathbf{P}_{ij}(t) = \mathbf{I}$, where I is the identity matrix, i.e., $\mathbf{I}_{ij} = 1$ if $i = j$, $\mathbf{I}_{ij} = 0$ if $i \neq j$). Reversibility means that $\pi_i \mathbf{P}_{ij}(t) = \pi_j \mathbf{P}_{ji}(t)$ for all i,j and t. The rate matrix for a reversible process has only real eigenvalues and eigenvectors. Stationarity means that the process is at equilibrium, i.e., nucleotide frequencies have remained more or less the same during the course of evolution. Base frequencies are generally different in different species; therefore, these assumptions are clearly violated. In particular, bacterial genomes show large differences in base compositions.

Instead of considering time measured in discrete generations we can work with continuous time. We can write:

$$\mathbf{P}(t+dt) = \mathbf{P}(t) + \mathbf{P}(t)\mathbf{Q}dt = \mathbf{P}(t)(\mathbf{I} + \mathbf{Q}dt)$$

where $\mathbf{Q}$ is the instantaneous rate matrix of transition probabilities. Simple matrix manipulations and spectral diagonalization in order to calculate $\mathbf{P}(t) = e^{t\mathbf{Q}}$ can be summarized as follows:

$$\frac{d\mathbf{P}(t)}{dT} = \mathbf{Q}\mathbf{P}(t)$$

$$\mathbf{P}(t) = e^{t\mathbf{Q}} = \mathbf{I} + \mathbf{Q}t + \frac{(\mathbf{Q}t)^2}{2!} + ...$$

$$\mathbf{Q} = \mathbf{U}.diag\{\lambda_1,.....\lambda_n\}.\mathbf{U}^{-1}$$

where

$$diag\{\lambda_1,....,\lambda_n\} \equiv \begin{bmatrix} \lambda_1 & 0 & 0 & 0 & . \\ 0 & \lambda_2 & 0 & . & 0 \\ 0 & 0 & . & 0 & 0 \\ 0 & . & 0 & \lambda_{n-1} & 0 \\ . & 0 & 0 & 0 & \lambda_n \end{bmatrix}.$$

Therefore, each component of $\mathbf{P}(t) = \mathbf{U}.diag\{e^{t\lambda_1},....,e^{t\lambda_n}\}.\mathbf{U}^{-1}$ can be written as: $P_{ij}(t) = \sum c_{ijk}e^{t\lambda_k}$, where $i, j, k = 1,....,4$ for DNA sequences; $i, j, k = 1,....,20$ for proteins; and c_{ijk} is a function of $\mathbf{U}$ and $\mathbf{U}^{-1}$. The row sums of the transition probability matrix $\mathbf{P}(t)$ at time t are all ones. It is noteworthy that t and $\mathbf{Q}$ are confounded, $\mathbf{Q}t = (\gamma\mathbf{Q})(t/\gamma)$ for any $\gamma \neq 0$; twice the rate at half the time has the same results.

3. DNA Substitution Models

In 1969 Jukes and Cantor proposed a model in which all the π_is are set equal to $1/4$, and one base changes into any of the others with equal probability α *(7)*. Kimura, in 1980 proposed a two-parameter model that considered the transitions versus transversions bias *(8)*. The substitution matrix probability for this model can be represented as:

$$\mathbf{Q} = \begin{bmatrix} -\alpha - 2\beta & \beta & \alpha & \beta \\ \beta & -\alpha - 2\beta & \beta & \alpha \\ \alpha & \beta & -\alpha - 2\beta & \beta \\ \beta & \alpha & \beta & -\alpha - 2\beta \end{bmatrix}$$

In this matrix, bases are in alphabetic order, i.e., A, C, G, T. After Kimura, several authors proposed models with an increasing number of parameters. For instance, Blaisdell introduced an asymmetry for some reciprocal changes: $i \to j$ has a different substitution rate than $j \to i$ *(9)*. In contrast to Kimura's two-parameter model, the four-parameter model proposed by Blaisdell does not have the property of time reversibility. It is noteworthy that, beyond the biological rationale, time reversibility simplifies the calculations.

Felsenstein proposed a model in which the rate of substitution of a base depends on the equilibrium frequency of the nucleotide; given that the sum of equilibrium frequencies must be 1, this adds 3 more parameters *(10)*. The **Q** matrix for this model can be represented as:

$$\mathbf{Q} = \begin{bmatrix} -\mu(1-\pi_A) & \mu\pi_C & \mu\pi_G & \mu\pi_T \\ \mu\pi_A & -\mu(1-\pi_C) & \mu\pi_G & \mu\pi_T \\ \mu\pi_A & \mu\pi_C & -\mu(1-\pi_G) & \mu\pi_T \\ \mu\pi_A & \mu\pi_C & \mu\pi_G & -\mu(1-\pi_T) \end{bmatrix}$$

Hasegawa and co-workers improved Felsenstein's model by considering transition and transversion bias *(11)*. They considered $\mathbf{Q}_{ij} = \alpha\pi_j$ for transitions, $\mathbf{Q}_{ij} = \beta\pi_j$ for transversions (where $\Sigma\,\pi_j Q_{ij} = -1$). Moreover they considered two classes of DNA sites: Class 1 represents the third codon position; class 2 sites are the first and second codon position and also positions in ribosomal DNA and tRNA. This distinction is based on the observation that the majority of mutations occurring at the third site do not change the coded amino acid, i.e., they are synonymous. For both these classes they calculated the π_j. The rate matrix is:

$$\mathbf{Q} = \begin{bmatrix} -\alpha\pi_G - \beta\pi_Y & \beta\pi_C & \alpha\pi_G & \beta\pi_T \\ \beta\pi_A & -\alpha\pi_T - \beta\pi_R & \beta\pi_G & \alpha\pi_T \\ \alpha\pi_A & \beta\pi_C & -\alpha\pi_A - \beta\pi_Y & \beta\pi_T \\ \beta\pi_A & \alpha\pi_C & \beta\pi_G & -\alpha\pi_C - \beta\pi_R \end{bmatrix}$$

where $\pi_R = \pi_A + \pi_G$ and $\pi_Y = \pi_C + \pi_T$. Thus, with respect to Hasegawa's model, Kimura's model corresponds to the case in which all π_i are equal. Felsenstein's model corresponds to the case of $\beta = \alpha$. When both these simplifications are made we obtain the Jukes-Cantor model. The most general model can have at most 12 independent parameters; insisting on reversibility reduces this to 9 and can be parameterized as follows:

$$\mathbf{Q} = \begin{bmatrix} - & \alpha\pi_C & \beta\pi_G & \gamma\pi_T \\ \alpha\pi_A & - & \rho\pi_G & \sigma\pi_T \\ \beta\pi_A & \rho\pi_C & - & \tau\pi_T \\ \gamma\pi_A & \sigma\pi_C & \tau\pi_G & - \end{bmatrix}$$

where the diagonal elements should be replaced by terms that make row sums equal to zero. The models described above are parametric, in the sense that they are defined in terms of parameters (π_i, α, β, etc.) inspired by our understanding of biology. Empirical models of nucleotide substitution also have been studied. These models are derived from the analysis of inferred substitutions in reference sequences, perhaps the sequence under current study or from databases. Advantages of this approach can be a better description of the evolution of the sequences under study, if a suitable reference set is used, particularly if this reference set is large. Disadvantages can be inaccuracy due to an inappropriate reference set, and a lack of a broader biological interpretability of purely empirical findings. More general models of nucleotide substitution have been studied by other authors as for instance Lanave et al. *(12)*, Zharkikh *(13)*, and Li *(14)*.

3.1. Modeling Rate Heterogeneity Along the Sequence

The incorporation of heterogeneity of evolution rates among sites has led to a new set of models that generally provide a better fit to observed data, and phylogeny reconstruction has improved *(15)*. Hasegawa considered three categories of rates (invariable, low rate, and high rate) for human mitochondrial DNA *(16)*. Although a continuous distribution in which every site has a different rate seems to be the most biological plausible model, Yang has shown that four categories of evolutionary rates with equal probability, chosen to approximate a gamma distribution (the discrete Gamma model), perform very well. This model is also considerably more practical computationally *(15)*. The gamma distribution, Γ, has two parameters: a shape parameter, Γ, and a scale parameter, β. If we assume $\beta = \alpha$, then the mean becomes 1 and the variance $1/\alpha$. The shape parameter is inversely proportional to the mutation rate. If α is less than 1, there is a relatively large amount of rate variation, with many sites evolving very slowly, but some sites evolving at a high rate. For values of α greater than 1, the shape of the distribution changes qualitatively, with less variation and most sites having roughly similar rates. Yang *(16)* and Felsenstein and Churchill have implemented methods in which several categories of evolutionary rates can be defined *(17)*. Both methods use hidden Markov model techniques *(18–20)* to describe the organization of areas of unequal and unknown rates at different sites along sequences. All possible assignments of evolutionary rate category at each site contributed to the phylogenetic analysis of sequences, and algorithms are also available to infer the most probable rate category for each site.

3.2. Models Based on Nearest Neighbor Dependence

It is well known that neighboring nucleotides in DNA sequences do not mutate independently of each other. The assumption of independent evolution at neighboring sites simplifies calculations since, under this assumption, the likelihood is the product of individual site likelihoods. A good example of violation is provided by the methylation-induced rate increase of C to T (and G to A) substitutions in vertebrate CpG dinucleotides, which results in 1,718-fold increased CpG to TpG/CpA rates compared with other substitutions. Several authors have investigated the relative importance of observed changes in two adjacent nucleotide sites due to a single event *(21)*. Seipel and Haussler (2003) *(22)* have shown that a Markov chain along a pair of sequences fits sequence data substantially better than a series of independent pairwise nucleotide distributions (a zero-order Markov chain). They also used a Markov model on a phylogenetic tree, parameterized by a di-nucleotide rate matrix and an independent-site equilibrium sequence distribution, and estimated substitution parameters using an expectation maximization (EM) procedure *(23)*.

Whelan and Goldman *(24)* have developed the singlet-doublet-triplet (SDT) model, which incorporated events that change one, two, or three adjacent nucleotides. This model allows for neighbor- or context-dependent models of base substitutions, which consider the N-bases preceding each base and are capable of capturing the dependence of substitution patterns on neighboring bases. They found that the inclusion of doublet and triplet mutations in the model gives statistically significant improvements in fit of model to data, indicating that larger-scale mutation events do occur. There are indications that higher-order states, autocorrected rates, and multiple functional categories all improve the fit of the model and that the improvements are roughly additive. The effect of higher-order states (context dependence) is particularly pronounced.

4. Codon Models

In an attempt to introduce greater biological reality through knowledge of the genetic code and the consequent effect of nucleotide substitutions in protein coding sequences on the encoded amino acid sequences, Goldman and Yang *(25, 26)* described a codon mutation model. They considered the 61 sense codons i consisting of nucleotides $i_1 i_2 i_3$. The rate matrix **Q** consisted of elements Q_{ij} describing the rate of change of codon $i = i_1 i_2 i_3$ to $j = j_1 j_2 j_3$ ($i \neq j$) depending on the number and type of differences between i_1 and j_1, i_2 and j_2, and i_3 and j_3 as follows:

$$Q_{ij} = \begin{cases} 0 & \text{if 2 or 3 of the pairs } i_k, j_k \text{ are different} \\ \mu \pi_j e^{-d_{aa_i, aa_j}/V} & \text{if one pair differ by a transversion} \\ \mu \kappa \pi_j e^{-d_{aa_i, aa_j}/V} & \text{if one pair differ by a transition} \end{cases}$$

where d_{aa_i, aa_j} is the distance between the amino acid coded by the codon i (aa_i) and the amino acid coded by the codon j (aa_j) as calculated by Grantham *(27)* on the basis of the physicochemical properties of the amino acids. This model takes account of codon frequencies (through π_j), transition/transversion bias (through κ), differences in amino acid properties between different codons

(d_{aa_i,aa_j}), and levels of sequence variability (V). Recent works by Yang and Nielsen *(28)* and Pedersen et al. *(29)* have developed and improved this model.

Note that the use of likelihood ratio tests, LRTs, to compare pairs of models using different statistical distributions to describe the variation in x has proved an effective way of identifying adaptive evolution. For these tests, the null model describes the evolution of a protein as a distribution containing only neutral and purifying selection ($x ≤ 1$), and the alternative model describes a similar distribution that also allows positive selection (x can take all values). A popular choice of models for forming these hypotheses are M7 and M8 *(30)*. M7 (the null model) describes variation in x between 0 and 1 with a beta distribution, which can take a variety of shapes that describe a wide range of potential selective scenarios and requires only two simple parameters to be estimated from the data. M8 (the alternate model) contains the beta distribution of M7, but also includes a single variable category of x to describe positive selection. When statistical tests show that M8 explains the evolution of the protein significantly better than M7 and the additional category of x is >1, positive selection is inferred. Many new incidences of adaptive evolution have been found using this approach and extensions of these methods allow the detection of the specific sites in a protein that are undergoing positive selection.

5. Amino Acid Mutation Models

In contrast to DNA substitution models, amino acid replacement models have concentrated on the empirical approach. In reality it is hardly practical to use a model with hundreds of parameters. Dayhoff and co-workers developed a model of protein evolution that resulted in the development of a set of widely used replacement matrices *(31, 32)*. In the Dayhoff approach, replacement rates are derived from alignments of protein sequences that are at least 85% identical and used to build phylogenetic trees. The internal nodes of the tree give the inferred ancestral sequences; the count of amino acid changes between the ancestral and the actual sequences gives the relative mutability of each of the 20 amino acids. The 85% threshold of amino acid identity between aligned sequences ensures that the likelihood of a particular mutation (e.g., $L \to V$) being the result of a set of successive mutations (e.g., $L \to x \to y \to V$) is low. An implicit instantaneous rate matrix was estimated, and replacement probability matrices $\mathbf{P}(t)$ generated for different values of t. Dayhoff and co-workers calculated the observed amino acid replacement pattern in a set of closely related sequences showing that an amino acid is very frequently replaced by an amino acid of similar physiochemical properties. Jones et al. *(33)* and Gonnet et al. *(34)* have also calculated an amino acid replacement matrix specifically for membrane spanning segments (**Fig. 13.3**). This matrix has remarkably different values from the Dayhoff matrices, which are known to be biased toward water-soluble globular proteins. One of the main uses of the Dayhoff matrices has been in database search methods where, for example, the matrices **P**(0.5), **P***(1)*, and **P**(2.5) (known as the PAM50, PAM100, and PAM250 matrices) are used to assess the significance of proposed matches between target and database sequences. Since relatively few families were considered, the resulting matrix of accepted point mutations included a large number of entries equal to 0 or 1. The counts of the amino acid changes between the inferred ancestral sequences and the actual sequences give an estimate of the relative mutability of each of the 20 amino acids. The number of the matrix (PAM50 and PAM100) refers to the evolutionary distance; greater numbers are greater distances. **Figure 13.2** shows a comparison between the most widely used amino acid substitution models. Matrices for greater evolutionary distances are extrapolated from

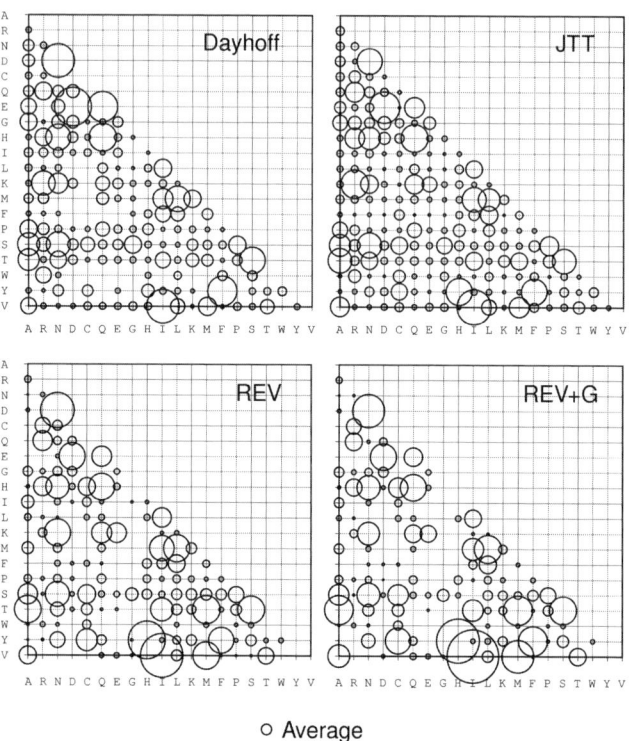

○ Average

Fig. 13.2. Comparison of amino acid replacement models: Dayhoff, JTT, REV, REV+ Gamma.

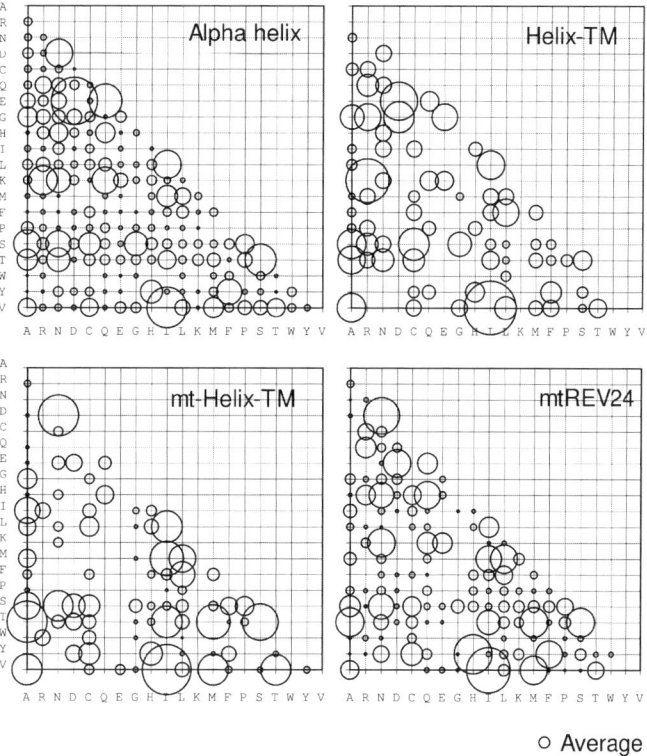

Fig. 13.3. Comparison of structural replacement matrices: alpha helix, transmembrane helix, mitochondrial transmembrane helix, mtREV24.

those for lesser distances. The mutation probability matrix is specific for a particular evolutionary distance, but may be used to generate matrices for greater evolutionary distances by multiplying it repeatedly by itself. The BLOSUM series of matrices generally perform better than PAM matrices for local similarity searches *(35)*. Claverie has developed a set of substitution matrices designed explicitly for finding possible frame shifts in protein sequences *(36)*.

5.1. Generating Mutation Matrices

All the methods for generating mutational data matrices are similar to that described by Dayhoff et al. *(32)*. The method involves three steps: *(1)* clustering the sequences into homologous families, *(2)* tallying the observed mutation between highly similar sequences, and *(3)* relating the observed mutation frequencies to those expected by pure chances.

A method proposed by David Jones, Willie Taylor, and Janet Thornton uses inferred phylogenetic relationships among the sequence data *(33)*.

The first step is to generate a mutation probability matrix; Elements of this matrix give the probability that a residue in column j will mutate to the residue in row i in a specific unit of evolutionary time. The diagonal elements of the matrix represent the probability that residue $i = j$ remains unchanged, $M_{ij} = 1 - \lambda m_j$, where m_j is the average relative mutability of residue j, and λ is a proportional constant. Non-diagonal elements are given by $M_{ji} = \lambda m_i A_{ij} / \Sigma_i A_{ij}$, where A_{ij} is a non-diagonal element of the raw PAM matrix. The value of λ relates to the evolutionary distance accordingly: $\Sigma_i f_i M_{ij} = 1 - 0.01 \cdot P$, where f_i is the normalized frequency of occurrence of residue i, and P approximates the evolutionary distance (in PAMs) represented by the matrix. The relationship breaks down for $P \gg 5$. P is usually given the value of 1 so that the basic mutation matrix represents a distance of 1 PAM. Matrices representing larger evolutionary distances may be derived from the 1 PAM matrix by matrix multiplication. When used for the comparison of protein sequences, the mutation probability matrix is usually normalized by dividing each element by the relative frequency of exposure to mutation of the amino acid. This operation results in the symmetrical "relatedness odds matrix," in which each element gives the probability of amino acid replacement. A second step is the calculation of log-odds $M_{ij} = 10 \log_{10} R_{ij}$, where R_{ij} are elements of the relatedness odds matrix. The logarithm of each element is taken to allow probabilities to be summed over a series of amino acids rather than requiring multiplication. The resulting matrix is the "log-odds matrix," which is frequently referred to as "Dayhoff's matrix" and often used at a distance of close to 256 PAM since this lies near to the limit of detection of distant relationships. Henikoff and Henikoff (35), using local, ungapped alignments of distantly related sequences, derived the BLOSUM series of matrices. The number after the matrix (BLOSUM62) refers to the minimum percent identity of the blocks used to construct the matrix; as a rule of thumb, greater numbers are lesser distances. It is noteworthy that these matrices are directly calculated without extrapolations. There is no such thing as a perfect substitution matrix; each matrix has its own limitations. The general consensus is that different matrices are better adapted to different purposes and matrices derived from observed substitution data (e.g., the Dayhoff or BLOSUM matrices) are superior to identity, genetic code, or physical property matrices. However, there are Dayhoff matrices of different PAM values and BLOSUM matrices of different percentage identities. The most widely used matrix for protein sequence comparison has been the PAM-250 matrix. This matrix was selected since in Monte Carlo studies matrices reflecting the evolutionary distance gave a consistently higher significance score than other matrices in the range 0–750 PAM. When using a local alignment method, Altschul suggests that three matrices should ideally be used: PAM40, PAM120, and PAM250. The lower matrices will tend to find short alignments of highly similar sequences, whereas higher PAM matrices will find longer, weaker local alignments *(37)*.

It is now possible to recreate inferred ancestral proteins in the laboratory and study the functions of these molecules. Tracing changes in protein structure along the branches of a phylogenetic tree can provide important insights into molecular function. How confident can we be of correctly reconstructing ancestral proteins? For example, if you were 95% sure of which nucleotide was present at each site in a sequence 100 nucleotide long, then your overall probability for having the correct sequence would be $(0.95)^{100}$, which results in a <1% chance of reconstructing the correct sequence. Thus the development of good models of evolution is of utmost importance.

5.2. Amino Acid Models Incorporating Structural Features

The relationship between phenotype and survival of the genotype is central to both genetics and evolution. The idea is that protein sequences are close to the genotype, whereas protein structures are a fundamental unit of the phenotype. The fact that protein structure changes more slowly over time than does protein sequence allows one to explore the constraints on protein sequence evolution that serve to maintain protein structure. Early investigations were mainly concerned with characterizing how patterns of amino acid replacement at a site in a protein are associated with the structural environment of the site *(38)*. In addition to the obvious relevance of this research for a better understanding of the process of molecular evolution, it is also pertinent to prediction of protein structure, which is a central problem in biotechnology.

Phylogenetic inference has been improved by incorporating structural and functional properties into inferential models. This information can be used to refine phylogenetic models and provide structural biologists with additional clues to natural selection and protein structure. The first approach is to consider details that influence structure but are not immediately related to it, such as physicochemical properties of amino acids (hydrophobicity, charge, and size) *(38)*. Approaches closer to structural biology have been implemented by Rzhetsky et al. *(39)* and Goldman et al. *(40)*. **Figure 13.3** shows the comparison between structure-specific or organelle specific amino acid substitution models.

Goldman and co-workers have introduced a set of evolutionary models that combine protein secondary structure and amino acid replacement. Their approach is related to that of Dayhoff and co-workers but considers different categories of structural environment: α, helix, trans-membrane helix; β, sheet, turn, and loop; and further classifies each site according to whether it is exposed to solvent or is buried; the Dayhoff approach simply considers an average environment for each amino acid. These matrices are organized in a composite evolutionary model through the means of a hidden Markov model algorithm *(28–31)*. Maximum likelihood estimation of phylogeny is then possible using a multiple sequence alignment. An important characteristic of the algorithm is that it can work without any specific structural information for the protein under study. In this case it uses the series of structure-specific replacement matrices and a set of transition probabilities between them to represent a model of the typical structure of similar proteins, which will incorporate prior knowledge gained from the analysis of other proteins of known structure. The algorithm can compute likelihoods integrated over all possible structures, weighted in accordance to their probability under the structural model. Information relating to the statistical dependence of the sequences in the dataset and their patterns of amino acid replacements can be used to derive both phylogeny and predicted secondary structure organization. This section presents a brief description of the general algorithm implemented; full description of existing models are given by Goldman, Thorne, and Jones *(41, 42)* and Liò and Goldman *(43, 44)*.

Let the aligned dataset be denoted by S, its length by N, the first i columns of the dataset by S_i, and the ith column itself by s_i. Gaps in the alignments are considered as missing information, as in most common phylogenetic programs. The likelihood of the tree T is given by $\Pr(S \mid T)$, and this is calculated via the terms $\Pr(S_i, c_i \mid T)$ for each possible secondary structure category c_i at site i using the iteration:

$$\Pr(S_i, c_i \mid T) = \sum_{c_{i-1}} \Pr(S_{i-1}, c_{i-1} \mid T) \rho_{c_{i-1} c_i} \Pr(s_i \mid c_i, T)$$

for $i > 1$. The terms $\Pr(S_i, c_i \mid T)$ are evaluated using Markov process replacement models appropriate for each secondary structure c_i and the "pruning" algorithm of Felsenstein *(10)*. The ρ_{ij} are the HMM transition probabilities between states. For $i = 1$, the iteration is started using:

$$\Pr(S_1, c_1 \mid T) = \Pr(s_1 \mid c_1, T) . \psi_{c_1}$$

where ψ_{c_1} is the stationary distribution of c_i. When completed, the iteration gives the required $\Pr(S \mid T)$ because:

$$\Pr(S \mid T) = \sum_{c_N} \Pr(S_N, c_N \mid T).$$

If secondary structure and accessibility information are available, modified likelihood calculations can be performed that do not involve the ρ_{ij}. Once the tree topology and branch lengths ($\hat{T}$) that have maximum likelihood have been found, the calculation of *a posteriori* probabilities of secondary structures for each site of the protein, $\Pr(c_i \mid S, \hat{T})$, allows prediction of the secondary structure for each site. This maximum likelihood approach gives a solid base for hypothesis testing and parameter estimation.

Echave and collaborators have developed an approach to modeling structurally constrained protein evolution (SCPE) in which trial sequences generated by random mutations at the gene level are selected against departure from a reference three-dimensional structure *(45)*. The model is based on a sequence-structure distance score, S_{dist}, which depends on a reference native structure and a parameter, S_{div}, which measures the degree of structural divergence tolerated by natural selection. The sequence-structure distance measure S_{dist} is calculated as follows. First, the trial sequence is forced to adopt the three-dimensional reference structure. Then, mean field energies per position $E_{trial}(p)$ and $E_{ref}(p)$ are calculated for the trial and reference sequences, respectively. Finally, $S_{dist} = \sqrt{\left[E_{trial}(p)^2 - E_{ref}(p)^2\right]}$ is obtained. To calculate the mean-field energies, the authors used the PROSA II potential *(46)*, which includes additive pair contributions that depend on the amino acid types and the geometric distance between the Cb atoms of the interacting amino acids, as well as a surface term that models the protein solvent interactions. An SCPE simulation starts with a reference DNA sequence that codes for a reference protein of known three-dimensional structure. Then, each run involves the repetition of evolutionary time steps, which consist of the application of the following four operations. First, the DNA sequence of the previous time step is mutated by introducing a random nucleotide substitution into randomly chosen sequence positions (Jukes-Cantor model). Second, if the mutation introduces a stop codon, the mutated DNA is rejected; otherwise, the muted DNA is translated, using the genetic code, to obtain a trial protein sequence. Third, the sequence-structure distance score, S_{dist}, is computed. The trial sequence is accepted only if S_{dist} is below the specified cut-off, S_{div}, which represents the degree of structural divergence allowed by natural selection. A similar approach was proposed by Bastolla and colleagues, in which possible mutations are tested for conservation of structural stability using a computational model of protein folding *(47)*.

5.3. Amino Acid Models Incorporating Correlation

In all the previous amino acid substitution models, there is an assumption that the sites evolve independently. The evolution of interacting sites is less easy to model. Correlations between sites distant in the linear sequence of a protein often reflect effects on parts of the protein that are very close in the folded (three-dimensional) structure. As such analyses become more specialized, however, there is some concern over whether there will ever be enough data to find these correlations reliably. Pollock and co-workers *(48)* considered a Markov process model similar to that of Felsenstein *(10)* for a single site that may have two states A and a, where A might be a set of large residues and a the complementary set of small residues, or residues with a different charge. There is a rate parameter, λ, and an equilibrium frequency parameter, π_A ($\pi_A + \pi_a = 1$), such that the instantaneous rate of substituting state j for a different state i is equal to $\lambda \pi_j$. The matrix of transition probabilities at time t is then:

$$\mathbf{P}_{ij}(t) = \begin{cases} \exp(-\lambda t) + \pi_i \exp(1-\lambda t) & i = j \\ \pi_i \exp(1-\lambda t) & i \neq j \end{cases}$$

This substitution process is reversible, i.e., $\pi_i P_{ij}(t) = \pi_j P_{ji}(t)$. A further extension is to model correlated change in pairs of sites. This was first introduced by Pagel *(49)* for comparative analysis of discrete characters. Consider a second site with two states, B and b, with equilibrium frequencies π_B and π_b (where $\pi_B + \pi_b = 1$). Then the matrix of instantaneous transition rates is:

$$\mathbf{Q} = \begin{bmatrix} -\sum_{AB} & \lambda_B \pi_{Ab}/\pi_A & \lambda_A \pi_{aB}/\pi_b & 0 \\ \lambda_B \pi_{AB}/\pi_A & -\sum_{Ab} & 0 & \lambda_A \pi_{ab}/\pi_b \\ \lambda_A \pi_{AB}/\pi_B & 0 & -\sum_{aB} & \lambda_B \pi_{ab}/\pi_a \\ 0 & \lambda_A \pi_{Ab}/\pi_b & \lambda_B \pi_{aB}/\pi_a & -\sum_{ab} \end{bmatrix}$$

where Σ_{ij} is the sum of off-diagonal elements for row ij and λ_A and λ_B are the two rate parameters governing substitution at the two loci, A and B. Rows and columns are ordered as AB, Ab, aB, ab. The number of free parameters is five: two rate parameters, and because the π_{ij} sum to one, three independent values of π_{ij}. There is an extra degree of freedom that can be represented by the quantity $RD = \pi_{AB}\pi_{ab} - \pi_{Ab}\pi_{aB}$; this quantity is analogous to the linkage disequilibrium. If the quantity RD is different from zero, there is some degree of dependence between the two sites. RD can be negative or positive and this corresponds to either compensation or anti-compensation of the residues. Again, the substitution probabilities for the co-evolving model can be calculated using $\mathbf{P}(t) = \exp[\mathbf{Q}t]$. Rather than using this model to construct a phylogenetic tree (which would be possible in principle), if there is a given phylogenetic tree, it is possible to use it to test the evolutionary model based on likelihood calculations.

As a final target, the understanding of protein evolution may allow one to distinguish between analogous and homologous proteins, i.e., detect similarities in those proteins that have very low sequence homology and have probably diverged from a common ancestor into the so-called twilight zone.

6. RNA Model of Evolution

Ribosomal RNA tertiary structure has been so far very hard to infer because the combinatorics of interaction grows very rapidly and strongly depends on the kinetics. Fairly reliable predictions of stable folded secondary structures can be made instead; it is possible to calculate free energies and other thermodynamic quantities for stem and loop regions. These two elements have different rates of base substitution because the stems are double-stranded regions and the loops are single-stranded RNA and thus selection pressures are different. The time reversibility constraint and the average mutation rate are set as they were for DNA models. The pairing within the stems involves the Watson-Crick A:U, G:C pairs, and the non-canonical G:U pair; other pairings exist, but they are rare enough to be disregarded in the current context. The stem regions are modeled using a 16-state rate matrix (because of all the possible pairings), whereas the loop regions are modeled using a four-state rate matrix. Rzhetsky and co-workers introduced a model to estimate base substitution in ribosomal RNA genes and infer phylogenetic relationships *(50)*. The model takes into account rRNA secondary structure elements: stem regions and loop regions. Other stimulating references are *(51)* and *(52)*. All the models described so far operate at the level of individual nucleotides.

7. Models of DNA Segment Evolution

Genome sequencing projects have revealed that much of the increase in genome size from bacteria to humans has been a result of DNA duplication. Different extents of duplication are possible: a small part of a gene, a module, an exon, an intron, a full coding region, a full gene, a chromosome segment containing a cluster of genes, a full chromosome, or the entire genome. In eukaryotes, gene duplications seem to have occurred often, for example, the olfactory *(4)*, HOX, or globin genes in animals. Moreover, the remnants of whole genome duplications have been identified, for example, for frogs, fishes, different yeast strains, and Arabidopsis *(6)*. The three main distinct mechanisms that generate tandem duplication of DNA stretches in eukaryotes are slipped-strand mis-pairing, gene conversion, and unequal recombination. Not every gene duplication results in the acquisition of a new function by one of the two duplicates: Most families, for example the globin and olfactory gene clusters, also contain many duplicates that have lost function (pseudogenes). Other duplicates can retain the original function and be maintained in a given genome for an indefinite time. The duplication of a single gene is an event whose average rate is on the order of 0.01 per gene per million years, ranging between 0.02 and 0.002 in different species. Moreover, duplicates exhibit an average half-life of about 4 million years, a time that appears to be congruent with the evolution of novel functions or the specialization of those that are ancestral and inefficient. Half of the genes in a genome are expected to duplicate and increase to high frequency at least once on time scales of 35 to 350 million years. Therefore, the rate of duplication of a gene is of the same order of magnitude as the rate of mutation per nucleotide site *(4)*.

A present debate on vertebrate evolution concentrates on the relative contribution of the large-scale genome duplication (the so-called big bang model) after the echinoderms/chordates split and before the vertebrate radiation, and on continuous origin by small-scale fragment duplications. After duplication, a gene starts diverging from the ancestral sequence. This is true not only of coding sequences, but also of regulatory sites.

The classical model of gene duplication states that, after the duplication event, one duplicate may become functionless, whereas the other copy retains the original function *(53)*. Complete duplicates (when the duplication involves both the coding and all the regulatory sequences of the original gene) are expected to be redundant in function (at least in the immediate beginning). In this case, one duplicate may represent a backup copy shielding the other from natural selection. This implies (as it is likely that a mutation will have a negative effect on function) that one duplicate will probably lose its function, whereas the other will retain it. Very rarely, an advantageous mutation may change the function of one duplicate and both duplicates may be retained. The most plausible fate in the light of the classical model is that one of the two duplicates will become a pseudogene. This fails to explain the amount of functional divergence and the long-term preservation of the large numbers of paralogous genes that constitute the eukaryotic multigene families, which often retain the original function for a long time. Accordingly, Walsh *(54)* used a Markov model of the evolutionary fate of duplicates immediately after the duplication event (when the duplicates are perfectly redundant). This Markov model has two absorbing states, fixation and pseudogenization. The main result is that if the population size is large enough, the fate of most duplicated genes is to gain a new function rather than become pseudogenes. Nadeau and Sankoff *(55)*, studying human and mice genes, estimated that about 50% of gene duplications undergo functional divergence. Other researches showed that the frequency of preservation of paralogous genes following ancient polyploidization events are in the neighborhood of 30–50% over periods of tens to hundreds of millions of years. To overcome these limitations, new models have been proposed that are a better fit to empirical data.

8. Sub-Functionalization Model

In eukaryotes, gene expression patterns are typically controlled by complex regulatory regions that finely tune the expression of a gene in a specific tissue, developmental stage, or cell lineage. Of particular interest is the combinatorial nature of most eukaryotic promoters, which are composed of different and partially autonomous regions with a positive or negative effect on downstream gene transcription, with the overall expression pattern being determined by their concerted (synergistic) action.

Similarly, proteins can contain different functional and/or structural domains that may interact with different substrates and regulatory ligands, or other proteins. Every transcriptionally important site or protein domain can be considered as a sub-functional module for a gene or protein, each one contributing to the global function of the gene or protein. Starting from this idea, Lynch and Force first proposed that multiple sub-functions of the original gene may play an important role in the preservation of gene duplicates. They focused on the role of degenerative mutations in different regulatory elements of an ancestral gene expressed at rates which depend on a certain number of different transcriptional modules (sub-functions) located in its promoter region. After the duplication event, deleterious mutations can reduce the number of active sub-functions on one or both the duplicates, but the sum of the sub-functions of the duplicates will be equal to the number of original functions before duplication (i.e., the original functions have been partitioned between the two duplicates). Similarly, considering both duplicates, they are together able to complement all the original sub-functions; moreover, they may have partially redundant functions.

This example of sub-functionalization considers both functions affecting the expression patterns dependent on promoter sequences recognized by different transcription factors, and also "hub" proteins with different and partially independent domains. The sub-functionalization, or duplication-degeneration-complementation model (DDC) of Lynch and Force, differs from the classical model because the preservation of both gene copies mainly depends on the partitioning of sub-functions between duplicates, rather than the occurrence of advantageous mutations.

A limitation of the sub-functionalization model is the requirement for multiple independent regulatory elements and/or functional domains. The classical model is still valid if gene functions cannot be partitioned; for example, when selection pressure acts to conserve all the sub-functions together. This is often the case when multiple sub-functions are dependent on one another.

9. Sub-Neo-Functionalization

A recent improvement in gene duplication models has been proposed by He and Zhang *(58)* starting from the result of a work on yeast protein interaction data (from MIPS) and human expression data that have been tested both under the neo-functionalization and sub-functionalization models. Neither model alone satisfied the experimental results for duplicates. Further progress is formalized as the sub-neo-functionalization model, which is a mix of the previous models; sub-functionalization appears to be a rapid process, whereas neo-functionalization requires more time and continues long after duplication *(59)*.

The rapid sub-functionalization observed by He and Zhang can be viewed as the acquisition of expression divergence between duplicates. After sub-functionalization has occurred, both duplicates are essential (because only together can they maintain the original expression patterns); hence, they are kept. Once a gene is established in a genome, it may retain its function or evolve to a new specialization (i.e., undergo neo-functionalization).

10. Tests for Sub-Functionalization

Dermitzakis and Clark *(53)* developed a sub-functionalization test, and identified several paralogs common to both humans and mice in which this has occurred. The basic idea is that the substitution rate in a given functional region will be low if a gene possesses a function that relies on those residues; otherwise, the substitution rate in the same region will be higher. The statistical method used by the authors allows the identification of regions with significantly different substitution rates in two paralogous genes. Following Tang and Lewontin *(60)*, Dermitzakis and Clark represented the pattern of change across a gene by the cumulative proportion of differences between a pair of orthologs (termed "pattern graph"). The pattern graph shows sharp increases when the substitution rate is high and almost no change in regions of sparse changes. Dermitzakis and Clark proposed the "paralog heterogeneity test," comparing the pattern graphs for human and mouse paralogs of several interesting genes. If one paralog has had a higher rate of substitution than the other in a given region, the difference between the pattern graphs shows a sharp rise or fall. If the paralogs have evolved at the same rate within a region, the difference between the pattern graphs will change slowly across the region. The "paralog heterogeneity test" compares the longest stretch of increasing or decreasing difference between the two pattern graphs to what would be expected if they had evolved similarly. The null distribution is simulated by the repeated random permutation of the two genes, so neither gene has distinct regions. If a region contains significant differences when compared with the null distribution, then the paralogs have evolved at different rates in that region. In that case, the sub-functionalization model predicts that it is important to the function of the paralog in which it is conserved. Sub-functionalization assumes there is more than one region of functional importance and it is possible for more than one region to be involved in a given function. By testing the sum of the two or more largest stretches, the significance of multiple regions can be determined; this can lead to predictions of functional regions for both paralogs.

11. Tests for Functional Divergence After Duplication

A possible fate of a duplicated gene is the acquisition of a new function or the specialization of a pre-existing but inefficient function. Again, statistics permit the study of functional divergence in a rigorous manner. Changes in protein function generally lead to variations in the selective forces acting on specific residues. An enzyme can evolve to recognize a different substrate that plausibly interacts with residues not involved in docking the original ligand, leaving some of them free. Similar changes can often be detected as evolutionary rate changes at the sites in question. It is known that a good estimator of the sign and strength of the selective force acting on a coding sequence is the ratio between non-synonymous and synonymous rates: If the values exceed 1, the gene is said to be under positive selection; otherwise, the gene is said to be under negative selection.

This divergence of protein functions often is revealed by a rate change in those amino acid residues of the protein that are more directly responsible for its new function. To investigate this change in evolution, a likelihood ratio test (LRT) is developed for detecting significant rate shifts at specific sites in proteins. A slow evolutionary rate at a given site indicates that this position is functionally important for the protein. Conversely, a high evolutionary rate indicates that the position is not involved in an important protein function. Therefore, a significant rate difference between two sub-families at a given site means that the function of this position is probably different in the two groups.

Recent work takes into account the phylogeny and different substitution rates among amino acids to detect functional divergence among paralogous genes. Gu *(61, 62)* has developed a quantitative measure for testing the function divergence within a family of duplicated genes. The method is based on measuring the decrease in mutation rate correlation between gene clusters of a gene family; the hidden Markov model (HMM) procedures allow the amino acid residues responsible for the functional divergence to be identified.

12. Protein Domain Evolution

Evolutionary studies of proteins can be performed at the level of individual protein domains. First, this is because many proteins share only a subset of common domains; and second, because domains can be detected conveniently and with a reasonable accuracy using available collections of domain-specific sequence profiles *(6)*. Comparisons of domain repertoires revealed both substantial similarities among different species, particularly with respect to the relative abundance of housekeeping domains, as well as major differences. The most notable manifestation of such differences is lineage-specific expansion of protein/domain families, which probably points to unique adaptations. Furthermore, it has been demonstrated that more complex organisms, such as vertebrates, have a greater variety of domains and, in general, more complex domain architectures of proteins than simpler life forms *(6)*. The quantitative comparative analysis of the frequency distributions of proteins or domains in different proteomes shows that these distributions appeared to fit the power law: $P(i) = ci^{-y}$ where $P(i)$ is the frequency of domain families, including exactly i members, c is a normalization constant and y is a parameter that typically assumes values between 1 and 3. Recent studies suggest that power laws apply to the distributions of a remarkably wide range of genome-associated quantities, including the number of transcripts per gene, the number of interactions per protein, the number of genes or pseudogenes in paralogous families, and others *(63)*. Power law distributions are scale free; that is, the shape of the distribution remains the same regardless of scaling of the analyzed variable. In particular, scale-free behavior has been described for networks of diverse nature, for example, the metabolic pathways of an organism or infectious contacts during an epidemic spread. **Figure 13.5** shows the statistical properties of random, scale-free, and hierarchical networks. The principal pattern of network evolution that ensures the emergence of power distributions (and accordingly, scale-free properties) is preferential attachment, whereby the probability of a node acquiring a new connection increases with the number of this node's connections. A simple model of evolution of the domain composition of proteomes was developed by Karev, Koonin, and collaborators with the following elementary processes: *(1)* domain birth (duplication with divergence), *(2)* death (inactivation and/or deletion), and *(3)* innovation (emergence from non-coding or non-globular sequences or acquisition via horizontal gene transfer). This model of evolution can be described as a birth, death, and innovation model (BDIM; *see* **Fig. 13.4a**) and is based on the following independence assumption: *(1)* all elementary events are independent of each other; and *(2)* the rates of individual domain birth (λ) and death (δ) do not depend on i (number of domains in a family). In a finite genome, the maximal number of domains in a family cannot exceed the total number of domains and, in reality, is probably much smaller; let N be the maximum possible number of domain family members. The model, described in **Fig. 13.4b**, considers classes of domain families that have only one common feature, namely the number of members. Let f_i be the number of domain families in the ith class, that is, families that are represented by exactly i domains in the given genome, $i = 1,2,...,N$. Birth of a domain in a family of class i results in the

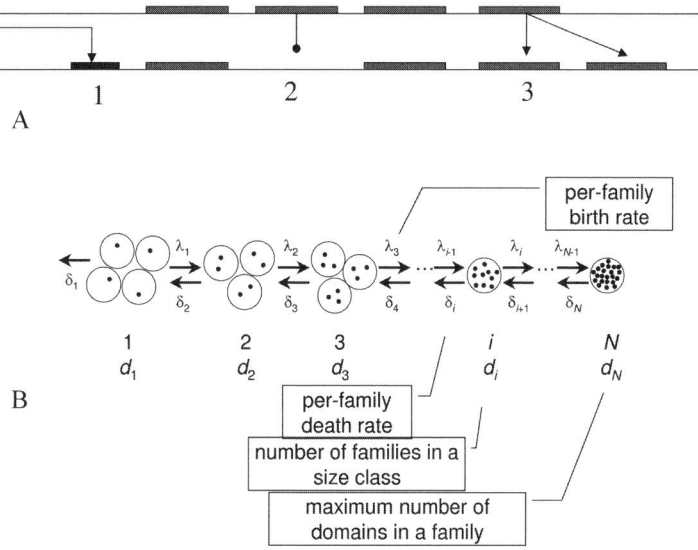

Fig. 13.4. (**A**) The layout of the Birth (3) + death (2) + innovation (1) model. (**B**) Domain dynamics and elementary evolutionary events under BDIM model. A dot is a domain; a circle represents a domain family.

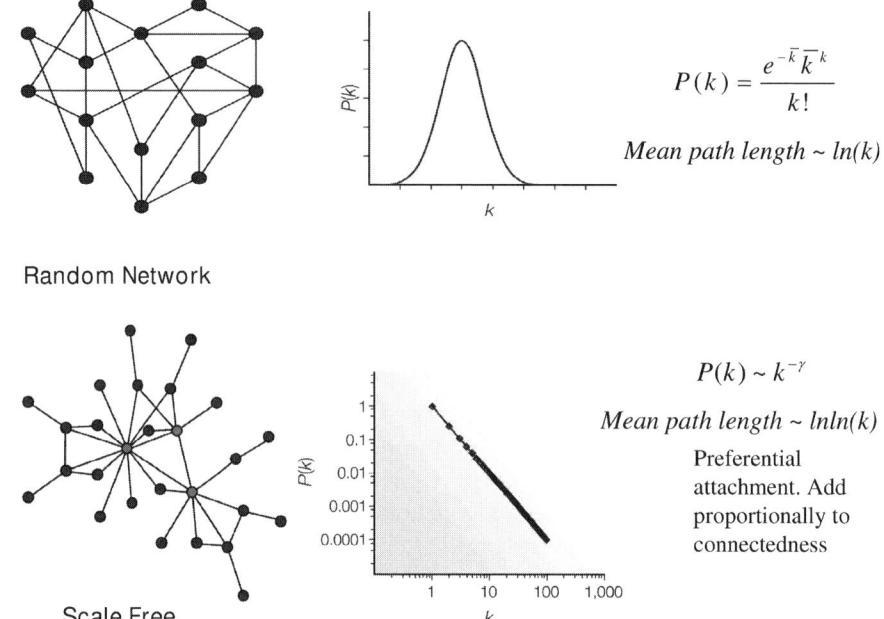

Fig. 13.5. Comparison of random, scale-free and hierarchical networks statistical properties.

relocation of this family from class i to class $i + 1$ (decrease of f_i and increase of f_{i+1} by 1). Conversely, death of a domain in a family of class i relocates the family to class $i - 1$; death of a domain in class 1 results in the elimination of the corresponding family from the given genome, this being the only considered mechanism of family death. They consider time to be continuous and suppose it very unlikely that more than one elementary event will occur during a short time interval; formally, the probability that more than one event occurs during an interval is very small.

Under these assumptions, the instantaneous rate at which a domain family leaves class i is proportional to the following simple BDIM, which describes the evolution of such a system of domain family classes:

$$df_1(t)/dt = -(\lambda_1 + \delta_1)f_1(t) + f_2(t) + v$$
$$df_i(t)/dt = \lambda_{i-1}f_i(t) - (\lambda + \delta)f_{i-1}(t) + \delta_{i+1}f_{i+1}(t) \text{ for } 1 < i < N$$
$$df_N(t)/dt = \lambda_{N-1}f_{N-1}(t) + \delta_N f_N(t)$$

The equilibrium for the number of domain families in each size class is $df_i(t)/dt = 0$, whereas the equilibrium for the total number of families gives $dF_i(t)/dt = 0$.

The solution of the model evolves to equilibrium, with a unique distribution of domain family sizes, $f_i(\lambda/\delta)^i/i$; in particular, if $\lambda = \delta$, this is $f_i\ 1/i$. Thus, under the simple BDIM, if the birth rate equals death rate, the abundance of a domain class is inversely proportional to the size of the families in this class. When the observations do not fit this particular asymptotic (as observed in several studies on distributions of protein family sizes), a different, more general model needs to be developed. Let $F(t) = \sum_{t=1}^{N} f_i(t)$ be the total number of domain families at instant t; then $dF(t)/dt = v - \delta - \delta_1 f_1(t)$. The system has an equilibrium solution $f_1, \ldots, f_N$ defined by the equality $df_i(t)/dt = 0$ for all i. Accordingly, there exists an equilibrium solution that we will designate F_{eq} (the total number of domain families at equilibrium). At equilibrium, $v = \delta_1 f_1$, that is, the processes of innovation and death of single domains (more precisely, the death of domain families of class 1, i.e., singletons) are balanced. It has been proved that the power law asymptotic appears if, and only if, the model is balanced, that is domain duplication and deletion rates are asymptotically equal up to the second order. It has been further proved that any power asymptotic with the degree not equal to –1 can appear only if the hypothesis of independence of the duplication/deletion rates on the size of a domain family is rejected. Specific cases of BDIMs, namely simple, linear, polynomial, and rational models, have been considered in detail and the distributions of the equilibrium frequencies of domain families of different size have been determined for each case. The authors have applied the BDIM formalism to the analysis of the domain family size distributions in prokaryotic and eukaryotic proteomes and show an excellent fit between these empirical data and a particular form of the model; namely, the second-order balanced linear BDIM. Calculation of the parameters of these models suggests surprisingly high innovation rates, comparable to the total domain birth (duplication) and elimination rates, particularly for prokaryotic genomes. In conclusion, the authors show that a straightforward model of genome evolution, which does not explicitly include selection, is sufficient to explain the observed distributions of domain family sizes, in which power laws appear as asymptotic. However, for the model to be compatible with the data, there has to be a precise balance among domain birth, death and innovation rates, and this is likely to be maintained by selection.

12.1. The Evolution of Introns

Eukaryotic genes usually contain introns, whereas bacterial genes are uninterrupted. Rzhetsky and co-workers analyzed the intron content of aldehyde dehydrogenase genes in several species in order to infer the validity of the introns-late or introns-early hypotheses *(65)*. They defined an instantaneous transition rate matrix corresponding to a first-order Markov chain description of intron evolution. They considered three rate parameters, intron insertion (λ), deletion (μ), and slippage (φ). Consider a hypothetical gene with only two sites potentially hosting introns: There are four possible intron configurations: 00, 01, 10, and 11, where zero and one stand for intron absence and presence respectively; transition 00 → 01 corresponds to an intron insertion; transition 01 → 00 is an intron deletion, and 01 → 10 is an intron slippage. The resulting instantaneous transition rate matrix, **Q**, shows all the configurations for a gene containing two sites with 2 a state of absence/presence of introns (the configuration order is, from left to right, 00, 01, 10, and 11):

$$\mathbf{Q} = \begin{bmatrix} -2\lambda & \lambda & \lambda & 0 \\ \mu & -\lambda-\mu-\varphi & \varphi & \lambda \\ \mu & \varphi & -\mu-\lambda-\varphi & \lambda \\ 0 & \mu & \mu & -2\mu \end{bmatrix}$$

The matrix of transition probabilities between gene arrangement states during time t is computed numerically as matrix exponentials ($\exp[\mathbf{Q}t]$) of the corresponding instantaneous transition rate matrix. Then the likelihood value can be calculated as described by Felsenstein *(10)*. The authors found that, using this model, their data support the introns-late theory.

13. Evolution of Microsatellite and ALU Repeats

The human genome is particularly rich in homo- and di-nucleotide repeats (*see* **Figs. 13.1 and 13.6**) *(6)* and in families of interspersed, mobile elements hundreds of base pairs (bp) long, among which are the Alu families. A first approach in modeling microsatellite evolution would consider the statistics of separations between consecutive words to distinguish content-bearing terms from generic terms. In general, the former words tend to cluster themselves as a consequence of their high specificity (attraction or repulsion), while the latter ones have a tendency to be evenly distributed across the whole text. In order to eliminate the dependency on frequency for different words, it is convenient to analyze the sequences of normalized separations between consecutive words of length L, $s = x(L)/\langle x(L)\rangle$. If homogeneous tracts were distributed at random in the genome, the inter-tract distribution $P_L(s)$ for words of length L would be: $P_L(s) = e^{-s}$.

As a consequence, we expect that non-specific words will run close to a Poisson law, whereas larger deviations should occur for highly specific content-bearing words. Such analysis may be implemented systematically in a quantitative fashion by studying the standard deviations $\sigma_L = \sqrt{\overline{s^2} - \overline{s}^2}$ of the distributions $P_L(s)$,

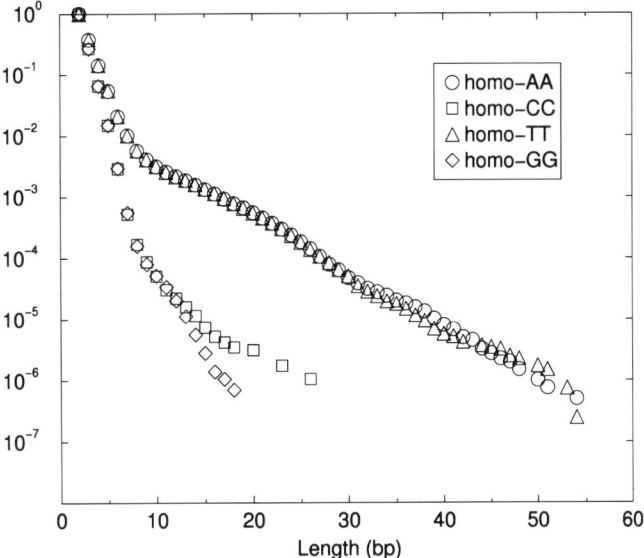

Fig. 13.6. Microsatellite length distribution in human genome.

since this is the simplest variable for characterizing a normalized distribution and its fluctuations. For a Poisson distribution $\sigma_L = 1$, whereas if there is attraction $\sigma_L > 1$. In the case of repulsion among words one should expect $\sigma_L < 1$. Moreover, further interesting information may be gathered by studying the skewness $\gamma = \langle (s-\bar{s})\rangle^3/\sigma^{3/2}$ and kurtosis $\kappa = \langle (s-\bar{s})^4\rangle/\sigma^2$ of the distribution $P_L(s)$. Interestingly, the same over-abundance of poly(X) tracts longer than a threshold of approximately 10 bp is found in all higher Eukaryotes analyzed. Moreover, *H. sapiens* shows a lower abundance of poly(G) and poly(C) tracts as compared with dog, chicken, and mouse genomes. A link between homo- and di-polymeric tracts and mobile elements recently has been highlighted. A statistical analysis of the genome-wide distribution of lengths and inter-tract separations of poly(X) and poly(XY) tracts in the human genome shows that such tracts are positioned in a non-random fashion, with an apparent periodicity of 150 bases (**Fig. 13.7**) *(66)*. In particular, the mobility of Alu repeats, which form 10% of the human genome, has been correlated with the length of poly(A) tracts located at one end of the Alu. These tracts have a rigid and non-bendable structure and have an inhibitory effect on nucleosomes, which normally compact the DNA.

Interestingly, all mobile elements, such as SINEs and LINEs, and processed pseudogenes, contain A-rich regions of different length. In particular, the Alu elements, present exclusively in the primates, are the most abundant repeat elements in terms

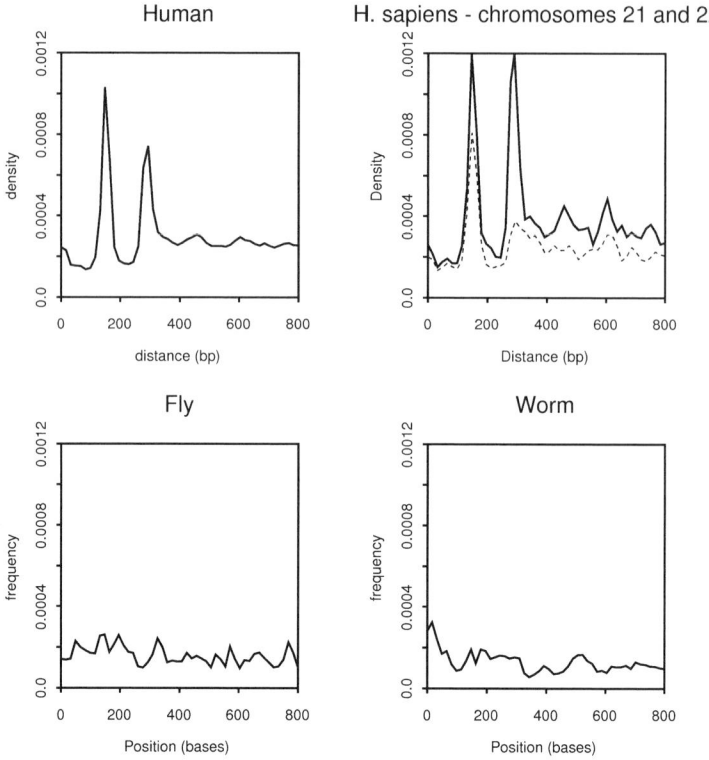

Fig. 13.7. The 150 and 300 bp periodic patterns found in analyzing the distances between poly(A)-poly(T) tracts.

of copy number (>10^6 in the human genome) and account for >10% of the human genome. They are typically 300 nucleotides in length, often form clusters, and are mainly present in non-coding regions. Higher Alu densities were observed in chromosomes with a greater number of genes and vice versa. Alus have a dimeric structure and are ancestrally derived from the gene 7SL RNA. They amplify in the genome by using a RNA polymerase III-derived transcript as template, in a process termed retroposition. The mobility is facilitated by a variable-length stretch of an A-rich region located at the 3′ end.

Although all Alu elements have poly(A) stretches, only a very few are able to retropose. Therefore, the mere presence of a poly(A) stretch is not sufficient to confer on an Alu element the ability to retropose efficiently. However, the length of the A stretch correlates positively with the mobility of the Alu.

The abundance of poly(X) and poly(XY) tracts shows an apparent correlation with organism complexity. For example, simple repeats are absent in viruses, rather rare in bacteria and low eukaryotes and very abundant in high vertebrate genomes.

References

1. Hein, J. (1994) TreeAlign. *Methods Mol Biol* 25, 349–364.
2. Whelan, S., Goldman, N. (2001) A general empirical model of protein evolution derived from multiple protein families using a maximum-likelihood approach. *Mol Biol Evol* 18, 691–699.
3. Liò, P., Vannucci, M. (2003) Investigating the evolution and structure of chemokine receptors. *Gene* 317, 29–37.
4. Glusman, G., Yanai, I., Rubin, I., et al. (2001) The complete human olfactory subgenome. *Genome Res* 11, 685–702.
5. Weiner, A. M. (2002) SINEs and LINEs: the art of biting the hand that feeds you. *Curr Opin Cell Biol* 14, 343–350.
6. Li, W. H. (2006) *Molecular Evolution*. Sinauer Associates, Sunderland, MA.
7. Jukes, T. H., Cantor, C. R. (1969), Evolution of protein molecules in (Munro, H. N., ed.). *Mammalian Protein Metabolism*. Academic Press, New York.
8. Kimura, M. (1980) Estimation of evolutionary distances between homologous nucleotide sequences. *Proc Natl Acad Sci U S A* 78, 454–458.
9. Blaisdell, J. (1985) A method of estimating from two aligned present-day DNA sequences their ancestral composition and subsequent rates of substitution, possibly different in the two lineages, corrected for multiple and parallel substitutions at the same site. *J Mol Evol* 22, 69–81.
10. Felsenstein, J. (1981) Evolutionary trees from DNA sequences: a maximum likelihood approach. *J Mol Evol* 17, 368–376.
11. Hasegawa, M., Kishino, H., Yano, T. (1985) Dating of the human-ape splitting by a molecular clock of mitochondrial DNA. *J Mol Evol* 22, 160–174.
12. Lanave, C., Preparata, G., Saccone, C., et al. (1984) A new method for calculating evolutionary substitution rates. *J Mol Evol* 20, 86–93.
13. Zarkikh, A. (1994) Estimation of evolutionary distances between nucleotide sequences. *J Mol Evol* 39, 315–329.
14. Li, W.-H. (1997) *Molecular Evolution*. Sinauer Associates, Sunderland, MA.
15. Yang, Z. (1994) Maximum likelihood phylogenetic estimation from DNA sequences with variable rates over sites: approximate methods. *J Mol Evol* 39, 306–314.
16. Hasegawa, M., Di Rienzo, A., Kocher, T. D., et al. (1993) Toward a more accurate time scale for the human mitochondrial DNA tree. *J Mol Evol* 37, 347–354.
17. Yang, Z., Goldman, N., Friday, A. (1995) Maximum likelihood trees from DNA sequences: a peculiar statistical estimation problem. *Syst Biol* 44, 384–399.
18. Felsenstein, J., Churchill, G. A. (1996) A Hidden Markov Model approach to variation among sites in rate of evolution. *Mol Biol Evol* 13, 93–104.
19. Rabiner, L. R. (1989) A tutorial on hidden Markov models and selected applications in speech recognition. *Proc IEEE* 77, 257–286.
20. Eddy, S. (1996) Hidden Markov models. *Curr Opinion Struct Biol* 6, 361–365.
21. Averof, M., Rokas, A., Wolfe, K. H., et al. (2000) Evidence for a high frequency of simultaneous double-nucleotide substitutions. *Science* 287, 1283–1286.
22. Siepel, A., Haussler, D. (2003) Combining phylogenetic and hidden Markov models in biosequence analysis. *Proceedings of the Seventh Annual international Conference on Research in Computational Molecular Biology (RECOMB'03)*. ACM Press, Berlin, Germany, 10–13 April. pp. 277–286.
23. Siepel, A., Haussler, D. (2004) Phylogenetic estimation of context dependent substitution rates by maximum likelihood. *Mol Biol Evol* 21, 468–488.
24. Whelan, S., Goldman, N. (2004) Estimating the frequency of events that cause multiple-nucleotide changes. *Genetics* 167, 2027–2043.
25. Goldman, N., Yang, Z. (1994) A codon-based model of nucleotide substitution for protein-coding DNA sequences. *Mol Biol Evol* 11, 725–736.
26. Yang, Z., Nielsen, R. (1998) Estimating synonymous and nonsynonymous substitution rates under realistic evolutionary models. *Mol Biol Evol* 46, 409–418.
27. Grantham, R. (1974) Amino acid difference formula to help explain protein evolution. *Science* 185(4154), 862–864.
28. Yang, Z., Nielsen, R., Goldman, N., et al. (2000) Codon-substitution models for heterogeneous selection pressure at amino acid sites. *Genetics* 155, 431–449.
29. Pedersen, A.-M. K., Wiuf, C., Christiansen, F. B. (1998) A codon-based model designed to describe lentiviral evolution. *Mol Biol Evol* 15, 1069–1081.
30. Yang, Z., Nielsen, R., Goldman, N., et al. (2000) Codon-substitution models for

heterogeneous selection pressure at amino acid sites. *Genetics* 155, 431–449.

31. Dayhoff, M. O., Eck, R. V., Park, C. M. (1972) A model of evolutionary change in proteins, in (Dayhoff, M. O., ed.), *Atlas of Protein Sequence and Structure*. vol. 5. National Biomedical Research Foundation, Washington, DC.

32. Dayhoff, M. O., Schwartz, R. M., Orcutt, B. C. (1978) A model of evolutionary change in proteins, in (Dayhoff, M. O., ed.), *Atlas of Protein Sequence and Structure*. vol. 5. National Biomedical Research Foundation, Washington, DC.

33. Jones, D. T., Taylor, W. R., Thornton, J. M. (1992). The rapid generation of mutation data matrices from protein sequence. *CABIOS* 8, 275–282.

34. Gonnet, G. H., Cohen, M. A., Benner, S. A. (1992). Exhaustive matching of the entire protein sequence database. *Science* 256, 1443–1445.

35. Henikoff, S., Henikoff, J. G. (1992) Amino acid substitution matrices from protein blocks. *Proc Natl Acad U S A* 89, 10915–10919.

36. Claverie, J. M. (1993) Detecting frame shifts by amino acid sequence comparison. *J Mol Biol* 234, 1140–1157.

37. Altschul, S. F. (1993) A protein alignment scoring system sensitive at all evolutionary distances. *J Mol Evol* 36, 290–300.

38. Naylor, G., Brown, W. M. (1997) Structural biology and phylogenetic estimation. *Nature* 388, 527–528.

39. Rzhetsky, A. (1995) Estimating substitution rates in ribosomal RNA genes. *Genetics* 141, 771–783.

40. Goldman, N., Thorne, J. L., Jones, D. T. (1996) Using evolutionary trees in protein secondary structure prediction and other comparative sequence analyses. *J Mol Biol* 263, 196–208.

41. Thorne, J. L., Goldman, N., Jones, D. T. (1996) Combining protein evolution and secondary structure. *Mol Biol Evol* 13, 666–673.

42. Goldman, N., Thorne, J. L., Jones, D. T. (1998) Assessing the impact of secondary structure and solvent accessibility on protein evolution. *Genetics* 149, 445–458.

43. Liò, P., Goldman, N., Thorne, J. L., et al. (1998) PASSML: combining evolutionary inference and protein secondary structure prediction. *Bioinformatics* 14, 726–733.

44. Liò, P., Goldman, N. (1999) Using protein structural information in evolutionary inference: transmembrane proteins. *Mol Biol Evol* 16, 1696–1710.

45. Fornasari, M. S., Parisi, G., Echave, J. (2002) Site-specific amino acid replacement matrices from structurally constrained protein evolution simulations. *Mol Biol Evol* 19, 352–356.

46. Sippl, M. J. (1993) Recognition of errors in three-dimensional structures of proteins. *Proteins* 17, 355–362.

47. Bastolla, U., Porto, M., Roman, H. E., et al. (2005) The principal eigenvector of contact matrices and hydrophobicity profiles in proteins. *Proteins* 58, 22–30.

48. Pollock, D. D., Taylor, W. R., Goldman, N. (1999) Coevolving protein residues: maximum likelihood identification and relationship to structure. *J Mol Biol* 287, 187–198.

49. Pagel, M. (1994) Detecting correlated evolution on phylogenies: a general method for the comparative analysis of discrete characters. *Proc R Soc (B)* 255, 37–45.

50. Rzhetsky, A. (1995) Estimating substitution rates in ribosomal RNA genes. *Genetics* 141, 771–783.

51. Telford, M. J., Wise, M. J., Gowri-Shankar, V. (2005) Consideration of RNA secondary structure significantly improves likelihood-based estimates of phylogeny: examples from the Bilateria. *Mol Biol Evol* 22, 1129–1136.

52. Hudelot, C., Gowri-Shankar, V., Jow, H., et al. (2003) RNA-based phylogenetic methods: application to mammalian mitochondrial RNA sequences. *Mol Phyl Evol* 28, 241–252.

53. Dermitzakis, E. T., Clark, A. G. (2001) Differential selection after duplication in mammalian developmental genes. *Mol Biol Evol* 18, 557–562.

54. Walsh, J. B. (1995) How often do duplicated genes evolve new functions? *Genetics* 139, 421–428.

55. Nadeau, J. H., Sankoff, D. (1997) Comparable rates of gene loss and functional divergence after genome duplications early in vertebrate evolution. *Genetics* 147, 1259–1266.

56. Force, A., Cresko, W. A., Pickett, F. B., et al. (2005) The origin of sub-functions and modular gene regulation. *Genetics* 170, 433–446.

57. Lynch, M., O'Hely, M., Walsh, B., et al. (2001) The probability of preservation of a newly arisen gene duplicate. *Genetics* 159, 1789–1804.

58. He, X., Zhang, J. (2005) Rapid sub-functionalization accompanied by prolonged and substantial neo-functionalization in duplicate gene evolution. *Genetics* 169, 1157.

59. von Mering, C., Krause, R., Snel, B., et al. (2002) Comparative assessment of large-scale datasets of protein-protein interactions. *Nature* 417(6887), 399–403.

60. Tang, H., Lewontin, R. C. (1999) Locating regions of differential variability in DNA and protein sequences. *Genetics* 153, 485–495.

61. Gu, X. (1999) Statistical methods for testing functional divergence after gene duplication. *Mol Biol Evol* 16, 1664–1674.

62. Gu, X. (2001) Maximum-likelihood approach for gene family evolution under functional divergence. *Mol Biol Evol* 18, 453.

63. Karev, G. P, Wolf, Y. I., Koonin, E. V. (2003) Simple stochastic birth and death models of genome evolution: was there enough time for us to evolve? *Bioinformatics* 19, 1889–1900.

64. Karev, G., et al. (2002) Birth and death of protein domains: a simple model of evolution explains power law behavior. *BMC Evol Biol* 2, 18–24.

65. Rzhetsky, A., Ayala, F. J., Hsu, L. C., et al. (1997) Exon/intron structure of aldehyde dehydrogenase genes supports the "introns-late" theory. *Proc Natl Acad Sci USA* 94, 6820–6825.

66. Piazza, F., Liò, P. Statistical analysis of simple repeats in the human genome. *Physica A* 347, 472–488.

67. Odom, G. L., Robichaux, J. L., Deininger, P. L. (2004) Predicting mammalian SINE subfamily activity from A-tail length. *Mol Biol Evol* 21, 2140–2148.

68. Roy-Engel, A. M., Salem, A. H., Oyeniran, O. O., et al. (2002) Active Alu element "A-tails": size does matter. *Genome Res* 12, 1333–1344.

Chapter 14

Inferring Trees

Simon Whelan

Abstract

Molecular phylogenetics examines how biological sequences evolve and the historical relationships between them. An important aspect of many such studies is the estimation of a phylogenetic tree, which explicitly describes evolutionary relationships between the sequences. This chapter provides an introduction to evolutionary trees and some commonly used inferential methodology, focusing on the assumptions made and how they affect an analysis. Detailed discussion is also provided about some common algorithms used for phylogenetic tree estimation. Finally, there are a few practical guidelines, including how to combine multiple software packages to improve inference, and a comparison between Bayesian and maximum likelihood phylogenetics.

Key words: Phylogenetic inference, evolutionary trees, maximum likelihood, parsimony, distance methods, review.

1. Introduction

Phylogenetics and comparative genomics use multiple sequence alignments to study how the genetic material changes over evolutionary time and draw biologically interesting inferences. The primary aim of many studies, and an important byproduct of others, is finding the phylogenetic tree that best describes the evolutionary relationship of the sequences. Trees have proved useful in many areas of molecular biology. In studies of pathogens, they have offered a wealth of insights into the interaction between viruses and their hosts during evolution. Pre-eminent among these have been studies on HIV in which, for example, trees were crucial in demonstrating that HIV was the result of at least two different zoonoses from chimpanzees (1). Trees have also provided valuable insights in molecular and physiological studies, including how the protein repertoire evolved *(2, 3)*, genome evolution *(4, 5)*, and the development of taxonomic classification and species concepts *(6, 7)*. Accurate reconstruction of evolutionary relationships is also important in studies in which the primary aim is not tree inference, such as investigating the selective pressures acting on proteins (*see* **Chapter 15**) or the identification of conserved elements in genomes *(8)*. Failure to correctly account for the historical relationship of sequences can lead to inaccurate hypothesis testing and impaired biological conclusions *(9)*.

Despite the importance of phylogenetic trees, obtaining an accurate estimate of one is not a straightforward process. There are many phylogenetic software packages available, each with unique advantages and disadvantages. An up-to-date list of such packages is held at the URL: http://evolution.genetics.washington.edu/phylip/software.html. This chapter aims to provide an introductory guide to enable users to make informed decisions about the phylogenetic software they use, by describing what phylogenetic trees are, why they are useful, and some of the underlying principles of phylogenetic inference.

2. Underlying Principles

Phylogenetic trees make implicit assumptions about the data. Most importantly, these include that all sequences share a common ancestor, and that sequences evolving along all branches in the tree evolve independently. Violations of the former assumption occur when unrelated regions are included in the data. This occurs, for example, when only subsets of protein domains are shared between sequences or when data have entered the tree from other sources, such as sequence contamination, lateral gene transfer, or transposons. Violations of the second assumption occur when information in one part of a tree affects sequence in another. This occurs in gene families under gene conversion or lateral gene transfer. Before assuming a bifurcating tree for phylogenetic analyses, one should try to ensure that these implicit assumptions are not violated.

The trees estimated in phylogenetics usually come in two flavors, rooted and unrooted, differing in the assumptions they make about the most recent common ancestor of the sequences, the root of the tree. Knowing the root of the tree enables the order of divergence events to be determined, which is valuable, for example, when investigating the evolution of phenotypic traits by examining their location on a tree.

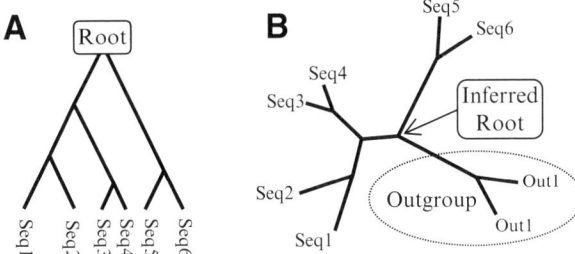

Fig. 14.1. Two common forms of bifurcating tree are used in phylogenetics. (**A**) Rooted trees make explicit assumptions about the most recent common ancestor of sequences and can imply directionality of the evolutionary process. (**B**) Unrooted trees assume a time-reversible evolutionary process. The root of sequences 1–6 can be inferred by adding an outgroup.

Rooted trees explicitly identify the location of this ancestor (**Fig. 14.1A**), whereas unrooted trees do not (**Fig. 14.1B**). In practice, the majority of studies use unrooted trees and, if rooting is required, it can be inferred through the use of an outgroup (**Fig. 14.1B**).

2.1. Scoring Trees

To discriminate between trees, a score function is required that quantifies how well a phylogenetic tree describes the observed sequences. The last 40 years have led to the proposal of many different scoring functions, summarized briefly as statistical, parsimony, and distance-based. There has been considerable debate in the literature over which methodology is the most appropriate for inferring trees. A brief discussion of each is provided in the following, although a full discussion is beyond this chapter's scope (but *see* **Note 1**; *see* also *(6, 9–11)* for an introduction).

Statistical methods are currently in the ascendancy and use likelihood-based scoring functions. Likelihood is a measure proportional to the probability of observing the data given the parameters specifying an evolutionary model and branch lengths in the tree. These describe how sequences change over time and how much change has occurred on particular lineages, respectively (*see* **Chapter 13**) *(9)*. Statistical methods come in two varieties: maximum likelihood (ML), which searches for the tree that maximizes the likelihood function, and Bayesian inference, which samples trees in proportion to the likelihood function and prior expectations. The primary strength behind statistical methods is that they are based on established and reliable methodology that has been applied to many areas of research, from classical population genetics to modeling world economies *(12)*. They are also statistically consistent: Under an accurate evolutionary model they tend to converge to the "true tree" as longer sequences are used *(13, 14)*. This, and other associated properties, enables statistical methodology to produce high-quality phylogenetic estimates with a minimum of bias under a wide range of conditions. The primary criticism of statistical methods is that they are computationally intensive. Progress in phylogenetic software and computing resources is steadily meeting this challenge.

Parsimony counts the minimum number of changes required on a tree to describe the observed data. Parsimony is intuitive to understand and computationally fast, relative to statistical methods. It is often criticized for being statistically inconsistent: Increasing sequence length in certain conditions can lead to greater confidence in the wrong tree. Parsimony does not include an explicit evolutionary model (although see *15*). In some quarters, this is seen as an advantage because of a belief that evolution cannot be modeled *(16)*. Others view this as a disadvantage because it does not account for widely acknowledged variation in the evolutionary process. In practice, parsimony may approximate statistical methods when the branch lengths on a tree are short *(15)*.

Distance-based criteria use pairwise estimates of evolutionary divergence to infer a tree, either by an algorithmic clustering approach (e.g., neighbor joining) *(17)* or assessing the fit of the distances to particular tree topologies (e.g., least-squares) *(18)*. Distance methods are exceptionally quick: A phylogeny can be produced from thousands of sequences within minutes. When an adequate evolutionary model is used to obtain distances, the methodology can be consistent, although it is probable that it converges to the "true tree" at a slower rate than full statistical methods. An additional weakness of distance methods is that they use only pairwise comparisons of sequences to construct a tree. These comparisons are long distances relative to the branches on a tree, and are consequently harder to estimate and prone to larger variances. In contrast, statistical methods estimate evolutionary distances on a branch-by-branch basis through a tree, exploiting evolutionary information more efficiently by using all sequences in a tree to inform about branch lengths and reduce the variance of each estimate. Purely algorithmic distance methods, such as neighbor-joining and some of its derivatives, do not use a statistical measure to fit distances to a tree. The estimate is taken as the outcome of a predefined algorithm, making it unclear what criterion is used to estimate the tree. In practice, however, algorithmic methods appear to function as good approximations to other more robust methods, such as least-squares or the minimum-evolution criterion.

2.2. Why Estimating Trees Is Difficult

The effective and accurate estimation of trees remains difficult, despite their wide-ranging importance in experimental and computational studies. It is difficult to find the optimal tree in statistical and parsimony methods because the size of tree space rapidly increases with the number of sequences, making an exhaustive analysis impractical for even modest numbers of sequences. For 50 sequences, there are approximately 10^{76} possible trees, a number comparable to the estimated number of atoms in the observable universe. This necessitates heuristic approaches for searching tree space that speed up computation, usually at the expense of accuracy. The phylogenetic tree estimation problem is unusual and there are few well-studied examples from other research disciplines to draw on for heuristics (20). Consequently, there has been a lot of active research into methodology to find the optimal tree using a variety of novel heuristic algorithms. Many of these approaches progressively optimize the tree estimate by iteratively examining the score of nearby trees and making the highest scoring the new best estimate, and stopping when no further improvements can be found. The nature of the heuristics mean there is often no way of deciding whether the newly discovered optimum is the globally best tree or whether it is one of many other local optima in tree space (although see (19), for a description of a Branch and Bound algorithm). Through acknowledging this problem and applying phylogenetic software to its full potential it is possible to produce good estimates of trees that exhibit many characteristics of a sequence's evolutionary history.

3. Point Estimation of Trees

The majority of software for inferring trees results in a single (point) estimate of the tree that best describes the evolutionary relationships of the data. This is ultimately what most researchers are interested in and what is usually included in any published work. In order to obtain the best possible estimate it is valuable to understand how phylogenetic software functions and the strengths and weaknesses of different approaches (see **Note 2**). Phylogenetic heuristics can be summarized by the following four-step approach:

1. Propose a tree.
2. Refine the tree using a traversal scheme until no further improvement found.
3. Check stopping criterion.
4. Resample from tree space and go to 2.

Not all four steps are employed by all phylogenetic software. Many distance methods, for example, use only step 1, whereas many others stop after refining a tree estimate.

3.1. Proposing an Initial Tree

Initial tree proposal is fundamental to all phylogenetic tree estimation and there is no substitute for a good starting topology when inferring trees. The most popular approach to producing an initial tree is to use distance-based clustering methods. This is usually chosen for computational speed, so purely algorithmic approaches are common. Occasionally, more sophisticated approaches, such as quartet puzzling (21), are used. These can be highly effective for smaller datasets, but often do not scale well to larger numbers of sequences. An alternative approach is to choose a tree based on external information, such as the fossil record or prior studies, which can be appropriate when examining the relationship of well-characterized species and/or molecules.

An alternative, widely used approach is to use sequence-based clustering algorithms (9, 19). These are similar to distance-based methods, but instead of constructing a tree using pairwise distances, they use something akin to full statistical or parsimony approaches. The two most popular are the stepwise addition and star-decomposition algorithms (**Fig. 14.2**). Stepwise addition (**Fig. 14.2A**) starts with a tree of three sequences and adds the remaining sequences to the tree in a random order to the location that maximizes the scoring criterion. The order that the sequences are added to the tree can affect the final proposed topology, also allowing a random order of addition to be used as a re-sampling step. Star-decomposition (**Fig. 14.2B**) starts with all of the sequences related by a star phylogeny: a tree with no defined internal branches. The algorithm progressively adds branches to this tree by resolving multi-furcations (undefined regions of a tree) that increase the scoring criteria by the largest amount. Providing there are no tied optimal scores the algorithm is deterministic, resulting in the same proposed tree each time.

3.2. Refining the Tree Estimate

Refining the tree estimate is the heuristic optimization step. The tree is improved using an iterative procedure that stops when no further improvement can be found. For each iteration, a traversal scheme is used to move around tree space and propose a set of candidate trees from the current tree estimate. Each tree is

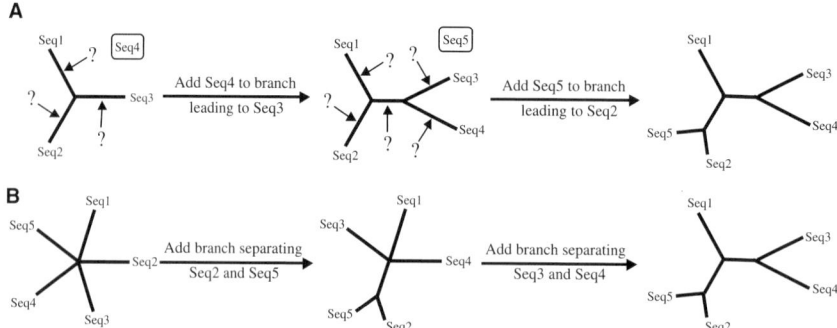

Fig. 14.2. Sequence-based clustering algorithms are frequently used to propose trees. (**A**) Stepwise addition progressively adds sequences to a tree at a location that maximizes the score function. (**B**) Star-decomposition starts with a topology with no defined internal branches and serially adds branches that maximize the score function. For example, in the first step branches could be added that separate all possible pairs of sequences.

assessed using the score function and one (or more) trees are chosen as the starting point for the next round of iteration. This is usually the best tree, although some approaches, such as Bayesian inference and simulated annealing, accept suboptimal trees (see the following). The popular traversal schemes discussed in the following share a common feature: they propose candidate trees by making small rearrangements on the current tree, examining each internal branch of a tree in turn and varying the way they propose trees from it. Three of the most popular methods are, in order of complexity and the number of candidate trees they produce: nearest neighbor interchange (NNI), subtree pruning and regrafting (SPR), and tree bisection and reconnection (TBR).

NNI (**Fig. 14.3A**) breaks an internal branch to produce four subtrees. Each of the three possible combinations for arranging these subtrees is added to the list of candidate trees. SPR (**Fig. 14.3B**) is a generalization of NNI that generates candidate trees by breaking the internal branch under consideration and proposing new trees by putting together the resultant subtrees in different ways, one of which is the original topology. The set of candidate trees is generated by regrafting the broken branch of subtree A to each of subtree B's branches; **Fig. 14.3B** shows

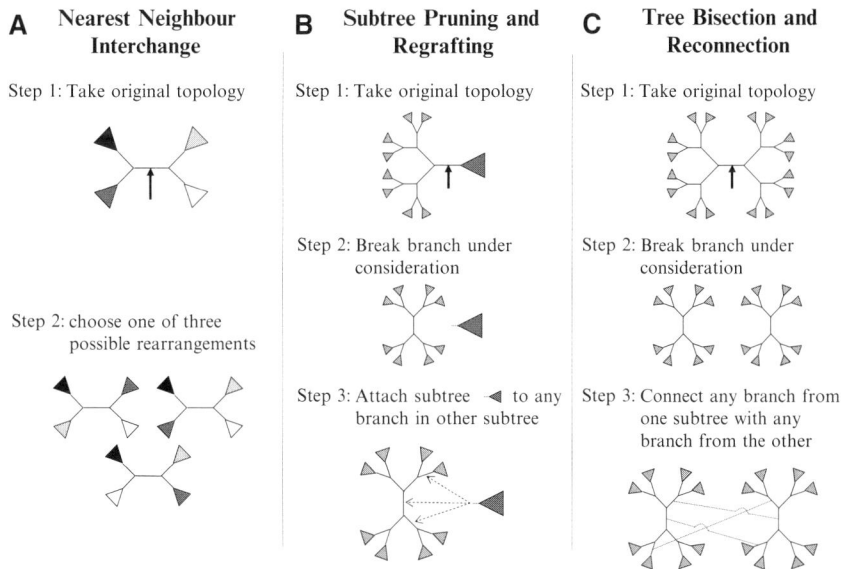

Fig. 14.3. Three related schemes for traversing tree space when searching for an optimal tree. (**A**) Nearest Neighbor Interchange: an internal branch is broken and the three potential arrangements of the four subtrees are examined. (**B**) Subtree Pruning and Regrafting: an internal branch is removed and all ways of regrafting one of the resulting subtrees to the other are examined. Dotted arrows demonstrate some potential regrafting points. (**C**) Tree Bisection and Reconnection: an internal branch is removed and all possible ways of reconnecting the subtrees are examined. Dotted lines demonstrate potential reconnection points.

three example regraftings *(dotted arrows)*. To complete the set of candidate trees the order of the subtrees is reversed and the process repeated, this time regrafting subtree B to each of subtree A's branches. TBR is similar to SPR, but generalizes it further to produce even more candidate trees per branch. When TBR breaks a branch, all possible ways of joining the two subtrees are added to the list. Some example reconnections are illustrated in **Fig. 14.3C** as dotted lines. NNI, SPR, and TBR are hierarchical in structure because the set of trees proposed by the more complex approach completely contains those proposed by the simpler approach *(22)*. This hierarchical structure is not general: There are other approaches based on removing multiple branches from a tree that do not follow this pattern, but these are not widely implemented *(23, 24)*.

The advantages and disadvantages of these methods emerge from the number of candidate trees they produce. NNI produces a modest number of proposed trees, growing linearly with the number of sequences in the phylogeny. This limits its effectiveness by allowing only small steps in tree space and a greater susceptibility to local optima than more expansive schemes. The number of trees proposed by SPR rises rapidly as the number of sequences increases, making it computationally impractical for large numbers of sequences, although the greater number of candidate trees results in a larger step size and fewer local optima. Innovations based around SPR limit the number of candidate trees by bounding the number of steps away that a subtree can move from its original position. The subtree in **Fig. 14.3B**, for example, could be bounded in its movement to a maximum of two branches (all branches not represented by a triangular subtree in the figure). Each branch has a maximum number of subtrees it can produce, returning a linear relationship between the number of candidate trees and sequences. This approach offers a promising direction for future algorithm research, but currently is not widely implemented *(25, 26)*. The characteristics of TBR are similar to SPR, but amplified because the number of candidate topologies per branch increases even more rapidly with the number of sequences.

3.3. Stopping Criteria

Many phylogenetic software packages do not resample tree space and stop after a single round of refinement. When resampling is used, a stopping rule is required. These are usually arbitrary, allowing only a pre-specified number of resamples or refinements. A recent innovation offers an alternative based on how frequently improvements in the overall optimal tree are observed. This dynamically estimates the number of iterations required before no further improvement in tree topology will be found and stops the algorithm when this has been exceeded *(27)*.

3.4. Resampling from Tree Space

Sampling from one place in tree space and refining still may not find the globally optimal tree. The goal of resampling from tree space is to expand the area of tree space searched by the heuristic and uncover new, potentially better optima. This is achieved by starting the refinement procedure from another point in tree space. This approach was originally used in some of the earliest phylogenetic software *(28)*, but is only lightly studied relative to improvements in the refinement methodology. Three of the many possible resampling schemes are discussed here: uniform resampling, stepwise addition, and importance quartet puzzling (IQP) *(27)*.

Uniform resampling is the simplest resampling strategy and the probability of picking each possible tree is one divided by the total number of trees. Although rarely used in practice, its deficiencies are edifying to the tree estimation problem. Each optimum has an area of tree space associated with it that will lead back to it during the refinement process, referred to as a center of attraction. In the majority of phylogenetic problems there are potentially large numbers of optima and the centers of attraction can be relatively small. Uniform sampling is prone to ending up in poor regions of tree space where nearby optima are unlikely to be particularly high. Finding centers with high optima can be difficult and requires an intelligent sampling process.

Stepwise addition with random sequence ordering is a viable resampling strategy because adding sequences in a different order is liable to produce a different, but equally good, starting tree. Both stepwise addition and uniform resampling effectively throw away information from the current best estimate of a tree. IQP keeps some of this information and can be viewed as a partial stepwise addition process. It resamples by randomly removing a number of sequences from the current best tree, then consecutively adding them back to the tree in a random order using the IQP algorithm. This identifies good locations to insert sequences by examining a set of four-species subtrees that all include the newly added sequence (*see* ref. *27* for more details). IQP resampling has been demonstrated to be reasonably effective for tree estimation when coupled with NNI.

An alternative approach to this purist resampling is to combine two or more refinement heuristics, the quicker of which (e.g., NNI) is used for the refinement step, and an alternative with a slower more expansive scheme (e.g., TBR) is used rarely to make larger steps in tree space.

3.5. Other Approaches to Point Estimation

Other popular approaches to phylogenetic tree estimation include genetic algorithms, simulated annealing (SA), and supertree reconstruction. Genetic algorithms are a general approach for numerical optimization that use evolutionary principles to allow a population of potential trees to adapt by improving their fitness (score function) according to a refinement scheme defined by the software designer. As the algorithm progresses, a proxy for natural selection weeds out trees with a lower fitness, and better trees tend to become more highly represented. After a period of time, the algorithm is stopped and the best topology discovered is the point estimate. The construction of genetic algorithms is very much an art and highly dependent on the designer's ability to construct a coherent and effective fitness scale, and the application of quasi-natural selection. Some approaches for estimating trees using genetic algorithms have been noticeably successful *(29, 30)*.

SA bases its optimization strategy on the observation that natural materials find their optimal energy state when allowed to cool slowly. Usual approaches to SA propose a tree at random from a traversal scheme, and the probability of accepting this as the current tree depends on whether it improves the score function. A tree that improves the score is accepted. Trees with lower scores are accepted with a probability related to the score difference between the current and new tree, and a "heat'" variable that decreases slowly during time. This random element allows the optimization process to move between different centers of attraction. SA starts "hot," frequently accepting poor trees and covering large tracts of tree space. As it gradually "cools," it becomes increasingly focused on accepting only trees with a higher score and the algorithm settles on a best tree. SA has proved very useful in other difficult optimization problems, but has yet to be widely used in phylogenetics (but *see (31–33)*).

4. Confidence Intervals on Point Estimates

In studies in which the phylogeny is of prime importance, it is necessary to attach a degree of confidence to the point estimate. This typically involves using computer simulations to generate new datasets from features of the original. The underlying principle is that simulated data represent independent draws from the same distribution over tree space (and evolutionary model space) as the process that generated the real data. This allows an assessment of the variability of the estimate and the construction of a confidence interval. This form of simulation is often known as a bootstrap *(34)*, after the phrase "pulling oneself up by the bootstraps," as it is employed when a problem is too difficult for classical statistics to solve. There are two broad approaches to bootstrap data widely used in phylogenetics: non-parametric bootstrapping and parametric bootstrapping. **Figure 14.4** contains examples of all the methods discussed in the following.

Inferring Trees 297

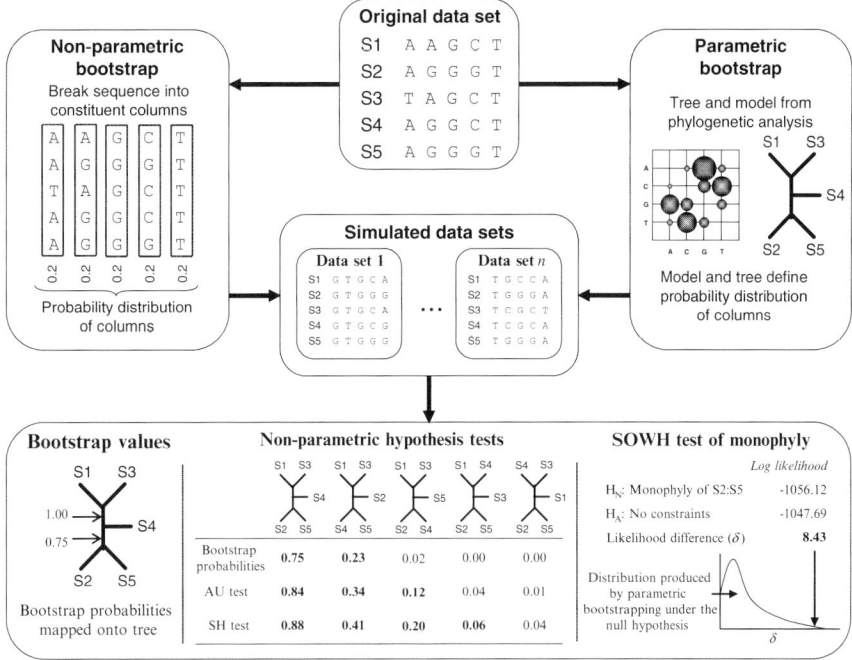

Fig. 14.4. Confidence in phylogenetic trees is often assessed through simulation approaches called bootstrapping. The different forms of bootstrap use the original data in different ways to simulate new datasets *(center)*. Non-parametric bootstrapping *(left)* produces new data by resampling with replacement from the alignment columns from the original data. Parametric bootstrapping *(right)* uses parameters estimated from the original data to produce new datasets. Trees are estimated for each of the simulated sets and their results summarized using a variety of measures *(bottom)*. The values shown are for demonstrative purposes only. See text for more details of the procedure.

4.1. The Non-Parametric Bootstrap

The non-parametric bootstrap is applicable to all methods of phylogenetic inference. It assumes that the probability distribution of observed columns in the original sequence alignment is representative of the complex and unknown evolutionary process. In other words, it assumes that if evolution had produced multiple copies of the original data, the average frequency of each alignment column would be exactly that observed in the original data. This philosophy underpins the simulation strategy. Sampling with replacement is used to produce a simulated dataset by repeatedly drawing from the original data to make a new dataset of suitable length. Each column in the original alignment has an equal probability of contributing to the simulated data. This approach allows non-parametric bootstrapping to encompass some of the complexities of sequence evolution that are not easily modeled, such as complex substitution patterns and rate variation, but introduces a finite sampling problem: Only a small proportion of all possible data columns could possibly be represented in the real data. For a complete DNA alignment covering 20 species there are $\sim 10^{12}$ possible data columns (four DNA bases raised to the 20th power). It would be unreasonable to expect any real dataset to provide a detailed representation of the probability distribution over this space.

The most ubiquitous non-parametric bootstrap test of confidence is simple bootstrapping, which assesses confidence in a tree on a branch-by-branch basis *(35)*. Tree estimates are obtained for a large number of simulated datasets. Bootstrap values are placed on branches of the original point estimate of the tree as the frequency that implied bipartitions are observed in trees estimated from the simulated data. This is useful for examining the evidence for particular subtrees, but becomes difficult to interpret for a whole tree because it hides information about how frequently other trees are estimated. This can be addressed by describing the bootstrap probability of different trees. In **Fig. 14.4**, the two bootstrap values are expanded to five bootstrap probabilities and, using hypothesis testing, three of the five can be rejected. These forms of simple bootstrapping are demonstrably biased and often place too much confidence in a small number of trees *(36, 37)*, but due to their simplicity they remain practical and useful tools for exploring confidence in a tree estimate.

Two other useful forms of the non-parametric bootstrap are employed in the Shimodaira-Hasegawa (SH) *(38)* and Approximately Unbiased (AU) *(39)* tests. These tests require a list of candidate trees to be proposed, representing a set of alternate hypotheses, such as species grouping, and usually containing the optimal tree estimate. The tests form a confidence set by calculating a value related to the probability of each tree being the best tree, and then rejecting those that fall below the critical value. A well-chosen list allows researchers to reject and support biologically interesting hypotheses based on tree shape, such as monophyly and gene duplication. Both tests control the level of type I (false-positive) error successfully. In other words, the confidence interval is conservative and does not place unwarranted confidence in a small number of trees. This was a particular problem for the tests' predecessor, the Kishino-Hasegawa (KH) test *(40)* that, due to a common misapplication, placed undue confidence in a small number of trees. The AU test is constructed in a subtly different manner than the SH test, which removes a potential bias and increases statistical power. This allows the AU test to reject more trees than the SH test and produce tighter confidence intervals (demonstrated in **Fig. 14.4**).

4.2. The Parametric Bootstrap

The parametric bootstrap is applicable to statistical methods of phylogenetic inference and is widely used to compare phylogenetic models as well as trees. The simulation is performed by generating new sequences and allowing them to evolve on a tree topology, according to the parameters in the statistical model estimated from the original data, such as replacement rates between bases/residues and branch lengths. This completely defines the probability distribution over all possible data columns, even those not observed in the original data, which avoids potential problems introduced by sampling in the non-parametric bootstrap. This introduces a potential source of bias because errors in the evolutionary model and its assumptions are propagated in the simulation. As evolutionary models become more realistic and as the amount of data analyzed grows, the distributions across data columns produced by parametric and non-parametric bootstrapping may be expected to become increasingly similar. When the model is sufficiently accurate and the sequences become infinitely long the two distributions may be expected to be the same, although it is unclear whether anything close to this situation occurs in real data.

The most popular test using the parametric bootstrap is the Swofford-Olsen-Waddell-Hillis (SOWH) *(19)* test. This also addresses the comparison of trees from a hypothesis testing perspective, by constructing a null hypothesis (H_N) of interest and an alternative, more general hypothesis (H_A). **Figure 14.4** demonstrates this through a simple test of monophyly. A likelihood is calculated under the null hypothesis of monophyly, which restricts tree space by enforcing that a subset of sequences always constitutes a single clade (group) on a tree, in this case S2 and S5 always being together. In practice this means performing a tree search on the subset of tree space in which the clade exists. A second likelihood is calculated under the alternative hypothesis of no monophyly, which allows tree estimation from the entirety of tree space. The SOWH test examines the improvement in likelihood, δ, observed by allowing the alternative hypothesis. To perform a statistical test, δ needs to be compared with some critical value on the null distribution. No standard distribution is appropriate and parametric bootstrapping is used to estimate it, with the parameters required for the simulation taken from H_N. Each simulated dataset is assessed under the null and alternate hypotheses and the value of δ is taken as a sample from the null distribution. When repeated large numbers of times, this produces a distribution of δ that is appropriate for significance testing. In the example, the observed value of δ falls outside the 95% mark of the distribution, meaning that the null hypothesis of monophyly is rejected. This example demonstrates some of the strengths and weaknesses of parametric bootstrapping. The SOWH test does not require a limited list of trees to be defined because hypotheses can be constructed by placing simple restrictions on tree space. This allows the SOWH test to assess complex questions that are not otherwise easily addressed, but the computational burden of each simulated dataset may be as extreme as the original dataset. This can make large numbers of bootstrap replicates unfeasible.

4.3. Limitations of Current Bootstrap Simulation

Both of the discussed bootstrapping approaches have theoretical and practical limitations. The primary practical limitation of both bootstrapping methods is that they are computationally very intensive, although there are many texts detailing computational approximations to make them computationally more efficient *(19, 41)*. The theoretical limitations of current bootstrapping methods stem from the simplifying assumptions they make when describing sequence evolution. These, most seriously, include poor choice of evolutionary model, neglecting insertion-deletion mutations (indels), and the effect of neighboring sites on sequence evolution. In statistical methods, a poor choice of model can adversely affect all forms of bootstrapping. In non-parametric bootstrapping, inaccuracies in the model can lead to biases in the tree estimate that may manifest as overconfidence in an incorrect tree topology. Modeling errors in parametric bootstrapping may result in differences between the probability distribution of columns generated during the simulation and the distribution of the "true" evolutionary process, which may make the simulated distribution inappropriate for statistical testing. The relative effects of model mis-specification on bootstrapping are generally poorly characterized and every care should be taken when choosing a model for phylogenetic inference (*see* **Chapter 16**).

Indels are common mutations that can introduce alignment errors, which can impact phylogenetic analysis (*see* **Chapters 7 and 15**). These effects are generally poorly characterized and, to their detriment, the majority of phylogenetic methodologies ignore them. The context of a site in a sequence may have significant effect on its evolution, and there are numerous well-characterized biological dependencies that are not covered by standard simulation models. In vertebrate genomes sequences, for example, methylation of the cytosine in CG dinucleotides results in rapid mutation to TG. Non-parametric bootstrap techniques using block resampling *(42)* or the use of more complex evolutionary models (e.g., hidden Markov models) *(43)* in the parametric bootstrap would alleviate some of these problems, but they are rarely used in practice.

5. Bayesian Inference of Trees

Bayesian inference of phylogenetic trees is a relatively recent innovation that simultaneously estimates the tree and the parameters in the evolutionary model, while providing a measure of confidence in those estimates. The following provides a limited introduction to Bayesian phylogenetics, highlighting some of the principles behind the methods, its advantages, disadvantages, and similarities to other methodology. More comprehensive guides to Bayesian phylogenetics can be found in Huelsenbeck et al. *(44)*, Holder and Lewis *(45)*, and online in the documentation for the BAMBE and MrBayes software. The major theoretical difference between Bayesian inference and likelihood approaches is that the former includes a factor describing prior expectations about a problem (*see* **Note 3**). More precisely, the prior is a probability distribution over all parameters, including the tree and model, describing how frequently one would expect values to be observed before evaluating evidence from the data. Bayesian inference examines the posterior distribution of the parameters of interest, such as the relative probabilities of different trees. The posterior is a probability distribution formed as a function of the prior and likelihood, which represents the information held within the data. When Bayesian inference is successful, a large dataset would ideally produce a posterior distribution focused tightly around a small number of good trees.

5.1. Bayesian Estimation of Trees

In order to make an inference about the phylogenetic tree, the posterior distribution needs to be processed in some way. To obtain the posterior probability of trees, the parameters not of direct interest to the analysis need to be integrated out of the posterior distribution. These are often referred to as "nuisance parameters," and include components of the evolutionary model and branch lengths. A common summary of the list of trees that Bayesian phylogenetics produces is the maximum *a posteriori* probability (MAP) tree, which is the tree that contains the largest mass of probability in tree space. The integration required to obtain the MAP tree is represented in the transition from left to right in the posterior distribution section of **Fig. 14.5**, where the area under the curve for each tree on the left equates to the posterior probability for each tree on the right. The confidence in the MAP tree can be estimated naturally from the posterior probabilities and requires no additional computation. In Bayesian parlance, this is achieved by constructing a credibility interval, which is similar to the confidence interval of classical statistics, and is constructed by adding trees to the credible set in order of decreasing probability. For example, the credibility interval for the data in **Fig. 14.5** would be constructed by adding the trees in the order of C, A, and B. The small posterior probability contained in B means that it is likely to be rejected from the credibility interval. Readers should be aware that there are other equally valid ways of summarizing the results of a Bayesian analysis, including the majority rule consensus tree *(46)*.

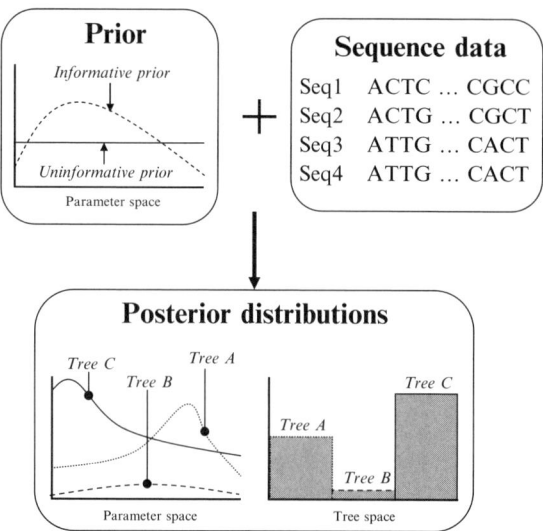

Fig. 14.5. A schematic of Bayesian tree inference. The prior *(left)* contains the information or beliefs one has about the parameters contained in the tree and the evolutionary model before seeing the data. During Bayesian inference, this is combined with the information about the tree and model parameter values held in the original data *(right)* to produce the posterior distribution. These may be summarized to provide estimates of parameters *(bottom left)* and trees *(bottom right)*.

5.2. Sampling the Posterior Using Markov Chain Monte Carlo

In phylogenetics, a precise analysis of the posterior distribution is usually not computationally possible because it requires a summation across all possible tree topologies. Markov chain Monte Carlo (MCMC) rescues Bayesian inference by forming a series (chain) of pseudo-random samples from the posterior distribution as an approximation to it. Understanding how this sampling works is useful to further explain Bayesian tree inference. A simplified description of the MCMC algorithm for tree estimation follows:

1. Get initial estimate of tree.
2. Propose a new tree (often by methods similar to traversal schemes in Section 3.2.).
3. Accept the tree according to a probability function.
4. Go to Step 2.

The options for obtaining an initial tree estimate: *(1)* are the same as for point estimation, although a random starting place can also be a good choice because starting different MCMC chains from very different places can be useful for assessing their convergence (see the following). The tree proposal mechanism in *(2)* needs to satisfy at least three criteria: (a) the proposal process is random; (b) every tree is connected to every other tree, and (c) the chain does not periodically repeat itself. The necessity for (a) and (b) allows the chain potential access to all points in tree space, which in principle allows complete sampling if the chain is allowed to run long enough. The final point (c) is a technical requirement that ensures the chain does not repeatedly visit the areas of tree space in the same order; it is aperiodic. The ability of MCMC to effectively sample tree space is highly dependent on the proposal scheme, and the most popular schemes are similar to those used in point estimation (see the preceding).

The probability of a new tree being accepted (Step 3) is the function that enables the MCMC algorithm to correctly sample from the posterior distribution. The probability of acceptance depends on the difference in likelihood scores of the current and new tree, their chances of occurring under the prior, and an additional correction factor dependent on the sampling approach. A good sampling scheme coupled with this acceptance probability enables the chain to frequently accept trees that offer an improvement, whereas occasionally accepting mildly poorer trees. Trees with very low posterior probability are rarely visited. The overall result is that the amount of time a chain spends in regions of tree space is directly proportional to the posterior distribution. This allows the posterior probability of trees to be easily calculated as the frequency of time that the chain spends visiting different topologies.

5.3. How Long to Run a Markov Chain Monte Carlo

The number of samples required for MCMC to successfully sample the posterior distribution is dependent on two factors: convergence and mixing. A chain is said to have converged when it begins to accurately sample from the posterior distribution, and the period before this happens is called burn in. The mixing of a chain is important because it controls how quickly a chain converges and its ability to sample effectively from the posterior distribution afterward. When a chain mixes well, all trees can be quickly reached from all other trees and MCMC is a highly effective method. When mixing is poor, the chain's ability to sample effectively from the posterior is compromised.

It is notoriously difficult to confirm that the chain has converged and is successfully mixing, but there are diagnostic tools available to help. A powerful way to examine these conditions is to run multiple chains and compare them. If a majority of chains starting from substantially different points in tree space concentrate their sampling in the same region, it is indicative that the chains have converged. Evidence for successful mixing can be found by comparing samples between converged chains. When samples are clearly different, it is strong evidence that the chain is not mixing well. These comparative approaches can go awry; for example, when a small number of good tree topologies with large centers of attraction are separated by long and deep troughs in the surface of posterior probabilities. If by chance all the chains start in the same center of attraction, they can misleadingly appear to have converged and mixed well even when they have poorly sampled the posterior. This behavior has been induced for small trees under artificial mis-specifications of the evolutionary model, although the general prevalence of this problem is currently unknown.

An alternative diagnostic is to examine a plot of the likelihood and/or model parameter values, such as rate variation parameters and sum of branches in the tree, against sample number. Before convergence these values may tend to show discernible patterns of change. The likelihood function, for example, may appear on average to steadily increase, as the chain moves to progressively better areas of tree-space. When the chain converges, these values may appear to have quite large random fluctuations with no apparent trend. Fast fluctuation accompanied by quite large differences in likelihood, for example, would be indicative of successful mixing. This character alone is a weak indicator of convergence because chains commonly fluctuate before they find better regions of tree space. New sampling procedures, such as Metropolis Coupled MCMC (MC^3), are being introduced that can address more difficult sampling and mixing problems and are likely to feature more frequently in phylogenetic inference. *See* also **Note 3**.

5.4. The Specification of Priors

The subjectivity and applicability of priors is one of the thorniest subjects in the use of Bayesian inference. Their pros and cons are widely discussed elsewhere (e.g., pros *(44)*; cons *(9)*) and I shall concentrate on practical details of their use in tree inference. If there is sufficient information in a dataset and the priors adequately cover tree space and parameter space, then the choice of prior should have only minimal impact on an analysis. There are broadly two types of prior: informative priors and uninformative priors. Informative priors describe a definite belief in the evolutionary relationships in sequences prior to analyzing the data, potentially utilizing material from a broad spectrum of areas, from previous molecular and morphological studies to an individual researcher's views and opinions. This information is processed to form a probability distribution over tree space. Strong belief in a subset of branches in a phylogeny can be translated as a higher prior probability for the subsection of tree space that contains them. This utility of informative priors has been demonstrated *(44)*, but is rarely used in the literature. This is partly because the choice of prior in Bayesian inference often rests with those who produce tree estimation software, not the researcher using it. Implementing an opinion as a prior can be an arduous process if you are not familiar with computer programming, which limits a potentially interesting and powerful tool.

Uninformative priors are commonly used in phylogenetics and are intended to describe the position of no previous knowledge about the evolutionary relationship between the sequences. In tree inference this can be interpreted as each tree being equally likely, which is philosophically similar to how other methods, such as likelihood and parsimony, treat tree estimation. Producing uninformative priors has proved problematic because of the interaction of tree priors with those of other parameters. This problem has been demonstrated to manifest itself as high confidence in incorrect tree topologies *(47)* and overly high support for particular branches in a tree *(48, 49)*. Further research demonstrated that this is likely to be the result of how Bayesian analysis deals with trees, in which the length of some branches are very small or zero *(47, 50)*. There have been several suggestions to deal with these problems, including bootstrapping *(49)* and describing trees with zero branches in the prior *(51)*. There is currently no settled opinion on the effectiveness of these methods. Users of Bayesian phylogenetics should keep abreast of developments in the area and employ due care and diligence, just as with any other tree estimation method.

12. Notes

1. Methodology and statistical models: The first step in any study requiring the estimation of a phylogenetic tree is to decide which methodology to use. Statistical methods are arguably the most robust for inferring trees. Their computational limitations are outweighed by benefits, including favorable statistical properties and explicit modeling of the evolutionary process. Choice of evolutionary model is also important and other chapters offer more details about the considerations one should make prior to phylogenetic inference. In general, models that more realistically describe evolution are thought more likely to produce accurate estimates of the tree. Models should include at least two components: a factor defining variation in the overall rate between sites, and an adequate description of the replacement rates between nucleotides/amino acids. Phylogenetic descriptions of rate variation allow each site in an alignment to evolve at a different overall rate with a defined probability. The distribution of potential rates is often described as a Γ-distribution, defined through a single parameter α that is inversely proportional to amount of rate variation *(52)*.

   For DNA models, replacement rates between bases should minimally consist of the parameters of the widely implemented HKY model *(53)*, which describe the relative frequency of the different bases and the bias toward transition mutation. The replacement rates between amino acids in protein models are usually not directly estimated from the data of interest. Instead, substantial amounts of representative data are used to produce generally applicable models, including the empirical models of Dayhoff *(54)* and Whelan and Goldman (WAG) *(55)* for globular proteins, and mtREV *(56)* and mtmam *(57)* for mitochondrial proteins. It is also common practice to adjust the relative frequency of the amino acids in these models to better reflect the data under consideration *(58, 59)*. Models describing the evolution of codons can also be used for phylogeny estimation *(60)*, although their primary use in phylogenetics remains the study of selection in proteins (*see* **Chapter 15**).

2. Choosing phylogenetic software: After deciding upon an appropriate methodology, a set of phylogenetic software must be chosen to estimate the tree. Phylogenetic studies require the best possible estimate of the tree, albeit tempered by computational limitations, and choosing software to maximize the coverage of tree space is advantageous. This should ideally involve creating a list of potentially best trees from a range of powerful phylogenetic software packages using complementary methods of tree searching. When using software that does not resample from tree space, it is useful to manually start the estimation procedure from different points in tree space, mimicking resampling and expanding coverage. In practice, not all phylogenetic software may implement the chosen model. In these cases, the model most closely resembling the chosen model should be used.

The final list of potentially best trees is informative about the difficulty of the phylogenetic inference problem on a particular dataset. If the trees estimated by different software and starting points frequently agree it is evidence that the tree estimate is good. If few of the estimates agree, inferring a tree from those data is probably hard and continued effort may reveal even better trees. Direct comparisons between the scores of different phylogenetic software packages are difficult because the models used and the method for calculating scores can vary. The final list should be assessed using a single consistent software package that implements the chosen model. The tree with the highest score is taken to be the optimal estimate and confidence intervals can then be calculated.

3. Bayesian inference versus ML: Bayesian and ML approaches to statistical tree inference are highly complementary, sharing the same likelihood function and their use of evolutionary models. The MAP tree in Bayesian inference and the optimal tree under ML are likely to be comparable and this can be exploited by running them in parallel and comparing their optimal tree estimates. This is also a useful

diagnostic to assess whether the MCMC chain has converged: If the trees under Bayesian inference and ML agree, it is good evidence that the chain has successfully burned in and all is well. When the trees do not agree, one or both of the estimation procedures may have gone awry. When the ML estimate is better (in terms of likelihood or posterior probability) and does not feature in the MCMC chain, it may demonstrate that the MCMC chain did not converge. When this occurs, making any form of inference from the chain is unwise.

When the Bayesian estimate is better, it is indicative that the MCMC tree search was more successful than the ML point estimation. This demonstrates a potential use of MCMC as a tool for ML tree estimation. If an MCMC sampler is functioning well, its tree search can potentially outperform the point estimation algorithms used under ML. The tree with the highest likelihood from the chain is therefore a strong candidate as a starting point for point estimation, or even as an optimal tree itself. This approach is not currently widely used for phylogenetics inference.

Acknowledgments

S.W. is funded by EMBL. Comments and suggestions from Nick Goldman, Lars Jermiin, Ari Loytynoja, and Fabio Pardi all helped improve previous versions of the manuscript.

References

1. Hahn, B. H., Shaw, G. M., de Cock, K.M., et al. (2000) AIDS as a zoonosis: Scientific and public health implications. *Science* 287, 607–614.

2. Pellegrini, M., Marcotte, E. M., Thompson, M. J., et al. (1999) Assigning protein functions by comparative genome analysis: protein phylogenetic profiles. *Proc Natl Acad Sci U S A* 96, 4285–4288.

3. Tatusov, R. L., Natale, D. A., Garkavtsev, I. V., et al. (2001) The COG database: new developments in phylogenetic classification of proteins from complete genomes. *Nucleic Acids Res* 29, 22–28.

4. Mouse Genome Sequencing Consortium. (2002) Initial sequencing of the mouse genome. *Nature* 420, 520–562.

5. The ENCODE Project Consortium. (2004) The ENCODE (Encyclopedia of DNA Elements) project. *Science* 306, 636–640.

6. Page, R. D. M., Holmes, E. C. (1998) *Molecular Evolution: A Phylogenetic Approach*. Blackwell Science, Oxford, UK.

7. Gogarten, J. P., Doolittle, W. F., Lawrence, J. G. (2002) Prokaryotic evolution in light of gene transfer. *Mol Biol Evol* 19, 2226–2238.

8. Siepel, A., Bejerano, G., Pedersen, J. S., et al. (2005) Evolutionarily conserved elements in vertebrate, insect, worm, and yeast genomes. *Genome Res* 15, 1034–1050.

9. Felsenstein, J. (2004) *Inferring Phylogenies*. Sinauer Associates, Sunderland, MA.

10. Nei, M., Kumar, S. (2000) *Molecular Evolution and Phylogenetics*. Oxford University Press, New York.

11. Whelan, S., Lio, P., Goldman, N. (2001). Molecular phylogenetics: state-of-the-art methods for looking into the past. *Trends Genet* 17, 262–272.

13. Chang, J. T. (1996) Full reconstruction of Markov models on evolutionary trees: Identifiability and consistency. *Math Biosci* 137, 51–73.
14. Rogers, J. S. (1997) On the consistency of maximum likelihood estimation of phylogenetic trees from nucleotide sequences. *Syst Biol* 46, 354–357.
15. Steel, M. A., Penny, D. (2000) Parsimony, likelihood, and the role of models in molecular phylogenetics. *Mol Biol Evol* 17, 839–850.
16. Siddall M. E., Kluge A. G. (1997) Probabilism and phylogenetic inference. *Cladistics* 13, 313–336.
17. Saitou, N., Nei, M. (1987) The neighbor-joining method: a new method for reconstructing phylogenetic trees. *Mol Biol Evol* 4, 406–425.
18. Fitch, W. M., Margoliash, E. (1967) Construction of phylogenetic trees. A method based on mutation distances as estimated from cytochrome c sequences is of general applicability. *Science* 155, 279–284.
19. Swofford, D. L., Olsen, G. J., Waddell, P. J., et al. (1996) Phylogenetic inference, in (Hillis, D.M., Moritz, C., and Mable B. K., eds.), *Molecular Systematics*, 2nd ed. Sinauer, Sunderland, MA.
20. Yang, Z., Goldman, N., Friday, A. (1995) Maximum likelihood trees from DNA sequences: a peculiar statistical estimation problem. *Syst Biol* 44, 384–399.
21. Strimmer, K., von Haeseler, A. (1996) Quartet puzzling: A quartet maximum likelihood method for reconstructing tree topologies. *Mol Biol Evol* 13, 964–969.
22. Bryant, D. The splits in the neighbourhood of a tree. *Ann Combinat* 8, 1–11.
23. Sankoff, D., Abel Y., Hein, J. (1994) A tree, a window, a hill; generalisation of nearest neighbor interchange in phylogenetic optimisation. *J Classif* 11, 209–232.
24. Ganapathy, G., Ramachandran, V., Warnow, T. (2004) On contract-and-refine transformations between phylogenetic trees. *Proc Fifteenth ACM-SIAM Symp Discrete Algorithms (SODA)*, 893–902.
25. Wolf, M. J., Easteal, S., Kahn, M., et al. (2000) TrExML: a maximum-likelihood approach for extensive tree-space exploration. *Bioinformatics* 16, 383–394.
26. Stamatakis, A., Ludwig, T., Meier, H. (2005) RAxML-III: a fast program for maximum likelihood-based inference of large phylogenetic trees. *Bioinformatics* 21, 456–463.
27. Vinh, L. S., von Haeseler, A. (2004) IQPNNI: moving fast through tree space and stopping in time. *Mol Biol Evol* 21, 1565–1571.
28. Felsenstein, J. (1993) *PHYLIP (Phylogeny Inference Package)*. Distributed by the author. Department of Genetics, University of Washington, Seattle.
29. Lewis, P. O. (1998) A genetic algorithm for maximum-likelihood phylogeny inference using nucleotide sequence data. *Mol Biol Evol* 15, 277–283.
30. Lemmon, A. R., Milinkovich, M. C. (2002) The metapopulation genetic algorithm: an efficient solution for the problem of large phylogeny estimation. *Proc Natl Acad Sci U S A* 99, 10516–10521.
31. Lundy, M. (1985) Applications of the annealing algorithm to combinatorial problems in statistics. *Biometrika* 72, 191–198.
32. Salter, L., Pearl., D. K. (2001) Stochastic search strategy for estimation of maximum likelihood phylogenetic trees. *Syst Biol* 50, 7–17.
33. Keith J. M., Adams P., Ragan M. A., et al. (2005) Sampling phylogenetic tree space with the generalized Gibbs sampler. *Mol Phy Evol* 34, 459–468.
34. Efron, B., Tibshirani, R. J. (1993) *An Introduction to the Bootstrap*. Chapman and Hall, New York.
35. Felsenstein, J. (1985) Confidence limits on phylogenies: an approach using the bootstrap. *Evolution* 39, 783–791.
36. Hillis, D., Bull, J. (1993) An empirical test of bootstrapping as a method for assessing conference in phylogenetic analysis. *Syst Biol* 42, 182–192.
37. Efron, B., Halloran, E., Holmes, S. (1996) Bootstrap confidence levels for phylogenetic trees. *Proc Natl Acad Sci U S A* 93, 13429–13434.
38. Shimodaira, H., Hasegawa, M. (1999) Multiple comparisons of log-likelihoods with applications to phylogenetic inference. *Mol Biol Evol* 16, 1114–1116.
39. Shimodaira, H. (2002) An approximately unbiased test of phylogenetic tree selection. *Syst Biol* 51, 492–508.
40. Kishino, H., Hasegawa, M. (1989) Evaluation of the maximum-likelihood estimate of the evolutionary tree topologies from DNA-sequence data, and the branching order in Hominoidea. *J Mol Evol* 29, 170–179.
41. Hasegawa, M., Kishino, H. (1994) Accuracies of the simple methods for estimating

the bootstrap probability of a maximum-likelihood tree. *Mol Biol Evol* 11, 142–145.
42. Davison, A. C., Hinkley, D. V. (1997) *Bootstrap Methods and Their Application*. Cambridge University Press, Cambridge, MA.
43. Siepel, A., Haussler, D. (2005) Phylogenetic hidden Markov models, in (Nielsen, R., ed.), *Statistical Methods in Molecular Evolution*. Springer, New York.
44. Huelsenbeck, J. P., Larget, B., Miller, R. E., et al. (2002) Potential applications and pitfalls of Bayesian inference of phylogeny. *Syst Biol* 51, 673–688.
45. Holder, M., Lewis, P. O. (2003) Phylogeny estimation: traditional and Bayesian approaches. *Nat Rev Genet* 4, 275–284.
46. Larget, B., Simon, D. (1999) Markov chain Monte Carlo algorithms for the Bayesian analysis of phylogenetic trees. *Mol Biol Evol* 16, 750–759.
47. Suzuki, Y., Glazko G. V., Nei, M. (2002) Overcredibility of molecular phylogenies obtained by Bayesian phylogenetics. *Proc Natl Acad Sci U S A* 99, 16138–16143.
48. Alfaro, M. E., Zoller, S., Lutzoni, F. (2003) Bayes or bootstrap? A simulation study comparing the performance of Bayesian Markov chain Monte Carlo sampling and bootstrapping in assessing phylogenetic confidence. *Mol Biol Evol* 20, 255–266.
49. Douady, C. J., Delsuc, F., Boucher, Y., et al. (2003) Comparison of Bayesian and maximum likelihood bootstrap measures of phylogenetic reliability. *Mol Biol Evol* 20, 248–254.
50. Yang, Z., Rannala, B. (2005) Branch-length prior influences Bayesian posterior probability of phylogeny. *Syst Biol* 54, 455–470.
51. Lewis, P. O., Holder, M. T., Holsinger, K. E. (2005) Polytomies and Bayesian phylogenetic inference. *Syst Biol* 54, 241–253.
52. Yang, Z. (1996) Among-site rate variation and its impact on phylogenetic analysis. *Trends Ecol Evol* 11, 367–372.
53. Hasegawa, M., Kishino, H., Yano, T. (1985) Dating of the human-ape splitting by a molecular clock of mitochondrial DNA. *J Mol Evol* 22, 160–174.
54. Dayhoff, M. O., Eck, R. V., Park, C. M. (1972) A model of evolutionary change in proteins, in (Dayhoff, M. O., ed.), *Atlas of Protein Sequence and Structure, vol.* 5. National Biomedical Research Foundation, Washington, DC.
55. Whelan, S., Goldman, N. (2001) A general empirical model of protein evolution derived from multiple protein families using a maximum likelihood approach. *Mol Biol Evol* 18, 691–699.
56. Adachi, J., Hasegawa M. (1996) Model of amino acid substitution in proteins encoded by mitochondrial DNA. *J Mol Evol* 42, 459–468.
57. Yang, Z., Nielsen, R., Hasegawa, M. (1998) Models of amino acid substitution and applications to mitochondrial protein evolution. *Mol Biol Evol* 15, 1600–1611.
58. Cao, Y., Adachi, J., Janke, A., et al. (1994) Phylogenetic relationships among eutherian orders estimated from inferred sequences of mitochondrial proteins: instability of a tree based on a single gene. *J Mol Evol* 39, 519–527.
59. Goldman, N., Whelan, S. (2002) A novel use of equilibrium frequencies in models of sequence evolution. *Mol Biol Evol* 19, 1821–1831.
60. Ren, F., Tanaka, H., Yang, Z. (2005) An empirical examination of the utility of codon-substitution models in phylogeny reconstruction. *Syst Biol* 54, 808–818.

Chapter 15

Detecting the Presence and Location of Selection in Proteins

Tim Massingham

Abstract

Methods to detect the action of selection on proteins can now make strong predictions about its strength and location, but are becoming increasingly technical. The complexity of the methods makes it difficult to determine and interpret the significance of any selection detected. With more information being extracted from the data, the quality of the protein alignment and phylogeny used becomes increasingly important in assessing whether or not a prediction is merely a statistical artifact. Both data quality issues and statistical assessment of the results are considered.

Key words: Positive selection, maximum likelihood, molecular evolution, protein evolution, neutral theory, adaptation, phylogeny.

1. Introduction

The appeal of detecting the role that selection has played in a protein's evolution lies in the connection to function: sites showing an unusual reluctance to change residue may be important for retained function, whereas those that change more than is expected by chance might be responsible for divergence in the protein's function and so adaptation. The patterns of conserved and diversified sites along a protein are informative about protein–protein and protein–ligand interactions, and expose the molecular fossils of past evolutionary arms races.

Traditionally, selection has been modeled in a population genetics context in terms of the difference in fitness between individual alleles, which affect how rapidly alleles propagate though a population.

The methods considered in this chapter detect selection by looking at the average behavior of many mutations over relatively long periods of time—whether there has been a prevalence of amino acid changing nucleotide mutations (non-synonymous) compared to silent (synonymous) mutations in a protein's coding sequence. Methods of this type are best suited to detecting cases in which selection has acted to cause unusually frequent changes in the amino acid composition of the protein, such as might be observed if some regions have been involved in an evolutionary arms race or when ligand-specificity has changed in a highly duplicated gene family.

Where change in function has been caused by a few particularly fit mutations and followed by a long period of conservation, the frequency of amino acid change is relatively low and the methods have little power to detect that those mutations that did occur were subjected to diversifying selection.

2. Methods

2.1. McDonald-Krietman Test

Under strictly neutral evolution, there is no difference between the fitness of a non-synonymous or synonymous mutation or in how rapidly they propagate through a population. The probability that a neutral mutation ultimately becomes fixed in all members of the population is independent of whether that mutation was non-synonymous or synonymous, and so the ratio of non-synonymous to synonymous mutations that ultimately become fixed is equal to the ratio of non-synonymous to synonymous polymorphisms within the population.

The McDonald-Kreitman test *(1)* considers the relative proportions of non-synonymous and synonymous mutants that occur within and between populations. Sites that exhibit variation in any population are classed as polymorphic, whereas sites that have no variation within a population but are different in at least one population represent fixed mutations; monomorphic sites are ignored since they are not informative about the types of mutation that occur (but may be informative about the rate at which mutations arise). Mutations in these two categories are then tabulated according to whether they are synonymous or non-synonymous, ambiguous sites being discarded, to form a 2-by-2 contingency table that can be used to test whether the type of mutation (synonymous or non-synonymous) is independent of where it is observed (within or between populations).

Table 15.1 shows the results reported by McDonald and Kreitman *(1)* for the *adh* locus, encoding alcohol dehydrogenase, in three species from the *Drosophila* subgroup. A one-sided application of Fisher's exact test *(2)* gives a p value of 0.007 for the

Table 15.1
Contingency table showing the different types of mutations observed within and between three populations in the *Drosophila* subgroup

	Between	Within
Non-synonymous	7	2
Synonymous	17	42

As originally reported by McDonald and Kreitman (1991). No sites were discarded because of ambiguity, although two sites showed more than one synonymous polymorphism.

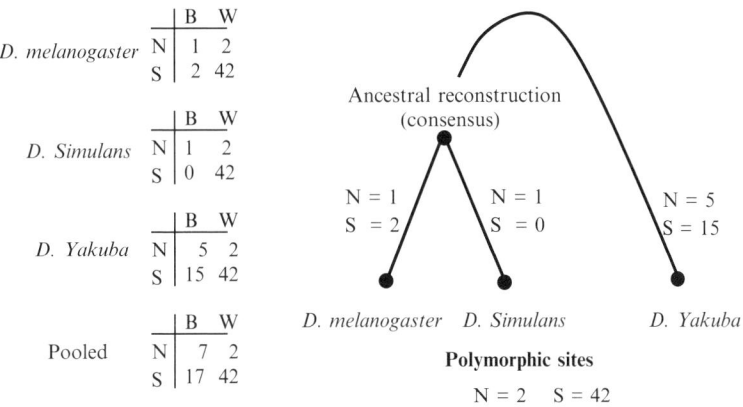

Fig. 15.1. Mutations observed in *Drosophila* adh tabulated for a McDonald-Kreitman test. The number of synonymous, *S*, and non-synonymous, *N*, mutations polymorphic within, *W*, and fixed between, *B*, populations in the *Drosophila* subgroup. Mutations are broken down by branch, using the consensus ancestral reconstruction, and contingency tables are presented both for the original MK test (data for all branches pooled) and for each branch using the polarized variant. Data taken from McDonald and Kreitman (1991).

presence of diversifying selection, but there is no indication as to where it took place.

The "polarized" variant of the McDonald-Kreitman test enables selection to be more precisely located in the phylogeny so that one can conclude, for example, that diversifying selection acted on the protein on the lineage leading to *D. melanogaster* rather than just somewhere between *D. melanogaster* and *D. simulans*. The polarized test uses a consensus ancestral sequence to assign fixed mutations to a particular lineage; non-polymorphic sites in which the ancestral sequence is ambiguous are ignored. The contingency table for non-neutral evolution on a single lineage is constructed by comparing fixed mutations on that particular lineage to mutations that are polymorphic in any lineage; the difference between this and the original test is illustrated in **Fig. 15.1**. One-sided application of Fisher's exact test gives p values of 0.026 for diversifying selection leading to *D. yakuba*, 0.183 for the lineage leading to *D. melanogaster*, and 0.364 for that leading to *D. simulans*. If multiple lineages are considered in this manner, p values must be adjusted to allow for the number of tests performed (see "Correcting for Multiple Tests"); the adjusted p value for the diversifying selection indicated on the *D. yakuba* branch is 0.08 and so would not be considered significant.

The relative number of synonymous and non-synonymous mutations that might be the result of a single nucleotide mutation depends on the protein's amino acid composition and codon bias, so the MK test should only be applied to species in which these are similar. A contingency table of species against codons can be used to test whether there is a significant difference in codon usage between the two sequences considered.

Because it does not take into account uncertainty in the ancestral reconstruction, the McDonald-Kreitman test is limited in the number of sequences and the degree of divergence for which it is valid. For example, there are sites in the *adh* locus data that show fixed differences between *D. melanogaster* and *D. simulans*, but are polymorphic in *D. yakuba*; potential problems with ambiguity of the ancestral reconstruction are avoided by counting these sites as polymorphic but information has been lost since there are fixed differences that have not been counted. If all ancestors can be reconstructed perfectly, then fixed mutations on all branches of a phylogeny are known and can be pooled, if not then most data have to be discarded as ambiguous.

A more severe limitation of the MK test is that mutations at all sites are pooled in the same contingency table, so the procedure measures the average selective pressure across the entire sequence and a few sites under strong diversifying selection may be masked by a lot of conserved sites. The likelihood-based methods for detecting selection rectify most of the drawbacks of the MK test at the expense of employing more technical statistical machinery.

2.2. Probabilistic Models of Codon Evolution

The likelihood-based approaches for detecting diversifying selection rely on a probabilistic model of nucleotide substitution that is composed of two steps: background mutation followed by selection. New mutations arise in the population and become fixed with some probability dependent on the increase or decrease in fitness they bring the organism. The process is described by a matrix, containing the relative rate that one codon is substituted by another, mutations to and from stop codons being ignored. Although more complex versions can be produced, the following simple substitution matrix *(3)* describes many important features of molecular evolution: differing fixation probabilities for non-synonymous and synonymous changes (p_N and p_S, respectively), transition/transversion bias (κ) in nucleotide mutation, and compositional differences through the frequency (π_i) that codon i is observed. The i,j entry of this matrix is determined by:

$$q_{ij} \propto \begin{cases} 0 & i \to j \text{ more than one nucleotide substitution} \\ \pi_j & p_S \quad i \to j \text{ synonymous transversion} \\ \pi_j \quad \kappa & p_S \quad i \to j \text{ synonymous transition} \\ \pi_j & p_N \quad i \to j \text{ nonsynonymous transversion} \\ \pi_j \quad \kappa & p_N \quad i \to j \text{ nonsynonymous transition} \end{cases}$$

with the diagonal entries (q_{ii}) set so that the substitution matrix is a valid rate matrix for a continuous-time Markov chain. Since time and rate are confounded, the process is scaled so one substitution is expected per unit time.

The probabilities of fixation for synonymous and non-synonymous mutations are often written as their ratio $\omega = \frac{p_N}{p_S}$, the relative rate ratio of non-synonymous to synonymous substitution. The quantity ω is also known as dN/dS or K_A/K_S in the literature. Interpreting the non-synonymous/synonymous ratio as the ratio of two probabilities of fixation connects it with the population genetic notion of fitness, with implications for how any observation of diversifying selection should be interpreted: under the Wright-Fisher model, the probability that a newly arisen mutation ultimately becomes fixed in the population depends on both the relative fitness of the new allele and the effective size of the population.

If there are no differences between the probability that non-synonymous and synonymous mutations are fixed, then $\omega = 1$, independently of the population size and hence independently of the effects of fluctuations in the size of the population. Although the point where purifying selection switches to being diversifying selection does not depend on the population size, the relationship between relative fitness and dN/dS for non-neutral mutations does. Shifts in the size of the population mean that mutations resulting in the same increase or decrease in fitness may have different probabilities of fixation, depending on where they occurred in the phylogenetic tree—for this reason caution should be exercised when comparing dN/dS at two different sites and declaring that one is under stronger selection than the other. The mutations that lead to two sites having the same ω may not have caused the same average change in the fitness of the organism.

These models of substitution make several assumptions that may be violated in real data; for example, they assume that all mutations have become fixed in the population, whereas in reality individual sequences may contain mutations that will ultimately be lost from the population (*see* **Note 1**). Multi-nucleotide mutations, such as those that can be caused by UV damage or by a second mutation occurring in the same codon before the first is fixed, are assumed not to occur, although there is considerable evidence that they do in practice. (*See (4)* for a review and suggestions of how these effects can be incorporated in substitution matrices.) Large-scale evolutionary events like frame-shifts, gene conversions or any other form of horizontal transfer are not described by these models and their presence in data can lead to erroneous indications of diversifying selection. The substitution model allows for differing codon composition through the codon frequency but the mechanisms that create and sustain a codon bias are not well understood and such a simple correction may not reflect reality (**Note 2**).

Although insertion events are not explicitly modeled, they are often incorporated into calculations as "missing data": codons are used when they are present, and all possibilities are considered otherwise. This procedure ignores any information that the presence of an insertion event conveys about the selection pressures on the bordering sequence, but does take into account how the inserted sequence evolves afterward and so may reveal the evolutionary constraints that lead to the event. Deletions are dealt with similarly.

Given an alignment of protein-coding sequences and a phylogeny connecting them, the probability of observing column i of the alignment under this model of evolution can be calculated using the standard algorithms *(5)*. Let the codons in column z, D_i, be under selective pressure ω, and all other parameters such as κ and branch length be represented by θ. The log-likelihood quantifies how well a particular set of parameters describes the data, and is related to the probability by:

$$l_i(\omega, \theta | D_i) = \ln P(D_i | \omega, \theta).$$

Assuming all sites are independent, the log-likelihood is additive and so the log-likelihood of the parameters given all columns in the alignment is:

$$l(\omega, \theta | D) = \sum_i l_i(\omega, \theta | D_i).$$

Parameter values that maximize the log-likelihood are good estimates and can be found using standard numerical optimization techniques, although there are some practical difficulties in ensuring that the values found are indeed maximal (*see* **Note 3**).

The maximum log-likelihood can also be used for hypothesis testing, for example, to test whether the data support a more general probabilistic model over a simpler one. When one model is a special case of the other, the models are said to be nested. The likelihood-ratio test statistic is twice the difference in the maximum log-likelihood between the more general and simpler models; the LRT statistic is small when there is no evidence that the general model is a better description of the data, with larger values indicating stronger evidence in favor of the more general model. The exact distribution of the LRT statistic, which is used for the calculation of p values, may depend on the models used, although it tends to a χ^2 distribution, or mixture of χ^2 distributions, under very general conditions *(6)*.

A "strictly neutral" model of evolution (as defined in the preceding with $\omega = 1$) is nested within one in which ω is free to vary, so the hypothesis test of neutral verses non-neutral evolution can be framed in terms of nested models. The likelihood ratio test statistic, Δ, for this example is defined by:

$$\tfrac{1}{2}\Delta = \max_{\omega,\theta} l(\omega,\theta|D) - \max_{\theta} l(1,\theta|D)$$

and is distributed as χ_1^2, critical values for which are available in most statistical tables. The likelihood for each model is maximized separately, so the ML estimate of θ when $\omega = 1$ may be different than when ω is free to vary.

For neutrally evolving sequences, the maximum likelihood value of ω is equally likely to be above or below one. If only diversifying selection is an interesting alternative and so the more general model has the restriction $\omega \geq 1$ then the likelihood-ratio statistic is zero half the time—rather than conforming to a χ_1^2 distribution, the likelihood ratio statistic is distributed as an equal mix of a point mass on zero and a χ_1^2 distribution, written $\tfrac{1}{2}I_0 \wedge \tfrac{1}{2}\chi_1^2$. The critical values of this mixture distribution are equal to those from a χ_1^2 distribution that has half the size.

Like the McDonald-Kreitman test, this procedure assumes that all sites are under the same level of non-synonymous selection and, although useful for determining whether or not the sequences evolved like pseudo-genes, does not realistically describe how selection might operate on a functional protein: some regions may be so important that any change would be deleterious, whereas others may be involved in an evolutionary arms race and change constantly.

The random-site methods of detecting selection *(3, 7)* add an extra layer to the model of sequence evolution to describe variation in ω along the sequence. Suppose it was known that each site along a protein had been subjected to one of n different levels of selective pressures, with proportion p_j sites evolving with selection pressure ω_j, then without knowing anything further about each aligned site, the probability of the observed data is:

$$P(D_i) = \sum_{j=1}^{n} p_j P(D_i|\omega_j)$$

and the likelihood for each site, and hence for the entire sequence, can be calculated as previously. By altering the number of categories and constraining the values of ω_j, nested tests for diversifying selection can be constructed.

Models with only a few categories of sites cannot describe the full range of selection that operates on different sites of a protein but each additional category adds two more parameters. Instead of allowing complete freedom of choice, the categories are often constrained to those of a particular form ("family"), where a few parameters determine the strength of selection for all categories. Many such families have been produced (7) by breaking a continuous distribution into equal chunks and representing each chunk by a category with ω equal to the median value of the chunk. The categories produced can either be thought of as a distribution in their own right or as an approximation to a continuous distribution which improves as the number of categories increases. The more popular models are described in **Note 4**.

The use of ω to detect selection implicitly assumes that the probability that a synonymous mutation becomes fixed in the population is constant across sites, and so can be used as a surrogate for the probability of neutral fixation. *See* **Note 5** for alternatives if synonymous variation is suspected.

As is often the case in phylogenetics, hypothesis tests for the presence of diversifying selection are more complex than is covered by the standard theory for likelihood ratio tests. Rather than taking fixed values under the null model, parameters that differentiate the null and alternative models may become inestimable as categories merge or their weight is reduced to zero. There is also the additional complication that not all cases in which the alternative model has a greater likelihood than the null have categories of sites with $\omega_j > 1$, and so do not represent the possibility of diversifying selection. These problems can be overcome by using parametric bootstrap techniques, comparing the test statistic for the set of data being studied to those from many random "pseudo-sets" of data, generated under the null model. Using the appropriate set of maximum likelihood parameters, a large number of pseudo-sets of data are generated using the null model (8) and the maximum likelihood values found under both the null and alternative models for each pseudo-set. Considering the jth pseudo-set, if the ML parameters under the alternative model indicate diversifying selection, then define B_j to be equal to the likelihood-ratio test statistic, otherwise it is set to zero. If the ML parameters under the alternative model for the original set of data do not indicate the presence of diversifying selection, then we conclude none is present. Otherwise the p value for the presence of diversifying selection can be estimated by the proportion of pseudo-sets for which $B_j > \Delta$.

A protein that has undergone a recent insertion event contains residues that are not homologous to any other in the alignment and so provide no information about the phylogeny or selection pressures that have acted along it. However, artificially simulated data sets are often produced ungapped and so are not fair representations of the original data, each site being more informative on average. This can be remedied by transferring the pattern of gaps from the original alignment onto each of the artificial sets of data, although this incorrectly assumes that there is no relationship between the gap occurrence and the strength of selection.

The non-parametric bootstrap provides an alternative approach to testing for the presence of diversifying selection. Rather than generating alignment columns using a model, columns from the original data are sampled with replacement to create a new alignment of equal length. This new alignment reflects the biases of the original, containing about the same number of sites under diversifying selection and the same number of gaps, but ML parameter estimates will differ from the original, depending on how confidently they were estimated. The proportion of resampled sets for which the ML parameter estimates indicate the presence of diversifying selection (that is, having one or more categories with non-zero weight and $\omega > 1$) indicates the confidence that the original data contained sites under diversifying selection. When the diversifying selection detected in any of the resampled sets is borderline, having a small weight or close to the conserving/diversifying boundary, extra care should be taken to ensure that the result is not due to numerical error in finding the ML estimates.

The use of the nonparametric bootstrap was first introduced into phylogenetics to assess the confidence when comparing many potential topologies *(9)*. The parametric bootstrap has been used to test whether a given substitution model adequately describes a sequence alignment *(8)*. Efron and Tibshirani *(10)* provide a very readable introduction to bootstrap techniques. In practice, the amount of computing power needed makes bootstrap techniques infeasible for many studies and approximations are used instead (*see* **Note 6**).

2.3. Location of Diversifying Selection

From a Bayesian perspective, the weight on category j can also be thought of as the prior probability that a randomly chosen site is in that category. The posterior probability that site i is in category j can be calculated as:

$$\frac{p_j P(D_i|\omega_j)}{\sum_k p_j P(D_i|\omega_j)}$$

and the sum of the posterior probabilities from all categories where $\omega_j > 1$ gives the posterior probability that a site is under diversifying selection. This procedure is known as naïve empirical Bayes estimation and it does not allow for uncertainties in the estimates of any parameters, which are particularly important when some of the categories lie close to the conserving/diversifying boundary and a small change can flip them from one side to the other. A much improved procedure that corrects this defect, Bayes empirical Bayes *(11)* and these posterior probabilities should be reported in preference.

A more direct approach to detecting deviations from neutral evolution at a particular site is to use a likelihood ratio test at each individual site, the Site-wise Likelihood Ratio (SLR) approach *(12, 13)*. All parameters, including ω, are estimated by maximum likelihood, assuming the strength of selection is the same at every site. Holding all parameters other than ω fixed, a separate value of ω is re-estimated at every site to produce a new likelihood ratio test statistic:

$$\tfrac{1}{2}\Delta_i = \max\, l(\omega|D_i) - l(1|D_i).$$

Unlike the random-sites models, the site-wise likelihood ratio statistic has a simple distribution when the null model is true and so it is not necessary to use the bootstrap techniques discussed earlier. The test statistic Δ_i is distributed as χ_1^2 for neutral versus non-neutral selection, and $\tfrac{1}{2}I_0 \wedge \tfrac{1}{2}\chi_1^2$ for neutral versus diversifying or neutral versus conserving selection. Since each site requires a separate hypothesis test, the probability that at least one gives a false-positive result is greater than for a single test and so the results need to be corrected for multiple comparisons. Methods are discussed in "Correcting for Multiple Tests."

The random-sites and SLR methods are complementary, the former providing a hypothesis test for the presence or absence of diversifying selection, whereas the latter tests for location. Detecting and locating diversifying selection demands much more from the data than just determining presence or absence, so it is possible for random-sites methods to detect diversifying selection, even though no single residue is found to be changing significantly faster than neutral.

3. Preparation of Materials

The likelihood methods for detecting diversifying selection rely on the accuracy of the phylogenetic tree and sequence alignment used, and problems with either could lead to erroneous indications of selection. Confidence in individual regions of the phylogeny can be assessed using bootstrap techniques *(9)* (*see* **Chapter 14**) and badly supported clades should be removed from the data, perhaps to be analyzed separately if they are internally well supported.

Estimating confidence in different regions of an alignment is a more difficult problem than for phylogeny and general techniques are not available. Although far from ideal, the following procedure provides useful information about confidence in the alignment without making strong assumptions about how the sequences evolved:

1. Align the sequences.
2. Train a profile Hidden Markov Model (profile HMM) on the alignment.
3. Align all sequences with the profile HMM.
4. For each sequence, calculate the (posterior) probability that each site is aligned.

A profile HMM is a model of what a typical protein in the alignment looks like, consisting of a sequence of states that each describe what residues may occur at each site along its length. Proteins can be aligned with this model and sites in two different proteins that align with the same state are homologous. Columns of the alignment that represent insertions relative to the profile HMM, called "mismatch" states, do not represent homologous residues and so should not be included in further analyses. The theory behind the use of profile HMMs in biology is described in *(14)* and the necessary calculations can be performed using the hmmbuild, hmmalign, and hmmpostal programs from the HMMER package *(15)*, which can be obtained from http://hmmer.janelia.org/.

The variation in the posterior probability of correct alignment along three different sequences of the HIV/SIV envelope glycoprotein (GP120) is shown in **Fig. 15.2**. There is a pronounced dip between residues 101 and 151 corresponding to the hypervariable V1/V2 loop regions of the protein. Examination of this region of the alignment shows that it contains many gaps, which is consistent with an intuitive notion of a poorly aligned region.

Closer inspection of the alignment reveals that HIV1-EL contains an apparent frame-shift affecting residues 425–428. Since the mutation-selection model only describes evolution arising from single nucleotide substitutions, sequence changes caused by other modes of evolution (e.g., frame-shifts, horizontal transfer, gene conversions, etc.) should be removed before applying tests for diversifying selection. Rather than removing an entire sequence or column of the alignment, the sites affected can be replaced by gaps and so treated in the calculations as unknown "missing" data.

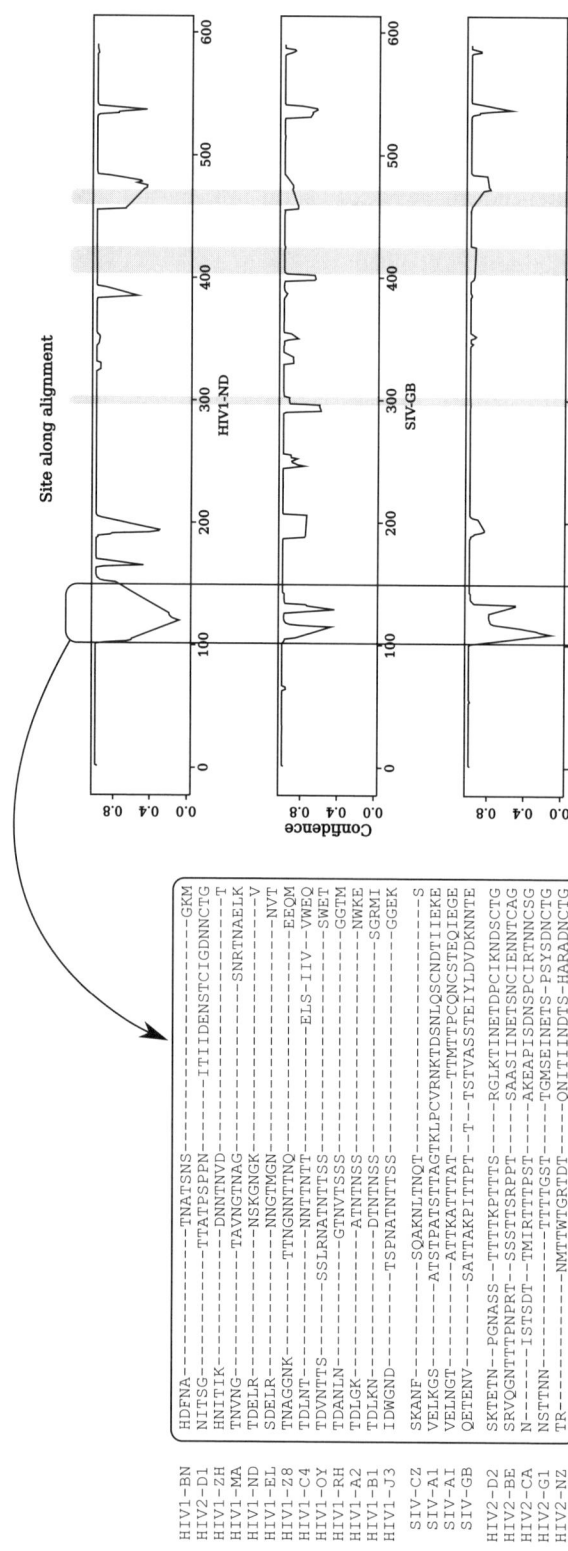

Fig. 15.2. Confidence in the alignment of 22 HIV and SIV amino acid sequences encoding GP120. The graphs give the posterior probability that each site in the sequence is correctly aligned, the shaded area being regions in which the sequences do not align with the HMM.

4. Correcting for Multiple Tests

An often neglected aspect of detecting diversifying selection is the need to correct for multiple tests. Results are often reported as a p value for the presence of diversifying selection and a list of sites at which it was most likely to have occurred. Making a prediction of diversifying selection represents an additional statistical test for each site considered and several may be false-positives.

Commonly used multiple comparison techniques seek to control the overall error by bounding one of two quantities: the Family-Wise Error Rate (FWER) is the probability that there are one or more false-positive results, whereas the False Discovery Rate is the proportion of positive results that are expected to be false. The classic Bonferroni procedure *(16)* falls into the former category, whereas more recent techniques, e.g., Benjamini and Hochberg *(17)*, control the latter.

One of the simplest methods of adjusting for multiple tests is the Bonferroni procedure: The significance level of each individual test is reduced so that the FWER is controlled. After removing sites from insertions only present in one sequence, suppose there are n sites at which diversifying could potentially be detected. If these sites are classified as being under diversifying selection only if they have a p value less than $\frac{\alpha}{n}$, then the Bonferroni inequality guarantees that the FWER is $\leq \alpha$. For example, to guarantee an FWER of at most 0.01 when considering 276 sites, only those with a p value $< \frac{0.01}{276} = 3.6 \times 10^{-5}$ should be considered significant.

Insertions that are only observed in one species do not count toward the total number of potential false-positives because it is not possible to detect diversifying selection at such sites as the likelihood ratio statistic is zero. The Bonferroni procedure assumes the worst possible case—that all the sites under consideration may be false-positives—and consequently the test is conservative. Consider, for example, the extreme case in which all sites bar one are under diversifying selection. Other procedures try to adjust the value of n depending on how much selection they predict to be present.

Although not the most powerful method of adjusting p values to control the FWER, Holm's procedure *(18)* is always at least as powerful as the Bonferroni procedure and is valid under the same general conditions. Holm's method is based on the observation that if diversifying selection detected at a site is a true-positive, then it should not count toward the number of potential false-positives. The Bonferroni procedure is applied iteratively, removing sites that are "definitely" under diversifying selection from those that may potentially be false-positives.

1. Mark all sites as "not significant."
2. Calculate Bonferroni correction (b) based on the number of sites currently marked as "not significant."
3. If all sites marked as "not significant" have p values larger than b, then STOP.
4. Mark all sites with a p value $\leq b$ as "significant."
5. If any sites are marked "not significant," go to 2; otherwise STOP.

The first application of iteration 1–3 is equivalent to applying the Bonferroni procedure to all sites; Holm's procedure only differs if there are sites that are significant after Bonferroni correction. Either these are all true-positives, in which case the Bonferroni correction was too severe, or at least one is a false-positive and would also have been a false-positive under the Bonferroni procedure. Therefore, Holm's procedure controls the FWER at least as well as the Bonferroni procedure and finds at least as many true-positives for a given nominal FWER. The price for this improved power is that when one false-positive result is found, Holm's procedure is more likely than Bonferroni to find another: The two methods both control the probability of observing at least one false-positive result, but the expected number of such results is higher for Holm's procedure.

For the GP-120 alignment discussed earlier, 341 sites have a p value for non-neutral evolution <0.01, including 19 with apparent diversifying selection. **Table 15.2** contains the results after applying Holm's procedure with an FWER of 0.01, finding five more sites than the Bonferroni method (equivalent to iteration 1). Unfortunately, 10 of the sites apparently under diversifying selection, including both the sites still significant after applying Holm's procedure, fall in the region indicated in **Fig. 15.2**, in which the alignment is uncertain and so could be artifacts due to alignment

Table 15.2
Number of sites in which significant non-neutral evolution was detected using Holm's procedure to control the FWER rate to less than 0.01 over 590 individual tests

Iteration	Significant sites
1	168 (2)
2	173 (2)
3	173 (2)

The number of sites with significant diversifying selection is in parentheses.

error. Further investigation would be needed here before it can safely be concluded that diversifying selection is present.

Many other procedures are available to adjust p values for multiple comparisons; the R statistical software (http://www.r-project.org/) is a good starting point, in particular the help page for the routine *p.adjust*, which implements several methods to control both FWER and FDR. The Hochberg's FWER method *(19)* was used by Massingham and Goldman *(12)* and that of Simes *(20)* has also been used *(21)*. Although not the most powerful technique, Holm's procedure has the advantage of making fewer assumptions about how the tests are correlated with each other.

In real proteins, the need to retain the ability to fold and remain soluble causes dependencies between sites, which could lead to negative association between the strength of selection. For example, some sites may have to be conserved in order to stabilize the disruption caused by frequent changes occurring at other sites.

For random effects models, it is common to claim all sites with a posterior probability of diversifying selection greater than some threshold are significant. This criterion treats the Bayesian posterior like any other statistic and so should be corrected for multiple tests, although converting the posterior into a p value is complex and dependent on the set of data under study *(21)*. The incorporation of parameter uncertainty causes complex dependence between the posterior probabilities of different sites, which makes correction difficult.

5. Notes

1. In future, models of substitution may be constructed that allow for sequences to have been sampled from a population and so may contain partially selected mutations that are not yet fixed in the population as a whole. This would be advantageous since the McDonald-Kreitman test shows that polymorphic sites contain information about neutral evolution that could improve estimates of selection. Currently, population sampling is not allowed for and the presence of a polymorphism will bias the estimates of non-synonymous and synonymous rates at a site in similar fashion to a sequencing error. Although not ideal, a practical measure to guard against the adverse effects of sampling from a population is to see if diversifying selection at a site can still be detected after the removal of individual sequences. This procedure relies on unselected mutations being sufficiently rare that they only occur in at most one sequence at each site. Unfortunately, this situation is akin to the birthday paradox:

The probability of an unselected mutation occurring at the same site in any two of the sequences sampled can be quite high, even though the probability of one occurring in any individual sequence is small. Removing sequences dilutes population sampling effects, but does not remove them.

2. The evolutionary origins of codon bias and how it interacts with selection are not fully understood. Although the codon substitution matrix includes a correction for composition, it does not describe any mechanistic reason why such a bias could arise or how it perpetuates. Consequently, the matrix does not accurately reflect the full effects that strong codon bias has had on the evolution of a protein. Measures of codon bias include the "relative synonymous codon usage" *(22)* and the "effective number of codons" *(23)*, and care should be taken when highly biased data is indicated.

3. Finding the parameter values at which the log-likelihood is maximal is a difficult problem in phylogenetics and requires the use of numerical techniques. The likelihood ratio test requires the highest possible point of the likelihood function, the global maximum, whereas those found by numerical techniques may only be local maxima. This situation is analogous to finding the highest peak in a range of hills while blindfolded: You do not know which hill you are on, or how many hills there are, but you can walk uphill until you find the peak of whichever hill you are on. The only remedy is to start the optimizer at many different points and use the best result. The difference in log-likelihood between two nested models can never be less than zero, since one is just a special case of the other, and this is a useful check on whether a local maximum has been found. If Δ is ever found to be negative for nested models, then the numerical optimization has failed to find the global maximum in at least the alternative model. The advice from the PAML manual *(24)* is that the results from several different pairs of models should be compared, and many different starting points should be used for each optimization. In addition, the results from both fixed and random sites models should be compared before concluding that any selection found is not an artifact of bad optimization.

4. Random effects models require some distribution to be used to model how selection varies across sites in the sequence. Identifying the best choice for such a distribution is an open and perhaps insoluble problem. A number of different choices have been suggested *(7)*, some of which are nested and so permit a likelihood-ratio test. **Table 15.3** describes some common pairs used in the literature. However, many of the possible pairs of nested models do not properly represent

Table 15.3
Some common distributions used in random-sites models of how selection varies across sites in a sequence

Name	Formula	Restrictions	Number parameters
$\begin{cases} M_0 \\ M_3 \end{cases}$	I_ω $p_0 I_{\omega_0} \wedge \ldots \wedge p_n I_{\omega_n}$		1 $2n - 1$
$\begin{cases} M_7 \\ M_8 \end{cases}$	$\beta(a,b)$ $(1 - p)\beta(a,b) \wedge p I_\omega$		2 4
$\begin{cases} M_{1a} \\ M_{2a} \end{cases}$	$p I_{\omega-} \wedge (1 - p) I_1$ $p I_{\omega-} \wedge p_1 I_1 \wedge p_2 I_{1+}$	$\omega_- \leq 1$ $\omega_- \leq 1, \omega_+ \geq 1$	2 4
$\begin{cases} M_{8a} \\ M_8 \end{cases}$	$(1 - p)\beta(a,b) \wedge p I_1$ as previous	$\omega \geq 1$	3 4

I_ω is a point mass at ω, $\beta(a,b)$ is the density of a two-parameter beta function, and $\wedge$ means "mixed with" in the given proportions, so $(1 - p)\beta(a,b) \wedge I_c$ is a mixture of a beta distribution and a point mass at c in the ratio $1 - p$:p. The models are grouped into nested pairs. All models must satisfy the restriction $\omega \geq 0$.

tests for the presence of diversifying selection and can report significant results when none are present. A good example is the pair *M*0 (all sites are under the same selective pressure) and *M*3 (selective pressure at a site takes one of n values), which is more properly a test of homogeneity than of diversifying selection. The most common failing for other pairs of models is that there are special cases of the alternative model in which no site is under diversifying selection but cannot be described by the null model, for example, a beta distribution with a point mass representing strictly neutral evolution is a special case of *M*8 but cannot be described by *M*7 (just a beta distribution). This was remedied by introducing the model *M*8*a* and restricting the point mass in *M*8 so it can only represent strictly neutral or diversifying selection *(25)*. The comparisons *M*8*a* vs. *M*8 and *M*1*a* vs. *M*2*a* *(21)* should be used in preference to those previously published.

5. If synonymous rate variation is suspected, there are variations of the probabilistic methods that take it into account *(26)*. These variants of the methods described here allow both the rate of non-synonymous and synonymous substitution to vary at each site, testing whether the two are different. These models can be interpreted in two ways: either the rate of neutral evolution is changing on a site-by-site basis, or there is selection for or against synonymous change. In the latter case, non-synonymous mutations might be fixed more

readily than synonymous ones but still slower than a truly neutral mutation: a different definition of positive selection to that used in this chapter. When synonymous rate variation is not present, these methods are less powerful than their equivalents because they must estimate more from the same data.

6. One of the attractions of the likelihood ratio test that has led to its widespread use in phylogenetics is that the distribution of the test statistic usually tends toward a simple distribution, independent of the particular problem. Unfortunately, several of the assumptions required for this convergence do not hold for the mixture models used to test for diversifying selection, and other techniques must be used. The parametric and non-parametric bootstrap techniques described here can be computationally onerous, and often we would be willing to sacrifice some statistical power for expediency. In practice, the likelihood ratio statistic is compared to a simple known distribution that, although not exact, is thought to produce a conservative test for a given pair of models. Simulation studies *(21, 27)* suggest that chi-squared distributions, with the degrees of freedom listed in **Table 15.4**, are a reasonable alternative to bootstrap methods and do not result in an excessive false-positive rate. For the comparison *M0* vs. *M3*, the chi-squared distribution listed is extremely conservative and so may fail to detect heterogeneity of selection in many cases in which it is present. In this particular case, the penalized likelihood-ratio test *(28)* provides an alternative method of comparison that has extremely good power.

Table 15.4
Degrees of freedom for the chi-squared distributions that have been suggested for testing for diversifying selection by comparing the stated models

Comparison	Degrees of freedom
M0 vs. *M3*	$2n - 2$
M1a vs. *M2a*	2
M7 vs. *M8*	2
M8a vs. *M8*	1

The comparisons M0 vs. M3 and M7 vs. M8 are included as they appear often in the literature but are not correct tests for diversifying selection and should not be used. n is the number of categories in the model.

References

1. McDonald, J., Kreitman, M. (1991) Adaptive protein evolution at the adh locus in *drosophila*. *Nature* 351, 652–654.

2. Sokal, R. R., Rohlf, F. J. (1995) *Biometry*, 3rd ed. W. H. Freeman and Company, New York.

3. Nielsen, R., Yang, Z. (1998) Likelihood models for detecting positively selected amino acid and applications to the HIV-1 envelope gene. *Genetics* 148, 929–936.

4. Whelan, S., Goldman, N. (2004) Estimating the frequency of events that cause multiple nucleotide changes. *Genetics* 167, 2027–2043.

5. Felsenstein, J. (1981) Evolutionary trees from DNA sequences: a maximum likelihood approach. *J Mol Evol* 17, 368–376.

6. Self, S. G., Liang, K.-Y. (1987) Asymptotic properties of maximum likelihood estimators and likelihood ratio tests under nonstandard conditions. *J Amer Stat Assoc* 82, 605–610.

7. Yang, Z., Nielsen, R., Goldman, N., et al. (2000) Codon-substitution models for heterogeneous selection pressure at amino acid sites. *Genetics* 155, 431–449.

8. Goldman, N. (1993) Statistical tests of models of DNA substitution. *J Mol Evol* 36, 182–198.

9. Felsenstein, J. (1985) Confidence limits on phylogenies: an approach using the bootstrap. *Evolution* 39, 783–791.

10. Efron, B., Tibshirani, R. J. (1993) *An Introduction to the Bootstrap*. Chapman and Hall/CRC, Florida.

11. Yang, Z., Wong, W. S. W., Nielsen, R. (2005) Bayes empirical Bayes inference of amino acid sites under positive selection. *Mol Biol Evol* 22, 1107–1118.

12. Massingham, T., Goldman, N. (2005) Detecting amino acid sites under positive selection and purifying selection. *Genetics* 169, 1753–1762.

13. Suzuki, Y. (2004) New methods for detecting positive selection at single amino acid sites. *J Mol Evol* 59, 11–19.

14. Durbin, R., Eddy, S., Krogh, A., et al. (1998) *Biological Sequence Analysis: Probabilistic Models of Proteins and Nucleic Acids*. Cambridge University Press, Cambridge, England.

15. Eddy, S. R. (1998) Profile hidden Markov models. *Bioinformatics* 14, 755–763.

16. Hsu, J. C. (1996) *Multiple Comparisons: Theory and Methods*. Chapman and Hall, London.

17. Benjamini, Y., Hochberg, Y. (1995) Controlling the false discovery rate: a practical and powerful approach to multiple testing. *J Roy Stat Soc B* 57, 289–300.

18. Holm, S. (1979) A simple sequentially rejective multiple test procedure. *Scand J Stat* 6, 65–70.

19. Hochberg, Y. (1988) A sharper Bonferroni procedure for multiple tests of significance. *Biometrika* 75, 800–803.

20. Simes, R. J. (1986) An improved Bonferroni procedure for multiple tests of significance. *Biometrika* 73, 751–754.

21. Wong, W. S. W., Yang, Z., Goldman, N., et al. (2004) Accuracy and power of statistical methods for detecting positive adaptive evolution in protein coding sequences and for identifying positively selected sites. *Genetics* 168, 1041–1051.

22. Sharp, P. M., Li, W.-H. (1987) The codon adaptation index — a measure of directional synonymous usage bias, and its potential applications. *Nucl Acids Res* 15, 1281–1295.

23. Wright, F. (1990) The "effective number of codons" used in a gene. *Gene* 87, 23–29.

24. Yang, Z. (2000) *Phylogenetic Analysis by Maximum Likelihood (PAML)*, version 3.0. University College London. http://abacus.gene.ucl.ac.uk/software/paml.html.

25. Anisimova, M., Bielawski, J. P., Yang, Z. (2001) Accuracy and power of the likelihood ratio test in detecting adaptive molecular evolution. *Mol Biol Evol* 18, 1585–1592.

26. Kosakovsky-Pond, S. L., Frost, S. D. W. (2005) Not so different after all: a comparison of methods for detecting amino acid sites under selection. *Mol Biol Evol* 22, 1208–1222.

27. Swanson, W. J., Nielsen, R., Yang, Q. (2003) Pervasive adaptive evolution in mammalian fertilization proteins. *Mol Biol Evol* 20, 18–20.

28. Chen, H., Chen, J., Kalbfleisch, J. D. (2001) A modified likelihood ratio test for homogeneity in finite mixture models. *J Roy Stat Soc B* 63, 19–29.

Chapter 16

Phylogenetic Model Evaluation

Lars Sommer Jermiin, Vivek Jayaswal, Faisal Ababneh, and John Robinson

Abstract

Most phylogenetic methods are model-based and depend on Markov models designed to approximate the evolutionary rates between nucleotides or amino acids. When Markov models are selected for analysis of alignments of these characters, it is assumed that they are close approximations of the evolutionary processes that gave rise to the data. A variety of methods have been developed for estimating the fit of Markov models, and some of these methods are now frequently used for the selection of Markov models. In a growing number of cases, however, it appears that the investigators have used the model-selection methods without acknowledging their inherent shortcomings. This chapter reviews the issue of model selection and model evaluation.

Key words: Evolutionary processes, Markov models, phylogenetic assumptions, model selection, model evaluation.

1. Introduction

Molecular phylogenetics is a fascinating aspect of bioinformatics with an increasing impact on a variety of life science areas. Not only does it allow us to infer the historical relationships of species [1], genomes [2], and genes [3], but it also provides a framework for classifying organisms [4] and genes [5], and for studying co-evolution of traits [6]. Phylogenies are the final products of some studies and the starting points of others. Charleston and Robertson [7], for example, compared a phylogeny of pathogens with that of their hosts and found that the pathogens' evolution had involved co-divergence and host-switching. Jermann et al. [8], on the other hand, used a phylogeny of artiodactyls to manufacture enzymes thought to have been expressed by the genomes of their 8- to 50-million-year-old common ancestors. In the majority of cases, including the examples cited in the preceding, the phylogeny is unknown, so it is useful to know how to infer a phylogeny.

Phylogenetic inference is often regarded as a challenge, and many scientists still shy away from approaching important questions from the evolutionary point of view because they consider the phylogenetic approach too hard. Admittedly, phylogenetic methods are underpinned by mathematics, statistics, and computer science, so a basic knowledge of these sciences goes a long way toward establishing a sound theoretical and practical basis for phylogenetic research. It need not be that difficult, though, because user-friendly phylogenetic programs are available for most computer systems. Instead, the challenges lie in: (i) choosing appropriate phylogenetic data for the question in mind, (ii) choosing a phylogenetic method to analyze the data, and (iii) determining the extent to which the phylogenetic results are reliable.

Most molecular phylogenetic methods rely on substitution models that are designed to approximate the processes of change from one nucleotide (or amino acid) to another. The models are usually selected by the investigator, in an increasing number of cases with the assistance of methods for selecting such models. The methods for selecting a substitution model are available for both nucleotide sequences *(9)* and amino acid sequences *(10)*, and they are now commonly used in phylogenetic research. However, the substitution models considered by many of the model-selecting methods implicitly assume that the sequences evolved under stationary, reversible, and homogeneous conditions (defined in the following). Based on the evidence from a growing body of research *(11–26)*, it appears that many sequences of nucleotides or amino acids have evolved under more complex conditions, implying that it would be: (i) inappropriate to use model-selecting methods that assume that the sequences evolved under stationary, reversible, and homogeneous conditions; and (ii) wise to use phylogenetic methods that incorporate more general Markov models of molecular evolution *(27–44)*.

The choice of substitution model is obviously important for phylogenetic studies, but many investigators are still: (i) unaware that choosing an inappropriate substitution model may lead to errors in the phylogenetic estimates, (ii) unsure about how to select an appropriate substitution model, or (iii) unaware that the substitution model selected for a phylogenetic study is in fact not the most appropriate model for the analysis of their data. A reason for the uncertainty associated with the choice of appropriate substitution models may be that an easy-to-understand explanation of the problem and its potential solutions is not yet available in the literature. The following attempts to provide such an explanation.

Underlying the approach taken in this chapter is the idea that the *evolutionary pattern* and *evolutionary process* are two sides of the same coin: The former is the phylogeny, a rooted binary tree that describes the time and order of different divergence events, and the latter is the mechanism by which mutations in germ-line DNA accumulate over time along the diverging lineages. It makes no sense to consider the pattern and process separately, even though only one of the two might be of interest, because the estimate of evolutionary pattern depends on the evolutionary process, and vice versa. Underpinning this chapter is also a hope of raising the awareness of the meaning of the *rate of molecular evolution*: It is not a single variable, as commonly portrayed, but a matrix of variables. Finally, although many types of mutations can occur in DNA, the focus is on point mutations. The reasons for limiting the approach to those mutations is that phylogenetic studies mostly rely on the products of point mutations as the main source of phylogenetic information, and the substitution models used in model-based phylogenetic methods focus mostly on those types of changes.

The following sections first describe the phylogenetic assumptions and outline some relevant aspects of the Markov models commonly used in phylogenetic studies. The terminology used to characterize phylogenetic data is revised and several of the methods that can be used to select substitution models are described. Finally, the general need to use data surveying methods before and after phylogenetic analyses is discussed. Although using such methods before phylogenetic analyses is becoming more common, it remains rare to see phylogenetic results being properly evaluated; for example, using the parametric bootstrap.

2. Underlying Principles

The evolutionary processes that lead to the accumulation of substitutions in nucleotide sequences are most conveniently described in statistical terms. From a biologist's point of view, the descriptions may appear both complex and removed from what is known about the biochemistry of DNA. However, research using the parametric bootstrap (*see* **Chapter 14**) has shown that the variation found among real sequences can be modeled remarkably well using statistical descriptions of the evolutionary processes *(44–47)*, so there is reason to be confident about the use of a statistical approach to describe the evolutionary processes.

From a statistical point of view, it is convenient to describe the evolutionary process operating along each edge of a phylogeny in terms of a Markov process, that is, a process in which the conditional probability of change at a given site in a sequence depends only on the current state and is independent of its earlier states. The Markov model, commonly described as a substitution model, is an approximation of the evolutionary process. It is also an integral part of all model-based phylogenetic methods, so it is convenient to know about phylogenetic assumptions and Markov models if selecting appropriate substitution models for a phylogenetic study is on the agenda.

The following three subsections describe the phylogenetic assumptions and Markov process in the context of alignments of nucleotides. The description also applies to alignments of amino acids, except that it then would be necessary to accommodate 20 amino acids. The description of the phylogenetic assumptions and the Markov process is based primarily on two papers by Tavaré *(48)* and Ababneh et al. *(49)*. For a good alternative description, see Bryant et al. *(50)*.

2.1. Phylogenetic Assumptions

In the context of the evolutionary pattern, it is commonly assumed that the sequences evolved along a bifurcating tree, where each edge in the tree represents the period of time over which point mutations accumulate and each bifurcation represents a speciation event. Sequences that evolve in this manner are considered useful for studies of many aspects of evolution. (Edges in a phylogeny are sometimes called arcs, internodes, or branches. In the interest of clarity, we recommend that the commonly used term branch be avoided, as it is ambiguous. It is sometimes used in reference to a sub-tree (i.e., a set of edges) and other times to a single edge *(51)*.) A violation of this assumption occurs when the evolutionary process also covers gene duplication, recombination between homologous chromosomes, and/or lateral gene transfer between different genomes. In phylogenetic trees, gene duplications look like speciation events and they may be interpreted as such unless both descendant copies of each gene duplication are accounted for, which is neither always possible nor always the case. Recombination between homologous chromosomes is most visible in phylogenetic data of recent origin, and the phylogenetically confounding effect of recombination diminishes with the age of the recombination event (due to the subsequent accumulation of point mutations). Lateral gene transfer is more difficult to detect than gene duplication and recombination between homologous chromosomes but is thought to affect studies of phylogenetic data with a recent as well as ancient origin. A variety of methods to detect recombination *(52–56)* and lateral gene transfer *(57–62)* are available, but their review is beyond the scope of this chapter. This chapter assumes that the sequences evolved along a bifurcating tree, without gene duplication, recombination, and lateral gene transfer.

In the context of the evolutionary process, it is most commonly assumed that the sites evolved independently under the same Markov process, the advantage being that only one Markov model is required to approximate the evolutionary process along an edge. In general, the sites are said to be *independent and identically distributed*. The simplest exception to this model assumes that some sites are *permanently invariant*; that is, unable to change over the entire period of time under consideration, whereas the other sites evolve independently under a single Markov model *(44, 63, 64)*. Other possible exceptions assume that: (i) the variable sites evolve independently under different Markov models *(65–67)*; (ii) the variable sites evolve in a correlated manner *(68–79)*; or (iii) the sites may be *temporarily invariant*; that is, alternate between being variable and invariant over the period of time under consideration *(80)*.

In addition, it is commonly assumed that the process at each site is locally *stationary, reversible*, and *homogeneous* (where locally refers to an edge in the tree). In phylogenetic terms, the stationary condition implies that the marginal probability of the nucleotides is the same over the edge. Reversibility implies that the probability of sampling nucleotide i from the stationary distribution and going to nucleotide j is the same as that of sampling nucleotide j from the stationary distribution and going to nucleotide i. Homogeneity implies that the conditional probabilities of change are constant over the edge *(43, 49, 50)*. The advantage of assuming stationary, reversible, and homogeneous conditions is that reversibility allows us to ignore the direction of evolution during the phylogenetic estimation (by definition, stationarity is a necessary condition for reversibility), and homogeneity permits us to use one Markov model to approximate the evolutionary process between consecutive speciation events.

A further simplification of the three conditions is that the process at the site is globally stationary, reversible, and homogeneous, where globally refers to all edges in the tree. However, the relationship among these three conditions remains complex (**Table 16.1**), with six possible scenarios. (Two additional scenarios are impossible because a reversible process, by definition, also is a stationary process.) Given that the phylogenetic methods in most cases assume that the sequences evolved under stationary, reversible, and homogeneous conditions (Scenario 1) and that features in the sequence alignments (e.g., compositional heterogeneity) suggest that the other scenarios more appropriately describe the conditions under which the sequences evolved, it is reasonable to question whether the sequences evolved under conditions that are more complex than those of Scenario 1. We will return to this issue in a following section.

When the processes are globally stationary, reversible, and homogeneous, we can use a single time-reversible Markov model to approximate the instantaneous rates of change for all edges in the un-rooted phylogeny. The Markov models available for these restrictive conditions range from the one-parameter model *(81)* to the general time-reversible model *(82)*. When these conditions are

Table 16.1
The spectrum of conditions relating to the phylogenetic assumptions

Scenario	Stationarity	Reversibility	Homo-geneity	Comment on each scenario
1	+	+	+	Possible
2	−	+	+	Impossible (by definition)
3	+	−	+	Possible
4	−	−	+	Possible
5	+	+	−	Possible
6	−	+	−	Impossible (by definition)
7	+	−	−	Possible
8	−	−	−	Possible

Note: "+" implies the condition is met; "−" implies the condition is not met.

not met by the data, various methods are available *(27–44)*. When the assumptions of locally stationary, reversible, and homogeneous conditions are not met by the data, only a few methods are available *(27, 28, 30–32, 36, 37, 42–44)*.

Given that the function of many proteins and RNA molecules is maintained by natural selection, there is reason to assume that the evolutionary process at different sites is more heterogeneous than described in the preceding. For example, it is quite likely that some sites could have evolved under time-reversible conditions, whereas other sites could have evolved under more general conditions, thus creating a complex signal in the alignment, which we might not be able to fully appreciate using the phylogenetic methods currently available. Accordingly, it is highly recommended that phylogenetic results be evaluated using the parametric bootstrap.

2.2. Modeling the Process at a Single Site in a Sequence

Consider a site in a nucleotide sequence, and allow the site to contain one of four possible states (A, C, G, and T, indexed as 1, 2, 3, and 4 for the sake of convenience). As the sequence evolves, the site may change from state i to state j, where $i, j = 1, 2, 3, 4$. Consider a random process, X, that takes the value 1, 2, 3, or 4 at any point in continuous time. The Markov process $X(t)$, $t \geq 0$, can be described by the transition function:

$$p_{ij}(t) = \Pr\left[X(t) = j \mid X(0) = i\right], \qquad [1]$$

where $p_{ij}(t)$ is the probability that the nucleotide is j at time t, given that it was i at time 0. Assuming a homogeneous Markov process, let r_{ij} be the instantaneous rate of change from nucleotide i to nucleotide j, and let **R** be the matrix of these rates of change. Then, representing $p_{ij}(t)$ in matrix notation as $\mathbf{P}(t)$, we can write equation [1] as:

$$\mathbf{P}(t) = \mathbf{I} + \mathbf{R}t + \frac{1}{2!}(\mathbf{R}t)^2 + \frac{1}{3!}(\mathbf{R}t)^3 + \ldots$$
$$= \sum_{k=0}^{\infty} \frac{(\mathbf{R}t)^k}{k!} \quad [2]$$
$$= e^{\mathbf{R}t}$$

where **R** is a time-independent rate matrix satisfying three conditions:

1. $r_{ij} > 0$ for $i \neq j$;
2. $r_{ii} = -\sum_{j \neq i} r_{ij}$, implying that $\mathbf{R}\mathbf{1} = \mathbf{0}$, where $\mathbf{1}^T = (1, 1, 1, 1)$ and $\mathbf{0}^T = (0, 0, 0, 0)$—this condition is needed to ensure that $\mathbf{P}(t)$ is a valid transition matrix for $t \geq 0$;
3. $\pi^T \mathbf{R} = \mathbf{0}^T$, where $\pi^T = (\pi_1, \pi_2, \pi_3, \pi_4)$ is the stationary distribution, $0 < \pi_j < 1$, and $\sum_{j=1}^{4} \pi_j = 1$.

In addition, if f_{0j} denotes the frequency of the jth nucleotide in the ancestral sequence, then the Markov process governing the evolution of a site along a single edge is:

1. Stationary, if $\Pr(X(t) = j) = f_{0j} = \pi_j$, for $j = 1, 2, 3, 4$, where π is the stationary distribution, and
2. Reversible, if the balance equation $\pi_i r_{ij} = \pi_j r_{ji}$ is met for $1 \leq i, j \leq 4$, where π is the stationary distribution.

In the context of modeling the accumulation of point mutations at a single site of a nucleotide sequence, **R** is an essential component of the Markov models that are used to do so. Each element of **R** has a role in determining what state the site will be in at time t, so it is useful to understand the implications of changing the values of the elements in **R**. For this reason, it is unwise to consider the rate of evolution as a single variable when, in fact, it is a matrix of rates of evolution.

2.3. Modeling the Process at a Single Site in Two Sequences

Consider a site in a pair of nucleotide sequences that evolve from a common ancestor by independent Markov processes. Let X and Y denote the Markov processes operating at the site, one along each edge, and let $\mathbf{P}^X(t)$ and $\mathbf{P}^Y(t)$ be the transition functions that describe the Markov processes $X(t)$ and $Y(t)$. The joint probability that the sequences contain nucleotide i and j, respectively, is then given by:

$$f_{ij}(t) = \Pr[X(t) = i, Y(t) = j \mid X(0) = Y(0)], \quad [3]$$

where $i, j = 1, 2, 3, 4$. Given that $X(t)$ and $Y(t)$ are independent Markov processes, the joint probability at time t can be expressed in matrix notation as:

$$\mathbf{F}(t) = \mathbf{P}^X(t)^T \mathbf{F}(0) \mathbf{P}^Y(t), \qquad [4]$$

where $\mathbf{F}(0) = diag(f_{01}, f_{02}, f_{03}, f_{04})$, $f_{0k} = P[X(0) = Y(0) = k]$ and $k = 1, \ldots, 4$. Equation [4] can be extended to n sequences, as described in Ababneh et al. *(49)*. The joint probability function has useful properties that will be relied upon in the next section.

3. Choosing a Substitution Model

Before describing the methods available to select substitution models, it is necessary to discuss the terminology used to describe some of the properties of sequence data.

3.1. Bias

The term *bias* has been used to describe: (i) a systematic distortion of a statistical result due to a factor not allowed for in its derivation; (ii) a non-uniform distribution of the frequencies of nucleotides, codons, or amino acids; and (iii) compositional heterogeneity among homologous sequences. In some contexts, there is little doubt about the meaning but in a growing number of cases, the authors have inadvertently provided grounds for confusion. Because of this, we recommend that the term bias be reserved for statistical purposes, and that the following four terms be used to describe the observed nucleotide content:

- The nucleotide content of a sequence is *uniform* if the nucleotide frequencies are identical; otherwise, it is *non-uniform*.
- The nucleotide content of two sequences is *compositionally homogeneous* if they have the same nucleotide content; otherwise, it is *compositionally heterogeneous*.

The advantages of adopting this terminology are that we can discuss model selection without the ambiguity that we otherwise might have had to deal with, and that we are not forced to state what the unbiased condition might be; it need not be a uniform nucleotide content, as implied in many instances (the five terms are applicable to codons and amino acids without loss of clarity).

3.2. Signal

In discussing the complexity of an alignment of nucleotides, it is often convenient to consider the variation in the alignment in terms of the sources that led to the complexity. One such source is the order and time of divergence events, which leaves a signal in the alignment; it is this *historical signal (83)* that is the target of most phylogenetic studies. The historical signal is detectable because the sequences have a tendency to accumulate point mutations over time. Other sources of complexity include: the *rate signal*, which may arise when the sites and/or lineages evolve at non-uniform rates; the *compositional signal*, which may arise when the sites and/or lineages evolve under different stationary conditions; and the *covarion signal*, which may emerge when the sites evolve non-independently (e.g., by alternating between being variable and invariable along different edges). The term *phylogenetic signal* is used sometimes, either synonymously with the historical signal or to represent the signals that the phylogenetic methods use during inference of a phylogeny. Due to this ambiguity and the fact that the most frequently used phylogenetic methods appear to misinterpret the *non-historical signals* (i.e., by treating non-historical signals as if they were the historical signals), we recommend that the term phylogenetic signal be used with caution.

Separating the different signals is difficult because their manifestations are similar, so inspection of the inferred phylogeny is unlikely to be the best solution to this problem. Recent simulation studies of nucleotide sequences generated under stationary, reversible, and homogeneous conditions as well as under more general conditions have highlighted a complex relationship among the historical signal, the rate signal, and the compositional signal *(84)*. The results show that the historical signal decays over time, whereas the other signals may increase over time, depending on the evolutionary processes operating at any point in time. The results also show that the relative magnitude of the signals determines whether the phylogenetic methods are likely to infer the correct tree. Finally, the results show that it is possible to infer the correct tree irrespective of the fact that the historical signal has been lost. Hence, there is good reason to be cautious when studying ancient evolutionary events—what might appear to be a well-supported phylogeny may in fact be a tree representing the non-historical signals.

3.3. Testing the Stationary, Reversible, and Homogeneous Condition

Alignments of nucleotides may vary compositionally across sequences and/or sites. In the first case, the sites would not have evolved under stationary, reversible, and homogeneous conditions, and in the second case, the sites would have evolved under different stationary, reversible, and homogeneous conditions. In either case, it would be inappropriate to estimate the phylogeny by using a single time-reversible Markov model to approximate the evolutionary processes.

A solution to this problem is to determine the type of compositional heterogeneity that is found in the sequence data. If compositional heterogeneity is across sites but not across sequences, then the evolutionary process may be approximated by a combination of time-reversible Markov models applied to different sections of the alignment. If, on the other hand, compositional heterogeneity is across sequences, then it is not possible to approximate the evolutionary process by any time-reversible Markov model.

Methods to detect compositional heterogeneity in alignments of nucleotides fall into four categories, with those of the first category using graphs and tables to visualize the nucleotide content of individual sequences, and those of the other categories producing test statistics that can be compared to expected distributions (for a review, see *(85)*). However, the methods are either statistically invalid, of limited use for surveys of species-rich data sets, or not yet accommodated by the wider scientific community. Faced with these challenges, Ababneh et al. *(86)* and Ho et al. *(87)* developed statistical and visual methods to examine under what conditions nucleotide sequence data might have evolved. The methods provide an opportunity to gain a better understanding of the evolutionary processes that underpin variation in alignments of nucleotides. We will now describe these methods.

3.3.1. Matched-Pairs Tests of Symmetry, Marginal Symmetry, and Internal Symmetry

Suppose we have l matched sequences of m independent and identically distributed variables taking values in n categories. An example of such data would be an alignment of $l = 8$ sequences of $m = 500$ nucleotides, implying that $n = 4$. Data of this nature can be summarized in an l-dimensional divergence matrix, **D**, which has n^l categories and is the observed equivalent of $m\mathbf{F}(t)$. The hypotheses of interest concern the symmetry, marginal symmetry, and internal symmetry of **D**. In the context of diverging nucleotide sequences, the matched-pairs tests can be used to examine the goodness of fit of Markov models thought to approximate the evolutionary processes. The relevance of using these tests prior to phylogenetic analysis of aligned nucleotide sequences has long been recognized *(48, 66, 88–90)*, but the tests are yet to be accommodated by the wider scientific community.

The matched-pairs tests can be divided into two groups depending on the number of sequences considered. In the simple case, in which only two sequences are considered, the matched-pairs tests can be used to test for symmetry *(91)*, marginal symmetry *(92)*, and internal symmetry *(86)* of a divergence matrix derived from the nucleotide sequences. In more complex cases, in which more than two sequences are under consideration, two tests of marginal symmetry *(86, 89)* are available. Ababneh et al. *(86)* reviewed the matched-pairs tests cited in the preceding and placed them in the context of other matched-pairs tests, so rather than describing them again, we will show how the tests can be used to determine whether a set of sequences have evolved under stationary, reversible, and homogeneous conditions.

3.3.2. Matched-Pairs Tests with Two Sequences

Suppose we have a pair of homologous nucleotide sequences that diverged from their common ancestor by independent Markov processes, and that the character states within each sequence are independent and identically distributed. Then the question is: Which scenario in **Table 16.1** represents the most plausible spectrum of conditions under which the sequences arose? To answer this question, we need to examine the divergence matrix, **D** (each element in **D**, i.e., d_{ij}, is the number of sites in which the sequences have nucleotides i and j, respectively). At the beginning of the sequences' evolution, the divergence matrix will be a 4 × 4 diagonal matrix. Later on, at time $t > 0$, the divergence matrix might look like this:

$$\mathbf{D}(t) = \begin{bmatrix} 1478 & 362 & 328 & 334 \\ 315 & 1526 & 356 & 319 \\ 338 & 327 & 1511 & 334 \\ 338 & 330 & 324 & 1480 \end{bmatrix}. \quad [5]$$

In order to conduct Bowker's *(91)* matched-pairs test of symmetry, we simply need to enter the off-diagonal elements of $\mathbf{D}(t)$ into the following equation:

$$S_B^2 = \sum_{i<j} \frac{(d_{ij} - d_{ji})^2}{d_{ij} + d_{ji}}, \quad [6]$$

where the test statistic, S_B^2, is asymptotically distributed as a χ^2-variate on $v = n(n-1)/2$ degrees of freedom. In the present example, $S_B^2 = 5.007$ and $v = 6$. Under the assumption of symmetry, the probability of a test statistic ≥ 5.007 is 0.5430; hence, we conclude that these data are consistent with the null hypothesis for symmetry.

In order to conduct Stuart's *(92)* matched-pairs test of marginal symmetry, we need to calculate the vector of marginal differences, **u**, and the variance-covariance matrix, **V**, of those differences. Here $\mathbf{u} = (d_{1\bullet} - d_{\bullet 1}, d_{2\bullet} - d_{\bullet 2}, d_{3\bullet} - d_{\bullet 3})^T$ and **V** equals a 3 × 3 matrix with the elements:

$$v_{ij} = \begin{cases} d_{i\bullet} + d_{\bullet i} - 2d_{ii}, & i = j, \\ -(d_{ij} + d_{ji}), & i \neq j, \end{cases} \quad [7]$$

where $d_{i\bullet}$ is the sum of the ith column of **D**, $d_{\bullet i}$ is the sum of the ith row of **D**, and so forth. Given **u** and **V**, we can obtain:

$$S_S^2 = \mathbf{u}^T \mathbf{V}^{-1} \mathbf{u}, \quad [8]$$

where the test statistic, S_S^2, is asymptotically distributed as a χ^2-variate on $v = n-1$ degrees of freedom. In the present example, $\mathbf{u} = (33, -29, -9)^T$ and:

$$\mathbf{V} = \begin{bmatrix} 2015 & -677 & -666 \\ -677 & 2009 & -683 \\ -666 & -683 & 2007 \end{bmatrix}, \quad [9]$$

so $S_S^2 = 0.7598$ and $v = 3$. Under the assumption of marginal symmetry, the probability of a test statistic ≥ 0.7598 is 0.8591;

hence, we conclude that these data are consistent with the null hypothesis for marginal symmetry.

In order to conduct Ababneh et al.'s *(86)* matched-pairs test of internal symmetry, we draw on the fact that the test statistic $S_I^2 = S_B^2 - S_S^2$ is asymptotically distributed as a χ^2-variate on $v = (n-1)(n-2)/2$ degrees of freedom. In the present example, $S_I^2 = 4.2469$ and $v = 3$. Under the assumption of internal symmetry, the probability of a test statistic ≥ 4.2469 is 0.2360; hence, we conclude that these data are consistent with the null hypothesis for internal symmetry.

3.3.3. Matched-Pairs Tests with More Than Two Sequences

We now turn to the more complex matched-pairs test of marginal symmetry. Suppose we have four nucleotide sequences that have evolved independently along a binary tree with three bifurcations. The alignment may look like that in **Fig. 16.1**. Visual inspection of the alignment emphasizes how difficult it is to determine whether the sequences have evolved under stationary, reversible, and homogeneous conditions, and therefore why the statistical methods are needed to provide the answer. In the following, we illustrate how statistical tests may be used to provide a detailed answer.

First, we use the overall matched-pairs test of marginal symmetry by Ababneh et al. (86), which yields a test statistic, T_S, that is asymptotically distributed as a χ^2-variate on $v = (n-1)(l-1)$ degrees of freedom. In the present example, $T_S = 56.93$ and $v = 9$. Under the assumption of marginal symmetry, the probability of a test statistic ≥ 56.93 is $\sim 5.22 \times 10^{-09}$, implying that it is unlikely that these data have evolved under stationary, reversible, and homogeneous conditions.

```
Seq1    TTTCTGTAGACTACAGCCGAACTGATACAATACAAGCACAAACAATTCACCGCGTCGCGCACAGT
Seq2    CGTCTGGGATCTTTTGCCGGGCTGGGTCGCTACACGAACGCAGAGTTCTACTCCGGTCGCACTTG
Seq3    CTACAGTTAAGTTCTGCAGAGCTGCTTGACTATACGATCAACGAATACAAGACGGGGCGCACAGG
Seq4    CTTCGGTATAGTTCTGCCGAGCTGGTTCGCTACATGATCAATGATTACGACCCTGGGCCCTCTGG

CGTCAAAGCGGCATTCCATAAAAGTTCATCCATACCCCGAGGTAACCTCACGTCGTCACGGGCTGACGTAATCAC
CGGATGAGTTGGTTACGGAGAGTGCGGGTCTTTTCCCAAAGTTCATTTCCCGTCGTTTCGGCCTGTTGTAATCAT
CATATAAGTGGGATTCCGTAAGATCATGTCTCTACCCAAAGGGTACATGTTGTCTTCACGGCCAAACCTAATCAC
CGTATGAGTGGGATGGTGTCAAATTTCTTCTTGACCGGCAGGTCACCTCTTGTCCTGAGGGCCGGGCGGCAGCAG

GAAAGCACCGCCCGACCGGTCAAGCCTCAGAAGGGTCGAACACGGACTCAGTCTCAAGTGCTCCTCCACAAACGT
GTGTGCTCCGCCCCATCGGTGAAGCCCCGCTAGCGTATTACTCGGAATGTGTATCTAGTGCCAATTCATATACGT
GGGTCACCTGCCCAACAGTTGAAGGCGCGCCAGGCCGGCCCACGCATACAGACTCCAGAGCAACTCCATCAACGT
GTCTGTGCTGTTTTGCCTGTAATGCCTCGTCAGGCGGGAGCACGGTTTTAGTATCCTTGCCTACTCTATTATTCT

CATACTTAGTTCACCATCCCCGAGCCTATTTCCCTTAAAATGCGGTAACCCGGCCAGGGAGGAGAGAAAGAGTGG
ATAAGTTAGTTTATAATCTCCTCGCCTATTTCCTTAGAAATAGTATTATCGATCTTTGACGGAGTGAACTATGGG
CAAACTTACCTCAAAGTCTCCGCGGCTAGTCCCATTGAAATACGATTATCTCACCTTGCAAGAGTGAAAAAATCG
GAAATTTAGCTGATAATCTCTTCAGCTAATTCTTTAGAAATAGGCTTATCGTCCCGGGTTGGTGCGAAACATCCG
```

Fig. 16.1. An alignment of nucleotide sequences: the data were generated using Hetero *(93)* with default settings invoked, except that the stationary distributions for the terminal edges were non-uniform (Π = (0.4, 0.3, 0.2, 0.1) for Seq1 and Seq3 and Π = (0.1, 0.2, 0.3, 0.4) for Seq2 and Seq4).

Table 16.2
Results from tests of symmetry, marginal symmetry and internal symmetry for the four sequences in Fig. 16.1

	Seq1	Seq2	Seq3
Test of symmetry			
Seq2	0.0000		
Seq3	0.8064	0.0002	
Seq4	0.0000	0.6146	0.0000
Test of marginal symmetry			
Seq2	0.0000		
Seq3	0.6829	0.0000	
Seq4	0.0000	0.9092	0.0000
Test of internal symmetry			
Seq2	0.8500		
Seq3	0.6772	0.4374	
Seq4	0.3479	0.2706	0.4477

Note: Each number corresponds to the probability of obtaining the pair of sequences by chance under the assumptions of symmetry, marginal symmetry, and internal symmetry.

The overall matched-pairs test of marginal symmetry did not provide an indication of the sources of the data's higher than expected complexity. To better understand what the sources might be, the matched-pairs tests of symmetry, marginal symmetry, and internal symmetry are needed. **Table 16.2** demonstrates that the complexity is due to differences in the sequences' marginal distribution (i.e., each sequence has acquired a different nucleotide content over time). The matched-pairs test of symmetry produced high probabilities for two of the comparisons (Seq1–Seq3 and Seq2–Seq4), whereas the other comparisons produced very low probabilities. These results are consistent with the notion that the approximation of the four sequences' evolution requires at least two Markov models—one for Seq1 and Seq3 and another for Seq2 and Seq4. The two Markov models must be quite different to produce the low probabilities observed for four of the six comparisons.

The matched-pairs test of symmetry disclosed which pairs of sequences had evolved under different conditions but did not identify the sources of complexity in the data. To obtain such details, it is necessary to use the matched-pairs tests of marginal symmetry and internal symmetry. The matched-pairs test of marginal symmetry produced results that are similar to those obtained by using the matched-pairs test of symmetry, whereas the matched-pairs test of internal symmetry produced high probabilities for every pair of sequences (**Table 16.2**). The results are consistent with the notion that the sequences evolved under conditions that are not stationary, reversible, and homogeneous; that is, the conditions under which the sequences were generated (**Fig. 16.1**). The sources of complexity in the alignment seem to be a combination of the historical signal and the compositional signal. Given that the data's complexity is due not only to the historical signal, it would be wise to analyze the data phylogenetically using methods that accommodate the general Markov models of molecular evolution *(27–44)*, because the time-reversible Markov models would not suffice.

Based on Monte Carlo simulations, where the sequences evolved under conditions of the scenarios in **Table 16.1**, it is possible to summarize our knowledge of the three matched-pairs tests of symmetry, marginal symmetry, and internal symmetry. In general, they are able to detect that sequences have evolved according to the conditions of two of the six possible scenarios. Specifically, when the sequences evolved according to the conditions of:

Scenario 1: The probabilities from the three tests will be uniformly distributed, with 5% of the tests producing a probability ≤0.05.

Scenario 3: The probabilities from the three tests will be uniformly distributed because the homogeneity of the process masks the non-reversible aspect of the process. A new surveying method outlined in Jayaswal et al. *(43)* may be useful but it has not yet been tested in this context.

Scenario 4: The probabilities from the three tests will be uniformly distributed because the homogeneity of the process masks the non-stationary and non-reversible aspects of the process. A new method by Jayaswal et al. *(43)* may be useful but it has not yet been tested in this context.

Scenario 5: The result will be like that for Scenario 1. To detect that the sequences evolved according to this scenario would require a third sequence and the use of a relative-rates test *(94)*.

Scenario 7: The probabilities from the matched-pairs test of marginal symmetry will be uniformly distributed, whereas the probabilities from the matched-pairs tests of symmetry and marginal symmetry may be non-uniformly distributed.

Scenario 8: The probabilities from the three tests may be non-uniformly distributed, with >5% of the tests producing probabilities ≤0.05.

In summary, if the matched-pairs tests yield a non-uniform distribution of probabilities with a larger than expected proportion of probabilities ≤0.05, then the conclusion is that the sequences must have evolved under the conditions of Scenarios 7 or 8. If the tests produce a uniform distribution of probabilities with 5% of the probabilities ≤0.05, then the conclusion is that the sequences may have evolved under the conditions of Scenarios 1, 3, 4 or 5, or alternatively, the non-historical signals may be too weak to detect.

3.3.4. Using Matched-Pairs Tests to Analyze Complex Alignments

The matched-pairs tests of symmetry, marginal symmetry, and internal symmetry have several features that make them extremely useful. Due to the design of the tests, the results are unaffected by invariant sites, and how far the sequences have diverged (because the emphasis is on those sites in which the sequences differ). If sites are thought to evolve at different rates, or there is reason to assume that certain sites have evolved under different conditions, then it is recommended that the sites be binned in a manner that reflects the evolutionary process at those sites (before the matched-pairs tests are conducted). If there is reason to think that the approximation to the evolutionary process will require the use of a covarion model *(77)*, then the matched-pairs test may produce biased results because a site may be variable in some sequences and invariant in others; therefore, the interpretation of the test statistics should be done cautiously.

The procedure of binning those sites that may have evolved under the same conditions is sensible because it allows us to characterize more precisely under which conditions the sites evolved. In binning the sites, there is a trade-off to consider between the number of bins and the sample size of the bins: Is a single large sample better than many small samples? Increasing the sample size will increase the power of the matched-pairs tests of symmetry, marginal symmetry, and internal symmetry, but the ability to determine under which conditions the sites evolved may be compromised. Hence, binning should be done with attention to, for example, the structure of the gene and the gene product. In so doing, there are some problems to consider: (i) it is not always clear what sites should be binned—protein-coding DNA, for example, can be binned in several ways, depending on whether each site or codon is considered the unit of a sample, and depending on whether the gene product's structure is considered; (ii) the small size of some bins may render it impossible to conduct the matched-pairs tests of marginal symmetry and internal symmetry (because estimation of the test statistics involves inverting a matrix that is sometimes singular).

3.3.5. Visual Assessment of Compositional Heterogeneity

An alternative approach to the statistical methods described in the preceding is to survey the data using an approach that combines visual assessment of nucleotide content with the results from the matched-pairs test of symmetry *(87)*. The visualization of the nucleotide content builds in the idea that a de Finetti plot *(95)* can be extended to a tetrahedral plot with similar properties (i.e., each observation comprises four variables, where $a + b + c + d = 1$ and

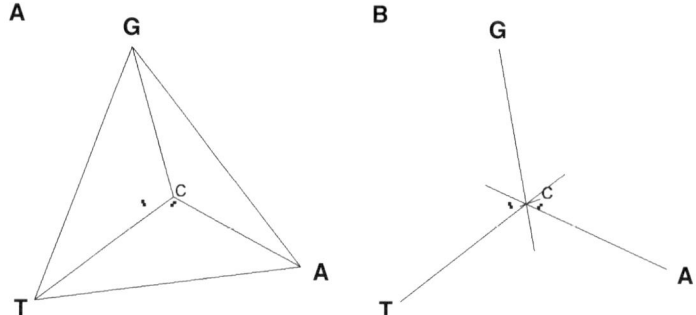

Fig. 16.2. The tetrahedral plot with the four sequences from Fig. 16.1, plotted (**A**) with the borders of the tetrahedron highlighted or (**B**) with the four axes highlighted.

$0 \leq a, b, c, d, \leq 1$). Each axis in the tetrahedral plot starts at the center of one of the surfaces at value 0 and finishes at the opposite corner at value 1. The nucleotide content of a given sequence is simply the set of shortest distances from its point within the tetrahedron to the surfaces of the tetrahedron (**Fig. 16.2**). Visual assessment of the spread of points shows the extent of compositional heterogeneity as well as the trends that may exist in the data. Rotation of the tetrahedron permits inspection of the scatter of points from all angles, thus enhancing the chance of detecting distributional trends or sequences that are outliers. Having characterized the distribution of points visually, it is recommended to conduct the matched-pairs test of symmetry to determine whether the sequences have evolved under stationary, reversible, and homogeneous conditions. The inclusion or exclusion of outliers, or a larger subset of the data, from the ensuing phylogenetic analysis can then be decided on the basis of results from the matched-pairs test of symmetry.

The tetrahedral plot is particularly useful for surveys of compositional heterogeneity in species-rich alignments. To illustrate this, we surveyed an alignment of the mitochondrial cytochrome oxidase subunit I gene from 74 species of butterflies—the data are part of a longer alignment analyzed by Zakharov et al. *(1)*. When all the sites are considered equal (i.e., all the sites were placed in the same bin) and the tetrahedron was allowed to rotate, the 74 points were found scattered tightly in an area where the proportion of A and T are >25% (**Fig. 16.3A**). The points are clearly spread within a confined area, implying that there may be more compositional heterogeneity in these data than the initial analysis suggested. To address this issue, we binned the nucleotides according to the positions within a codon—three bins were created. The distributions of points differ for the first, second, and third codon sites, with first codon sites displaying a small amount of scatter (**Fig. 16.3B**), second codon sites displaying hardly any scatter (**Fig. 16.3C**), and third codon sites displaying a lot of scatter (**Fig. 16.3D**). Rotating the three tetrahedral plots shows that

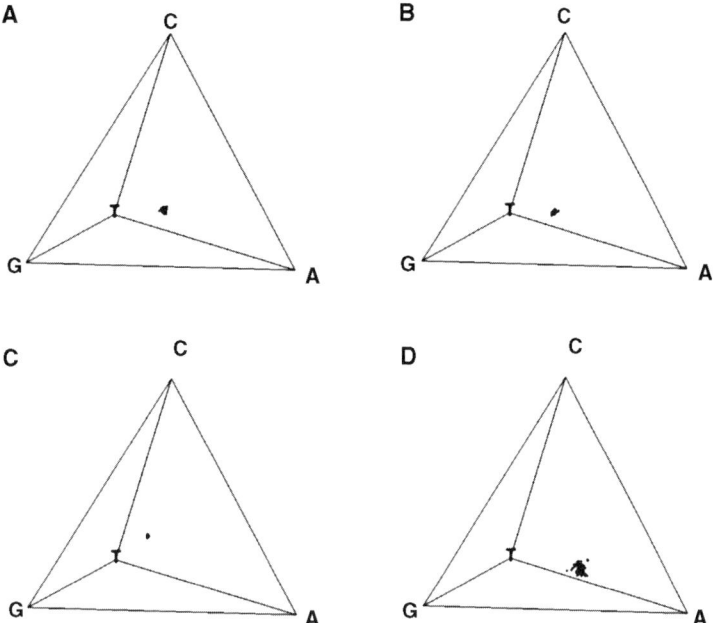

Fig. 16.3. The tetrahedral plot based on the cytochrome oxidase subunit I gene from 74 butterflies. The plots show the nucleotide composition at (**A**) all sites, (**B**) first codon sites, (**C**) second codon sites, and (**D**) third codon sites.

the centroids differ for the codon sites, thus suggesting that it would be necessary to apply three Markov models to these data in order to analyze them appropriately within a phylogenetic context. The reason for needing two Markov models for the first and second codon sites is that the tetrahedral plots imply that the first and second codon sites evolved under different stationary conditions.

Having examined the tetrahedral plots for the first, second and third codon sites, we completed the visual survey with a matched-pairs test of symmetry. Of the 2,775 pairwise tests conducted:

- One hundred forty-seven (147) tests (5.3%) for first codon site were found to produce a probability ≤0.05, implying first codon site is consistent with evolution under stationary, reversible, and homogeneous conditions.

- Two tests (0.1%) for second codon site were found to produce a probability ≤0.05.

- Nine hundred four (904) tests (32.6%) for third codon site were found to produce a probability ≤0.05, implying third codon site is inconsistent with evolution under stationary, reversible, and homogeneous conditions.

Given these results, a sensible approach to analyze these data phylogenetically would be to apply a time-reversible Markov model to the first codon sites, another such model to the second codon sites, and a more general Markov model to the third codon sites.

In conclusion, visual and statistical assessment of the nucleotide content in alignments of nucleotides provides considerable insight into the evolutionary processes that may have led to contemporary sequences—failure to acknowledge this, or to incorporate the information that the methods described in the preceding can provide, may lead to undetected errors in phylogenetic estimates.

3.4. Testing the Assumption of Independent and Identical Processes

It is commonly assumed that the sites in an alignment of nucleotides are independent and identically distributed, or at least independently distributed. The latter case includes the commonly occurring scenario where rate-heterogeneity across sites is modeled using a Γ distribution and a proportion of sites are assumed to be permanently invariant. However, the order and number of nucleotides usually determine the function of the gene product, implying that it would be unrealistic, and maybe even unwise, to assume that the sites in the alignment of nucleotides are independent and identically distributed.

To test whether sites in an alignment of nucleotides are independent and identically distributed, it is necessary to compare this simple model to the more complex models that describe the interrelationship among sites in the alignment of nucleotides. The description of the more complex models depends on prior knowledge of the genes and gene products, and the comparisons require the use of likelihood-ratio tests, sometimes combined with permutation tests or the parametric bootstrap.

The likelihood-ratio test provides a statistically sound framework for testing alternative hypotheses within an evolutionary context (96). In statistical terms, the likelihood-ratio test statistic, Δ, is defined as:

$$\Delta = \frac{\max(L(H_0 \mid \text{data}))}{\max(L(H_1 \mid \text{data}))} \qquad [10]$$

where the likelihood, L, of the null hypothesis (H_0), given the data, and the likelihood of the alternative hypothesis (H_1), given the data, both are maximized with respect to the parameters. When $\Delta > 1$, the data are more likely to have evolved under H_0—when $\Delta < 1$, the data favor the alternative hypothesis. When the two hypotheses are nested, that is H_0 is a special case of H_1, Δ is < 1, and $-2 \log \Delta$ is usually asymptotically distributed under H_0 as a χ^2 variate with v degrees of freedom (where v is the extra number of parameters in H_1)—for a detailed discussion of the likelihood-ratio test, see Whelan and Goldman (97) and Goldman and

Whelan *(98)*. When the hypotheses are not nested, it is necessary to use the parametric bootstrap *(96, 99)*, in which case pseudo-data are generated under H_0. In some instances, permutation tests may be used instead *(100)*.

The evolution of protein-coding genes and RNA-coding genes is likely to differ due to the structural and functional constraints of their gene products, so to determine whether sites in an alignment of such genes evolved under independent and identical conditions, it is necessary to draw on knowledge of the structure and function of these genes and their gene products. For example, although a protein-coding gene could be regarded as a sequence of independently evolving sites (**Fig. 16.4A**), it would be more appropriate to consider it as a sequence of independently evolving codons (**Fig. 16.4B**) or a sequence of independently evolving codon positions, with sites in the same codon position evolving under identical and independent conditions

A

```
Gene      ATGAACGAAAATCTGTTCGCTTCATTCATTGCCCCCACAATCCTAGGCCTACCCGCCGCA
Unit      ------------------------------------------------------------
```

B

```
Gene      ATGAACGAAAATCTGTTCGCTTCATTCATTGCCCCCACAATCCTAGGCCTACCCGCCGCA
Unit      ─── ─── ─── ─── ─── ─── ─── ─── ─── ─── ─── ─── ─── ─── ─── ─── ─── ─── ─── ───
```

C

```
Gene      ATGAACGAAAATCTGTTCGCTTCATTCATTGCCCCCACAATCCTAGGCCTACCCGCCGCA
Unit      ------------------------------------------------------------
Category  123123123123123123123123123123123123123123123123123123123123
```

D

```
Gene      ATGAACGAAAATCTGTTCGCTTCATTCATTGCCCCCACAATCCTAGGCCTACCCGCCGCA
Unit      ------------------------------------------------------------
Category  123123123123456456456456456456456456456456456456789789789789
```

E

```
Gene      ATGAACGAAAATCTGTTCGCTTCATTCATTGCCCCCACAATCCTAGGCCTACCCGCCGCA
Unit 1    ─── ─── ─── ─── ─── ─── ─── ─── ─── ─── ─── ─── ─── ─── ─── ─── ─── ─── ─── ───
Unit 2    - ─── ─── ─── ─── ─── ─── ─── ─── ─── ─── ─── ─── ─── ─── ───
```

Fig. 16.4. Models used to describe the relationship among sites in protein-coding genes. A protein-coding gene may be regarded as a sequence of independently evolving units, where each unit is (**A**) a site, (**B**) a codon, or (**C**) a site assigned its own evolutionary model, depending on its position within a codon. The more complex models include those that consider (**D**) information about the gene product's structure and function. (Here, categories 1, 2, and 3 correspond to models assigned to sites within the codons that encode amino acids in one structural domain, categories 4, 5, and 6 correspond to models assigned to sites in codons that encode amino acids in another structural domain, and so forth.) (**E**) Overlapping reading frames. (Here, unit 1 corresponds to one reading frame of one gene whereas unit 2 corresponds to that of the other gene.)

(**Fig. 16.4C**). There are advantages and disadvantages of using each of these approaches:

- The first approach (**Fig. 16.4A**) is fast because the number of parameters required to approximate the evolutionary processes is small (because **R** is a 4×4 matrix), and it is catered for by a large number of substitution models. However, the approach fails to consider that the evolution of neighboring sites may be correlated, which is highly likely for codon sites.

- The second approach (**Fig. 16.4B**) is slow because the number of parameters needed to model the evolutionary process is large (because **R** is a 64×64 matrix), and it is catered for by only a few substitution models *(74, 101, 102)*. However, the approach does address the issue of correlations among neighboring sites within a codon.

- The third approach (**Fig. 16.4C**) is a compromise between the previous approaches. It assigns a category to each codon position, after which each codon position is assigned its own substitution model. The advantage of this approach is that codon-site-specific characteristics may be accounted for; for example, the nucleotide content and the rates of evolution are often found to vary across codon positions. However, the issue of correlation among neighboring sites within each codon is not adequately addressed.

An alternative to the approaches described in the preceding would be to translate the codons into amino acids before subsequent analysis. Like the first approach, the polypeptides could be regarded as sequences of independently evolving sites and analyzed, as such using one of the available amino acid substitution models *(103–113)*. The approach is appealing because it accounts for correlations among neighboring codon sites, but it is slower than the first approach (because **R** is a 20×20 matrix) and does not account for correlations among neighboring amino acids.

Using protein-coding genes obtained from viral and eukaryote genomes, a recent study compared the first three approaches and found that the performance of the second approach is superior to that of the other two; however, the second approach comes at an extremely high computational cost *(114)*. The study also found that the third approach is an attractive alternative to the second approach.

The first three approaches can be extended to account for the structure and function of the gene product or the fact that a nucleotide sequence may encode several gene products. One way to account for the structure and function of the gene product is to incorporate extra categories (**Fig. 16.4D**). Hyman et al. *(115)*, for example, used six rate categories to account for differences between the first and second codon positions (the third codon position was ignored) as well as differences among the codons.

(The discriminating factor is whether the corresponding amino acid ultimately would be located in the: (i) lumen between the mitochondrial membranes, (ii) inner mitochondrial membrane, or (iii) mitochondrial matrix.) The same approach could easily be used in conjunction with codon-based substitution models. In some cases, a nucleotide has more than one function. For example, it may encode more than one product, in which case complex models are required to approximate the evolutionary process. In the case of mitochondrial and viral genomes, some protein-coding genes overlap (**Fig. 16.4E**), in which case complex models are available *(70, 77)*.

Occasionally, there are reasons to suspect that some sites may have been temporarily invariant. Under such conditions, it may be beneficial to use statistical tests developed by Lockhart et al. *(80)* and a phylogenetic method developed by Galtier *(116)*. However, the statistical tests rely on prior knowledge allowing the investigator to partition the sequences into evolutionarily sound groups, and such information is not always available.

In the context of RNA-coding genes, the sequences could be viewed as independently evolving sites (**Fig. 16.5A**), but that approach would ignore the structure and function of the gene product. A more appropriate method would be to: (i) partition the sites according to the features they encode in the gene product, and (ii) assign Markov models to the sites in accordance with this partition. For example, for alignments of transfer RNA-coding genes it would be necessary to partition the sites into three categories,

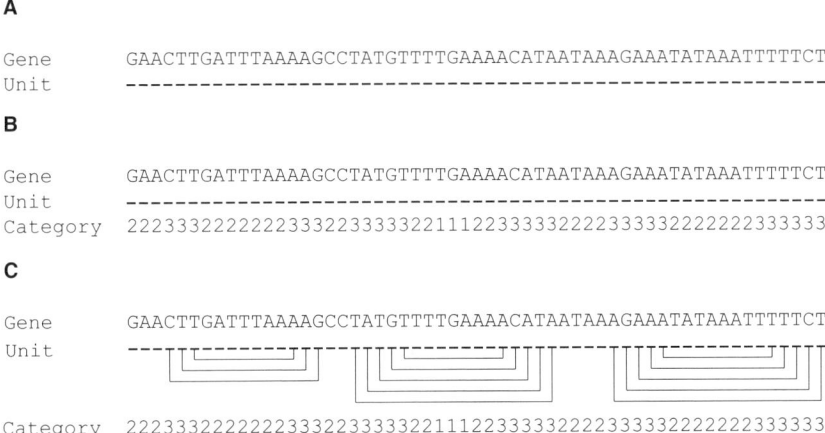

Fig. 16.5. Models used to describe the relationship among sites in RNA-coding genes. An RNA-coding gene may be regarded as a sequence of independently evolving units, where each unit is (**A**) a site, or (**B**) a site assigned its own evolutionary model, depending on what feature it encodes. (Here, category 1 corresponds to the model assigned to sites that encode the anticodon in a tRNA molecule, category 2 corresponds to the model assigned to sites that encode loops in the gene product, and category 3 corresponds to the model assigned to sites that encode the stems in the gene product.) An even more advanced approach uses information about stem-coding nucleotides that match each other in the gene product. (**C**) A thin line connects each pair of those nucleotides.

one for sites encoding the anticodon, another for sites encoding loops, and a third for sites encoding stems (**Fig. 16.5B**). This approach may be extended further by incorporating knowledge of sites that encode the stems of RNA molecules (**Fig. 16.5C**), although such sites may be at a distance from one another on the gene, their evolution is still likely to be correlated because of their important role in forming the stems of RNA molecules.

The sites that help forming the stems in RNA molecules have most likely evolved in a correlated manner, and for this reason the Markov models assigned to these sites should consider substitutions between pairs of nucleotides rather than single nucleotides (i.e., the Markov models should consider changes between 16 possible pairs of nucleotides: AA, AC, ..., TT). Several Markov models have been developed for pairs of nucleotides *(68, 69, 71–73, 75, 76, 78)* and used to address a variety of phylogenetic questions *(e.g., 79, 100, 115, 117, 118)*. Each of these Markov models, however, assumes that the sites evolved under stationary, reversible, and homogeneous conditions, a potential problem that was discussed in a previous section.

Regardless of whether the alignment is of protein-coding genes or RNA-coding genes, the inclusion of additional information on the structure and function of the gene products leads to phylogenetic analyses that are more complex than they would have been if the sites were assumed to be independent and identically distributed. However, the benefit of including this information is that the phylogenetic results are more likely to reflect the evolutionary pattern and processes that gave rise to the data. Finding the most appropriate Markov models for different sites in RNA-coding genes and protein-coding genes can be a laborious and error-prone task, especially if the sites have not evolved under stationary, reversible, and homogeneous conditions. Moreover, there is always the possibility that the extra parameters used to approximate the evolutionary processes simply fit noise in the data rather than the underlying trends *(78)*. To address these problems, it is often useful to compare alternative models using statistical methodology, such as the parametric bootstrap *(96, 99)* or permutation tests *(100)*, both of which are tree-dependent methods. We will return to this issue in a following section.

3.5. Choosing a Time-Reversible Substitution Model

If a partition of sites in an alignment were found to have evolved independently under the same stationary, reversible, and homogeneous conditions, then there is a large family of time-reversible Markov models available for analysis of these sites. Finding the most appropriate Markov model from this family of models is easy due to the fact that many of the models are nested, implying that a likelihood-ratio test *(96)* is suitable for determining whether the alternative hypothesis, H_1, provides a significantly better fit to the data than the null hypothesis, H_0.

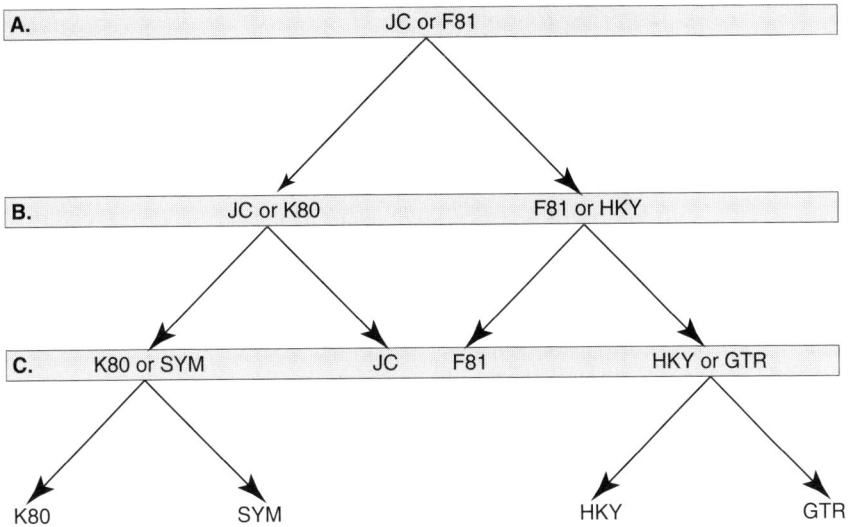

Fig. 16.6. Relationship among models within the family of the time-reversible Markov models. The models are the Jukes and Cantor model (JC: *(81)*), the Kimura 1980 model (K80: *(121)*), the symmetrical model (SYM: *(122)*), the Felsenstein 1981 model (F81: *(119)*), the Hasegawa-Kishino-Yano model (HKY: *(120)*), and the general time-reversible model (GTR: *(82)*). The questions that allow us to distinguish these models are: (**A**) Is the nucleotide content uniform or non-uniform? (**B**) Are there one or two conditional rates of change? (**C**) Are there two or six conditional rates of change? The degrees of freedom associated with questions (**A, B**) and **C** is 3, 1, and 4, respectively.

For nucleotide sequences, the choice between time-reversible Markov models may conform to a decision tree like that shown in **Fig. 16.6**. Assuming a tree and the alignment, the initial test is between the JC model *(81)*, which assumes uniform nucleotide content, and the F81 model *(119)*, which allows the nucleotide content to be non-uniform (here the first model corresponds to the null hypothesis, whereas the other model corresponds to the alternative hypothesis). From a statistical point of view, the difference is 3 degrees of freedom, so if $-2 \log \Delta > 7.81$ (i.e., the 95% quantile of the χ^2-distribution with 3 degrees of freedom), the F81 model is favored over the JC model. The next likelihood-ratio test is between the F81 model, which assumes one conditional rate of change, and the HKY model *(120)*, which assumes two conditional rates of change. From the statistical point of view, the difference is 1 degree of freedom, so if $-2 \log \Delta > 3.84$ (i.e., the 95% quantile of the χ^2-distribution with 1 degree of freedom), the HKY model is favored over the F81 model. On the other hand, if $-2 \log \Delta \leq 3.84$, the F81 model would be favored.

The approach outlined in the preceding has been extended to cover the family of time-reversible Markov models for nucleotide sequences *(9)*, and it is now available also for amino acid sequences *(10)*. The approach also allows for the presence of permanently invariant sites and rate-heterogeneity across sites, thus catering for some of the differences found among sites in phylogenetic data.

Although the approach described above appears attractive, there is reason for concern. For example, the likelihood-ratio test assumes that at least one of the models compared is correct, an assumption that would be violated in most cases in which we compare the fit of an alignment to all potentially suitable Markov models. Other problems include those arising when: (i) multiple tests are conducted on the same data and the tests are non-independent; (ii) sample sizes are small; and (iii) non-nested models are compared (for an informative discussion of the problems, see *(123, 124)*). Finally, it is implicitly assumed that the tree used in the comparison of models is the most likely tree for every model compared, which might not be the case.

Some of the problems encountered when using the hierarchical likelihood-ratio test are readily dealt with by other methods of model selection. Within the likelihood framework, alternative Markov models may be compared using Akaike's Information Criterion (*AIC*) *(125)*, where *AIC* for a given model, **R**, is a function of the maximized log-likelihood on **R** and the number of estimable parameters, K (e.g., the nucleotide frequency, conditional rates of change, proportion of invariant sites, rate variation among sites, and the number of edges in the tree):

$$AIC = -2\max\left(\log L(\mathbf{R}\,|\,\text{data})\right) + 2K. \qquad [11]$$

If the sample size, l, is small compared with the number of estimable parameters (e.g., $l/K < 40$), the corrected *AIC* (AIC_c) is recommended *(126)*. (The exact meaning of the sample size is currently unclear but it is occasionally thought to be approximately equal to the number of characters in the alignment.):

$$AIC_c = AIC + \frac{2K(K+1)}{l - K - 1}. \qquad [12]$$

The *AIC* may be regarded as the amount of information lost by using **R** to approximate the evolutionary processes, whereas $2K$ may be regarded as the penalty for allowing $2K$ parameters; hence, the best-fitting Markov model corresponds to the smallest value of *AIC* (or AIC_c).

Within the Bayesian context, alternative models may be compared using the Bayesian Information Criterion (*BIC*) *(127)*, Bayes factors (*BF*) *(128-130)*, posterior probabilities (*PP*) *(131, 132)* and decision theory (*DT*) *(133)*, where, for example:

$$BF_{ij} = \frac{P(\text{data}\,|\,\mathbf{R}_i)}{P(\text{data}\,|\,\mathbf{R}_j)} \qquad [13]$$

and

$$BIC = -2\max\left(\log L(\mathbf{R}\,|\,\text{data})\right) + K \log l. \qquad [14]$$

A common feature of the model-selection methods that calculate *BF* and *PP* is that the likelihood of a given model is calculated by integrating over parameter space; hence, the methods often rely on computationally intensive techniques to obtain the likelihood of the model. The model-selection method that calculates *PP* is more attractive than that which calculates *BF* because the former facilitates simultaneous comparison of multiple models, including non-nested models. However, the drawback of the model-selection method that calculates *PP* is that, although some Markov models may appear more realistic than other such models, it is difficult to quantify the prior probability of alternative models *(124)*.

The method for calculating *BIC* is more tractable than those for calculating *BF* and *PP*, and provided the prior probability is uniform for the models under consideration, the *BIC* statistics are also easier to interpret than the *BF* statistics *(124)*. The prior probability is unlikely to be uniform, however, implying that interpretation of the *BIC* and *DT* statistics, which is an extension of the *BIC* statistics *(133)*, may be more difficult.

There is evidence that different model-selection methods may lead to the selection of different Markov models for phylogenetic analyses of the same data *(134)*, so there is a need for more information on how to select model-selection methods. Based on practical and theoretical considerations, Posada and Buckley *(124)* suggested that model-selection methods should: (i) be able to compare non-nested models; (ii) allow for simultaneous comparison of multiple models; (iii) not depend on significance levels; (iv) incorporate topological uncertainty; (v) be tractable; (vi) allow for model averaging; (vii) provide the possibility of specifying priors for models and model parameters; and (viii) be designed to approximate, rather than to identify, truth. Based on these criteria and with reference to a large body of literature on philosophical and applied aspects of model selection, Posada and Buckley *(124)* concluded that the hierarchical likelihood-ratio test is not the optimal approach for model selection in phylogenetics and that the *AIC* and Bayesian approaches provide important advantages, including the ability to simultaneously compare multiple nested and non-nested models.

3.6. General Approaches to Model Selection

If a partition of sites in an alignment were found to have evolved under conditions that are more general than those considered by the general time-reversible Markov model, then there are only a few methods available for finding appropriate models to approximate the evolutionary process. The problem of finding such models is difficult and may be further exacerbated if variation in the alignment is the outcome of several evolutionary processes operating across different sites. For example, the survey of butterfly mitochondrial DNA disclosed that in order to analyze these data phylogenetically, it would be necessary to use two time-reversible Markov models and a general Markov model to approximate what may have led to the variation in these data. (The solution presented here is not necessarily the only suitable one. For example, it might be better to employ Markov models designed for pairs of nucleotides—in this case, first and second codon sites—codons or amino acids (in this case, after translation of the codons.)) Were the alignment to be analyzed with other alignments, possibly using the Pareto set approach *(135)*, it is likely that many Markov models would be required to analyze these data—one model for each set of binned sites—and critics of the multi-model approach might argue that the data were in danger of being over-fitted. (For a discussion of model selection and multi-model inference, see *(123)*).

To address the problems described in the preceding, it is necessary to use approaches that allow us to ascertain whether the models chosen for different partitions of the alignment are the most appropriate, and whether the combination of models is a sufficient approximation of the evolutionary processes that gave rise to the data. Relevant methods for addressing the issue of fitting multiple models to data involve parametric and non-parametric bootstrap procedures and therefore are computationally intensive.

The use of the parametric bootstrap to test the appropriateness of a particular Markov model was proposed by Goldman *(99)* and is a modification of Cox's *(136)* test, which considers non-nested models. The test can be performed as follows:

1. For a given model, **R**, use the original alignment to obtain the log-likelihood, log L, and the maximum-likelihood estimates of the free parameters.
2. Calculate the unconstrained log-likelihood, log L^*, for the original alignment using the following equation:

$$\log L^* = \sum_{i=1}^{N} \log\left(\frac{N_i}{N}\right) \quad [15]$$

   where N_i is the number of times the pattern at column i occurs in the alignment and N is the number of sites in the alignment.
3. Calculate $\delta_{obs} = \log L^* - \log L$.
4. Use the parameters estimated during Step 1 to generate 1,000 pseudo-data sets on the tree in question.
5. For each pseudo-data set, $j = 1, ..., 1000$, calculate log L_j (i.e., the log-likelihood under **R**), log L^*_j, and $\delta_j = \log L^*_j - \log L_j$.
6. Determine p, that is, the proportion of times where $\delta_j > \delta_{obs}$. A large p value supports the hypothesis that **R** is sufficient to explain the evolutionary process underpinning the data while a small p value provides evidence against this hypothesis.

For the family of time-reversible Markov models, a parametric bootstrap analysis may be done with the help of Seq-Gen *(137)*, whereas for the more general Markov models of nucleotide sequence evolution, the analysis may be performed with the help of purpose-developed programs *(41, 44, 49, 93)*. Foster *(41)* and Jayaswal et al. *(44)* used the parametric bootstrap to show that the models employed provide a good fit even though the sequences appear to have evolved under conditions that are not stationary, reversible, and homogeneous. **Fig. 16.7A** illustrates the results from parametric bootstrap analysis for the bacterial data set analyzed by Jayaswal et al. *(44)*. If the bacterial sequences had been analyzed using time-reversible models, ModelTest *(9)* would have chosen the GTR+Γ model as the most appropriate. However, the result in **Fig. 16.7A** shows that the difference between the unconstrained log-likelihood and log-likelihood under the GTR+Γ model is significant, thus showing that the GTR+Γ model fails to adequately describe the complex conditions under which the data evolved. On the other hand, a similar analysis involving the BH+I model *(44)* produced results that show that the BH+I model is an adequate approximation to the evolutionary processes that gave rise to these bacterial sequences (**Fig. 16.7B**).

To compare the hypothesis that the sites of an alignment have evolved under identical and independent conditions to the hypothesis that the nucleotides encode a molecule that necessitates the alignment be partitioned (in accordance with the structure and function of the gene product) and the partitions be analyzed phylogenetically using different Markov models, a permutation test *(138)* can be performed. A suitable strategy for comparing two such hypotheses is described as follows:

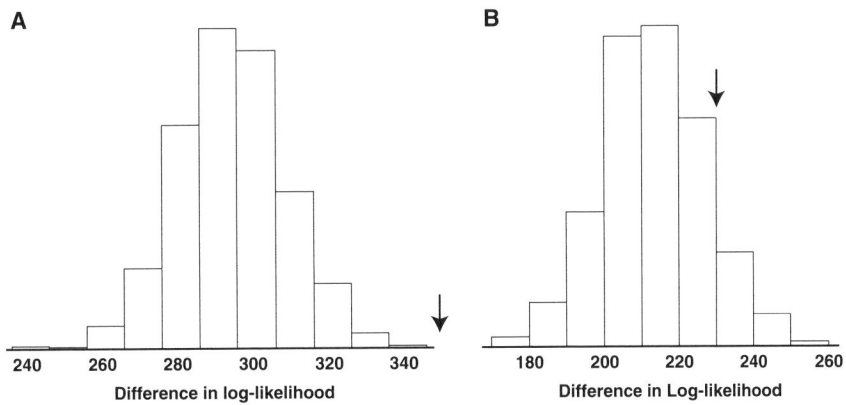

Fig. 16.7. (**A**) Examples of the results from two parametric bootstrap analyses. Parametric bootstrap results under the GTR+Γ model based on 1,000 simulations. For each bootstrap replicate, the difference in log-likelihoods was obtained by subtracting the log-likelihood under the GTR+Γ model from the unconstrained log-likelihood. The arrow indicates the difference in log-likelihood for the actual data under the GTR+Γ model. (**B**) Parametric bootstrap results under the BH+I model based on 1,000 simulations. For each bootstrap replicate, the difference in log-likelihoods was obtained by subtracting the log-likelihood under the BH+I model from the unconstrained log-likelihood. The arrow indicates the difference in log-likelihood for the actual data under the BH+I model.

1. For a given set of models, R_1, R_2, ..., each of which is applied to its own partition of the alignment, calculate the log-likelihood, log L_{obs}, of the original alignment.

2. Generate 1,000 pseudo-data sets by randomizing the order of columns in the data.

3. For each pseudo-data set, $j = 1, ..., 1000$, calculate log L_j (i.e., the log-likelihood of the pseudo-data under R_1, R_2, ...).

4. Determine p, that is, the proportion of times where log L_j > log L_{obs}. A small p-value supports the notion that the sites should be partitioned and analysed using separate models for each partition.

Telford et al. *(100)* used a modification of the approach described above to show that the evolution of the small subunit ribosomal RNA genes of Bilateria can be explained best by using two Markov models, one for the loop-coding sites and another for the stem-coding sites. They also found that paired-sites models *(68, 71, 72)* were significantly better at describing the evolution of stem-coding sites than a model that assumes independence of the stem-forming sites.

4. Discussion

It is clear that model selection plays an important role in molecular phylogenetics, in particular in the context of distance, maximum-likelihood, and Bayesian methods. Given that different model-selection methods may pick different models and that application of inappropriate models affects many aspects of phylogenetic studies, including estimates of phylogeny, substitution rates, posterior probabilities, and bootstrap values, it is important to be aware of the advantages and disadvantages of each of the available model-selection methods.

It is equally important to know that any model that we can construct to analyze a set of sequences is extremely unlikely to be the true model. Rather, the models that we infer by means of prior knowledge of the data and the model-selection methods available are at best good approximations of the underlying evolutionary processes. The process of selecting Markov models for phylogenetic studies, therefore, should be considered as a method of approximating rather than identifying the evolutionary processes *(123, 124)*. In identifying an adequate approximation to the evolutionary processes, it is important to strike a balance between bias and variance, where the problem of bias may arise when too few parameters are used to approximate the evolutionary processes and the problem of variance may arise if too many parameters are used *(123, 124)*. Using model-selection methods that include a penalty for including more than the necessary number of parameters, therefore, appears very appealing.

Having identified a suitable Markov model and subsequently inferred the phylogeny, it is also important to employ the parametric bootstrapping procedure to determine whether the estimates of evolutionary patterns and evolutionary processes are consistent with the data. This is not done as often as it ought to be, although the trend appears to be changing (see e.g., *(139)*). While using the parametric bootstrap, it is important to note that a good fit does not guarantee that the model is correct. For example, Jayaswal et al. *(44)* obtained a good fit between the 16S ribosomal RNA from bacteria and a general Markov model that does not take into account structural and functional information about the gene product.

Finally, it is clear that there is a need for more efficient and reliable methods to identify appropriate Markov models for phylogenetic studies, in particular for data that have not evolved under stationary, reversible, and homogeneous conditions. Likewise, there is a need for phylogenetic methods that (i) allow the partitions of alignments to be analyzed using a combination of Markov models and (ii) allow the parameters of each model to be optimized independently (except for the length of edges in each tree). Finally, there is a need for a large number of different Markov models to bridge the gap between the family of time-reversible models and the most general Markov models.

Acknowledgment

This research was partly funded by Discovery Grants (DP0453173 and DP0556820) from the Australian Research Council. Faisal Ababneh was supported by a postgraduate scholarship from the Al-Hussein Bin Talal University in Jordan. The authors wish to thank J.W.K. Ho, J. Keith, K.W. Lau, G.J.P. Naylor, S. Whelan, and Y. Zhang for their constructive thoughts, ideas, and criticism on this manuscript.

References

1. Zakharov, E. V., Caterino, M. S., Sperling, F. A. H. (2004) Molecular phylogeny, historical biogeography, and divergence time estimates for swallowtail butterflies of the genus *Papilio* (Lepidoptera: Papilionidae). *Syst Biol* 53, 193–215.

2. Brochier, C., Forterre, P., Gribaldo, S. (2005) An emerging phylogenetic core of Archaea: phylogenies of transcription and translation machineries converge following addition of new genome sequences. *BMC Evol Biol* 5, 36.

3. Hardy, M. P., Owczarek, C. M., Jermiin, L. S., et al. (2004) Characterization of the type I interferon locus and identification of novel genes. *Genomics* 84, 331–345.

4. de Queiroz, K., Gauthier, J. (1994) Toward a phylogenetic system of biological nomenclature. *Trends Ecol Evol* 9, 27–31.

5. Board, P. G., Coggan, M., Chelnavayagam, G., et al. (2000) Identification, characterization and crystal structure of the Omega class of glutathione transferases. *J Biol Chem* 275, 24798–24806.

6. Pagel, M. (1999) Inferring the historical patterns of biological evolution. *Nature* 401, 877–884.

7. Charleston, M. A., Robertson, D. L. (2002) Preferential host switching by primate lentiviruses can account for phylogenetic similarity with the primate phylogeny. *Syst Biol* 51, 528–535.

8. Jermann, T. M., Opitz, J. G., Stackhouse, J., et al. (1995) Reconstructing the evolutionary history of the artiodactyl ribonuclease superfamily. *Nature* 374, 57–59.

9. Posada, D., Crandall, K. A. (1998) MODELTEST: testing the model of DNA substitution. *Bioinformatics* 14, 817–818.

10. Abascal, F., Zardoya, R., Posada, D. (2005) ProtTest: selection of best-fit models of protein evolution. *Bioinformatics* 21, 2104–2105.

11. Weisburg, W. G., Giovannoni, S. J., Woese, C. R. (1989) The *Deinococcus* and *Thermus* phylum and the effect of ribosomal RNA composition on phylogenetic tree construction. *Syst Appl Microbiol* 11, 128–134.

12. Loomis, W. F., Smith, D. W. (1990) Molecular phylogeny of *Dictyostelium discoideum* by protein sequence comparison. *Proc Natl Acad Sci USA* 87, 9093–9097.

13. Penny, D., Hendy, M. D., Zimmer, E. A., et al. (1990) Trees from sequences: panacea or Pandora's box? *Aust Syst Biol* 3, 21–38.

14. Lockhart, P. J., Howe, C. J., Bryant, D. A., et al. (1992) Substitutional bias confounds inference of cyanelle origins from sequence data. *J Mol Evol* 34, 153–162.

15. Lockhart, P. J., Penny, D., Hendy, M. D., et al. (1992) Controversy on chloroplast origins. *FEBS Lett* 301, 127–131.

16. Hasegawa, M., Hashimoto, T. (1993) Ribosomal RNA trees misleading? *Nature* 361, 23.

17. Olsen, G. J., Woese, C. R. (1993) Ribosomal RNA: a key to phylogeny. *FASEB J* 7, 113–123.

18. Sogin, M. L., Hinkle, G., Leipe, D. D. (1993) Universal tree of life. *Nature* 362, 795.

19. Klenk, H. P., Palm, P., Zillig, W. (1994) DNA-dependent RNA polymerases as phylogenetic marker molecules. *Syst Appl Microbiol* 16, 638–647.

20. Foster, P. G., Jermiin, L. S., Hickey, D. A. (1997) Nucleotide composition bias affects amino acid content in proteins coded by animal mitochondria. *J Mol Evol* 44, 282–288.

21. van den Bussche, R. A., Baker, R. J., Huelsenbeck, J. P., et al. (1998) Base compositional bias and phylogenetic analyses: a test of the "flying DNA" hypothesis. *Mol Phylogenet Evol* 10, 408–416.

22. Foster, P. G., Hickey, D. A. (1999) Compositional bias may affect both DNA-based and protein-based phylogenetic reconstructions. *J Mol Evol* 48, 284–290.

23. Chang, B. S. W., Campbell, D. L. (2000) Bias in phylogenetic reconstruction of vertebrate rhodopsin sequences. *Mol Biol Evol* 17, 1220–1231.

24. Conant, G. C., Lewis, P. O. (2001) Effects of nucleotide composition bias on the success of the parsimony criterion on phylogenetic inference. *Mol Biol Evol* 18, 1024–1033.

25. Tarrío, R., Rodriguez-Trelles, F., Ayala, F. J. (2001) Shared nucleotide composition biases among species and their impact on phylogenetic reconstructions of the *Drosophilidae*. *Mol Biol Evol* 18, 1464–1473.

26. Goremykin, V. V., Hellwig, F. H. (2005) Evidence for the most basal split in land plants dividing bryophyte and tracheophyte lineages. *Plant Syst Evol* 254, 93–103.

27. Barry, D., Hartigan, J. A. (1987) Statistical analysis of hominoid molecular evolution. *Stat Sci* 2, 191–210.

28. Reeves, J. (1992) Heterogeneity in the substitution process of amino acid sites of proteins coded for by the mitochondrial DNA. *J Mol Evol* 35, 17–31.

29. Steel, M. A., Lockhart, P. J., Penny, D. (1993) Confidence in evolutionary trees from biological sequence data. *Nature* 364, 440–442.

30. Lake, J. A. (1994) Reconstructing evolutionary trees from DNA and protein sequences: paralinear distances. *Proc Natl Acad Sci USA* 91, 1455–1459.

31. Lockhart, P. J., Steel, M. A., Hendy, M. D., et al. (1994) Recovering evolutionary trees under a more realistic model of sequence evolution. *Mol Biol Evol* 11, 605–612.

32. Steel, M. A. (1994) Recovering a tree from the leaf colourations it generates under a Markov model. *Appl Math Lett* 7, 19–23.

33. Galtier, N., Gouy, M. (1995) Inferring phylogenies from DNA sequences of unequal base compositions. *Proc Natl Acad Sci USA* 92, 11317–11321.

34. Steel, M. A., Lockhart, P. J., Penny, D. (1995) A frequency-dependent significance test for parsimony. *Mol Phylogenet Evol* 4, 64–71.

35. Yang, Z., Roberts, D. (1995) On the use of nucleic acid sequences to infer early branches in the tree of life. *Mol Biol Evol* 12, 451–458.

36. Gu, X., Li, W.-H. (1996) Bias-corrected paralinear and logdet distances and tests of molecular clocks and phylogenies under nonstationary nucleotide frequencies. *Mol Biol Evol* 13, 1375–1383.

37. Gu, X., Li, W.-H. (1998) Estimation of evolutionary distances under stationary and nonstationary models of nucleotide substitution. *Proc Natl Acad Sci USA* 95, 5899–5905.

38. Galtier, N., Gouy, M. (1998) Inferring pattern and process: maximum-likelihood implementation of a nonhomogenous model of DNA sequence evolution for phylogenetic analysis. *Mol Biol Evol* 15, 871–879.

39. Galtier, N., Tourasse, N., Gouy, M. (1999) A nonhyperthermophilic common ancestor to extant life forms. *Science* 283, 220–221.

40. Tamura, K., Kumar, S. (2002) Evolutionary distance estimation under heterogeneous substitution pattern among lineages. *Mol Biol Evol* 19, 1727–1736.

41. Foster, P. G. (2004) Modeling compositional heterogeneity. *Syst Biol* 53, 485–495.

42. Thollesson, M. (2004) LDDist: a Perl module for calculating LogDet pair-wise distances for protein and nucleotide sequences. *Bioinformatics* 20, 416–418.

43. Jayaswal, V., Jermiin, L. S., Robinson, J. (2005) Estimation of phylogeny using a general Markov model. *Evol Bioinf Online* 1, 62–80.

44. Jayaswal, V., Robinson, J., Jermiin, L. S. (2007) Estimation of phylogeny and invariant sites under the General Markov model of nucleotide sequence evolution. *Syst Biol*, 56, 155–162.

45. Sullivan, J., Arellano, E. A., Rogers, D. S. (2000) Comparative phylogeography of Mesoamerican highland rodents: concerted versus independent responses to past climatic fluctuations. *Am Nat* 155, 755–768.

46. Demboski, J. R., Sullivan, J. (2003) Extensive mtDNA variation within the yellow-pine chipmunk, *Tamias amoenus* (Rodentia: Sciuridae), and phylogeographic inferences for northwestern North America. *Mol Phylogenet Evol* 26, 389–408.

47. Carstens, B. C., Stevenson, A. L., Degenhardt, J. D., et al. (2004) Testing nested phylogenetic and phylogeographic hypotheses in the *Plethodon vandykei* species group. *Syst Biol* 53, 781–792.

48. Tavaré, S. (1986) Some probabilistic and statistical problems on the analysis of DNA sequences. *Lect Math Life Sci* 17, 57–86.

49. Ababneh, F., Jermiin, L. S., Robinson, J. (2006) Generation of the exact distribution and simulation of matched nucleotide sequences on a phylogenetic tree. *J Math Model Algor* 5, 291–308.

50. Bryant, D., Galtier, N., Poursat, M.-A. (2005) Likelihood calculation in molecular phylogenetics, in (Gascuel, O., ed.), *Mathematics in Evolution and Phylogeny*. Oxford University Press, Oxford, UK, pp. 33–62.

51. Penny, D., Hendy, M. D., Steel, M. A. (1992) Progress with methods for constructing evolutionary trees. *Trends Ecol Evol* 7, 73–79.

52. Drouin, G., Prat, F., Ell, M., et al. (1999) Detecting and characterizing gene conversion between multigene family members. *Mol Biol Evol* 16, 1369–1390.

53. Posada, D., Crandall, K. A. (2001) Evaluation of methods for detecting recombination from DNA sequences: computer simulations. *Proc Natl Acad Sci USA* 98, 13757–13762.

54. Posada, D. (2002) Evaluation of methods for detecting recombination from DNA sequences: empirical data. *Mol Biol Evol* 19, 708–717.

55. Martin, D. P., Williamson, C., Posada, D. (2005) RDP2: Recombination detection and analysis from sequence alignments. *Bioinformatics* 21, 260–262.

56. Bruen, T. C., Philippe, H., Bryant, D. (2006) A simple and robust statistical test for detecting the presence of recombination. *Genetics* 172, 2665–2681.

57. Ragan, M. A. (2001) On surrogate methods for detecting lateral gene transfer. *FEMS Microbiol Lett* 201, 187–191.

58. Dufraigne, C., Fertil, B., Lespinats, S., et al. (2005) Detection and characterization of horizontal transfers in prokaryotes using genomic signature. *Nucl Acid Res* 33, e6.

59. Azad, R. K., Lawrence, J. G. (2005) Use of artificial genomes in assessing methods for atypical gene detection. *PLoS Comp Biol* 1, 461–473.

60. Tsirigos, A., Rigoutsos, I. (2005) A new computational method for the detection of horizontal gene transfer events. *Nucl Acid Res* 33, 922–933.

61. Ragan, M. A., Harlow, T. J., Beiko, R. G. (2006) Do different surrogate methods detect lateral genetic transfer events of different relative ages? *Trends Microbiol* 14, 4–8.

62. Beiko, R. G., Hamilton, N. (2006) Phylogenetic identification of lateral genetic transfer events. *BMC Evol Biol* 6, 15.

63. Fitch, W. M. (1986) An estimation of the number of invariable sites is necessary for the accurate estimation of the number of nucleotide substitutions since a common ancestor. *Prog Clin Biol Res* 218, 149–159.

64. Lockhart, P. J., Larkum, A. W. D., Steel, M. A., et al. (1996) Evolution of chlorophyll and bacteriochlorophyll: the problem of invariant sites in sequence analysis. *Proc Natl Acad Sci USA* 93, 1930–1934.

65. Yang, Z. (1996) Among-site rate variation and its impact on phylogenetic analysis. *Trends Ecol Evol* 11, 367–372.

66. Waddell, P. J., Steel, M. A. (1997) General time reversible distances with unequal rates across sites: mixing Γ and inverse Gaussian distributions with invariant sites. *Mol Phylogenet Evol* 8, 398–414.

67. Gowri-Shankar, V., Rattray, M. (2006) Compositional heterogeneity across sites: Effects on phylogenetic inference and modeling the correlations between base frequencies and substitution rate. *Mol Biol Evol* 23, 352–364.

68. Schöniger, M., von Haeseler, A. (1994) A stochastic model for the evolution of autocorrelated DNA sequences. *Mol Phylogenet Evol* 3, 240–247.

69. Tillier, E. R. M. (1994) Maximum likelihood with multiparameter models of substitution. *J Mol Evol* 39, 409–417.

70. Hein, J., Støvlbæk, J. (1995) A maximum-likelihood approach to analyzing nonoverlapping and overlapping reading frames. *J Mol Evol* 40, 181–190.

71. Muse, S. V. (1995) Evolutionary analyses of DNA sequences subject to constraints on secondary structure. *Genetics* 139, 1429–1439.

72. Rzhetsky, A. (1995) Estimating substitution rates in ribosomal RNA genes. *Genetics* 141, 771–783.

73. Tillier, E. R. M., Collins, R. A. (1995) Neighbor joining and maximum likelihood with RNA sequences: addressing the interdependence of sites. *Mol Biol Evol* 12, 7–15.

74. Pedersen, A.-M. K., Wiuf, C., Christiansen, F. B. (1998) A codon-based model designed to describe lentiviral evolution. *Mol Biol Evol* 15, 1069–1081.

75. Tillier, E. R. M., Collins, R. A. (1998) High apparent rate of simultaneous compensatory base-pair substitutions in ribosomal RNA. *Genetics* 148, 1993–2002.

76. Higgs, P. G. (2000) RNA secondary structure: physical and computational aspects. *Q Rev Biophys* 30, 199–253.

77. Pedersen, A.-M. K., Jensen, J. L. (2001) A dependent-rates model and an MCMC-based methodology for the maximum-likelihood analysis of sequences with overlapping frames. *Mol Biol Evol* 18, 763–776.

78. Savill, N. J., Hoyle, D. C., Higgs, P. G. (2001) RNA sequence evolution with secondary structure constraints: comparison of substitution rate models using maximum-likelihood methods. *Genetics* 157, 339–411.

79. Jow, H., Hudelot, C., Rattray, M., et al. (2002) Bayesian phylogenerics using an RNA substitution model applied to early mammalian evolution. *Mol Biol Evol* 19, 1591–1601.

80. Lockhart, P. J., Steel, M. A., Barbrook, A. C., et al. (1998) A covariotide model explains apparent phylogenetic structure of oxygenic photosynthetic lineages. *Mol Biol Evol* 15, 1183–1188.

81. Jukes, T. H., Cantor, C. R. (1969) Evolution of protein molecules, in (Munro, H. N., ed.), *Mammalian Protein Metabolism*. Academic Press, New York.

82. Lanave, C., Preparata, G., Saccone, C., et al. (1984) A new method for calculating evolutionary substitution rates. *J Mol Evol* 20, 86–93.

83. Naylor, G. P. J., Brown, W. M. (1998) Amphioxus mitochondrial DNA, chordate phylogeny, and the limits of inference based on comparisons of sequences. *Syst Biol* 47, 61–76.

84. Ho, S. Y. W., Jermiin, L. S. (2004) Tracing the decay of the historical signal in biological sequence data. *Syst Biol* 53, 623–637.

85. Jermiin, L. S., Ho, S. Y. W., Ababneh, F., et al. (2004) The biasing effect of compositional heterogeneity on phylogenetic estimates may be underestimated. *Syst Biol* 53, 638–643.

86. Ababneh, F., Jermiin, L. S., Ma, C., et al. (2006) Matched-pairs tests of homogeneity with applications to homologous nucleotide sequences. *Bioinformatics* 22, 1225–1231.

87. Ho, J. W. K., Adams, C. E., Lew, J. B., et al. (2006) SeqVis: Visualization of compositional heterogeneity in large alignments of nucleotides. *Bioinformatics* 22, 2162–2163.

88. Lanave, C., Pesole, G. (1993) Stationary MARKOV processes in the evolution of biological macromolecules. *Binary* 5, 191–195.

89. Rzhetsky, A., Nei, M. (1995) Tests of applicability of several substitution models for DNA sequence data. *Mol Biol Evol* 12, 131–151.

90. Waddell, P. J., Cao, Y., Hauf, J., et al. (1999) Using novel phylogenetic methods to evaluate mammalian mtDNA, including amino acid-invariant sites-LogDet plus site stripping, to detect internal conflicts in the data, with special reference to the positions of hedgehog, armadillo, and elephant. *Syst Biol* 48, 31–53.

91. Bowker, A. H. (1948) A test for symmetry in contingency tables. *J Am Stat Assoc* 43, 572–574.

92. Stuart, A. (1955) A test for homogeneity of the marginal distributions in a two-way classification. *Biometrika* 42, 412–416.

93. Jermiin, L. S., Ho, S. Y. W., Ababneh, F., et al. (2003) *Hetero*: a program to simulate the evolution of DNA on a four-taxon tree. *Appl Bioinf* 2, 159–163.

94. Muse, S. V., Weir, B. S. (1992) Testing for equality of evolutionary rates. *Genetics* 132, 269–276.

95. Cannings, C., Edwards, A. W. F. (1968) Natural selection and the de Finetti diagram. *Ann Hum Genet* 31, 421–428.

96. Huelsenbeck, J. P., Rannala, B. (1997) Phylogenetic methods come of age: Testing hypotheses in an evolutionary context. *Science* 276, 227–232.

97. Whelan, S., Goldman, N. (1999) Distributions of statistics used for the comparison of models of sequence evolution in phylogenetics. *Mol Biol Evol* 16, 11292–11299.

98. Goldman, N., Whelan, S. (2000) Statistical tests of gamma-distributed rate heterogeneity in models of sequence evolution in phylogenetics. *Mol Biol Evol* 17, 975–978.

99. Goldman, N. (1993) Statistical tests of models of DNA substitution. *J Mol Evol* 36, 182–198.

100. Telford, M. J., Wise, M. J., Gowri-Shankar, V. (2005) Consideration of RNA secondary structure significantly improves likelihood-based estimates of phylogeny: examples from the bilateria. *Mol Biol Evol* 22, 1129–1136.

101. Goldman, N., Yang, Z. (1994) A codon-based model of nucleotide substitution for protein-coding DNA sequences. *Mol Biol Evol* 11, 725–736.

102. Muse, S. V., Gaut, B. S. (1994) A likelihood approach for comparing synonymous and nonsynonymous nucleotide substitution rates, with application to the chloroplast genome. *Mol Biol Evol* 11, 715–724.

103. Dayhoff, M. O., Schwartz, R. M., Orcutt, B. C. (eds.) (1978) *A Model of Evolutionary Change in Proteins*. National Biomedical Research Foundation, National Biomedical Research Foundation, Washington, DC.

104. Jones, D. T., Taylor, W. R., Thornton, J. M. (1992) The rapid generation of mutation data matrices from protein sequences. *Comp Appl Biosci* 8, 275–282.

105. Henikoff, S., Henikoff, J. G. (1992) Amino acid substitution matrices from protein blocks. *Proc Natl Acad Sci USA* 89, 10915–10919.

106. Adachi, J., Hasegawa, M. (1996) Model of amino acid substitution in proteins encoded by mitochondrial DNA. *J Mol Evol* 42, 459–468.

107. Cao, Y., Janke, A., Waddell, P. J., et al. (1998) Conflict among individual mitochondrial proteins in resolving the phylogeny of eutherian orders. *J Mol Evol* 47, 307–322.

108. Yang, Z., Nielsen, R., Hasegawa, M. (1998) Models of amino acid substitution and applications to mitochondrial protein evolution. *Mol Biol Evol* 15, 1600–1611.

109. Müller, T., Vingron, M. (2000) Modeling amino acid replacement. *J Comp Biol* 7, 761–776.

110. Adachi, J., Waddell, P. J., Martin, W., et al. (2000) Plastid genome phylogeny and a model of amino acid substitution for proteins encoded by chloroplast DNA. *J Mol Evol* 50, 348–358.

111. Whelan, S., Goldman, N. (2001) A general empirical model of protein evolution derived from multiple protein families using a maximum likelihood approach. *Mol Biol Evol* 18, 691–699.

112. Dimmic, M. W., Rest, J. S., Mindell, D. P., Goldstein, R. A. (2002) RtREV: an amino acid substitution matrix for inference of retrovirus and reverse transcriptase phylogeny. *J Mol Evol* 55, 65–73.

113. Abascal, F., Posada, D., Zardoya, R. (2007) MtArt: a new model of amino acid replacement for Arthropoda. *Mol Biol Evol* 24, 1–5.

114. Shapiro, B., Rambaut, A., Drummond, A. J. (2005) Choosing appropriate substitution models for the phylogenetic analysis of protein-coding sequences. *Mol Biol Evol* 23, 7–9.

115. Hyman, I. T., Ho, S. Y. W., Jermiin, L. S. (2007) Molecular phylogeny of Australian Helicarionidae, Microcystidae and related groups (Gastropoda: Pulmonata: Stylommatophora) based on mitochondrial DNA. *Mol Phylogenet Evol*, 45, 792–812.

116. Galtier, N. (2001) Maximum-likelihood phylogenetic analysis under a covarion-like model. *Mol Biol Evol* 18, 866–873.

117. Hudelot, C., Gowri-Shankar, V., Jow, H., et al. (2003) RNA-based phylogenetic methods: Application to mammalian mitochondrial RNA sequences. *Mol Phylogenet Evol* 28, 241–252.

118. Murray, S., Flø Jørgensen, M., Ho, S. Y. W., et al. (2005) Improving the analysis of dinoflagelate phylogeny based on rDNA. *Protist* 156, 269–286.

119. Felsenstein, J. (1981) Evolutionary trees from DNA sequences: a maximum likelihood approach. *J Mol Evol* 17, 368–376.

120. Hasegawa, M., Kishino, H., Yano, T. (1985) Dating of the human-ape splitting by a molecular clock of mitochondrial DNA. *J Mol Evol* 22, 160–174.

121. Kimura, M. (1980) A simple method for estimating evolutionary rates of base substitutions through comparative studies of nucleotide sequences. *J Mol Evol* 16, 111–120.

122. Zharkikh, A. (1994) Estimation of evolutionary distances between nucleotide sequences. *J Mol Evol* 39, 315–329.

123. Burnham, K. P., Anderson, D. R. (2002) *Model Selection and Multimodel Inference: A Practical Information-Theoretic Approach.* Springer, New York.

124. Posada, D., Buckley, T. R. (2004) Model selection and model averaging in phylogenetics: advantages of akaike information criterion and bayesian approaches over likelihood ratio tests. *Syst Biol* 53, 793–808.

125. Akaike, H. (1974) A new look at the statistical model identification. *IEEE Trans Auto Cont* 19, 716–723.

126. Sugiura, N. (1978) Further analysis of the data by Akaike's information criterion and the finite corrections. *Commun Stat A: Theory Methods* 7, 13–26.

127. Schwarz, G. (1978) Estimating the dimension of a model. *Ann Stat* 6, 461–464.

128. Suchard, M. A., Weiss, R. E., Sinsheimer, J. S. (2001) Bayesian selection of continuous-time Markov chain evolutionary models. *Mol Biol Evol* 18, 1001–1013.

129. Aris-Brosou, S., Yang, Z. (2002) Effects of models of rate evolution on estimation of divergence dates with special reference to the metazoan 18S ribosomal RNA phylogeny. *Syst Biol* 51, 703–714.

130. Nylander, J. A., Ronquist, F., Huelsenbeck, J. P., et al. (2004) Bayesian phylogenetic analysis of combined data. *Syst Biol* 53, 47–67.

131. Kass, R. E., Raftery, A. E. (1995) Bayes factors. *J Am Stat Assoc* 90, 773–795.

132. Raftery, A. E. (1996) Hypothesis testing and model selection, in (Gilks, W. R., Richardson, S. and Spiegelhalter, D. J., eds.), *Markov Chain Monte Carlo in Practice.* Chapman & Hall, London.

133. Minin, V., Abdo, Z., Joyce, P., et al. (2003) Performance-based selection of likelihood models for phylogenetic estimation. *Syst Biol* 52, 674–683.

134. Posada, D., Crandall, K. A. (2001) Selecting methods of nucleotide substitution: An application to human immunodeficiency virus 1 (HIV-1). *Mol Biol Evol* 18, 897–906.

135. Poladian, L., Jermiin, L. S. (2006) Multi-objective evolutionary algorithms and phylogenetic inference with multiple data sets. *Soft Comp* 10, 358–368.

136. Cox, D. R. (1962) Further results on tests of separate families of hypotheses. *J Royal Stat Soc B* 24, 406–424.

137. Rambaut, A., Grassly, N. C. (1997) Seq-Gen: An application for the Monte Carlo simulation of DNA sequence evolution along phylogenetic trees. *Comp Appl Biosci* 13, 235–238.

138. Felsenstein, J. (2003) *Inferring phylogenies.* Sinauer Associates, Sunderland, MA.

139. Rokas, A., Krüger, D., Carroll, S. B. (2005) Animal evolution and the molecular signature of radiations compressed in time. *Science* 310, 1933–1938.

Chapter 17

Inferring Ancestral Gene Order

Julian M. Catchen, John S. Conery, and John H. Postlethwait

Abstract

To explain the evolutionary mechanisms by which populations of organisms change over time, it is necessary to first understand the pathways by which genomes have changed over time. Understanding genome evolution requires comparing modern genomes with ancestral genomes, which thus necessitates the reconstruction of those ancestral genomes. This chapter describes automated approaches to infer the nature of ancestral genomes from modern sequenced genomes. Because several rounds of whole genome duplication have punctuated the evolution of animals with backbones, and current methods for ortholog calling do not adequately account for such events, we developed ways to infer the nature of ancestral chromosomes after genome duplication. We apply this method here to reconstruct the ancestors of a specific chromosome in the zebrafish *Danio rerio*.

Key words: Ancestral reconstruction, chromosome evolution, genome duplication, automated workflow.

1. Introduction

Conservation of genome structure provides information about organismal origin and change over time. Conserved genomic islands rafting in a sea of genomic change identify features protected from variation by natural selection. Investigation of conserved non-coding regions, for example, has provided extensive data on genomic entities that control gene expression and has identified genes for small non-translated RNAs *(1–3)*. In contrast, genomic blocks that display conserved gene order have been investigated less fully.

A pair of organisms can show conservation of genome structure at several different levels. Level 1 is the conservation of syntenies (*syn*, same; *ten*, thread) between a pair of genomes. In a conserved synteny, two or more genes that are syntenic (on the same chromosome) in one genome have orthologs that are also on a single chromosome in another genome. (Orthologs are defined as genetic entities in different genomes descended from a single entity in the last common ancestor of those genomes. They can be defined operationally as reciprocal best hits (RBH) in BLAST searches (*4, 5*), although occasionally errors are made in the case of rapidly evolving genes.) Even if many genes intervene between the pair of genes in one or the other genome, the organization of two pairs of orthologs on a single chromosome in both genomes meets the criterion of conserved synteny. Conserved syntenies suggest the hypothesis that all members of the group of genes were syntenic in the last common ancestor of the two genomes. The alternative, less parsimonious (but undoubtedly sometimes true) hypothesis is that the genes were separate in the last common ancestor and chromosome rearrangements that occurred independently in both lineages brought the members of the group together fortuitously.

Somewhat more restrictive, in Level 2 a group of three or more genes that shows conserved syntenies could in addition display conserved gene order: the order of genes in one genome is the same as the order of their orthologs in the other genome (**Fig. 17.1**). Changes in gene order would reflect an inversion involving at least two genes in one of the genomes with respect to the other. Level 3 is the conservation of transcription orientation within a group of genes that possesses conserved gene order. Loss of conserved orientation reflects an inversion involving a single gene in the lineage of one genome relative to the other. The most restrictive condition, Level 4, involves a conserved genomic block in which all genes in the block in one species have orthologs that are in the same order and have no intervening or missing genes in the other species.

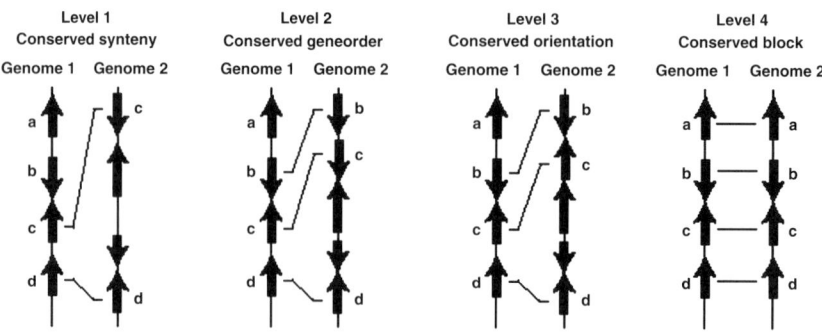

Fig. 17.1. Levels of genomic conservation. Levels 1 through 4 represent increasing amounts of conserved synteny. Whereas level 1 requires only that two orthologous genes occur on homoeologous chromosomes, level 2 requires conserved gene order; level 3 additionally requires conserved transcription orientation, and level 4 requires no intervening genes within the conserved block.

Conservation of gene order can reflect either constraint due to selection or simple failure by chance to fix chromosome rearrangements disrupting the block. If fully conserved genomic blocks persist in different lineages over increasing time periods, then selection becomes an increasingly probable mechanism for the maintenance of conserved blocks.

Probably the best-studied examples of conserved gene order in the human genome are the *HOX* clusters, which provide an example of gene order conserved due to functional constraints. Human *HOX* cluster genes show gene order conservation both within the human genome and among vertebrates. Within the human genome, the four partially degraded *HOX* clusters arose from two consecutive genome duplication events, called R1 and R2 (**Fig. 17.2**), that occurred in early vertebrate evolution about 500 million or more years ago *(6–9)*. The *HOXA*, *HOXB*, *HOXC*, and *HOXD* clusters are descended from a single cluster of 14 contiguous *HOX* genes present in an ancient

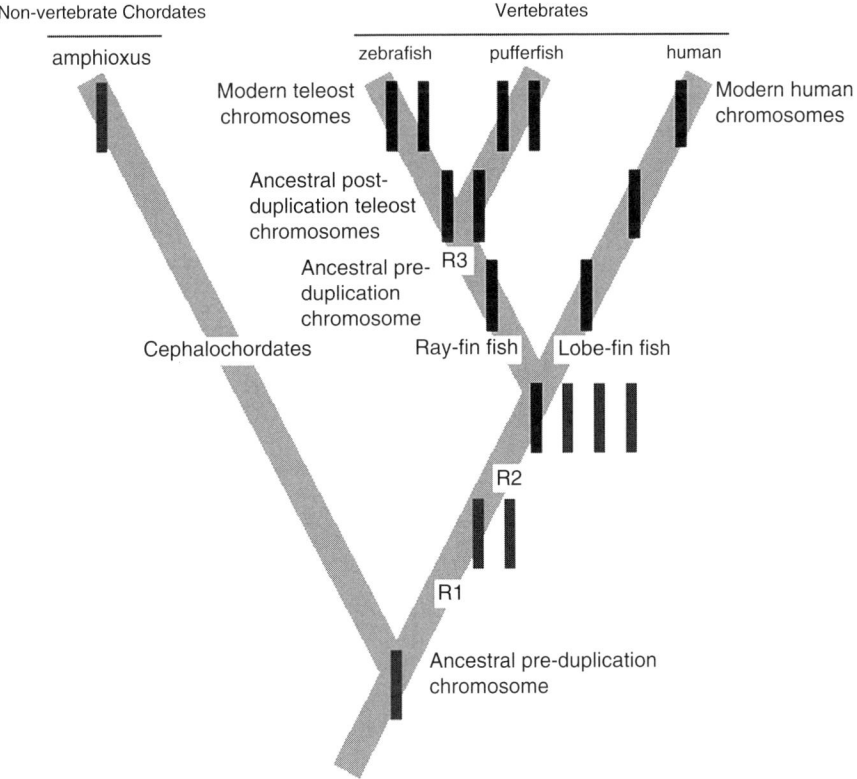

Fig. 17.2. Genome duplication events in chordate evolution. An ancestral pre-duplication chromosome duplicated twice (R1 and R2) to form four chromosome segments, only one of which is diagrammed higher in the figure. The R3 genome duplication event occurred in the ray-fin lineage, providing teleost fish with duplicates of human genes.

non-vertebrate chordate genome similar to the modern amphioxus genome *(10–13)*. Two genes derived from a gene duplication event within a lineage, such as the genes that make up the *HOX* clusters, are called paralogs. Moreover, each human *HOX* cluster is also conserved in gene order with the orthologous *HOX* cluster of other vertebrates, for example, the gene order of the human *HOXA* cluster is conserved with the *Hoxa* cluster of mouse as well as the *hoxaa* cluster of zebrafish *(14–18)*. *HOX* gene clusters have probably been preserved due to the sharing of regulatory elements by several genes in the cluster *(19)*.

In addition to the two rounds of genome duplication that occurred near the time of the vertebrate radiation (R1 and R2), an additional round of genome duplication (R3) occurred at the base of the radiation of teleost fish (the crown group of ray-fin fish, such as zebrafish, and pufferfish, distinct from basally diverging ray-fin fish, such as sturgeon and gar) *(15, 20–24)*. This R3 event (*see* **Fig. 17.2**) generated duplicate chromosome segments in teleosts corresponding to single chromosome segments in humans and other mammals. Duplicated chromosomes are called *homoeologous chromosomes*. (These are not to be confused with a pair of homologous chromosomes, which align and exchange during meiosis in a diploid; homoeologous chromosomes are a pair of chromosomes derived from a single, pre-duplication chromosome in an ancestral organism.) The R3 genome duplication event produced, for example, duplicate copies of the human *HOXA* cluster in teleosts (*hoxaa* and *hoxab*) surrounded by duplicated copies of many additional genes on the homoeologous chromosomes *(15, 25–28)*.

Other examples of gene clusters maintained by selection include hemoglobin genes and *DLX* genes, but the extent of conserved gene order is not yet fully investigated, and the relative importance of selection versus chance in regions of conserved gene order is currently unknown. Investigations of gene order in vertebrate lineages that diverged from the human lineage long ago could help to narrow the number of regions in which selection is a strong hypothesis for preserved gene order. To contribute to an understanding of these issues, we have begun to investigate the conservation of gene orders and inference of ancestral gene orders in fish genomes compared with the human genome. This section discusses software developed to identify paralogous chromosome segments within species as well as orthologous chromosome segments between species. We then use that software to identify conserved segments between teleost species, infer ancestral gene orders in the pre-duplication teleost genome, and infer genome content in the last common ancestor of teleost fish and mammals.

2. Methods

2.1. Bioinformatic Challenges

In early studies examining conserved synteny, such as the identification of *HOX* clusters, the burden of labor was weighted heavily toward laboratory work, with the objective of locating and sequencing genes. Initial analyses involved small numbers of genes, and the management of these data could be done in an ad hoc manner. Analysis involving homology searches and small-scale phylogenetic tree building could be accomplished on a case-by-case basis, and these were generally used to confirm hypotheses that had been determined *a priori*.

The availability of large-scale genomic data has inverted this equation and presents several new challenges with respect to the management, storage, and interpretation of genomic data. For example, consider a scenario to identify all the genes in mouse that are structurally similar to genes on Human chromosome 17 (Hsa17). If we were only interested in a single gene, we could quickly solve the problem using a Web-based tool to perform a remote homology search. But, when we scale the problem to consider a full chromosome, we face an explosion in both computational resources and data storage. Hsa17 contains approximately 1,800 genes. Since some of those genes produce multiple transcripts, the total number of sequences to examine is over 2,400. Each homology search can return tens of results containing numerical data and sequence alignments that are thousands of characters long. Each additional chromosome and organism we add to the analysis increases the problem size linearly, whereas each additional homology search increases the problem size by an order of magnitude.

In this environment, it becomes essential to employ a system that allows disparate parts of an analysis to be encapsulated, allowing each step to be run independently, resumed, and ideally, run in parallel with other steps. Most challenging of all is the presentation of the results generated by each step in a way that can be visualized and readily comprehended.

Reconstructing ancestral chromosomes requires us to examine modern genomes from several organisms and infer their ancestral states. At its core, this is an exercise in large-scale sequence matching, but we must be able to differentiate between several different types of evolutionary events. First, we must take into account the historical, phylogenetic relationships of organisms we are examining. For closely related vertebrates (e.g., human versus mouse), generally there should be a one-to-one correspondence between genes, because the R1 and R2 genome duplication events in early vertebrate evolution occurred prior to the divergence of mammals. Additionally, since human and mouse diverged about 100 million years ago, they are more closely related to each other than either is to fish, which diverged from the lineage leading to mammals about 450 million years ago. Thus, we expect sequence similarity in our homology searching to be much higher among mammals than between fish and mammals *(29)*.

When we include organisms such as the zebrafish *Danio rerio*, however, the picture becomes more complex. Although the relationship of mouse to human genes is usually one-to-one, because of the R3 genome duplication event at the beginning of the teleost radiation (*see* Fig. 17.2), the relationship of zebrafish genes to human genes is often two-to-one, although 70% to 80% of genes have returned to single copy after the genome duplication *(30)*. In addition, since the teleost/mammalian divergence occurred so much earlier than that of the rodent/primate, we should expect lower sequence similarity in homology searches, more lost genes in one or the other lineage, more genes that have diverged too far to be reliably classified as orthologs, and more genes duplicated independently in each lineage after they diverged.

Importantly, when including organisms that have experienced a third genome duplication, we must take care to differentiate between ancient paralogous genes that originated in the first and second genome duplications from those that originated in the most recent genome duplication. Further, we must be able to differentiate very recent tandemly duplicated genes from those duplicated in the third genome duplication.

The types of evolutionary relationships our software analysis must detect are shown in **Fig. 17.3**. The bottom half of the figure is a gene tree that shows evolutionary events that led to the current genomes for human (Hsa) and zebrafish (Dre): two rounds of genome duplication (R1 and R2), speciation (S), and a third round of genome duplication (R3) in the teleost lineage. In the top half of the figure, a vertical line represents a modern chromosome. The first row, labeled 4:8, is the ideal situation, and might be seen for genes that encode essential biological functions, e.g., for *HOX* clusters: all four Hsa descendants and seven of the eight Dre descendants of the original ancestral clusters have survived (the eighth cluster has survived in pufferfish). Assuming the methods that pair a gene with its closest relative are accurate, and that reciprocal closest sequence relatives are orthologs, the software will correctly match each human gene with the correct pair of zebrafish co-orthologs. The second row (1:4) shows a situation in which there are more zebrafish genes than surviving human genes, but again the software will make the correct choice because the zebrafish gene pair that has a human ortholog should be more similar to the surviving human gene than the zebrafish gene pair lacking a human ortholog. The third row is an example for which the software can identify only one ortholog for each of several human paralogs.

In some evolutionary scenarios, the software might mistakenly assign orthology relationships. The fourth row (labeled 1:2) illustrates a situation in which the software might erroneously

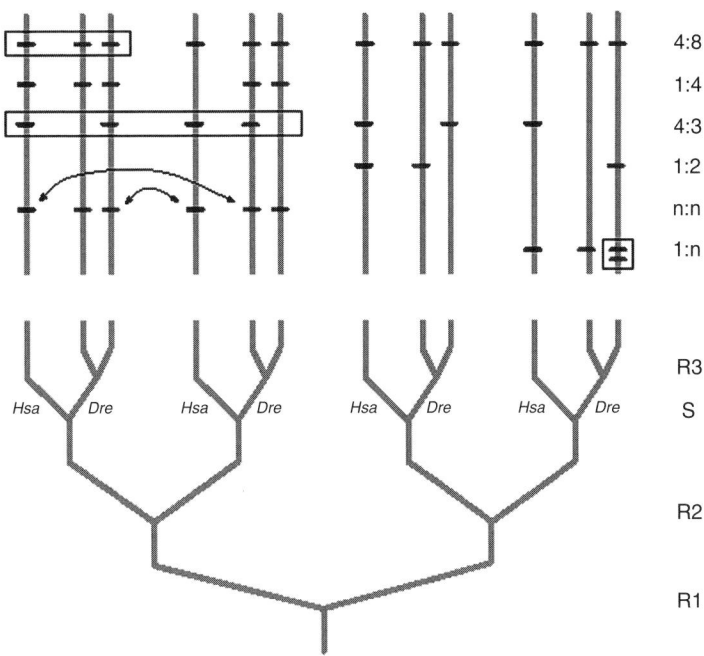

Fig. 17.3. Orthologs and paralogs. The box in line 4:8 indicates a human gene's zebrafish co-orthologs that arose in the R3 genome duplication event. The box in line 4:3 indicates two human paralogs arising from the R2 genome duplication event and their single-copy zebrafish orthologs. The box in line 1:n shows a pair of recent tandem duplicates of one of two zebrafish co-orthologs that arose in the R3 event.

identify two zebrafish genes as co-orthologs of a human gene: the two zebrafish genes are each other's "closest living relatives," but their last common ancestor preceded R3. The last two lines illustrate two additional ways in which the software could mistakenly assign orthology relations. In one case, the best result of our homology search might be to paralogs (e.g., during the period of relaxed selection immediately following R3 the zebrafish genes might have changed in a way that makes it difficult to distinguish them). In another case, a recent tandem duplication (here shown as a duplication in the zebrafish genome) can create an extra gene, and it is possible the software will erroneously identify the tandem duplicates as co-orthologs from a genome duplication, unless position in the genome is taken into account.

2.2. Reconstruction of Ancestral Chromosomes

Before we can reconstruct ancestral chromosomes, we must identify homoeologous chromosomes within a species, which requires identifying paralogs from the most recent genome duplication event. A major unsolved problem, however, is the assignment of zebrafish paralogs to their proper duplication event (R1, R2, or R3). The process uses three organisms: a *primary organism* (zebrafish in this case) and two outgroups. The *recent outgroup* is an organism that diverged from our primary organism after the most recent duplication event, and we will use the green-spotted pufferfish *Tetraodon nigroviridis*, whose genome sequence is nearly complete *(31)*. An organism that diverged from our primary organism prior to the most recent duplication can be used as an *ancient outgroup*; in this case we use the human genome because of its high quality of annotation.

Step 1 is to identify genes that are highly similar within our primary organism, defined as the reciprocal best hits (RBH) of a within-species BLASTn (**Fig. 17.4A1**). This preprocessing step allows us to identify pairs of genes within the organism that are most closely related (*see* **Fig. 17.4A**, P_1 and P_2), the majority of which should have been produced in the most recent genome duplication event R3, although some may be recent tandem duplicates or paralogs from the R1 or R2 events.

Next, we take the sister genes that are the output of Step 1 in our primary organism and perform a BLASTp analysis against the ancient, non-duplicated outgroup, the human genome (see **Fig. 17.4A**, arrows labeled 2a and 2b). If both sisters P_1 and P_2 in the primary organism match a single ancient outgroup gene (as for gene O_1 in **Fig. 17.4A**, Arrows labeled 2a), and gene O_1 has as two best hits genes P_1 and P_2 (*see* **Fig. 17.4A**, Arrows labeled 2b), then P_1 and P_2 are co-orthologs of outgroup gene O_1 and were produced after the divergence of the primary species and the ancient outgroup. This co-orthologous pair could have originated either in R3 (the boxed comparison of **Fig. 17.3**, line 4:8) or in a recent tandem duplication event (as the boxed pair in **Fig. 17.3**, line 1:n). These possibilities are distinguished by location: The R3 event will produce duplicates on different chromosomes, whereas tandem duplication produces duplicates that are adjacent, barring subsequent specific types of chromosome translocations.

If the two sister genes (P_1 and P_2) in the primary organism have as RBH different ancient outgroup genes (O_1 and O_2 in **Fig. 17.4B**), then the pipeline has identified a pair of ancient paralogs produced in the first or second round of genome duplication (as for the boxed genes in **Fig. 17.3** line 4:3). Finally, if only one member of the pair has an RBH with the outgroup, then the origins of this gene pair are ambiguous (*see* **Fig. 17.4C**).

Having completed the identification of co-orthologs in the primary species, we next identify orthologs of the primary species' genes in the recent outgroup, in this case the pufferfish, by RBH. This final identification concludes data collection, and we then proceed in the following way.

We compare the gene content of chromosomes in the primary species to the genome of the recent outgroup to infer the content of the ancestral post-duplication teleost chromosomes. This comparison reduces two pairs of modern chromosomes to a single, ancestral post-duplication pair.

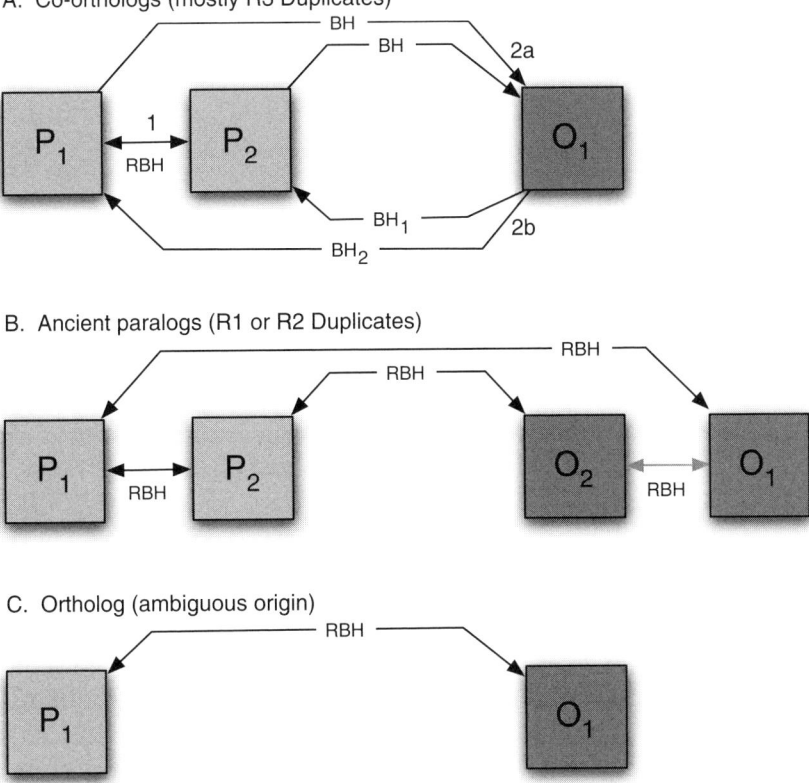

Fig. 17.4. Operational definitions of ancestral relationships in our pipeline.

We next infer the content of the ancestral pre-duplication chromosome of a ray-fin (Actinopterygian) fish, which existed about 300 million years ago, by collapsing the post-duplication pair of chromosomes.

Finally, we compare the pre-duplication ray-fin fish chromosome to our ancient outgroup, the lobe-fin (Sarcopterygian) fish called *Homo sapiens*. This final comparison allows us to infer the content of the ancestral bony fish (Osteichthyes) chromosome that existed about 450 million years ago.

2.3. The Analysis Pipeline

As described, to infer the content of ancestral chromosome sequences, we must conduct two major analyses, the identification of paralogs within a species and the identification of orthologs between species. A "software pipeline" conducts each analysis, manages the data, and runs the bioinformatic applications that process the data. The project described in this chapter uses a system known as PIP (pipeline interface program) *(32)*, a generic framework that allows us to create many different "pipelines" by combining arbitrary analysis Stages in different orders (*see* **Note 1**) (for a more in-depth look at computational pipelines, *see* **Chapter 24 of Volume 2**). The initial data set (gene sequences and annotations plus parameters for the various genomic applications) are stored in a relational database. PIP then runs each application in turn, passing results from one analysis Stage as inputs to the next analysis Stage (**Figs. 17.5** and **17.6**), archiving the results of each Stage in the database. Each of the two major analysis steps, identifying paralogs and identifying orthologs, is embodied in its own pipeline, the Paralog Identification pipeline (PARA) and the Ortholog Identification pipeline (ORTH) (*see* **Note 2**).

Prior to executing either pipeline, we acquire full genomic data for each of our organisms from an authoritative source, in this case Ensembl (http://www.ensembl.org/) and place it into a locally accessible database. Each gene has associated attributes, including its unique identifier, cDNA nucleotide sequence, and amino acid sequence of the protein. We prepare several searchable BLAST databases, one for each genome we will be examining. In both pipelines, BLAST searches were performed with the WU-BLAST program (http://blast.wustl.edu).

The purpose of the PARA pipeline is to identify paralogs in the primary species. In the first stage of the PARA pipeline (Forward BLAST, *see* **Fig. 17.5**), a nucleotide BLAST search is performed for each gene on our chromosome of interest (the query chromosome). For each query sequence, the PIP saves the five best matches from our subject genome. The second stage (Reverse BLAST) reverses the search: The new query sequences

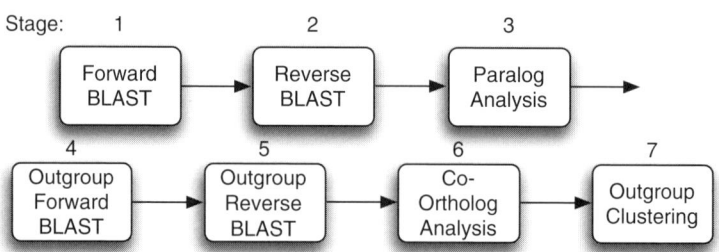

Fig. 17.5. Paralog Identification Pipeline (PARA). The PIP manages the input and output data of each Stage.

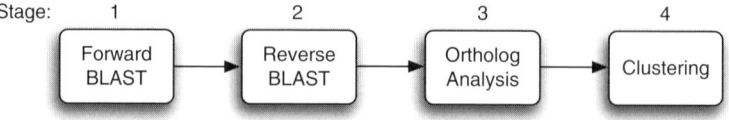

Fig. 17.6. The Ortholog Identification Pipeline (ORTH) consists of four major analytical stages.

now become the set of five best hits generated in the first stage and the original query genome now becomes the subject genome.

In Stage 3 (Paralog Analysis), the pipeline considers results from the intra-organism BLAST and identifies pairs of genes that may be recent duplicate genes, or sister genes, based on reciprocal best hit. For each gene X, the Paralog Analysis Stage considers the five top hits for the gene produced in the forward search. For each hit gene $Y_1 \ldots Y_5$, the stage checks to see if X is the top hit in the reverse search. If so, the searching stops and X and Y are considered paralogs. In the strictest sense, a reciprocal best hit algorithm should only accept the best hit in the forward direction and in the reverse direction, i.e., Y should be X's top hit and X should be Y's top hit. When dealing with large gene families and duplicated genomes, however, it is common to have a number of hits that all score highly and are almost indistinguishable from one another. Examining the top several hits allows us some flexibility. These initial gene pairs correspond to the predicted recent paralog genes, P_1 and P_2, in **Fig. 17.4**. Additionally, the Paralog Analysis Stage conducts a basic classification to identify gene pairs that fall on the same chromosome versus those that fall on different chromosomes to find potential tandem duplicate gene pairs.

An understanding of the relationship between the two members of the identified gene pair comes from Stages 4 and 5 (*see* **Fig. 17.5**), a reciprocal best hit analysis against an outgroup organism, represented by O_1 and O_2 in **Fig. 17.4**. In cases in which no gene pair appeared in Stages 1–3, Stages 4 and 5 also compare single genes against the outgroup to identify genes for which the paralog was lost at some point in the past. In Stage 6, the PARA pipeline examines the output from both reciprocal best hit analyses, the within-species and outgroup RBHs, and decides which duplication event, (R1/R2, or R3) produced the gene pairs. These relationships are specified in **Fig. 17.4**. Finally, the query gene pairs are clustered with the outgroup genes to find common chromosomal segments between them. The clustering process is described in more detail in the following.

The purpose of the ORTH pipeline is to understand the historical relationships of paralogs identified in the PARA pipeline by identifying orthologs in an appropriate outgroup. The ORTH pipeline starts off in a manner similar to the PARA pipeline, with forward and reverse BLAST search stages. In contrast to the PARA pipeline, however, the ORTH pipeline uses a protein BLAST to perform reciprocal searches between organisms (*see* **Fig. 17.6**, Stage 1 and 2). A BLAST of protein sequence rather than nucleotide sequence returns similarities in more distantly related organisms due to the degeneracy of the genetic code.

The third step (Ortholog Analysis) identifies pairs of genes between the query and subject genomes that are thought to be orthologs because they are reciprocal best hits. This identification is done in a manner similar to the Paralog Analysis and Co-Ortholog Analysis Stages in the PARA pipeline.

In the final stage (Clustering), we use a clustering technique that allows us to draw conclusions about the levels of conserved syntenic layout and conserved gene order and construct a set of comparative maps between two different species. To cluster the predicted orthologs, we construct a gene homology matrix *(33)*. To create the matrix, we plot the position of each gene on the query chromosome along the X axis and the position of the orthologous gene on the predicted chromosome along the Y axis (**Fig. 17.7a**). We traverse the length of the query chromosome, looking for sets of genes that occur near each other on the subject chromosome. We employ a sliding window to allow gaps of user-defined size between the pairs of ortholog genes. Each time we identify an ortholog pair, we slide the window up to that pair and begin searching for the next pair. If the next pair appears before we reach the end of the window, we continue the cluster and look for the next member. Otherwise, we record the current cluster and begin searching for the next one. Clusters appear as diagonal lines on the plots. An uninterrupted slope indicates conserved gene order, and the reversal of slope from positive to negative indicates an inversion.

Execution of the two analysis pipelines identified pufferfish orthologs of zebrafish genes, zebrafish orthologs and co-orthologs

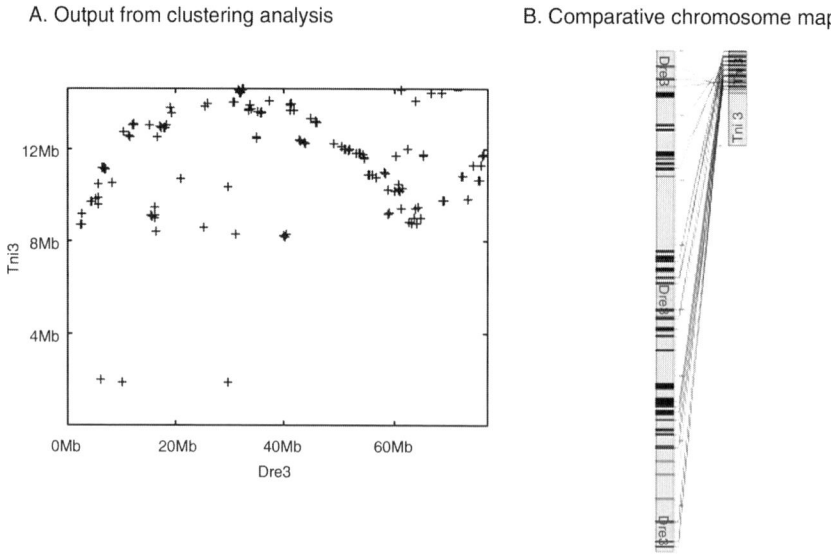

Fig. 17.7. Conservation of gene orders for zebrafish chromosome DreLG3 and the pufferfish chromosome Tni3. (**A**) Chromosome orthology matrix, each point represents a gene pair that is orthologous between zebrafish, plotted by location on the DreLG3 along the horizontal axis, and pufferfish chromosome Tni3, position plotted along the vertical axis. (**B**) Identified clusters on the orthologous chromosomes, DreLG3 and Tni3 respectively. Chromosomal segments that are inverted with respect to their local environments appear in light gray on Tni3.

of human genes, pufferfish orthologs and co-orthologs of human genes, gene pairs within each species derived from tandem duplications, and gene pairs derived from genome duplications. In addition, the analysis assigned each gene pair to a specific genome duplication event and constructed comparative chromosome maps. We used this analysis pipeline to reconstruct partially the ancestral chromosome history of one zebrafish chromosome.

3. Application

The initial focus was to reconstruct the ancestral chromosome and gene orders for the *Danio rerio* Linkage Group 3 (DreLG3), one of the 25 zebrafish chromosomes. The pipeline started with data from the zebrafish genome sequence version Zv6 available from the Sanger Institute Zebrafish Genome Project (http://www.ensembl.org/Danio_rerio/). The pipeline first identified paralogous genes within the *Danio rerio* genome to infer chromosome segments that constitute the most likely paralogon (chromosome segments resulting from duplication) *(34)* produced in the R3 duplication event. To confirm the predicted paralogs, the pipeline compared each duplicate against a pre-duplication outgroup *(Homo sapiens)* to ensure that the pair of genes originated with the R3 duplication. We operationally define two genes as paralogs in zebrafish if they both hit the same gene in the human genome during a reciprocal best hit analysis (*see* **Fig. 17.4A**). *Homo sapiens* genomic data was taken from NCBI build 36, version 1, available from Ensembl (http://www.ensembl.org/Homo_sapiens/). This analysis yielded *Danio rerio* Linkage Group 12 (DreLG12) as the most likely DreLG3 paralogon. **Figure 17.8** Step 1a shows that genes distributed along the full length of DreLG3 have duplicates distributed along the full length of DreLG12, but that the order of paralogs is quite different in the two homoeologous chromosomes, as evidenced by the crossing of lines that join paralogs. These types of differences in gene order would occur if many chromosome inversions occurred on both homoeologous chromosomes since the R3 genome duplication event.

Next, we took genes from DreLG3 and DreLG12 and searched for orthologous genes in pufferfish, using version 7 also from Ensembl (http://www.ensembl.org/Tetraodon_nigroviridis/). Because the identification of orthologs does not depend on which duplication event produced them, there is no need to compare the orthologous predictions against an outgroup. The rule followed in the current pipeline is simply that each gene is reciprocally the other's best BLAST hit. This analysis yielded Tni2 as

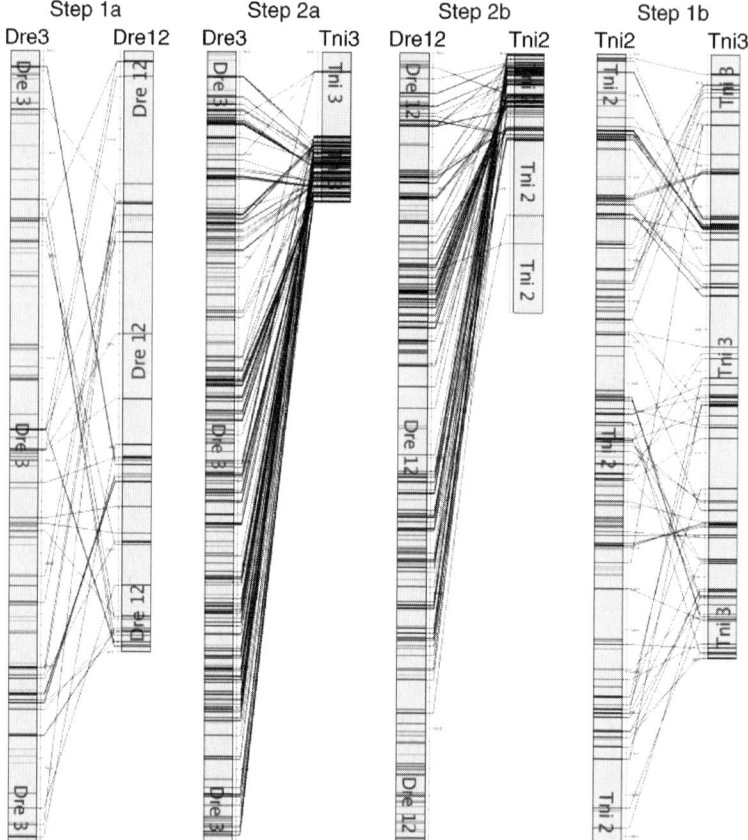

Fig. 17.8. Transitive search for homoeologous and orthologous chromosome segments. Homoeologous chromosomes are identified within either zebrafish or pufferfish (Steps 1a and 1b), and orthologous chromosomes between zebrafish and pufferfish (Steps 2a and 2b).

most closely related to DreLG3 and Tni3 as most closely related to DreLG12 (*see* **Fig. 17.8**, Steps 2a and 2b). The principle of transitive homology *(33)* demands that the chromosome homoeologous to Tni2 would be Tni3, and our data verified this prediction (*see* **Fig. 17.8**, Step 1b).

The distribution of orthologs revealed several features with implications regarding the mechanisms of chromosome evolution. First, zebrafish chromosomes appear to be stuffed into short regions on pufferfish chromosomes (*see* **Fig. 17.8**, Step 3a). This fits with the dramatic diminution of pufferfish genomes, a derived feature achieved by decreasing the length of introns and intergenic regions *(35)*.

The second result apparent from the analysis is that gene order on DreLG3 matches gene order on Tni3 far better than gene order on DreLG3 matches gene order on DreLG12. This result would be predicted by the hypothesis that fewer inversions occurred since the speciation event that produced the diverging zebrafish and pufferfish lineages (producing DreLG3 and Tni3) than occurred since the genome duplication event that produced DreLG3 and DreLG12. If one assumes that the rate of the fixation of inversions in populations is roughly constant over time and between lineages, then these results suggest that the R3 genome duplication event was substantially earlier than the zebrafish/pufferfish speciation event.

Third, the analysis shows that nearly all pufferfish orthologs of DreLG3 occupy only the lower portion of Tni3, and nearly all pufferfish orthologs of DreLG12 reside only in the upper part of Tni2 (*see* **Fig. 17.8**, Steps 3a and 3b). Two possible hypotheses can explain these distributions. According to the *pufferfish fusion hypothesis*, the last common ancestor of zebrafish and pufferfish had a chromosome like DreLG3 (or DreLG12), and in the pufferfish lineage, this chromosome became the lower part of Tni3 (or the upper part of Tni2), which joined by a translocation event an unrelated chromosome that became the top portion of Tni3 (or lower part of Tni2) (**Fig. 17.9A**). The alternative hypothesis, the *zebrafish fission hypothesis*, is that the last common ancestor of zebrafish and pufferfish had a chromosome like Tni3 (or Tni2), and that in the zebrafish lineage, this chromosome broke roughly in half, yielding DreLG3 from the lower half of Tni3, and DreLG12 from the upper half of Tni2 (*see* **Fig. 17.9B**).

The pufferfish fusion hypothesis and the zebrafish fission hypothesis make different predictions for the nature of the pufferfish chromosomes that are not related to DreLG3 and DreLG12. According to the pufferfish fusion hypothesis (*see* **Fig. 17.9A**),

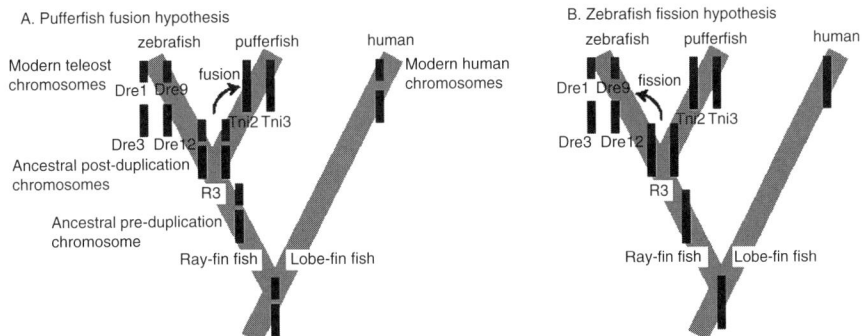

Fig. 17.9. Two hypotheses for the reconstruction of zebrafish and pufferfish chromosomes. Portions of DreLG3/12 and Tni2/3 chromosomes (in black) originated in a common, pre-duplication ancestor. (**A**) The pufferfish fusion hypothesis states that the remaining portions of Tni2 and Tni3 (in gray) are unrelated to one another because the fusion events that created them were independent and post-duplication. (**B**) Alternatively, the zebrafish fission hypothesis states that the remaining portions of Tni2 and Tni3 would be orthologous to the same portion of the human genome as they would have been part of the same ancestral pre-duplication chromosome. In this case the pipeline suggests the fission hypothesis to be more parsimonious.

the non-DreLG3/12 portion of pufferfish chromosomes Tni2 and Tni3 (gray) would most likely be unrelated to each other because the fusion events that created Tni2 and Tni3 would have occurred independently of each other. Under the zebrafish fission hypothesis, however (*see* **Fig. 17.9B**), the non-DreLG3/12 portions of pufferfish chromosomes Tni2 and Tni3 (gray) would be orthologous to the same portion of the human genome because they would have been part of the same ancestral pre-duplication chromosome.

The pipeline shows that the non-DreLG3 portion of Tni3 (corresponding to DreLG1) (*see* **Fig. 17.10A**), and the non-DreLG12 portion of Tni2 (orthologous to DreLG9) (*see* **Fig. 17.10B**) are both orthologous to the long arm of human chromosome two (Hsa2q, *see* **Fig. 17.10C,D**). This type of relationship would be expected according to the zebrafish fission hypothesis but not according to the pufferfish fusion hypothesis. Therefore, we conclude that the ancestral preduplication chromosome that was the ancestor to DreLG3 consisted of a chromosome that was substantially similar to the sum of the genetic content of pufferfish chromosomes Tni2 and Tni3. This result is somewhat surprising because *T. nigroviridis* has 21 chromosomes, whereas zebrafish and most other teleosts have 25 *(27)*, which is expected if chromosome fusion occurred more frequently in the pufferfish lineage than in most teleosts. Thus, although the zebrafish fission

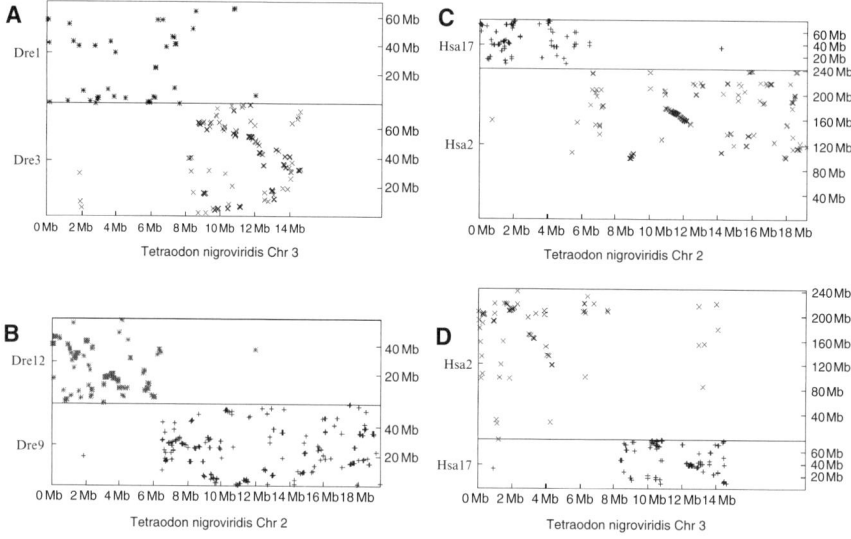

Fig. 17.10. Ancestral chromosome reconstruction. The portions of pufferfish chromosomes Tni2 and Tni3 that do not correspond to zebrafish chromosomes are orthologous to DreLG1 and DreLG9, respectively (**A** and **B**), but in both cases are orthologous to much of human chromosome arm Hsa2q (**C** and **D**). This suggests that the ancestral chromosome state was the sum of the two pufferfish chromosomes.

hypothesis works best for this case, the answer is likely to be quite different for other chromosomes.

The best way to finalize the inference of the ancestral chromosome is to analyze the situation in closely related outgroups, including a post-R3 teleost outgroup and a pre-R3 non-teleost ray-fin outgroup. Although appropriate outgroup lineages exist, including for post-R3 the Anguilliformes (eels) and the Osteoglossiformes (butterfly fish and bonytongues), and the pre-R3 outgroups Amiiformes (bowfin) and Semionotiformes (gars) *(36, 37)*. Unfortunately none have available genomic resources necessary to resolve the issue.

Finally, the analysis reveals two special regions of pufferfish chromosome Tni2 that have extensive regions of conserved gene order, one at about 9 Mb and one at about 11 Mb. The corresponding regions in human occupy about 9 Mb and about 30 Mb of Hsa2q, remarkably long conserved regions (at least 14 and 54 genes, respectively) preserved for a remarkably long time. Future challenges will be to understand the mechanisms for this preservation and identify other similar regions on other chromosomes.

4. Conclusions

The bioinformatic pipeline described here provides an initial step to reconstruct chromosome content and gene orders in ancient vertebrate chromosomes. Several additional features should be added to complete the analysis. First, the reciprocal best hit approach has several deficiencies, some of which can be overcome by the implementation of phylogenetic analysis to define orthologs and sister paralogs. Second, ancestral chromosomes reconstructed by the comparison of zebrafish and pufferfish chromosomes would be substantially more robust if data from newly sequenced teleost genomes is taken into account, namely medaka and stickleback. Third, genome sequence from an extant ray-fin fish whose lineage diverged from teleosts before the R3 event is essential as an outgroup to reconstruct the genomic constitution of an ancient ray-fin fish. The application of the completed pipeline will help us to describe what happened in vertebrate chromosome evolution and to identify chromosome regions whose gene order has been conserved much longer than is normal. These identified regions will allow us to develop hypotheses to explain what maintains gene order and hopefully to perform experiments to understand the evolutionary mechanisms that preserve the gene order of a chromosome segment over time.

5. Notes

1. PIP is agnostic when it comes to the origin of an analysis stage. It can be written in any language and simply needs to be an executable program that reads tab-separated records as input, and writes tab-separated records as output. In order to infer ancestral chromosomes, we have written PIP stages that are simple Perl programs, on the order of tens of lines, and we have written stages that are C++/Perl hybrids that run in parallel on a compute cluster.

2. When designing new Stages, be careful not to duplicate existing work; libraries exist to handle common tasks such as running BLAST or CLUSTALW as well as for parsing data files such as FASTA. For Perl, consider using the BioPerl modules (http://bioperl.org); for Python, consider BioPython (http://biopython.org); and for Ruby, consider BioRuby (http://bioruby.org).

Acknowledgments

This project was supported by grant no. 5R01RR020833-02 from the National Center for Research Resources (NCRR), a component of the National Institutes of Health (NIH). Its contents are solely the responsibility of the authors and do not necessarily represent the official views of NCRR or NIH. J. Catchen was supported in part by an IGERT grant from NSF in Evolution, Development, and Genomics (DGE 9972830).

References

1. Allende, M. L., Manzanares, M., Tena, J. J., et al. (2006) Cracking the genome's second code: enhancer detection by combined phylogenetic footprinting and transgenic fish and frog embryos. *Methods* 39, 212–219.
2. Tran, T., Havlak, P., Miller, J. (2006) MicroRNA enrichment among short 'ultraconserved' sequences in insects. *Nucleic Acids Res* 34, e65.
3. Sauer, T., Shelest, E., Wingender, E. (2006) Evaluating phylogenetic footprinting for human-rodent comparisons. *Bioinformatics* 22, 430–437.
4. Altschul, S. F., Madden, T. L., Schaffer, A. A., et al. (1997) Gapped BLAST and PSI-BLAST: a new generation of protein database search programs. *Nucleic Acids Res* 25, 3389–3402.
5. Wall, D. P., Fraser, H. B., Hirsh, A. E. (2003) Detecting putative orthologs. *Bioinformatics* 19, 1710–1711.
6. Ohno, S. (1970) *Evolution by Gene Duplication*. Springer-Verlag, New York.
7. Lundin, L. G. (1993) Evolution of the vertebrate genome as reflected in paralogous chromosomal regions in man and the house mouse. *Genomics* 16, 1–19.
8. Spring, J. (1997) Vertebrate evolution by interspecific hybridization—are we polyploid? *Fed Eur Biol Soc Lett* 400, 2–8.
9. Dehal, P., Boore, J. L. (2005) Two rounds of whole genome duplication in the ancestral vertebrate. *PLoS Biol* 3, e314.

10. Garcia-Fernàndez, J., Holland, P. W. (1994) Archetypal organization of the amphioxus Hox gene cluster. *Nature* 370, 563–66.
11. Ferrier, D. E., Minguillon, C., Holland, P. W., et al. (2000) The amphioxus Hox cluster: deuterostome posterior flexibility and Hox14. *Evol Dev* 2, 284–293.
12. Minguillon, C., Gardenyes, J., Serra, E., et al. (2005) No more than 14: the end of the amphioxus Hox cluster. *Int J Biol Sci* 1, 19–23.
13. Powers TP, A. C. (2004) Evidence for a Hox14 paralog group in vertebrates. *Curr Biol* 14, R183–184.
14. Koh, E. G., Lam, K., Christoffels, A., et al. (2003) Hox gene clusters in the Indonesian coelacanth, *Latimeria menadoensis*. *Proc Natl Acad Sci U S A* 100, 1084–1088.
15. Amores, A., Force, A., Yan, Y.-L., et al. (1998) Zebrafish *hox* clusters and vertebrate genome evolution. *Science* 282, 1711–1714.
16. Chiu, C. H., Amemiya, C., Dewar, K., et al. (2002) Molecular evolution of the HoxA cluster in the three major gnathostome lineages. *Proc Natl Acad Sci U S A* 99, 5492–5497.
17. Acampora, D., D'Esposito, M., Faiella, A., et al. (1989) The human HOX gene family. *Nucleic Acids Res* 17, 10385–10402.
18. Graham, A., Papalopulu, N., Krumlauf, R. (1989) The murine and *Drosophila* homeobox gene complexes have common features of organization and expression. *Cell* 57, 367–378.
19. Duboule, D. (1998) Vertebrate hox gene regulation: clustering and/or colinearity? *Curr Opin Genet Dev* 8, 514–518.
20. Postlethwait, J. H., Yan, Y.-L., Gates, M., et al. (1998) Vertebrate genome evolution and the zebrafish gene map. *Nat Genet* 18, 345–349.
21. Postlethwait, J. H., Amores, A., Yan, G., et al. (2002) Duplication of a portion of human chromosome 20q containing Topoisomerase (Top1) and snail genes provides evidence on genome expansion and the radiation of teleost fish., in (Shimizu, N., Aoki, T., Hirono, I., Takashima, F., eds.), *Aquatic Genomics: Steps Toward a Great Future*. Springer-Verlag, Tokyo.
22. Taylor, J., Braasch, I., Frickey, T., et al. (2003) Genome duplication, a trait shared by 22,000 species of ray-finned fish. *Genome Res.* 13, 382–390.
23. Van de Peer, Y., Taylor, J. S., Meyer, A. (2003) Are all fishes ancient polyploids? *J Struct Funct Genomics* 3, 65–73.
24. Hoegg, S., Brinkmann, H., Taylor, J. S., et al. (2004) Phylogenetic timing of the fish-specific genome duplication correlates with the diversification of teleost fish. *J Mol Evol* 59, 190–203.
25. Amores, A., Suzuki, T., Yan, Y. L., et al. (2004) Developmental roles of pufferfish Hox clusters and genome evolution in ray-fin fish. *Genome Res* 14, 1–10.
26. Hoegg, S., Meyer, A. (2005) Hox clusters as models for vertebrate genome evolution. *Trends Genet* 21, 421–424.
27. Naruse, K., Tanaka, M., Mita, K., et al. (2004) A medaka gene map: the trace of ancestral vertebrate proto-chromosomes revealed by comparative gene mapping. *Genome Res* 14, 820–828.
28. Aparicio, S., Hawker, K., Cottage, A., et al. (1997) Organization of the *Fugu rubripes* Hox clusters: evidence for continuing evolution of vertebrate Hox complexes. *Nat Genet* 16, 79–83.
29. Hedges, S. B. (2002) The origin and evolution of model organisms. *Nat Rev Genet* 3, 838–849.
30. Postlethwait, J. H., Woods, I. G., Ngo-Hazelett, P., et al. (2000) Zebrafish comparative genomics and the origins of vertebrate chromosomes. *Genome Res* 10, 1890–1902.
31. Jaillon, O., Aury, J. M., Brunet, F., et al. (2004) Genome duplication in the teleost fish *Tetraodon nigroviridis* reveals the early vertebrate proto-karyotype. *Nature* 431, 946–957.
32. Conery, J., Catchen, J., Lynch, M. (2005) Rule-based workflow management for bioinformatics. *VLDB J* 14, 318–329.
33. Van de Peer, Y. (2004) Computational approaches to unveiling ancient genome duplications. *Nat Rev Genet* 5, 752–763.
34. Coulier, F., Popovici, C., Villet, R., et al. (2000) MetaHox gene clusters. *J Exp Zool* 288, 345–351.
35. Elgar, G., Clark, M., Green, A., et al. (1997) How good a model is the *Fugu* genome? *Nature* 387, 140.
36. Inoue, J. G., Miya, M., Tsukamoto, K., et al. (2003) Basal actinopterygian relationships: a mitogenomic perspective on the phylogeny of the "ancient fish". *Mol Phylogenet Evol* 26, 110–120.
37. Miya, M., Takeshima, H., Endo, H., et al. (2003) Major patterns of higher teleostean phylogenies: a new perspective based on 100 complete mitochondrial DNA sequences. *Mol Phylogenet Evol* 26, 121–138.

Chapter 18

Genome Rearrangement by the Double Cut and Join Operation

Richard Friedberg, Aaron E. Darling, and Sophia Yancopoulos

Abstract

The *Double Cut and Join* is an operation acting locally at four chromosomal positions without regard to chromosomal context. This chapter discusses its application and the resulting menu of operations for genomes consisting of arbitrary numbers of circular chromosomes, as well as for a general mix of linear and circular chromosomes. In the general case the menu includes: inversion, translocation, transposition, formation and absorption of circular intermediates, conversion between linear and circular chromosomes, block interchange, fission, and fusion. This chapter discusses the well-known edge graph and its dual, the adjacency graph, recently introduced by Bergeron et al. Step-by-step procedures are given for constructing and manipulating these graphs. Simple algorithms are given in the adjacency graph for computing the minimal DCJ distance between two genomes and finding a minimal sorting; and use of an online tool (Mauve) to generate synteny blocks and apply DCJ is described.

Key Words: Genome rearrangements, gene order, Mauve, synteny, inversion, reversal, translocation, transposition, block interchange, fission, fusion.

1. Introduction

1.1. Background

Comparative analyses of genomes have identified regions of similarity or "conserved segments" *(1)* among species, which may be scrambled from one genome to another. Such regions range from a few nucleotides up to millions of base pairs, depending on the criteria. At the small end they cover only a fraction of a gene or a small stretch of an intergenic element, whereas at the large end they span vast genomic tracts containing a multitude of genes. This chapter does not focus on the identification of homologous regions or synteny blocks in this section (but *see* **Section 2.2.1**), and for simplicity, refers to these units as "genes." The reader is cautioned that this nomenclature connotes neither the size nor functional characterization of these genomic elements.

The problem of transforming one genome to another when the two genomes have the same gene content without paralogs can be reduced to that of sorting a signed permutation via a defined set of operations. The set of operations that are possible not only depends on, but can affect, the kind of genomes considered. For single chromosomes, a significant amount of effort has been given to the study of inversions (reversals) of segments of arbitrary length *(2)*. When multiple linear chromosomes are involved, these can be generalized to include translocations, which exchange end-segments between two chromosomes, as well as fissions and fusions. The study of evolutionary *"distance"* (minimum number of steps) between two genomes containing the same genetic material in different arrangements depends on the choice of elementary operations. It can be seen that with a menu of "generalized reversals" the distance cannot be less than $b - c$, where b = # *breakpoints*, and c = # *cycles* in the *breakpoint graph* (*see* **Section 1.2.3**). However, certain obstacles can prevent the lower bound from being achieved.

Recently, attention has been given to a single universal operation, the *double cut and join* (DCJ), which acts indiscriminately on gene ends without regard to the chromosomal disposition of the genes *(3, 4)*. Special cases of this operation are inversion, fission, fusion, translocation, and conversion between linear and circular chromosomes. When the basic operation is taken to be unrestricted DCJ, the lower bound, $b-c$, is always achieved and the finding of optimal paths is vastly simplified. One also obtains this minimal distance in the restricted case of linear chromosomes, with the caveat that once a circular intermediate is created (by one DCJ) it must be annihilated (by absorption or linearization) by the DCJ, which immediately follows. **Section 1.5** shows that such a dual operation is equivalent to a single operation given a weight of 2 also known as "block interchange" or generalized transposition *(3, 5, 6)*.

1.2. DCJ on Circular Chromosomes

1.2.1. Genome Graphs

A genome consisting of N genes is completely described by specifying the connections between adjacent gene ends. There are $2N$ gene ends and $(2N-1)!! = 1 * 3 * 5 * ... * (2N-1)$ ways to pair them. Each such pairing corresponds to a genome configuration. This allows chromosomes consisting of a single gene eating its tail. It is possible to represent the configuration of a genome by a graph (**Fig. 18.1**) in which the two ends of a single gene are connected by a white line, and the connection between two adjacent gene ends is represented by a black line. Such a graph may be called a genome graph; it must not be confused with the edge graph to be described later, which pertains to two genomes.

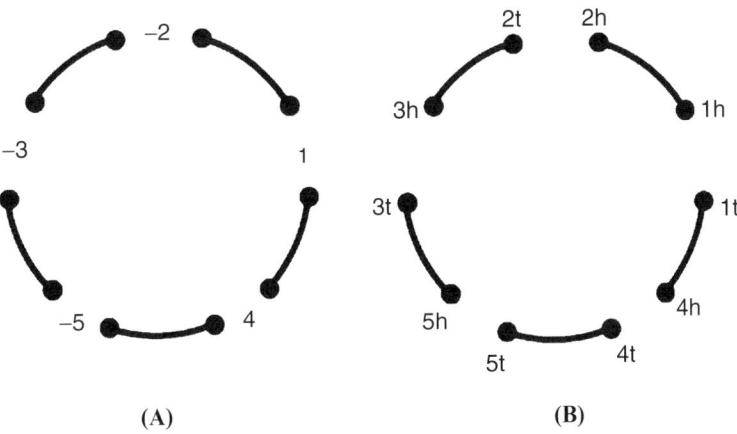

Fig. 18.1. Genome graph for a genome consisting of one circular chromosome. (**A**) Signed labeling of each gene: a positive gene points counterclockwise around the circle. (**B**) Labeling of gene ends: the gene points from its tail ("t") to its head ("h").

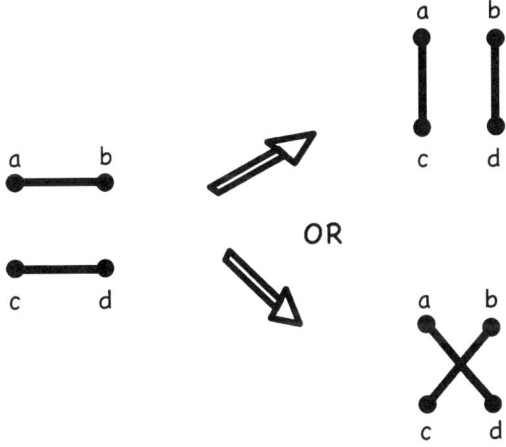

Fig. 18.2. Localized view of DCJ [1]. The letters a, b, c, d represent gene ends. The black lines represent attachments between genes. The rest of the genome is not shown.

We shall consistently speak of the path along a single gene as a "white line," although in the literature it has been variously represented as a solid line, a dashed line, or a space.

1.2.2. DCJ Operation

The DCJ operation can be defined entirely as a modification of the black lines without any reference to the white lines. One chooses two black lines and cuts them. This leaves four loose ends that can be rejoined in three different ways, one of which re-establishes the previous configuration. One completes the DCJ by choosing either of the remaining two ways of rejoining the ends (**Fig. 18.2**).

Sorting circular chromosomes by DCJ is equivalent to sorting by reversals, fusions, and fissions. If the two black lines to be cut are initially on different chromosomes, either of the two possible DCJs will accomplish a fusion. The two possible outcomes differ by relative reversal of the two fragments. If the two black lines to be cut are initially on the same chromosome, one of the possible DCJs will accomplish a reversal within the chromosome, and the other a fission into two chromosomes (**Fig. 18.3**).

The operations of block interchange *(3–6)* and of block interchange with reversal can each be accomplished by a fission followed by a fusion, and therefore can be reduced to two DCJs *(3)* (**Fig. 18.4**). Transposition and transversion are special cases of block interchange, in which the two blocks to be exchanged are contiguous *(7)*.

To distinguish these conventional operations from one another—for example, to distinguish reversal from fusion—requires knowledge of the placement of the gene ends on the chromosome, which in turn requires knowledge of which gene ends lie at opposite ends of the same gene. In other words, it requires access to the white lines. By studying DCJ without making these distinctions, one addresses a simpler mathematical problem in which the white lines play no part. Not only does this lead

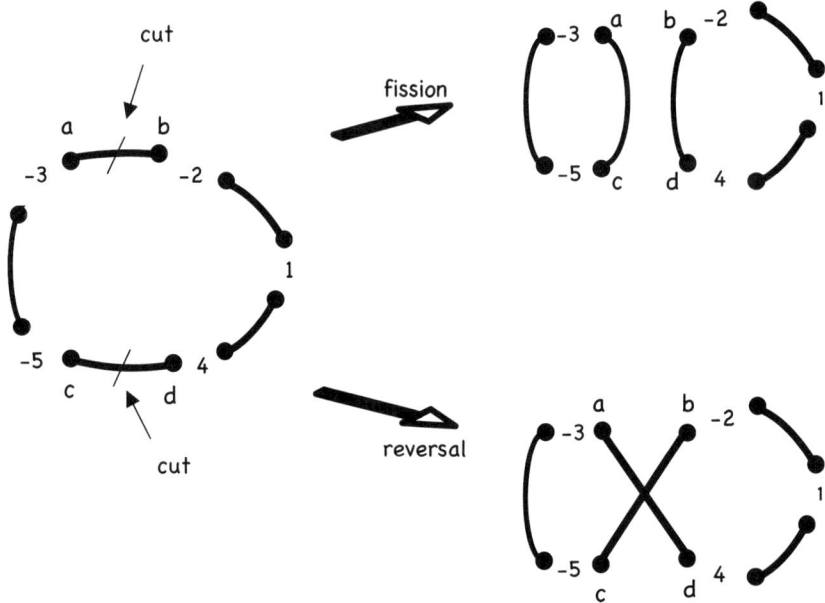

Fig. 18.3. DCJ on the genome shown in Fig. 18.1. Signed labeling of genes. The black lines cut are between 3h and 2t, above; and between 5t and 4t, below. The four gene ends 3h, 2t, 5t, 4t are labeled, respectively, a, b, c, and d to facilitate comparison with Fig. 18.2. In the upper outcome, the chromosome is split into two. In the lower outcome, the segment −3, −5 is reversed as can be seen by untwisting the chromosome so that the order of genes becomes 4, 1, −2, 5, 3. The reader may wish to follow the steps by relabeling the figure as in Fig. 18.1B.

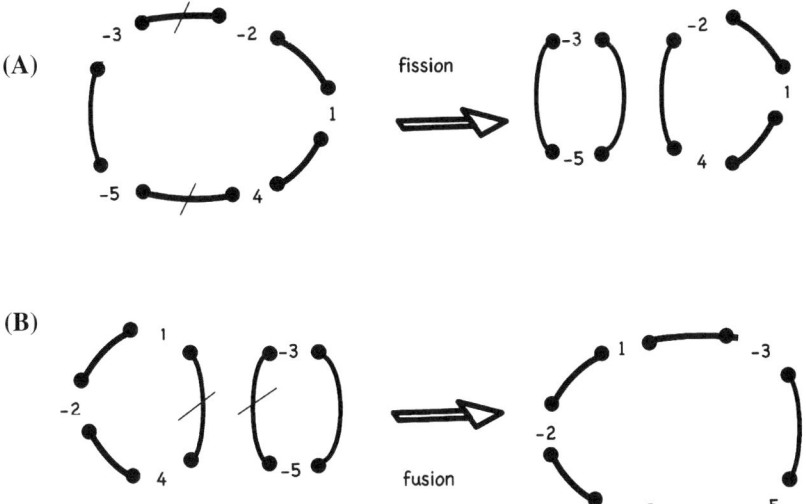

Fig. 18.4. Block interchange achieved by two successive DCJs. (**A**) Fission of one chromosome into two, taken from the upper outcome of Fig. 18.3. (**B**) The two chromosomes resulting from (**A**) are cut in a different way and fused into one circular chromosome. The result differs from the starting configuration of (**A**) by interchanging −3 and 4.

to a relatively simple distance formula, but also it may open the way to attack more complex rearrangement problems that would remain totally intractable in terms of the conventional operations that depend on the white lines.

1.2.3. Edge Graph

The problem of sorting one genome (the initial genome) into another (the target genome) by DCJ operations is simply the problem of progressing by such operations from the initial pairing of $2N$ gene ends to the target pairing. At any stage in this sequence of operations the intermediate configuration reached will be called the "current genome." It is conventional to represent the connections between adjacent gene ends by black lines in the initial as well as in the current pairing, and by gray lines in the target pairing.

The comparison between initial or current genome and target genome can be visualized by means of a graph (**Fig. 18.5C**) consisting of the $2N$ points joined by black and gray lines. The points represent gene ends; each point refers to that gene end in both genomes. The black lines tell how to connect the genes in the current genome, and the gray in the target genome. If nothing else is added, the graph gives no information about which gene ends belong to the same gene or which points belong to the same chromosome. It is often convenient, however, to supply this information by making the "white lines" (representing genes) visible or labeling the gene ends as in the genome graph. This graph, first introduced in *(8)*, has been variously called edge graph, breakpoint graph, and comparison graph in the literature (but *see* **Section 1.2.5**).

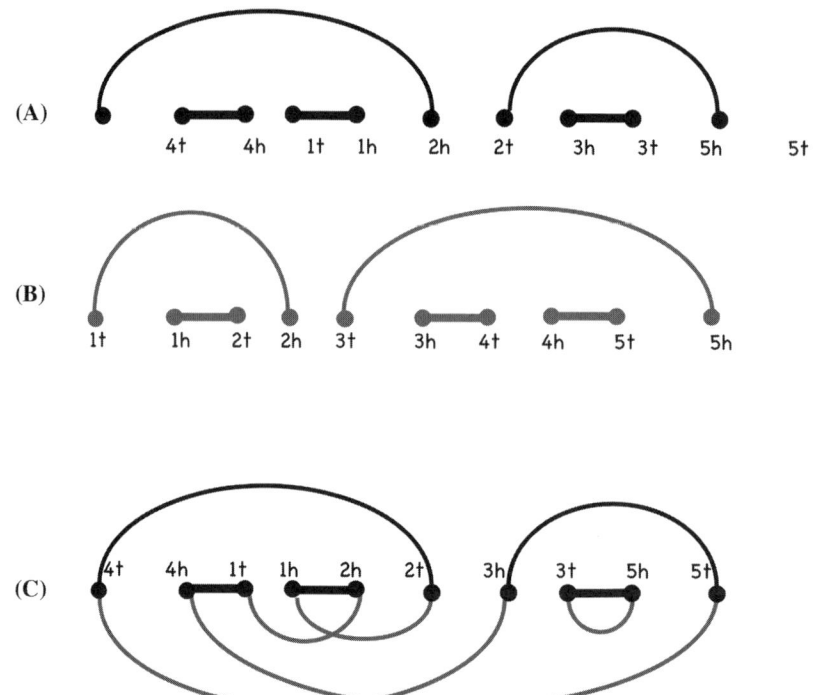

Fig. 18.5. Making an edge graph (only circular chromosomes). (**A**) Genome graph of the current genome, with gene ends labeled. This genome is identical with the upper outcome of Fig. 18.3, namely (4, 1, −2); (−3, −5), but displayed in such a way that all the black lines are horizontal except one in each chromosome. **B.** Genome graph of the target genome (1, 2); (3, 4, 5), similarly displayed. (**C**) The edge graph formed from (**A**) and (**B**). All of (**A**) is retained, and the labeled points are connected below by gray arcs in accordance with the connections shown in (**B**). Note that the large black-gray cycle visits both chromosomes of each genome.

1.2.4. Adjacency Graph

It has been suggested recently *(9)* that there are advantages to replacing the edge graph by its dual. To construct the dual graph, replace every line of the edge graph by a point and every point by a line. The resulting graph is called the adjacency graph *(4)*. Perhaps the most striking advantage of the adjacency graph is the ease with which it can be constructed, given the two genome graphs (initial and target) to be compared.

Begin with the initial (or current) genome graph, drawn with only white lines. That is, the black lines have been contracted to points. We label each point by the gene ends that meet there. Thus, if the head of gene 7 is adjacent to the tail of gene 3 one would label the point of adjacency 7h3t (**Fig. 18.6A**). In like fashion we draw the genome graph of the target genome (*see* **Fig. 18.6B**). This is placed underneath the first genome graph so that both appear in the same picture. Thus each gene is represented twice, once above and once below. To complete the adjacency graph, we simply join the endpoints of each gene in the current genome at the corresponding endpoints in the target genome below (*see* **Fig. 18.6C**). For ease of reference we shall refer to the joining lines as green

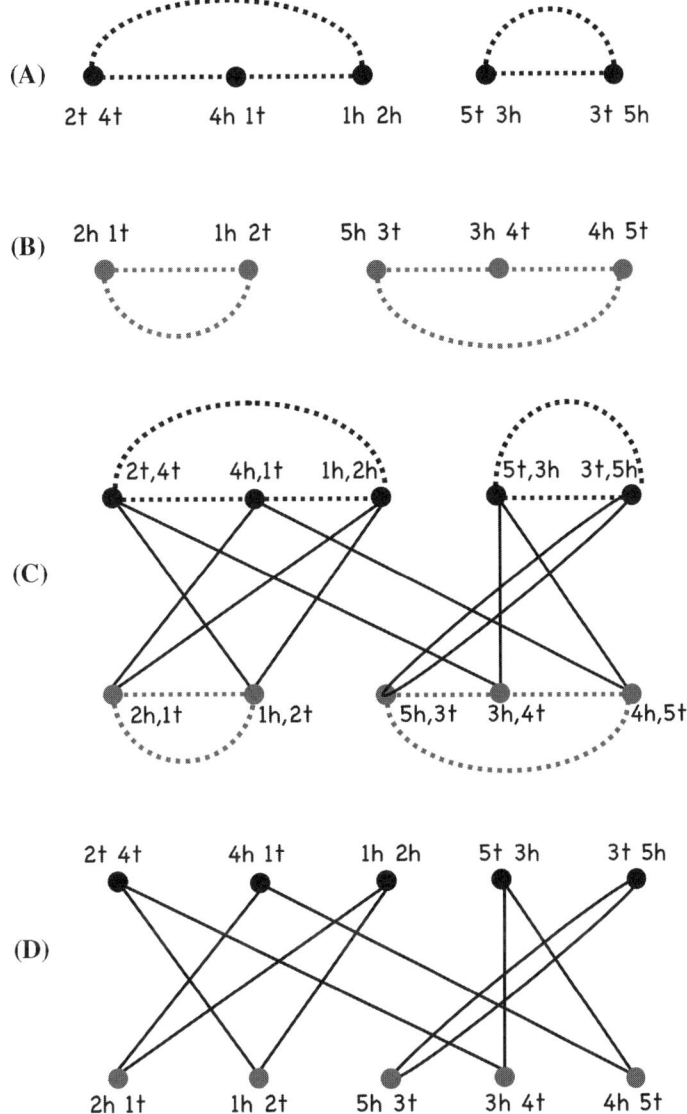

Fig. 18.6. Construction of an adjacency graph (only circular chromosomes). The current genome is (4, 1, −2); (−3, −5). The target genome is (1, 2); (3, 4, 5). (**A**) Current genome graph with black lines shrunk to points, white lines shown as dotted, and gene end labeling. (**B**) Target genome graph below (**A**). (**C**) Same as (**A, B**) with corresponding gene ends joined by green lines. (**D**) Same as (**C**) with white lines deleted.

lines. Thus, each adjacency is the terminus of two green lines. The adjacency graph proper is obtained by dropping the white lines and retaining only the green (*see* **Fig. 6D**).

The power of the adjacency graph is seen when it is recognized as the perfect dual (lines and points interchanged) of the edge graph described in the preceding section. Therefore, anything that can be found from one can be found from the other.

However, the adjacency graph has distinct advantages as visual display. First, it is considerably easier to construct by hand than the edge graph; the reader is encouraged to try both constructions starting from the same genome pair (*see* **Sections 3.3** and **3.4**), and the correctness of the construction is much easier to verify for the adjacency graph. Second, in the edge graph the two genomes are represented tangled together and at least one of them is necessarily difficult to visualize from an examination of the graph. In the adjacency graph the two genomes are visually distinct and each one can be represented in a way that closely resembles the set of chromosomes.

The duality between the adjacency and edge graphs can be visualized with the aid of a more complex graph, called the master graph (**Fig. 18.7**). The master graph contains all the black and gray lines of the edge graph as well as the green lines of the adjacency graph. Starting from the master graph, one obtains the edge graph by contracting the green lines to single points,

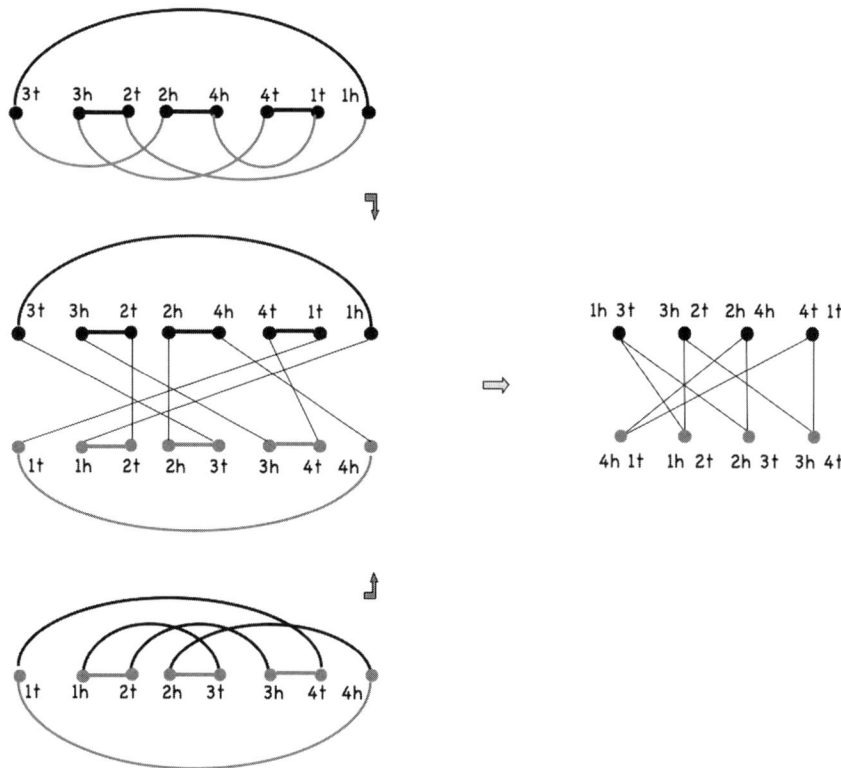

Fig. 18.7. Duality via the master graph. The current genome is (3, 2, −4, 1). The target genome is (1, 2, 3, 4). At left is shown the usual edge graph, above; the same edge graph with points reordered to render the gray lines horizontal rather than the black lines, below; and the master graph, middle. At right is shown the adjacency graph. "Green lines" in the master and adjacency graph are those that connect the upper half with the lower. To go from the master to either version of the edge graph, contract the green lines. To go from the master to the adjacency graph, contract the black and gray lines.

1.2.5. Distance

To solve the problem of sorting circular chromosomes by DCJ, we observe that in the edge graph each point has two connections and therefore the graph resolves itself into closed cycles consisting of alternating black and gray lines. (When linear chromosomes are allowed the situation is more complicated (*see* **Section 1.3**)). If every DCJ is begun by cutting two black lines in the same cycle, the ends can be rejoined in a way that causes the cycle to be split in two (*see* **Fig. 18.8**). At the end of the sorting, the current genome will be identical to the target genome, and the edge graph will consist of N cycles, each composed of one black and one gray line (**Fig. 18.9**), called 1-cycles. Since each DCJ increased the number of cycles by 1, the number of DCJ steps performed was $N-C$, where C is the number of cycles in the beginning edge graph.

From the preceding argument it is also clear that the number of cycles cannot be increased by more than one at a step, so that no sorting is possible in fewer than $N-C$ steps *(10)*. One thus arrives at the distance formula: $d = N-C$.

A frequent convention is to eliminate each 1-cycle as soon as it is formed by deleting the black line and the gray line of which

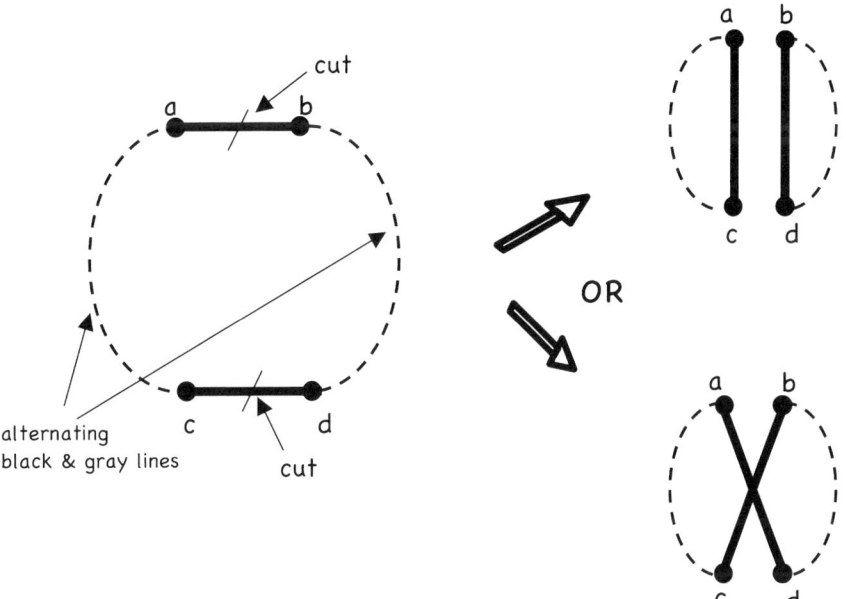

Fig. 18.8. DCJ acting on a cycle [1]. The cuts are shown as in Fig. 18.2, but the dashed arcs here represent the rest of the cycle, composed of black and gray lines, not the rest of the chromosome as is depicted in Fig. 18.3. This figure gives no indication of chromosomal structure. The upper outcome represents fission of a cycle, and the lower outcome reversal of part of a cycle, but there is no way to tell whether either outcome represents fission of a chromosome, fusion of two chromosomes, or reversal within a chromosome.

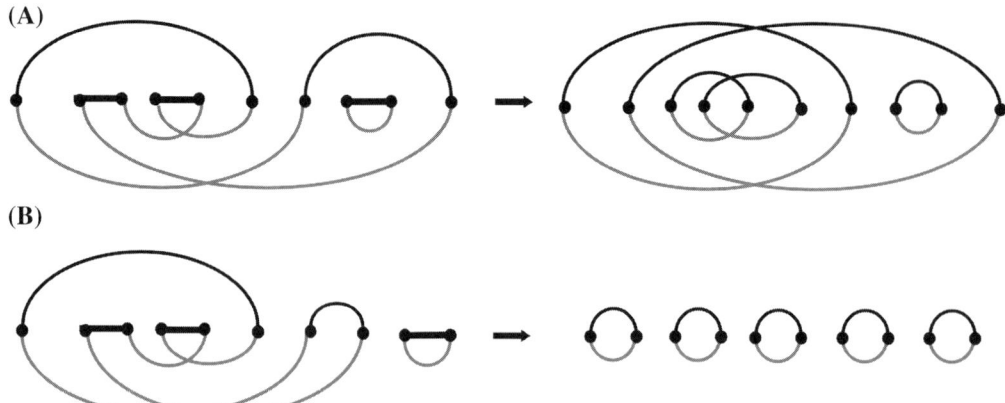

Fig. 18.9. (**A**) Sorting of an edge graph. At left, the edge graph of Fig. 18.5C with labels deleted. At right, the edge graph obtained by modifying the current genome until it matches the target genome. The gray arcs are the same as at left, but the black lines have been redrawn so as to mirror the gray arcs. All black lines are now shown as arcs above the points. (**B**) Cycle decomposition. The points of each graph in (**A**) have been shuffled so as to exhibit the cycles separately. There are two cycles in the graph at left, and five in the sorted graph at right; therefore the sorting requires a minimum of 5 − 2 = 3 DCJ steps.

it is comprised (**Fig. 18.10**). The lines of the edge graph with all 1-cycles deleted are called breakpoints. The number of either black or gray breakpoints is called b. We denote the number of cycles excluding the number of 1-cycles by c, as is usually done in this convention. Since the number of 1-cycles can be written either as $N-b$ or $C-c$, the distance formula can be written as $d = N - C = b - c$. The proof presented in the preceding for the distance can be presented equally well with $b-c$.

This chapter generally allows 1-cycles to be retained in the edge graph and writes the distance as $N-C$. Strictly, the term "breakpoint graph" is not applicable to our edge graph, since in most literature the term "breakpoint" refers only to a connection between two gene ends that is present in one genome but is broken in the other.

The formula $d = N - C$ can also be evaluated from the adjacency graph, since the green lines there form cycles that correspond one for one to the black-gray cycles of the edge graph. A 1-cycle then consists of an A point and a B point joined by two green lines.

In the adjacency graph the DCJ is defined in terms of two of the points belonging to the current genome. These points are deleted, leaving four green lines lacking a termination. These four lines are paired in either of the two remaining ways and for each pair a new point is provided to serve as the common terminus. (For linear chromosomes, see **Section 1.4**) This procedure is completely parallel to the one described in the preceding in terms of the edge graph, and one obtains the distance formula by essentially the same argument.

Genome Rearrangement

(A)

(B)

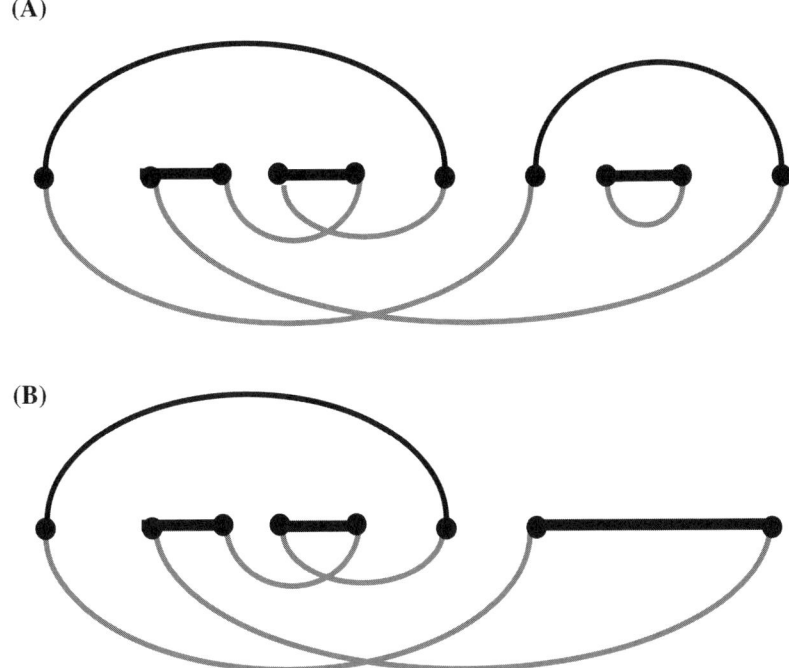

Fig. 18.10. Removal of a 1-cycle in an edge graph of circular chromosomes. **(A)** The edge graph of Fig. 18.5C, with labels removed. There are N = 5 genes (hence five black lines) and C = 2 cycles. **(B)** Same as **(A)** except that the 1-cycle has been removed. Now there are b = 4 black lines (breakpoints) and c = 1 cycle. The DCJ distance between the two genomes is 5 − 2 = 4 − 1 = 3.

1.3. DCJ on Circular and Linear Chromosomes (Applications of Edge Graph)

1.3.1. Caps and Null Chromosomes

This section allows both circular and linear chromosomes to be present in each genome. Now if we pair adjacent gene ends the two endpoints of each chromosome are not paired, so that there are only $2N-2L$ points to be paired where L is the number of linear chromosomes. In this way we obtain a genome graph consisting of $N-L$ black lines and N white lines (*see* **Fig. 18.11B**). It would be possible to use this as a basis for constructing the edge graph between two genomes, but the resulting DCJ operation would not be capable of handling endpoints of a linear chromosome. Thus, for example, the operation of reversing an end segment of a linear chromosome could not be subsumed under the DCJ operation. In order to include operations on endpoints including those of fission and fusion, it is convenient to consider each chromosomal endpoint in the current genome as the terminus of a black line whose other end is not attached to any gene. This unattached end is called a "cap" *(11)*. The resulting genome graph has $N+L$ black and N white lines (*see* **Fig. 18.11C**).

1.3.2. DCJ on Capped Lines

The DCJ operation is now defined just as in **Section 1.2.2**. The capped black lines are treated completely on a par with those that connect two genes. When one of the black lines to be cut is capped, the half attached to the cap provides one of the four loose ends to be rejoined. The rejoining then connects the cap

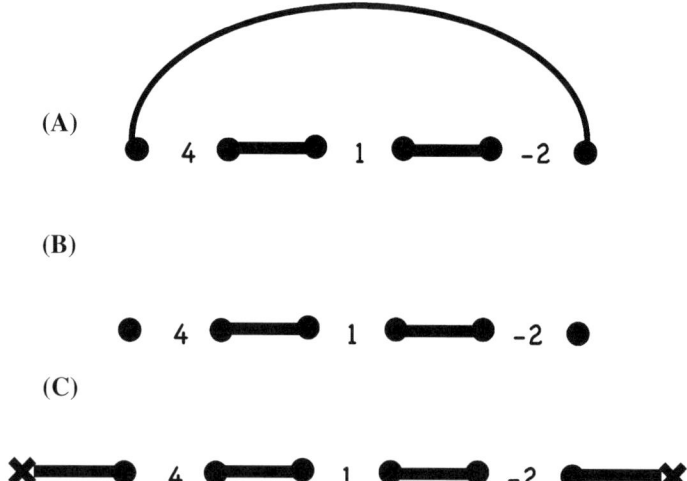

Fig. 18.11. Genome graphs for circular and linear chromosome. (**A**) The circular chromosome (4, 1, −2) displayed as in the left hand part of Fig. 18.5A, but with signed labeling of genes. There are three genes and three black lines. (**B**) The linear chromosome [4, 1, −2] (we use [brackets] for linear, (parentheses) for circular chromosomes) displayed in the same way. The difference from (**A**) is the absence of the arc above. There are three genes and two black lines. (**C**) The same chromosome as in (**B**), displayed with caps. There are three genes and four black lines.

to one of the two gene ends originally connected by the other black line; this gene end now becomes a chromosomal endpoint in place of the original one (**Fig. 18.12A**).

When two capped black lines are cut, it is possible to rejoin the loose ends so that the two caps are connected and the two chromosomal endpoints are joined. This operation decreases L by 1, either by converting a linear to a circular chromosome or fusing two linear chromosomes into one. The structure consisting of two caps joined by a black line is an artifact of the construction called a null chromosome (*see* **Fig. 18.12B**). Null chromosomes do not contain genes and can be added or deleted without affecting the genome.

By including any number of null chromosomes in the initial genome, one may achieve the inverse of the above process. A null chromosome is destroyed by cutting the black line it contains and rejoining the ends to those of another black line that was internal to a chromosome. Thus L is increased by 1, either by fission of a linear chromosome or conversion of a circular chromosome to a linear chromosome.

1.3.3. Menu of Operations

As noted in **Section 1.2.2**, when only circular chromosomes are present (and by tacit agreement no null chromosomes), the conventional operations mimicked by DCJ are reversals, fusions, and fissions. When linear chromosomes are allowed, we must add to this list translocations between linear chromosomes and conversions from linear to circular or circular to linear chromosomes.

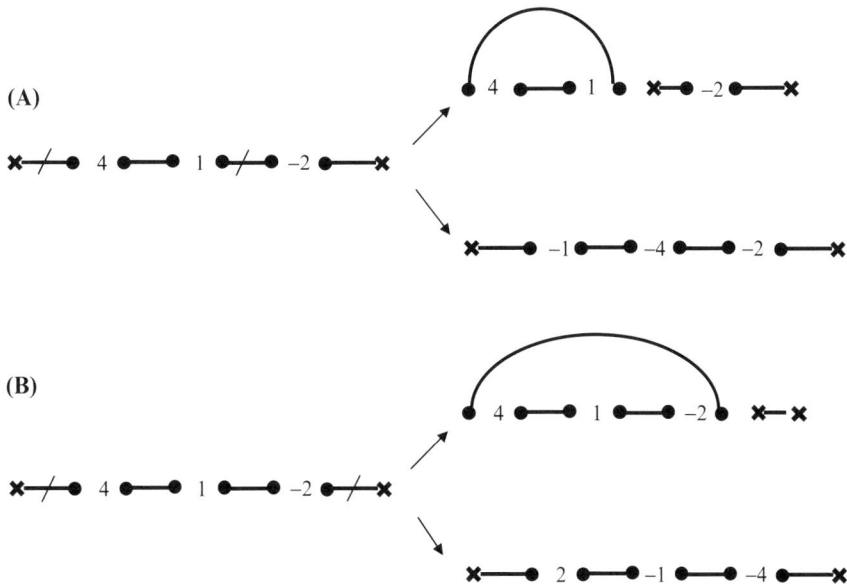

Fig. 18.12. DCJ on a linear chromosome (same as Fig. 18.11C) with cutting of capped line(s). (**A**) One capped line and one uncapped line are cut. Outcomes are fission into a linear and a circular, and reversal of the 4, 1 segment. (**B**) Two capped lines are cut. Outcomes are conversion to circular chromosome (same as Fig. 18.11A) and reversal of the entire chromosome; the latter is no change. There is a null chromosome *(black line bounded by two "x"-caps)* in the upper outcome.

A complete list of possible DCJ operations with their effect on genomic structure is presented in **Table 18.1**. The genomic effects are also summarized in **Fig. 18.13**.

1.3.4. Paths and Cycles

This section considers the DCJ distance between an initial genome A and a target genome B, when linear chromosomes are permitted. The edge graph may not consist entirely of cycles. Besides any cycles present there may be a number of paths, beginning at one cap and ending at another. In defining paths we regard the caps in genome A, which terminate black lines, as distinct from the caps in genome B, which terminate gray lines. A path will be called "odd" if it starts with a cap belonging to one genome and ends with a cap in the other. It will be called "even" if both its caps are in the same genome (**Fig. 18.14**). Odd paths may also be called AB paths; even paths may be called AA or BB paths, according to the location of the end caps.

1.3.5. Closure and Distance

The distance between two genomes can be found by closing all paths into cycles and applying the formula $d = N' - C'$, where N' and C' are found from the closed graph. To close an AB path, identify the two end caps. To close an AA path, introduce a null chromosome into the B genome and identify its caps with those of the AA path. To close a BB path, introduce a null chromosome into the A genome and identify its caps with those of the BB path. After closure the graph will contain an equal number N'

Table 18.1
Outcomes of DCJ on linear and circular chromosomes

	Lines cut	One or two chromosomes	Initial chromosome configuration	Two outcomes			
				Operation	Result	Operation	Result
1	int +int	1	C	Fission	C C	Reversal	C
2	int +int	1	L	(int) Fission	C L	(int) Reversal	L
3	int +int	2	C C	Fusion	C	Fusion	C
4	int +int	2	C L	(int) Fusion	L	(int) Fusion	L
5	int +int	2	L L	(reciprocal) Translocation	L L	(reciprocal) Translocation	L L
6	int + tel	1	L	(ext) Fission	L C	(ext) Reversal	L
7	int + tel	2	C L	(ext) Fusion	L	(ext) Fusion	L
8	int + tel	2	L L	(1-way) Translocation	L L	(1-way) Translocation	L L
9	tel + tel	1	L	Conversion	C N	No change	L
10	tel + tel	2	L L	Fusion	L N	No change	L L
11	int + null	2	C N	Conversion	L	Same conversion	L
12	int + null	2	L N	Fission	L L	Same fission	L L
13	tel + null	2	L N	No change	L N	No change	L N
14	null + null	2	N N	No change	N N	No change	N N

A DCJ on a given edge graph is defined by choosing two black lines to be cut and selecting one of two ways to rejoin the cut ends. The black lines are of three kinds: internal to a chromosome; telomere at the end of a chromosome, with one cap; and null, with two caps. Accordingly there are six possible cases for two black lines. For each case there are various subcases according to the number and type (circular, linear, null) of chromosomes initially containing the cut lines. For each subcase there are two outcomes, depending on the rejoining.

For each outcome we give the conventional name of the operation, and the subcase to which the final configuration belongs. For fission of L into CL, we distinguish between internal and external, depending on whether the ejected fragment was an interior segment or an end segment of the original chromosome. Likewise for fusion of CL to L, whether the circular material is inserted into the linear chromosome or appended to the end; and for reversal within L, whether an interior or an end portion is reversed. For translocation we distinguish between reciprocal (exchange of end portions) and 1-way (transfer of end portion from one to another).

Operating on the last two subcases, tel + null and null + null, leads only to the initial configuration because all caps are indistinguishable. For the same reason, operating on tel + tel or int + null yields only one outcome different from the initial state. C = circular chromosome; chr = chromosome; int = internal line; L = linear chromosome; N = null chromosome; tel = telomere.

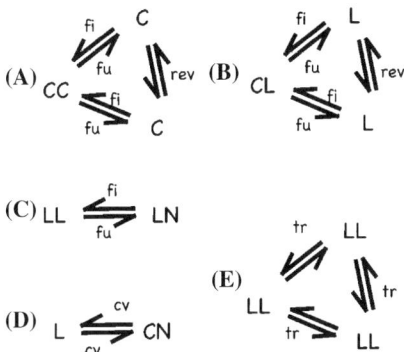

Fig. 18.13. The five possible "DCJ triangles" analogous to Fig. 18.3, sorted by number and type of chromosome participating. (Refer to Table 18.1 for **A–D**.) (**A**) Rows 1 and 3. (**B**) Rows 2, 4, 6, and 7. (**C**) Rows 10 and 12. (**D**) Rows 9 and 11. **E**. Rows 5 and 8. In (**C, D**) the third member of the triangle is not shown because it differs from the left hand member shown only by the exchange of two indistinguishable caps. C = circular; cv = conversion; fi = fission; fu = fusion; L = linear; N = null; tr = translocation, rv = reversal.

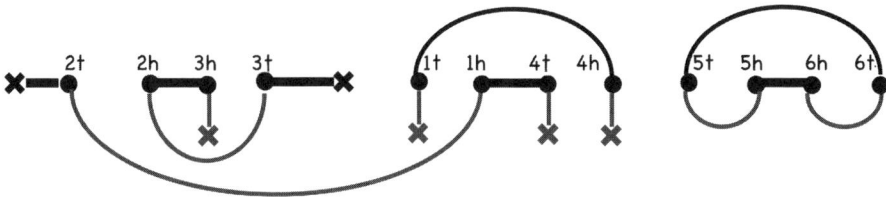

Fig. 18.14. A complex edge graph. The current genome is [2, −3]; (1, 4); (5, −6). The target genome is [1, 2, 3]; [4]; (5); (6). There are two AB paths (odd), one BB path (even), and one cycle.

of black lines and gray lines, and a number C' of cycles including those formed from paths.

1.4. DCJ on Circular and Linear Chromosomes (Applications of the Adjacency Graph)

1.4.1. Telomeres

If the adjacency graph is constructed in accordance with **Section 1.2.4**, each genome will in general contain not only adjacency points corresponding to the connections between adjacent genes in a chromosome, but also telomere points, which correspond to the endpoints of linear chromosomes. An adjacency has two labels and two green lines attached; a telomere has only one of each (*see* **Fig. 18.15D**). To perform a DCJ on two adjacencies, the four green lines incident on them are detached and reshuffled as described in **Section 1.2.5**.

To achieve the full menu of operations given in **Section 3**, it must be possible to perform DCJ also on telomeres. Then one has only three green lines to reshuffle, or only two if both points are telomeres; in the last case the result is a single adjacency point. Bergeron et al. (*4*) treat these possibilities as separate cases, along with another case in which a single adjacency is attacked, yielding two telomeres. Thus, the correct dualism is achieved between the

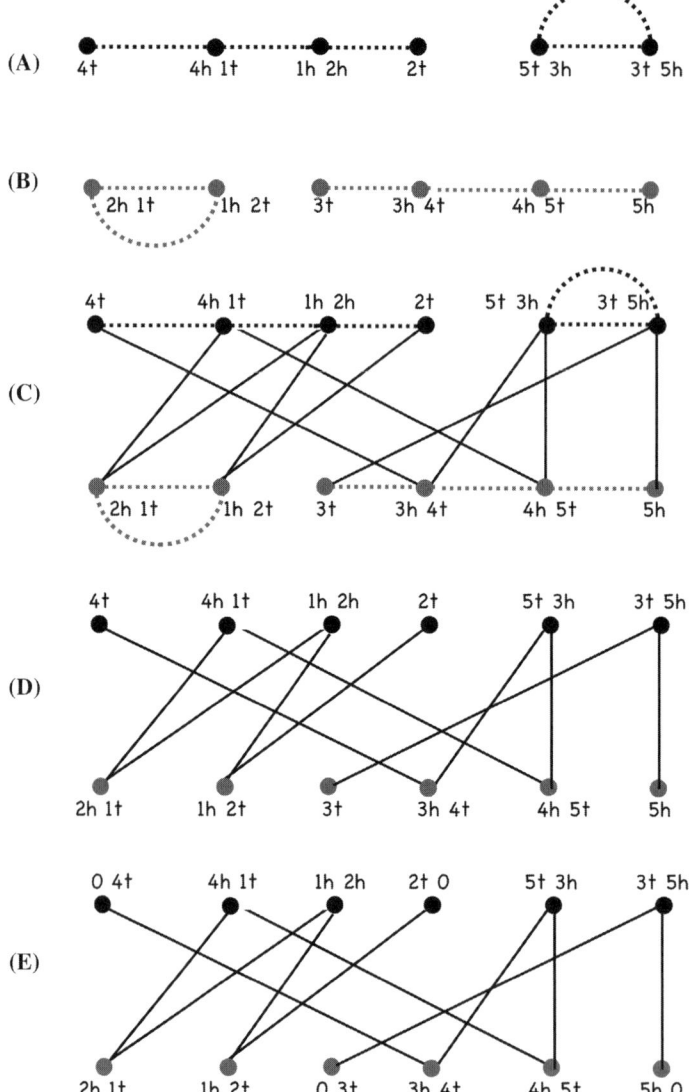

Fig. 18.15. Construction of an adjacency graph involving circular and linear chromosomes. This figure was obtained from Fig. 18.6 by making one chromosome in each genome linear. Current genome is [4, 1, −2]; (−3, −5). The target genome is (1, 2); [3, 4, 5]. (**A**) The current genome above. (**B**) The target genome below. (**C**) Same as (**A, B**) with green lines added. (**D**) Same as (**C**) with white lines deleted. (**E**) Same as (**D**) with "0" labels added to telomeres. The two cycles in Fig. 18.6 are replaced here by even paths.

edge and adjacency graph, but at the sacrifice of uniformity in the definition of DCJ.

1.4.2. Uniform Definition and 0 Labels

We can make the DCJ uniform in the adjacency graph by adding structures dual to the caps and null chromosomes. The dual of a cap would be a dangling green line attached to a telomere point, like a road with a beginning and no end. Although this structure can lead to a correct uniform definition of DCJ, it would mar the visual elegance of the adjacency graph. Therefore, we suggest the logically equivalent addition of a second label, "0," to each telomere, representing the absence of a green line (*see* **Fig. 18.15E**). The "0" labels play the role of caps. We also introduce null points, to play the role of null chromosomes. A null point belongs to one genome and has two "0" labels and no green line attached. Every point now has two labels. For further illustration we present in **Fig. 18.16** the adjacency graph corresponding to **Fig. 18.14**. The corresponding master graph is shown in **Fig. 18.17**.

The DCJ can now be performed by choosing any two points in the current genome, reshuffling the four labels attached to them, and reconnecting the green lines (however many there are) in accordance with the new labeling of the points. This is a uniform definition, and it yields all the special cases described in **Table 18.1**.

1.4.3. Paths and Cycles, Odd and Even Paths

Like the edge graph, the adjacency graph consists of paths and cycles. The paths begin and end on telomeres. An AB path (odd) begins in genome A and ends in genome B. An AA path (even) begins and ends in genome A. A BB path (even) begins and ends in genome B.

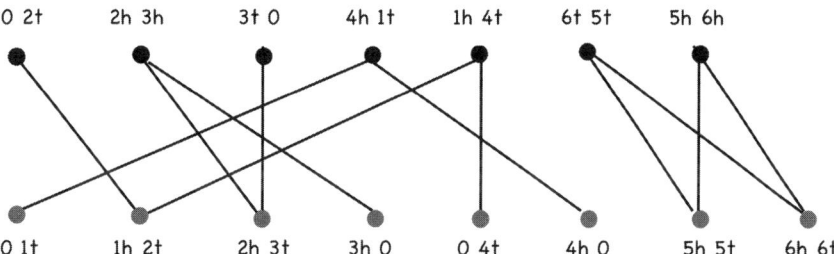

Fig. 18.16. The adjacency graph corresponding to Fig. 18.14. Current genome [2, −3]; (1, 4); (5, −6). Target genome [1, 2, 3]; [4]; (5); (6). "0" labels have been added.

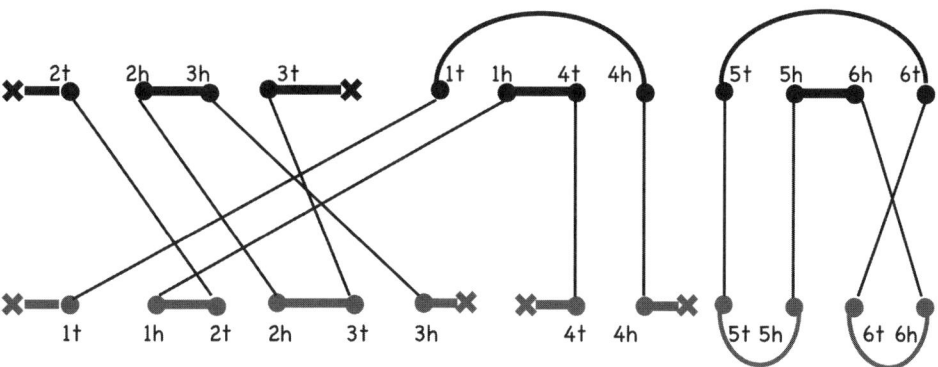

Fig. 18.17. The master graph for Figs. 18.14 and 18.16. Contract green lines to get Fig. 18.14. Contract black and gray lines to get Fig. 18.16.

Table 18.2
Pairwise DCJ distances between genomes of *Shigella* and *E. coli* as computed on the Mauve alignment with minimum LCB weight 1147

	Sb227	Sd197	5 str. 8401	2a 2457T	K12
S. boydii Sb227	—				
S. dysenteriae Sd197	86	—			
S. flexneri 5 str. 8401	40	79	—		
S. flexneri 2a 2457T	44	79	14	—	
E. coli K12	28	65	16	18	—

We see that in general, pairwise distances between *Shigella* spp. and *E. coli* K12 are lower than distances among pairs of *Shigella* genomes—a counterintuitive result for organisms from different genera. *Shigella* and *E. coli* were originally given different genus names because the diarrheal disease caused by *Shigella* had different symptoms than that caused by *E. coli*. *Shigella* and *E. coli* are now commonly considered members of the same bacterial "species." Some strains of *Shigella* appear to have acquired an elevated rate of genomic rearrangement, resulting in the high DCJ distances between strains.

It is possible to close the adjacency graph and use the distance formula $d = N' - C'$ given in **Section 1.3.5**. To close an AB path, join the two endpoints (telomeres) by a green line. To close an AA path, introduce a null point into genome B and connect it by green lines to both ends of the AA path. To close a BB path, introduce a null point into genome A and connect it by green lines to both ends of the BB path. Then C' is the number of cycles including those formed from paths, and N' is the number of points in each genome, including null points.

1.4.4. Distance Formula

Bergeron et al. *(4)* have also given a distance formula that can be applied directly to the adjacency graph without introducing "0" labels, null points, or closure. They arrive at $d = N - C - I/2$, where N is the number of genes, C the number of cycles, and I the number of odd paths. This is a lower bound on the distance because no DCJ can do better than increase C by 1 or increase I by 2, but not both. To see that this lower bound can be achieved, *see* **Section 3.10**, where we give a sorting procedure that achieves it.

This formula can be proved equivalent to the one based on closure as follows. One has $C' = C + P$, where P is the number of paths. One also has $2N' = 2N + L_A + L_B + Z_A + Z_B$, where L_A, L_B are the number of linear chromosomes in each genome and Z_A, Z_B are the number of null points introduced in each genome for closure. By counting telomeres one has $P = L_A + L_B$ and also $P = I + Z_B + Z_A$, since closure introduces one null point for each even path. Putting these equations together, one obtains $N' - C' = N - C - I/2$. The formula $d = N - C - I/2$ can also be applied to the edge graph since N, C, and I are the same in both graphs.

1.5. Linear Chromosomes with Restriction of Circular Intermediates

Of biological interest is the case in which only linear chromosomes are allowed in the initial and target genomes, and never more than one circular chromosome in the current genome. It has been shown *(3)* that in this case the DCJ distance is the same as though circular chromosomes were unrestricted. This case is equivalent to forbidding circulars altogether and allowing inversions, translocations, and block interchanges (with weight 2).

1.6. Outline of Procedures (see Section 3)

The interested reader may follow the procedures detailed in **Section 3** to construct the graphs that have been described or develop algorithms for the distance or sorting by DCJ:

Section 3.1. Construction of black-white genome graph

Section 3.2. Construction of white genome graph

Section 3.3. Construction of edge graph

Section 3.4. Construction of adjacency graph

Section 3.5. From edge graph to adjacency graph

Section 3.6. From adjacency graph to edge graph

Section 3.7. Distance without linear chromosomes

Section 3.8. Distance with or without linear chromosomes

Section 3.9. Sorting without linear chromosomes

Section 3.10. Sorting with or without linear chromosomes

2. Materials and Online Resources

The procedures in this chapter are of two kinds: constructing graphs (described in **Section 1**), and carrying out sorting algorithms to transform one genome into another by the minimum number of DCJs.

2.1. Constructing Graphs by Hand or Computer

The constructions can be done in two ways: by hand or computer. Hand construction is easier unless one has facility with computer software for creating diagrams. Some of the constructions involve alterations of a graph, and this requires an excellent eraser or else the willingness to draw the graph from scratch after each alteration.

2.2. Online Software for Performing Algorithms

2.2.1. Applying DCJ to Genome Sequence Data Using Mauve

The DCJ algorithm provides a measure of genomic rearrangement distance between genomes that have been coded as synteny blocks. Identification of synteny blocks among genomes remains a non-trivial task and is the subject of ongoing research and software development. As discussed in **Chapter 20**, GRIMM-Synteny provides one mechanism for identifying synteny blocks. This section describes the Mauve genome alignment software *(12)*, which can be used to both identify synteny blocks and compute pairwise DCJ distances among multiple genomes. Mauve is free, open-source software for Linux, Windows, and Mac OS X, available from http://gel.ahabs.wisc.edu/mauve.

Mauve creates synteny blocks using an algorithm that identifies homologous tracts of sequence that are unique in each genome. First, Mauve identifies putatively homologous local multiple alignments *(13)*. Next, the local multiple alignments are clustered into groups that are free from rearrangement, called locally collinear blocks (LCBs). Each LCB is assigned an *LCB weight* equal to the sum of lengths of the ungapped local alignments that comprise the LCB. Some LCBs may represent spurious homology predictions or paralogous sequence, and it is necessary to filter out such LCBs. Such LCBs typically have low LCB weight relative to LCBs representing non-paralogous homology. Mauve discards all LCBs that have an LCB weight less than a user-specified threshold value in a process called greedy breakpoint elimination. For more details, refer to the algorithm descriptions in *(12, 14)*.

2.2.2. Performing a DCJ Analysis of E. coli and Shigella Genomes with Mauve 1.3.0

Using the Mauve genome alignment software, one can generate synteny blocks and perform a DCJ analysis in a five-step process:

1. Download and run Mauve from http://gel.ahabs.wisc.edu/mauve.

2. Download whole genome sequence data in either Multi-FastA or GenBank format from NCBI. For this example we will use the genomes of five *Shigella* and *E. coli* strains. The genomes are available from:

   ftp://ftp.ncbi.nih.gov/genomes/Bacteria/Shigella_boydii_Sb227/NC_007613.gbk

   ftp://ftp.ncbi.nih.gov/genomes/Bacteria/Shigella_flexneri_2a_2457T/NC_004741.gbk

   ftp://ftp.ncbi.nih.gov/genomes/Bacteria/Shigella_flexneri_5_8401/NC_008258.gbk

   ftp://ftp.ncbi.nih.gov/genomes/Bacteria/Shigella_dysenteriae/NC_007606.gbk

   ftp://ftp.ncbi.nih.gov/genomes/Bacteria/Escherichia_coli_K12/NC_000913.gbk

3. Identify synteny blocks (LCBs) in Mauve.
   a. Select the "Align ..." option from the "File" menu. The "Align sequences..." dialog will appear, as shown in **Fig. 18.18**.
   b. In the dialog, add each genome sequence file that was downloaded from NCBI. Files can be dragged and dropped, or selected using the "Add sequence..." button, as shown in **Fig. 18.18A**. Also set the output file location. In this example, output will be written to the file C:\shigella\5way_comparison.
   c. Select the "Parameters" tab (*see* **Fig. 18.18B**) and disable the "Extend LCBs" option. Then disable the "Full Alignment" option. The synteny block generation process will run much more quickly without performing a full genome alignment.
   d. Click the "Align" button.
4. Choose the appropriate minimum LCB weight. A slider control at the top-right of the Mauve window allows one to change the minimum LCB weight, as shown in **Fig. 18.19**. The default value is often too low and thus many spurious LCBs appear. By sliding the control to the right, a higher LCB weight can be chosen that better represents the true set of LCBs. For the present data we select a minimum LCB weight of 1147, which roughly corresponds to the average gene size in bacteria.

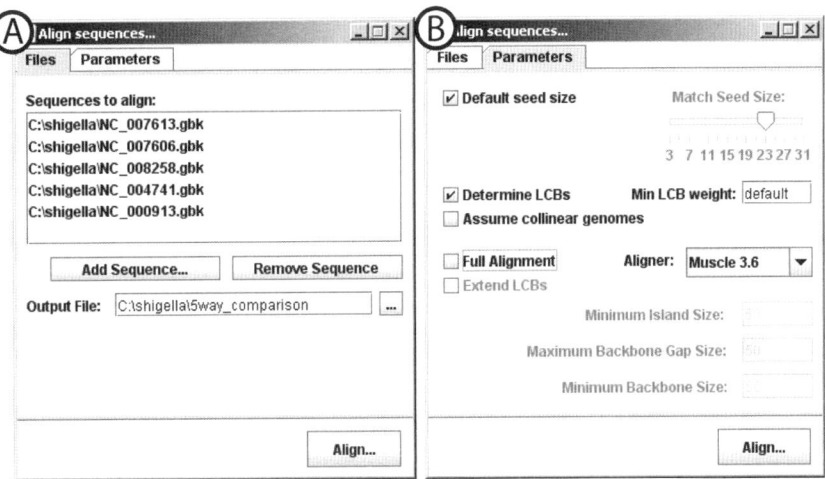

Fig. 18.18. The Mauve "Align sequences..." dialog window. (**A**) The selection of five GenBank sequence files that correspond to four genomes of *Shigella* spp. and one *E. coli* (NC_000913.gbk). (**B**) Selection of the alignment options. Note that the "Extend LCBs" option and the "Full Alignment" option have been unchecked. Once the appropriate sequence files and options have been selected, an alignment can be triggered by clicking the "Align..." button at bottom right. The series of DCJ operations used can be viewed by clicking the "Operations" button in the DCJ window.

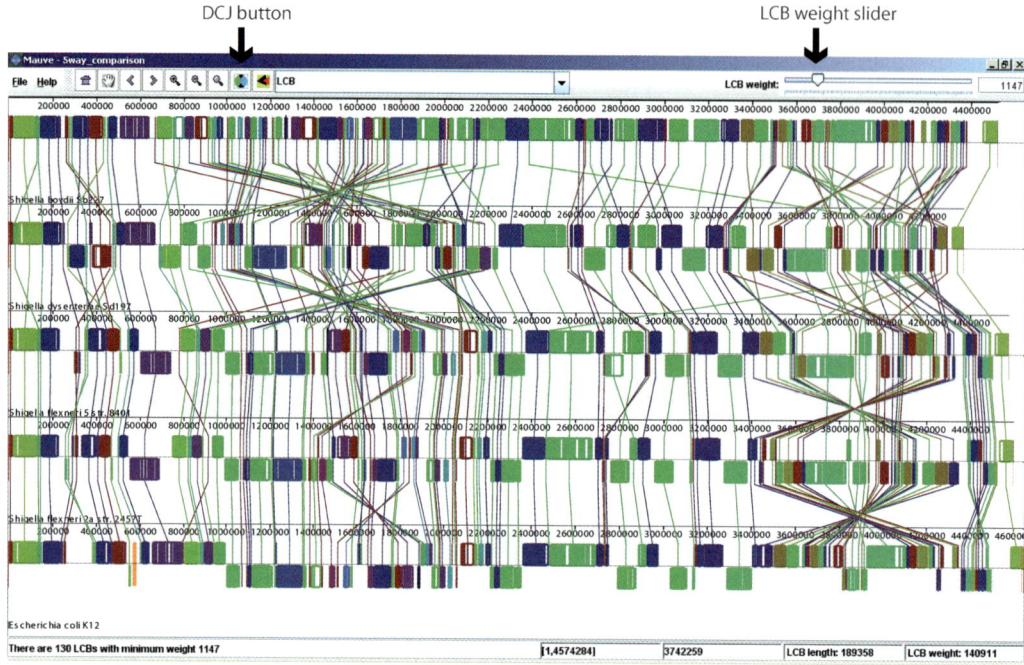

Fig. 18.19. Mauve alignment of four *Shigella* genomes with one *E. coli* genome. Genomes are laid out horizontally and linked colored blocks indicate the homologous segments in each genome. Blocks shifted downward in a given genome are inverted relative to *Shigella boydii* Sb227 *(top)*. The minimum LCB weight has been adjusted using the LCB weight slider *(top right)* to a value of 1147. The increase in minimum LCB weight removes spurious LCBs. A minimum weight value of 1,147 roughly coincides with the average length of prokaryotic genes, indicating that remaining LCBs (Synteny Blocks) are likely to consist of one or more complete genes. The Mauve software permits interactive browsing of Synteny Blocks in conjunction with annotated features such as genes. For further information on how to use Mauve, please refer to the online Mauve user guide at http://gel.ahabs.wisc.edu/mauve.

5. Perform a DCJ analysis by clicking the DCJ button in the Mauve window (shown at top in **Fig. 18.19**). The results appear as a textual table in a pop-up window. The table is tab-delimited text that can be copied-and-pasted into a program such as Microsoft Excel. The pairwise DCJ distances for the five *Shigella* and *E. coli* genomes are shown in **Table 18.2**.

3. Methods

3.1. Construction of a Black-White Genome Graph

Construct a black-white genome graph with a given gene content. Let the gene content be given in signed permutation form, thus [4, 1, –2] (–3, –5) signifies that there are two chromosomes, that the first chromosome is linear and reads from left to right as gene 4 pointing forward, gene 1 pointing forward, gene 2 pointing backward, and that the second chromosome is circular and reads counterclockwise as gene 3 pointing backward, gene 5 pointing backward.

1. For each circular chromosome of n genes, place 2n points in a horizontal line.

2. Join these 2n points by alternating white and black lines, starting with white. (The white lines may be left invisible.)

3. Join the first and last point by a black arc passing above.

4. For each linear chromosome of n genes, place 2n + 2 points in a horizontal line. (The first and last point may be distinguished as caps.)

5. Join these 2n + 2 points by alternating black and white lines, starting with black.

6. As a result of Steps 2 and 5, every chromosome now has as many white lines as genes. The white lines represent the genes, reading from left to right as in the input description. They may be labeled as follows.

7. If the gene is positive (pointing forward), label the right hand end of the white line with the gene number followed by "h," and the left hand end with the gene number followed by "t." If it is negative (pointing backward), label the left hand end with the gene number followed by "h," and the right hand end with the gene number followed by "t." Caps are not labeled.

With the input example given above, the first chromosome is given $2 \times 3 + 2 = 8$ points, the second is given $2 \times 2 = 4$ points, and the labels of the 12 points from left to right (caps indicated by x) are x, 4t, 4h, 1t, 1h, 2h, 2t, x, 3h, 3t, 5h, 5t.

3.2. Construction of a White Genome Graph

Construct a white genome graph with a given gene content. Let the input be given as in **Section 3.1**.

1. For each circular chromosome of n genes, place n points in a horizontal line.

2. Join these n points by n – 1 white lines, and join the first and last point by a white arc passing above. (This white arc must be visible, at least until its endpoints have been labeled in Step 6 below.)

3. For each linear chromosome of n genes, place n + 1 points in a horizontal line.

4. Join these n + 1 points by n white lines.

5. Each chromosome of n genes now has n white lines. Each white line will be labeled at each of its endpoints, so that the point of connection between two consecutive white lines will receive two labels. Label in accordance with the input description as follows.

6. For the white arc passing above a circular chromosome, label "in reverse." If the gene is positive, the left hand end receives

the gene number followed by "h," and the right hand end receives the gene number followed by "t." If the gene is negative, the right hand end receives the gene number followed by "h," and the left hand end receives the gene number followed by "t."

7. For all other white lines, label "directly." If the gene is positive, the right hand end receives the gene number followed by "h," and the left hand end receives the gene number followed by "t." If the gene is negative, the left hand end receives the gene number followed by "h," and the right hand end receives the gene number followed by "t."

8. All points now have two labels, except the endpoints of linear chromosomes.

With the input example given in **Section 3.1**, the first chromosome will have $3+1=4$ points, and the second will have two points. The labels of the six points will be (4t); (4h,1t); (1h,2h); (2t); (5t,3h); (3t,5h) (see **Fig. 18.15A**).

3.3. Construction of an Edge Graph

Construct an edge graph, given the initial and target genomes. Let the two genomes be given in signed permutation form, for example:

genome A (initial) = [4, 1, −2) (−3, −5)

genome B (target) = [1, 2) (3, 4, 5)

1. Construct the black-white genome graph of genome A, following **Section 3.1**.

2. Construct the black-white genome graph of genome B, but use gray lines instead of black.

3. For each gray line in the B graph, connecting two labeled points, find the corresponding labeled points in the A graph and connect them by a gray arc passing beneath the horizontal line.

4. For each gray line in the B graph connecting a labeled point to a cap, find the corresponding labeled point in the A graph and run a gray line vertically downward to a cap below.

5. Discard the B graph.

6. The white lines may now be deleted if desired.

The resulting elaborated A graph is the edge graph between genomes A and B. It should have $2N$ labeled points, $2L_A$ caps terminating horizontal black lines, and $2L_B$ caps terminating vertical gray lines, where N is the number of genes in each genome, L_A is the number of linear chromosomes in genome A, and L_B is the number of linear chromosomes in genome B.

Thus, for the input example given above, the edge graph has 10 labeled points, $2 \times 1 = 2$ caps terminating horizontal black lines, and $2 \times 1 = 2$ caps terminating vertical gray lines.

Genome Rearrangement 409

3.4. Construction of Adjacency Graph

Construct an adjacency graph, given the initial and target genomes. Let the input genomes be given as for Procedure 3.

1. Construct the white genome graph for genome A, using **Section 3.2**. (*see* **Fig. 18.15A**).

2. Construct the white genome graph for genome B, directly below the A graph (*see* **Fig. 18.15B**).

3. Join each label in the A graph to the corresponding label in the B graph by a green line (*see* **Fig. 18.15C**).

4. Delete white lines (*see* **Fig. 18.15D**).

5. If desired, add a "0" label to each telomere (point with only one label) (*see* **Fig. 18.15E**).

The resulting adjacency graph will have 2N green lines, where N is the number of genes in each genome. For the input given before **Section 3.3**, there should be $2 \times 5 = 10$ green lines.

3.5. From Edge Graph to Adjacency Graph

Given an edge graph, construct the corresponding adjacency graph. We shall assume that the white lines are invisible in both graphs.

1. The edge graph is built on a horizontal level called the upper level. Also visualize a lower level, some distance below the upper level.

2. For each gray arc in the edge graph, mark a point on the lower level. Join this point by two green lines to each endpoint of the gray arc. Delete the gray arc.

3. Replace each vertical gray line running from a point on the upper level down to a cap by a vertical green line running from that point on the upper level down to a point on the lower level. Delete the cap.

4. Label each point on the lower level by the same labels as appear at the upper ends of the green lines connected to it. If there is only one green line, add a label "0" if desired.

5. On the upper level, contract each horizontal black line to a point. Let this point inherit the labels of the previous endpoints of the black line. If one endpoint was a cap, add a label "0" if desired, otherwise give the point only one label.

6. Treat the black arcs according to the same principle. This may be done by deleting the arc and its right hand endpoint, and transferring the green line and label of the deleted endpoint to the left hand endpoint, which is now the point of contraction.

The resulting structure is the adjacency graph. Every point should have two labels, if "0" labels were added.

3.6. From Adjacency Graph to Edge Graph

Given an adjacency graph, construct the corresponding edge graph.

1. If the adjacency graph has "0" labels, delete them.

2. Label each green line in the adjacency graph with the single label that is found on both of its endpoints. If the two endpoints have both labels in common, there will be two green lines

joining them; one green line should receive one label and one the other.

3. Draw out each point on the upper level (in genome A) into two points connected by a horizontal black line. Give each point one of the green lines connected to the original point, with the corresponding label. If the original point has only one green line, let one of the new points inherit the green line with its label, and let the other be a cap.

4. Replace each pair of green lines having a common lower endpoint by a gray arc connecting the respective upper endpoints.

5. Replace each green line having only one lower endpoint by a vertical gray line running from the upper endpoint downward to a cap. The resulting structure is the edge graph.

3.7. Distance Without Linear Chromosomes

Find the DCJ distance between two genomes, given the adjacency graph. Restricted case: no linear chromosomes in either genome (no telomeres in the graph).

1. N is the number of genes in each genome, found by counting the points on either level.

2. C is the number of cycles in the graph, found by the following steps.

3. Start from any point and follow the green lines continuously, marking each point you reach, until you return to the starting point. You have traversed a cycle.

4. Start again from an unmarked point (if there is one) and traverse a second cycle by repeating Step 3.

5. Repeat until all points are marked. The number of cycles traversed is C.

6. The distance is $N-C$.

3.8. Distance With or Without Linear Chromosomes

Find the DCJ distance between two genomes, given the adjacency graph. In a general case, linear chromosomes may be present.

1. If the number N of genes in each genome is not known, determine it by counting the green lines and dividing by 2.

2. Explore all the paths by the following steps.

3. Start at a telomere; it has only one green line attached. Move along the green lines continuously, marking each point you reach, until you reach another telomere. You have traversed a path.

4. If you began on one level and ended on the other, the path is odd. Keep count of the odd paths.

5. If there remains an unmarked telomere, start again at that point and traverse a second path. Repeat until all paths are traversed (no unmarked telomere).

Genome Rearrangement   411

6. *I* is the number of odd paths you have found. This number must be even.

7. The part of the graph remaining unmarked consists completely of cycles. Find *C*, the number of cycles, by applying Steps 3 to 5 of **Section 3.7**.

8. The distance is $N-C-(I/2)$.

3.9. Sorting Without Linear Chromosomes

Perform a series of DCJs to transform a given initial genome consisting of circular chromosomes into a given target genome consisting of circular chromosomes.

1. Construct the adjacency graph between initial and target genome (**Section 3.4**). All points will be adjacencies (two green lines).

2. Proceed from left to right on the lower level. Choose the first point that is connected by green lines to two different points on the upper level. If there is none, sorting has been completed.

3. Cut the two points on the upper level that are connected to the chosen point.

4. Reconnect the four loose ends so that the two green lines from the chosen point are connected to one point on the upper level, and the other two green lines to the other point.

5. Repeat steps 2–4 until sorting is complete.

Each DCJ has increased the number of cycles by one. The final configuration consists completely of 1-cycles.

3.10. Sorting With or Without Linear Chromosomes

Perform a series of DCJs that transform a given initial genome of arbitrary type (circular and linear chromosomes both permitted) to a given target genome of arbitrary type.

1. Construct the adjacency graph between the initial and target genome (**Section 3.4**). Include "0" labels for telomere points (only one green line).

2. Proceed from left to right on the lower level. Choose the first point that is connected by green lines to two different points on the upper level. If there is none, proceed to Step 7.

3. Cut the two points on the upper level that are connected to the chosen point.

4. Shuffle the four labels on the two cut points so that the two labels corresponding to those on the chosen point are placed on one point on the upper level. Place the other two labels on the other point. If the latter are both "0," the resulting null point may be deleted.

5. Reconnect all loose ends of green lines according to the labels on the points.

6. Repeat Steps 2–5 until every point on the lower level is joined to no more than one point on the upper level. Every DCJ so far has increased the number of cycles by 1, and has not changed the number of odd paths, although it may have shortened some. The graph now consists of: (1) 1-cycles; (2) AB paths (odd) of length 1; (3) BB paths (even) of length 2.

7. Proceed from left to right on the upper level. Choose the first point that is connected by green lines to two different points on the lower level. (This configuration will be a BB path.) If there is none, sorting is complete.

8. Cut the chosen point and a null point introduced for this purpose on the upper level.

9. Shuffle the four labels so that the two "0" labels are on different points.

10. Reconnect the two loose ends of green lines. This has converted an adjacency (the chosen point) and a null point into two telomeres. The BB path has been split into two AB paths.

11. Repeat Steps 7–10 until sorting is complete. Each DCJ in this second loop has left the number of cycles unchanged and has increased the number of odd (AB) paths by 2. Each DCJ in the whole procedure has increased the number $C + I/2$ by 1, where C = number of cycles, I = number of odd paths. The final configuration consists of 1-cycles and AB paths of length 1. The labeled points on the upper level correspond exactly to those on the lower level.

An example of this procedure is shown in **Figs. 18.16** and **18.20**.

4. Notes

1. 1.: Homology is a binary characteristic. There is no such thing as 75% homologous, "more" homologous or "homologous depending on criteria". Given sequence data, we use computational methods to determine whether a given genomic segment is homologous to another genomic segment. These computational methods use parameters and "criteria" to make predictions about homology. The true relationship of homology does not depend on those criteria, but our predictions do.

2. 3, 3.1, 3.2: In all these constructions we represent the genome graph as far as possible as laid out in a straight horizontal row. This can be done completely for linear chromosomes, but a circular chromosome must contain one arc that returns from the last point to the first. In 3.1. we have the option to choose

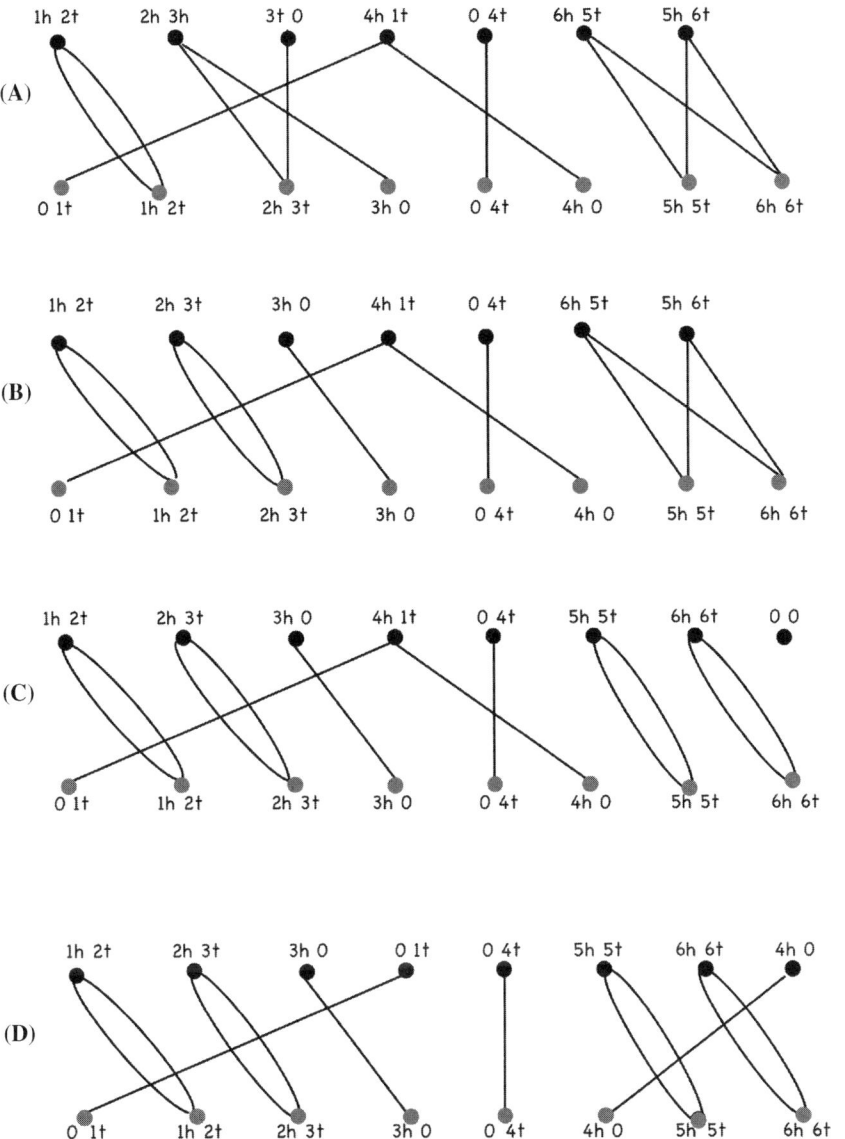

Fig. 18.20. Sorting of Fig. 18.16 by DCJ. Procedure as given in Section 3.10. (**A**) Start in Fig. 18.16 from 1h2t below, cut 02t, and 1h4t above. Operation is fusion of circular and linear. New genome is [4, 1, 2, −3]; (5, −6). $\Delta C = 1, \Delta l = 0$. (**B**) Start in (**A**) with 2h3t below, cut 2h3h and 3t0 above. Operation is reversal. New genome is [4, 1, 2, 3]; (5, −6). $\Delta C = 1, \Delta l = 0$. (**C**) Start in (**B**) from 5h5t below, cut 6h5t and 5h6t above. Operation is fission of circular. New genome is [4, 1, 2, 3]; (5); (6). $\Delta C = 1, \Delta l = 0$. (**D**) Start in (**C**) from 4h1t above, cut 4h1t, and 00 (null point has been introduced for this purpose). Operation is fission of linear to two linears. New genome is [1, 2, 3]; [4]; (5); (6), identical to target genome. $\Delta C = 0, \Delta l = 2$.

either a black or a white line to be this arc. We choose a black arc because a white arc, if invisible, would be difficult to infer from the appearance of the graph, and so circular chromosomes would not be easily distinguished from linear. (In principle the distinction could be made by studying the labels.)

In 3.2 we have no choice but to use a white arc. Therefore, the arc must not be deleted until its endpoints have been labeled.

3. 3.1: Caps seem to be necessary at least in the current genome if one wants a uniform definition of DCJ in terms of the edge graph. We have thought of making the linear chromosome "pseudocircular" by adding a black line to connect the endpoints, but marking this line to show that it is not really there. However, a DCJ made by cutting this line would have ambiguous effect, because it would remain unspecified which end of the chromosome was to receive new material. Another possibility *(11)* is to omit the caps in the target genome; we have preferred to use them in both genomes for the sake of symmetry between the two.

4. 3.2: The rules for labeling can be understood by imagining a toy train proceeding counterclockwise along a closed track, left to right in the foreground and right to left on the return segment in the background. Some of the cars (positive genes) have been put on correctly, heads forward and tails behind; some (negative genes) have been put on backwards. In the foreground the positive cars have their heads to the right and tails to the left, and the negative (backwards) cars have heads left and tails right. In the background the train is headed right to left; therefore, the positive cars have heads left and the negative cars have heads right. The white arc corresponds to the background portion of the track.

5. 3.4, 5; 3.5, 4 and 5; 3.6, 1; 3.10: The adjacency graph was presented by Bergeron et al. (3) without "0" labels, and for some purposes they are superfluous. Therefore, they are made optional in **Sections 3.4** and **3.5**. They only get in the way in **Section 3.6**, and so we delete them at the outset. They have no effect on the computation of distance, and so are not mentioned in **Section 3.8**. Their value is in sorting by DCJ, which becomes simpler if the operation is defined with only one universal case. Therefore we explicitly include them in **Section 3.10**.

6. 3.5: Steps 2 and 3 are ambiguous as to the placement of points on the lower level, and may therefore lead to an appearance of the adjacency graph different from that resulting from **Section 3.2**, but the connections will be the same.

7. 3.8: The number of genes can of course be determined by counting the white lines, but if they are not visible the easiest way may be by counting the green lines. Counting the points requires a distinction between adjacencies and telomeres.

8. 3.8: The number of AB paths, I, is always an even number; therefore, the distance formula $N-C-I/2$ always gives an

integer. *I* is even because the number of A-caps is equal to *I* plus twice the number of AA paths. Therefore, if *I* were odd, the number of A-caps would be odd, which is impossible because this number is double the number of linear chromosomes in genome A.

9. 3.11: In sorting with linear chromosomes present, the elements to be expected when sorting is complete are 1-cycles and AB paths of length 1. Each of these structures involves one point in each genome, with identical labeling. A 1-cycle represents an adjacency common to both genomes. An AB path of length 1 represents a chromosomal endpoint common to both genomes.

Acknowledgments

The authors are grateful to David Sankoff for his advice and encouragement; Anne Bergeron for communicating the idea of the adjacency graph in advance of publication; Mike Tsai for his online implementation of the DCJ; and Betty Harris for invaluable logistic support and encouragement. S.Y. thanks Nicholas Chiorazzi for his enthusiasm, encouragement, and support; and A.E.D is supported by NSF grant DBI-0630765.

References

1. Nadeau, J. H., Taylor, B. A. (1984) Lengths of chromosomal segments conserved since divergence of man and mouse. *Proc Natl Acad Sci U S A* 81, 814–818.
2. Pevzner, P. A. (2000) *Computational Molecular Biology: An Algorithmic Approach.* MIT Press, Cambridge, MA.
3. Yancopoulos, S., Attie, O., Friedberg, R. (2005) Efficient sorting of genomic permutations by translocation, inversion and block interchange. *Bioinformatics* 21, 3340–3346
4. Bergeron, A., Mixtacki, J., Stoye, J. (2006) A unifying view of genome rearrangements. *WABI 2006*, 163–173.
5. Christie, D. A. (1996) Sorting permutations by block interchanges. *Inform Proc Lett* 60, 165–169.
6. Lin, Y. C., et al. (2005) An efficient algorithm for sorting by block-interchanges and its application to the evolution of Vibrio species. *J Comput Biol* 12, 102–112.
7. Meidanis, J., Dias, Z. (2001) Genome rearrangements distance by fusion, fission, and transposition is easy. In *Proceedings of SPIRE'2001—String Processing and Information Retrieval*, Laguna de San Rafael, Chile.
8. Kececioglu, J., Sankoff, D. (1995) Exact and approximation algorithms for sorting by reversals, with application to genome rearrangement. *Algorithmica* 13, 180–210.
9. Bergeron, A., private communication.
10. Bafna, V., Pevzner, P.A. (1993) Genome rearrangements and sorting by reversals. In *Proceedings of the 34th Annual IEEE Symposium on Foundations of Computer Science*, IEEE Press, Los Alamitos, CA.
11. Hannenhalli, S., Pevzner, P. A. (1995) Transforming men into mice (polynomial algorithm for genomic distance problem). In *Proceedings of the 36th Annual IEEE Symposium on Foundations of Computer Science*, Milwaukee, WI.
12. Darling, A. C. E., Mau, B., Blattner, F. R., et al. (2004) Mauve: multiple alignment of conserved genomic sequence with rearrangements. *Genome Res* 14, 1394–1403.

13. Darling, A. E., Treangen, T. J., Zhang, L., et al. (2006) Procrastination leads to efficient filtration for local multiple alignment. *Lecture Notes in Bioinformatics.* Springer-Verlag, New York.

14. Darling, A. E., Treangen, T. J., Messeguer, X., et al. (2007) Analyzing patterns of microbial evolution using the Mauve genome alignment system, in (Bergman, ed.), *Comparative Genomics.* Humana Press, Totowa, NJ, in press.

Chapter 19

Inferring Ancestral Protein Interaction Networks

José M. Peregrín-Alvarez

Abstract

With the recent sequencing of numerous complete genomes and the advent of high throughput technologies (e.g., yeast two-hybrid assays or tandem-affinity purification experiments), it is now possible to estimate the ancestral form of protein interaction networks. This chapter combines protein interaction data and comparative genomics techniques in an attempt to reconstruct a network of core proteins and interactions in yeast that potentially represents an ancestral state of the budding yeast protein interaction network.

Key words: Ancestral state, protein interaction networks, phylogenetic profiles, comparative genomics, bioinformatics.

1. Introduction

With the advent of genomics, the focus of biology has changed from the individual functions of biomolecular components to the interactions between these components. Protein–protein interactions underlie many of the biological processes in living cells. In recent years, high throughput analyses have enabled us to obtain protein interaction data from a few model organisms (1–5). One of the key findings of such analyses is that these biological networks share many global topological properties such as scale-free behavior. The analysis of the topological properties of the interactions between essential proteins in the budding yeast interactome has also revealed a preferential attachment between these proteins, resulting in an almost fully connected sub-network (6). This sub-network includes proteins with a wide phylogenetic extent, suggesting that they may be of an earlier evolutionary origin (5, 7–9). In fact, the sub-network includes proteins involved in processes believed to have appeared early in evolution (e.g., transcription, translation, and replication) *(8–10)*. Thus, this sub-network of proteins with wide phylogenetic extent may represent an ancestral state of the yeast protein interaction network. In addition, this sub-network may represent a "core" common to other species, probably representing an ancestral form of the protein interaction network in Eukarya, around which other species- or taxa-specific interactions may merge *(6)*.

This chapter combines comparative genomics techniques and protein interaction data to infer the ancestral state of protein interaction networks (*see* **Note 1**). To illustrate the use of this approach we will use published yeast protein interaction data *(11)* aiming to reconstruct a potential ancestral form of the yeast protein interaction network. **Figure 19.1** shows a conceptual view of the proposed method. The approach presented here may help us to understand the overall structure and evolution of biological networks.

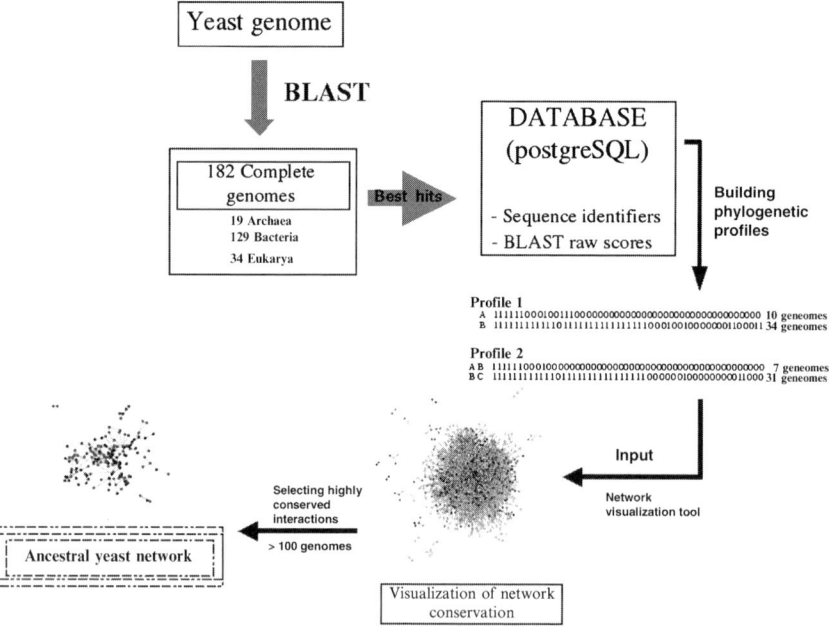

Fig. 19.1. Inferring potential ancestral protein interaction networks. This figure represents a conceptual view of the method presented here. In summary, the *Saccharomyces cerevisiae* genome is blasted against 182 other complete genomes, and the results are stored in a PostgreSQL database. The database is queried, phylogenetic profiles of proteins *(14)* and interactions *(5)* are produced, and the number of genomes in which yeast homologs are found is counted. These data are used as input to visualize the conservation of the interaction network using a network visualization tool. Finally, highly conserved (>100 genomes) nodes (proteins) and edges (protein interactions) are selected from the network. The resulting highly conserved sub-network may represent an ancestral state of the original yeast protein interaction network *(5–7)*.

2. Materials

Our approach assumes that the user has some basic computing/programming skills, such as using Unix/Linux command lines, writing Perl scripts, and building PostgreSQL (or MySQL) databases. Furthermore, it assumes a basic knowledge of bioinformatics tools, such as BLAST.

2.1. Equipment

1. The minimal computational infrastructure required is one Intel workstation with sufficient speed and disk space (*see* **Note 2**).

2.2. General Platform and Computing/Programming Tools

1. Linux (http://www.linux.org/), Perl (http://www.perl.org/), and PostgreSQL (http://www.postgresql.org/) installations.

2.3. Bioinformatics Tools

1. BLAST (ftp://ftp.ncbi.nih.gov/blast/) (12) and BioLayout Java v1.21 (http://cgg.ebi.ac.uk/services/biolayout/) (13) installations.

3. Methods

3.1. Input Data

3.1.1. Yeast Protein Interaction Data

1. Download and save in a file the yeast protein interaction data (core dataset) from the Database of Interacting Proteins (DIP) (http://dip.doe-mbi.ucla.edu/dip/). A total of 2,236 proteins involved in 5,952 interactions were obtained by January 17, 2006. Alternatively, you may want to use a different protein interaction dataset from another species or database/source.

2. Open the file with a file editor (e.g., Excel) and eliminate all columns but the ones corresponding to SwissProt identifiers. You may want to write a Perl script to do this step. At this point the file should have this format:
A B
B C
.. ..
which means that protein A interacts with protein B, B with C, and so on.

3. Remove self-interactions (e.g., A A). The resulting filtered file constitutes our initial protein interaction dataset. This step produced a total of 2,186 proteins involved in 5,164 interactions.

3.1.2. Sequence Data

1. Download and save in a file the ORFs encoding all yeast proteins from the Saccharomyces Genome Database (http://www.yeastgenome.org/). Using SGD, we downloaded a total of 6,714 yeast protein sequences by January 17, 2006 (*see* **Note 3**). These data constitute our initial yeast genome dataset.

2. Download and save in different files the ORFs encoding the protein sequences of the complete genomes present in the GOLD (Genomes OnLine) Database (http://www.genomesonline.org/) (*see* **Notes 3** and **4**). These data constitute our reference genome datasets.

3.2. Sequence Similarity Searches

1. For each *S. cerevisiae* protein sequence perform a BLASTP search, with a threshold bit score of 50 and an expectation value (Evalue) of 10^{-05} (*see* **Note 5**) and default parameters, against each of the reference genome datasets. (See point 2 under **Section 3.1.2**).

3.3. Parsing BLAST Outputs

1. From each BLAST output (*see* **Section 3.2**) we create an additional file (called, for example, yeast_vs_ecoli_parsed, the resulting BLAST output of yeast versus the *Escherichia coli* genome) with this information: yeast sequence identifiers and best BLAST bit scores. The file format should look like this:

   ```
   A 128
   B 55
   C 0
   .. ..
   ```

   which means that yeast protein A versus the reference genome (i.e., *E. coli* in our example) has a BLAST bit score of 128, protein B has a bit score of 55, and protein C has no sequence similarity to the reference genome (i.e., the BLAST E-value is higher than i.e., the BLAST bit score is less than the specified threshold, or there is no sequence similarity at all; *see* **Section 3.2**, and **Notes 5** and **6**).

3.4. Building a PostgreSQL Database

1. Use the file generated in **Section 3.3** as input to build a PostgreSQL database (*see* **Note 7**) where the first field (column) (PostgreSQL primary key) represents yeast sequence identifiers and the remaining fields represent the best BLAST bit scores of yeast versus the reference genomes (see point 2 in **Section 3.1.2**). *See* **Fig. 19.2** for an example of the result.

3.5. Querying the Database and Building Phylogenetic Profiles

1. Query each row of the database table (*see* **Section 3.4**) by first grouping the table fields (i.e., reference genomes) by taxonomy. One possible query involves grouping all genomes from Archaea together, followed by Bacteria, then Eukarya (*see* **Fig. 19.2**). This step is necessary for assessing sequence conservation (*see* **Section 3.6**).

```
select clus_id, M_thermoautotrophicum, M_jannaschii, M_maripaludis, V_cholerae, V_parahaemolyticus, W_pipientis, D_melanogaster, H_sapiens,
C_elegans from blast limit 10;
```

clus_id	M_thermoautotrophicum	M_jannaschii	M_maripaludis	V_cholerae	V_parahaemolyticus	W_pipientis	D_melanogaster	H_sapiens	C_elegans
YAL005C	439	0	0	503	506	513	848	863	745
YAL035W	348	323	333	134	130	133	390	607	207
YAL036C	309	302	274	70	67	59	366	438	227
YAR073W	207	220	196	243	248	224	236	407	165
YAR075W	81	82	83	80	77	85	145	195	95
YBL038W	0	0	0	84	84	80	0	0	0
YBL075C	434	0	0	498	494	516	868	873	767
YBL076C	565	522	488	322	306	644	811	1159	404
YBL099W	95	94	90	502	503	547	426	696	600
YBR025C	65	83	68	252	246	233	179	402	197

(10 rows)

Fig. 19.2. This figure shows a screenshot of a simplified PostgreSQL query (the number of fields was reduced to 12 and the number of rows to 10), in which the top line represents the database query; the first field (column) represents yeast sequence identifiers, and the remaining fields represent the best BLAST bit scores of yeast versus the reference genomes (field names) (see **Section 3.3**). BLAST bit scores < 50 were set to zero. Fields were queried in this order: 3 Archaea, 3 Bacteria, and 3 Eukarya genomes (see top line).

2. Using the output of this database query, generate a binary sequence for each protein, in which a "1" represents those genomes for which the protein has a BLAST bit score ≥50 and "0" represents those genomes in which the protein is potentially not present or has a bit score below the threshold (*see* **Section 3.2**). Save the resulting profiles in a file (*see* **Note 8**). The file format should look like this:

   ```
   A 111111000100111000000000000000000000000000000000
   B 111111111110111111111111111100010010000000110001111
   .. .. .. .. .. .. .. .. .. ..
   ```

   which means that protein A is potentially present in genomes 1, 2, 3, 4, 5, 6, 10, 13, 14, 15, and so on.

3. In addition, for each pair of yeast proteins from our interaction dataset (see point 3 in **Section 3.1.1**), query the database, as described, for those proteins that match the two interacting proteins from our interaction dataset, and generate a binary sequence in which a "1" represents those genomes for which both two proteins from our interaction dataset have a bit score ≥50 in our database, and a "0" represents genomes in which both proteins are not present in the different reference genomes (*see* **Note 9**). Save the result in a file. The file format should look like this:

   ```
   A B 111111000100000000000000000000000000000000000000000
   B C 11111111111101111111111111111110000001000000000110000
   .. .. .. .. .. .. .. .. .. ..
   ```

   which means that the interaction between protein A and B is potentially present in genome 1, 2, 3, 4, 5, 6, 10, and so on.

3.6. Analyzing Sequence Conservation: Number of Genomes and Phylogenetic Extent

1. Count the number of 1s in the two profile versions we have created (see points 2 and 3 in **Section 3.5**). The end result should be two files with these formats:

   a. For individual proteins:

   ```
   A 10
   B 36
   C 140
   .. ..
   ```

   which represents that protein A is potentially present in 35 reference genomes.

   b. For interacting proteins:

   ```
   A B 7
   B C 31
   D E 121
   .. .. ..
   ```

   which means that the interaction A B is potentially present in 7 reference genomes, and so on.

2. Transform the protein profiles (with the reference genomes—table columns—grouped by taxonomy; see point 1 in **Section 3.5**) into an abbreviated profile of three 0s/1s in this order: Archaea, Bacteria, and Eukarya (*see* **Note 10**). The file format should be:

```
A 011
B 110
.. ..
```

which means that protein A is absent in Archaea but is present in Bacteria and Eukarya, and so on. These abbreviated profiles represent the phylogenetic extent *(7)* of the yeast protein sequences.

3.7. Visualization of Protein Interaction Data Using BioLayout Java

BioLayout (http://cgg.ebi.ac.uk/services/biolayout/) is an automatic graph layout algorithm for similarity and network visualization, and is free for academic and non-profit research institutions (*see* **Note 11**).

3.7.1. Input Data

1. Change the format of the file obtained in point 1a in **Section 3.6** (i.e., protein A is potentially present in *n* reference genomes) by classifying the numeric values (i.e., the number of genomes in which homologs of a protein are found) into node classes (i.e., Nodes (proteins) with values ≤50 are assigned to class 1, values between 50 and 100 to class 2, values between 100 and 150 to class 3, and values >150 to class 4) (*see* **Note 12**).

2. Concatenate the resulting file to the bottom of the file obtained earlier (see point 1b in **Section 3.6**). The end file format should look like this (*see* **Note 13**):

```
A B 7
B C 31
C D 121
//NODECLASS A 1
//NODECLASS B 1
//NODECLASS C 3
.. .. .. .. .. .. .. .. .. ..
```

3.7.2. Loading and Visualizing Protein Interaction Data

1. Run BioLayout. Click on File → Open in the menu, and select the file to be loaded (see point 2 in **Section 3.7.1**). Click on Open. This step will produce a first look at the entire yeast protein interaction network.

2. Change the network layout (e.g., node size and edge directionality) by clicking on Edit → Selection → Select All → Tools → Properties → Vertices → Size [8], and Apply (to change node size); and click on Tools → Properties → General → Graph Options, uncheck Directional, and click on Apply (to remove edge directionality in the network).

3. Change the node color scheme by clicking on Tools → Properties → Classes, and changing the classes color (nodes) by clicking on the default colored boxes and choosing an appropriate color from the java pallet. This step is done in order to give a more meaningful color scheme to the nodes (proteins) in the network, i.e., we change node colors according to their degree of conservation. Here, we use a color grading from white (class 1, less conserved) to black (class 4, more highly conserved) (see point 1 in **Section 3.7.1**). Next, we add some text to the empty boxes to give a meaning to the different node classes (*see* **Fig. 19.3**). Finally, click on Apply and then Ok in the Edit Properties menu.

4. Click on View → Show Legend (*see* **Note 14**). At this point, the yeast interaction network should look like **Fig. 19.4**.

3.7.3. Inferring a Potential Ancestral State of the Yeast Protein Interaction Network

1. Click on Edit → Filter by Weight → Yes, add the value 100 in the box, and click on Ok (*see* **Note 15**).

2. Click on View → Show All Labels. This step will add yeast sequence identifiers to the nodes in the resulting sub-network.

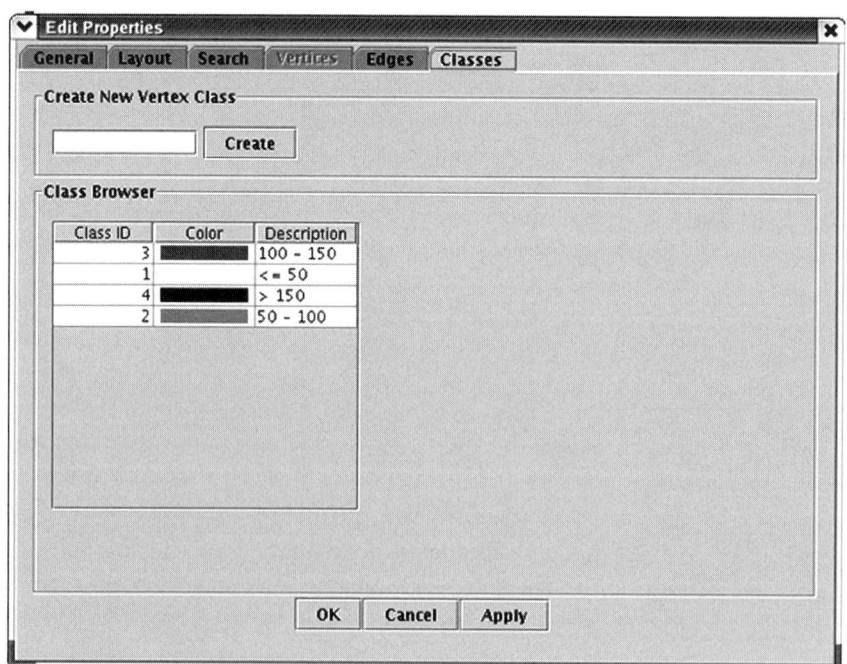

Fig. 19.3. BioLayout Edit Properties screenshot. Colors represent the conservation (i.e., the number of genomes in which yeast homologs are found, classified into four bin classes: ≤50; 50–100; 100–150; and >150 genomes).

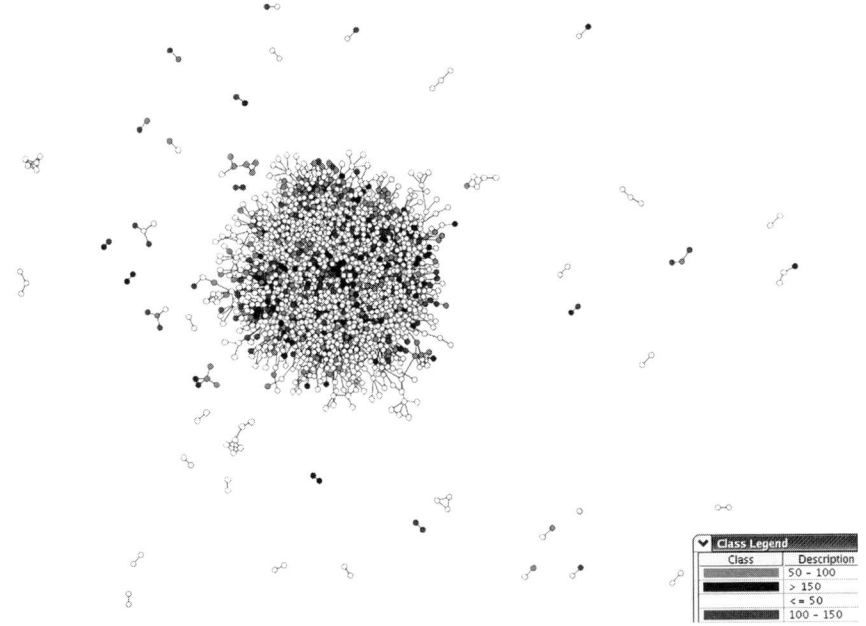

Fig. 19.4. The conservation of the yeast protein interaction network. This network is mainly formed by a giant cluster of interconnected nodes, surrounded by very small protein interaction clusters. The conservation of the network can be seen easily by looking at the predominant node colors. Part of the network appears to be highly conserved (which is suggested by the high number of gray and black nodes in the network). (See the legend inset and Fig. 19.3).

These steps will allow us to visualize our goal, a network of core proteins and interactions that potentially represents an ancestral state of the yeast protein interaction network (*see* **Fig. 19.5**).

4. Notes

1. The main advantage of the method presented here is its simplicity. Other much more complicated and time-consuming methods for the reconstruction of ancestral genomes have been reviewed recently *(17)*. These methods consider genomes as phylogenetic characters and reconstruct the phylogenetic history of the species and their ancestral states in terms of genome structure or function. The application of these methods in the context of protein interactions networks is an alternative approach for inferring ancestral protein interaction networks.

2. Computational infrastructure is a limiting factor when using the approach discussed here. Given the high number of similarity searches necessary to run our experiments, we have performed our analysis using the supercomputer facilities

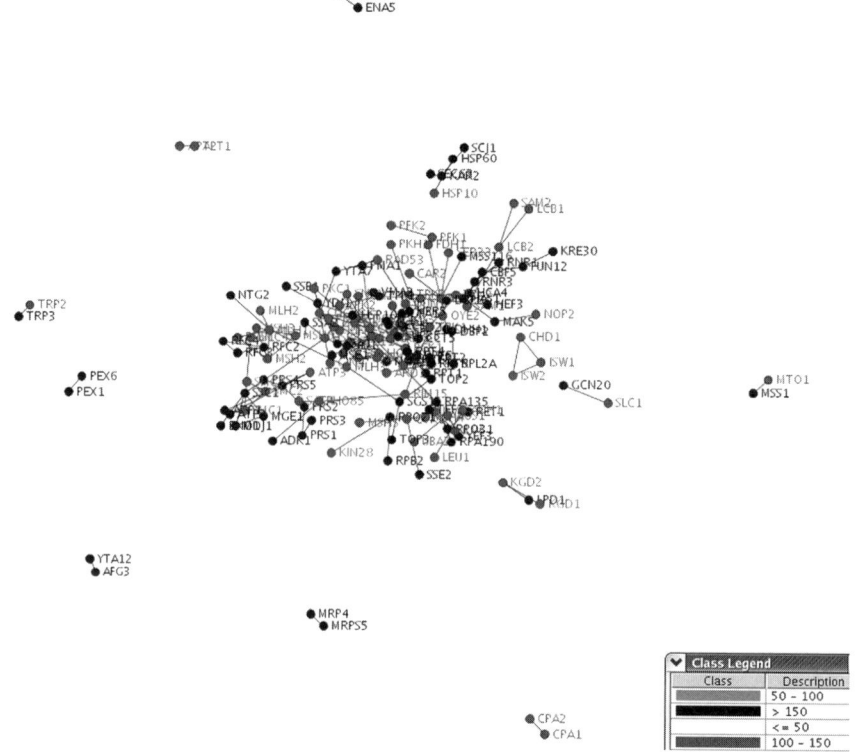

Fig. 19.5. Ancestral state inference of the yeast protein interaction network. This yeast sub-network (part of the original yeast network; see **Fig. 19.4**) is formed by highly conserved (present in >100 genomes) proteins and interactions. A closer examination of the reduced profiles generated (see point 2 in Section 3.6) shows that all proteins in this sub-network are universally conserved (i.e., present in Prokarya and Eukarya). This suggests that this sub-network may represent the ancestral form of the yeast protein interaction network (5–7). In addition, it may represent a "core" sub-network that is common to other interaction networks from other species. It may be an ancient form of the protein interaction network in Eukarya (6), or possibly a protein interaction network present in the universal ancestor (7, 15). In fact, a further characterization of the functional categories of the proteins participating in this potential ancestral state of the yeast network shows that many of these proteins participate in processes suggested to have appeared early in evolution, such as energy, metabolism, transcription, translation, transport, and replication (7–8, 10, 16).

located at the Centre for Computational Biology (Hospital for Sick Children, Toronto, Ontario, Canada).

3. Alternatively, the user may want to use other biological databases to retrieve these sequence data, such as COGENT (http://cgg.ebi.ac.uk/services/cogent/) or NCBI (ftp://ftp.ncbi.nih.gov/genomes/).

4. In this study, we used a filtered subset of the 335 complete genomes available in the GOLD database (see point 2 in **Section 3.1.2**) on January 17, 2006; that is, only one strain was used per species. This was done for simplicity and to avoid taxonomic redundancy. Thus, eventually a total of 182 complete genomes were downloaded (19 Archaea, 129 Bacteria, and 34 Eukarya). The user may want to download a smaller number of complete genomes in order to reduce

similarity search space using BLAST. This reduction should be done with caution to maximize taxonomic coverage.

5. A BLAST Evalue threshold of 10^{-05} is used initially to reduce the size of the BLAST output files. This initial Evalue threshold is complemented in later steps by applying an additional BLAST bit score threshold of 50, which has been found to be appropriate for the analysis of sequence similarity across genomes *(5)*. A BLAST bit score of 50 approximately corresponds to an Evalue of 10^{-06}. The user may want to use a more stringent threshold.

6. BLAST outputs can be parsed using your own Perl script or using Bioperl (http://www.bioperl.org/wiki/Main_Page). Alternatively, other BLAST output information could be stored in the parsed files (*see* **Section 3.3**), such as Evalue, % Sequence Identity, etc.

7. Alternatively, the user may want to create a MySQL database (http://www.mysql.com/), or even a flat file, instead of a PostgreSQL database. We chose to build a SQL database because it allows us to store the data at low disk space cost, and in a safe and convenient format that could be easily queried in possible more detailed future analyses.

8. These profiles are called phylogenetic profiles (*see* **Chapter 9** of **Volume 2**); they represent patterns of sequence co-occurrence across genomes *(14)*. These profiles usually have a value of 1 when a homolog is present, and 0 when it is absent.

9. We consider a protein interaction to be present across genomes if both interacting proteins have detectable homologs in any of the reference genomes analyzed. Otherwise, we consider the interaction to be not present in the reference genomes. The resulting profile is a modified version of a phylogenetic profile *(5)*.

10. In the abbreviated profiles, the values 0 and 1 represent the pattern of absence and presence, respectively, in different taxonomic groups *(7)*. These taxonomic groupings correspond to the three domains of life: Archaea, Bacteria, and Eukarya (see point 1 in **Section 3.5**). Different taxonomic classifications also could be adopted. For example, one could further split the Eukaryotic domain into Protist, Fungi, Metazoa, and Plantae *(7)*, or other more specific taxonomic classifications. Alternatively, a different format for the abbreviated profiles could be used, such as abbreviations instead of 0s/1s digits (e.g., protein1 BE, which means that protein1 is only present in Bacteria and Eukarya). This format requires correction since all yeast protein sequences are present in Eukarya by definition. Thus, for example, this hypothetical yeast abbreviated profile case "protein2 AB" should be corrected to "protein2 ABE."

11. Biolayout is a powerful and easy-to-use tool for network visualization. An alternative is Cytoscape (http://www.cytoscape.org/index.php), which is an open-source software package with additional features available as plug-ins (e.g., for network and molecular profiling analyses, new layouts, additional file format support, and connection with databases).

12. A file containing graph connections and properties is required by BioLayout to produce a graph layout. This file should have this format (see point 1B in **Section 3.6**):

    ```
    A B 7
    B C 31
    D E 121
    .. .. ..
    ```

    The first two columns represent two nodes (proteins) to be connected by an edge (protein interaction). The third column specifies an edge weight, which in our case represents the number of genomes in which homologs of two interacting proteins are found. The preceding example will create a graph with five nodes (A, B, C, D and E). According to the input, nodes A and B are connected, B and C are connected, and finally D and E are connected.

    BioLayout allows specification of "node classes" to distinguish nodes from each other, using this format:

    //NODECLASS A 1

    //NODECLASS B 1

    //NODECLASS C 3

    .. ..

    Node classes will have different colors in the final BioLayout graph and are specified using property lines in the input file. This means that in the preceding example node A and B will share the same color, whereas node C will have a different color. Node classes can be created and modified using the class properties browser and vertex properties in the BioLayout properties dialog.

13. Notice that the values on the first part of this formatted file (e.g., A B 7) refer to protein interactions (edges) (see point 1b in **Section 3.6**), whereas the values of the second part (e.g., //NODECLASS A 1) refer to individual proteins (nodes) (see point 1a in **Section 3.6**).

14. This step will display a class legend: a box with the meaning of node colors. Notice that by default the sentence "No Class" with a different color (usually green) will show up in the class legend. This will be the color that any node will have in the network if any data (in our case, sequence identifiers represented as interactions) do not match our class format.

15. Once a proper visualization of the network has been established, this step will filter out non-highly conserved nodes, keeping only the highly conserved ones, which potentially form the ancestral state of the yeast interaction network *(5–7)*. Since we have searched for sequence similarities across 182 other complete genomes, a conservation threshold of >100 genomes (which, in our case, includes protein and interactions widely distributed across Prokarya and Eukarya) is appropriate for this analysis. Alternatively, the user may want to choose a more flexible or stringent criterion of conservation. The more stringent the criteria of conservation used as filter, the smaller the resulting subnetwork size, and vice versa.

Acknowledgments

The author thanks John Parkinson for reading the manuscript and making useful comments. This work was supported by the Hospital for Sick Children (Toronto, Ontario, Canada) Research Training Center.

References

1. Uetz, P., Giot, L., Cagney, G., et al. (2000) A comprehensive analysis of protein-protein interactions in *Saccharomyces cerevisiae*. *Nature* 403, 623–627.
2. Gavin, A. C., Bosche, M., Krause, R., et al. (2002) Functional organization of the yeast proteome by systematic analysis of protein complexes. *Nature* 415, 141–147.
3. Giot, L., Bader, J. S., Brouwer, C., et al. (2003) A protein interaction map of *Drosophila melanogaster*. *Science* 302, 1727–1736.
4. Li, S., Armstrong, C. M., Bertin, N., et al. (2004) A map of the interactome network of the metazoan *C. elegans*. *Science* 303, 540–543.
5. Butland, G., Peregrin-Alvarez, J. M., Li, J., et al. (2005) Interaction network containing conserved and essential protein complexes in *Escherichia coli*. *Nature* 433, 531–537.
6. Pereira-Leal, J. B., Audit, B., Peregrin-Alvarez, J. M., et al. (2005) An exponential core in the heart of the yeast protein interaction network. *Mol Biol Evol* 22, 421–425.
7. Peregrin-Alvarez, J. M., Tsoka, S., Ouzounis, C. A. (2003) The phylogenetic extent of metabolic enzymes and pathways. *Genome Res* 13, 422–427.
8. Makarova, K. S., Aravind, L., Galperin, M. Y., et al. (1999) Comparative genomics of the Archaea (Euryarchaeota): evolution of conserved protein families, the stable core, and the variable shell. *Genome Res* 9, 608–628.
9. Harris, J. K., Kelley, S. T., Spiegelman, G. B., et al. (2003) The genetic core of the universal ancestor. *Genome Res* 13, 407–412.
10. Kyrpides, N., Overbeek, R., Ouzounis, C. (1999) Universal protein families and the functional content of the last universal common ancestor. *J Mol Evol* 49, 413–423.
11. Deane, C. M., Salwinski, L., Xenarios, I., et al. (2002) Protein interactions: two methods for assessment of the reliability of high-throughput observations. *Mol Cell Prot* 1, 349–356.
12. Altschul, S. F., Gish, W., Miller, W., et al. (1990) Basic local alignment search tool. *J Mol Biol* 215, 403–410.
13. Goldovsky, L., Cases, I., Enright, A. J., et al. (2005) BioLayout (Java): versatile network visualisation of structural and functional relationships. *Appl Bioinformatics* 4, 71–74.
14. Pellegrini, M., Marcotte, E. M., Thompson, M. J., et al. (1999) Assigning protein

functions by comparative genome analysis: protein phylogenetic profiles. *Proc Natl Acad Sci U S A* 96, 4285–4288.

15. Woese, C. (1998) The universal ancestor. *Proc Natl Acad Sci U S A* 9, 6854–6859.

16. Ouzounis, C. A., Kunin, V., Darzentas, N., et al. (2005) A minimal estimate for the gene content of the last common ancestor: exobiology from a terrestrial perspective. *Res Microbiol* 2005; Epub ahead of print.

17. Ouzounis, C. A. (2005) Ancestral state reconstructions for genomes. *Curr Opin Genet Dev* 15, 595–600.

Chapter 20

Computational Tools for the Analysis of Rearrangements in Mammalian Genomes

Guillaume Bourque and Glenn Tesler

Abstract

The chromosomes of mammalian genomes exhibit reasonably high levels of similarity that can be used to study small-scale sequence variations. A different approach is to study the evolutionary history of rearrangements in entire genomes based on the analysis of gene or segment orders. This chapter describes three computational tools (GRIMM-Synteny, GRIMM, and MGR) that can be used separately or in succession to contrast different organisms at the genome-level to exploit large-scale rearrangements as a phylogenetic character.

Key words: rearrangements, algorithms, homologous regions, phylogenetic tree, computational tool.

1. Introduction

The recent progress in whole genome sequencing provides an unprecedented level of detailed sequence data for comparative study of genome organizations beyond the level of individual genes. This chapter describes three programs that can be used in such studies:

1. GRIMM-Synteny: Identifies homologous synteny blocks across multiple genomes.
2. GRIMM: Identifies rearrangements between two genomes.
3. MGR: Reconstructs rearrangement scenarios among multiple genomes.

Genome rearrangement studies can be dissected into two steps: *(1)* identify corresponding orthologous regions in different genomes, and *(2)* analyze the possible rearrangement scenarios that can explain the different genomic organizations. The orthologous regions are typically numbered 1, 2, ... *n*, and each genome is represented as a signed permutation of these numbers, in which the signs indicate the relative orientation of the orthologous regions. In multi-chromosomal genomes, by convention, a "$" delimiter is inserted in the permutation to demarcate the different chromosomes.

The types of rearrangements that are considered by GRIMM and MGR are illustrated in **Fig. 20.1** using this signed permutation notation. In uni-chromosomal genomes, the most common rearrangements are reversals (also called inversions), shown in **Fig. 20.1A**; in signed permutation notation, a contiguous segment of numbers are put into reverse order and negated. In multi-chromosomal genomes, the most common rearrangements are reversals, translocations, fissions, and fusions. A fusion event concatenates two chromosomes into one, and a fission breaks one chromosome into two (*see* **Fig. 20.1B**). A translocation event transforms two chromosomes A B and C D into A D and C B, in which each letter stands for a sequence of signed genes (*see* **Fig. 20.1C**). There are other modes of evolution, such as large-scale insertions, deletions, and duplications, which are not addressed in the present chapter.

Two main categories of input data can be used to analyze genome rearrangements: *(1)* sequence-based data, relying on nucleotide alignments; and *(2)* gene-based data, relying on homologous genes or markers, typically determined by protein alignments or radiation-hybrid maps. The appropriate acquisition of such datasets is discussed in **Section 2.3**. Processing of this raw data to determine large-scale syntenic blocks is done using GRIMM-Synteny; this is discussed in **Section 3.1**. After constructing the syntenic blocks, GRIMM can be used to study the rearrangements between pairs of species *(1)*. GRIMM implements the Hannenhalli-Pevzner methodology *(2–7)*, and can efficiently compute the distance between two genomes and return an

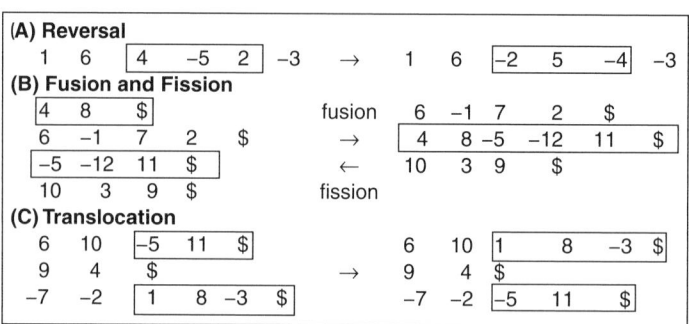

Fig. 20.1. Rearrangements in signed permutations showing impact of: (**A**) reversals; (**B**) fusions and fissions; and (**C**) translocations.

optimal path of rearrangements; it is described in **Section 3.2**. Finally, MGR is a tool to study these types of rearrangements in several genomes, resulting in a phylogenetic tree and a rearrangement scenario *(8)*; it is described in **Section 3.3**.

Bourque and colleagues presented a detailed application of these tools *(9)*. In that study, the global genomic architecture of four species (human, mouse, rat, and chicken) was contrasted using the two types of evidence: sequence- and gene-based data. That study is used as a reference point for many of the input and output files provided in this chapter.

Typically, the running time for GRIMM-Synteny is seconds to minutes. GRIMM takes a fraction of a second for most uses, and MGR takes minutes to days. However, it may take considerable time to prepare the inputs and analyze the outputs of each program.

2. Materials

2.1. Computer Requirements

At the time this is written, the software requires a UNIX system with shell access and a C compiler. PERL is also recommended for writing custom scripts to convert the output from other software to a format suitable for input to the software.

2.2. Obtaining the Software

The GRIMM-Synteny, GRIMM, and MGR software, and demonstration data used in this chapter, should be downloaded from http://www.cse.ucsd.edu/groups/bioinformatics/GRIMM/ The web site also has a web-based version of GRIMM and MGR *(1)*. The Web-based version is for small datasets only and cannot handle the datasets described in this chapter.

After downloading the files, run the following commands. Note that % is the Unix prompt, the commands may differ on your computer, and the version numbers might have changed since publication.

For GRIMM-Synteny:

```
% gzip -d GRIMM_SYNTENY-2.01.tar.gz
% tar xvf GRIMM_SYNTENY-2.01.tar
% cd GRIMM_SYNTENY-2.01
% make
```

For GRIMM-Synteny demonstration data hmrc_align_data:

```
% gzip -d hmrc_align_data.tar.gz
% tar xvf hmrc_align_data.tar
```

For GRIMM:

```
% gzip -d GRIMM-2.01.tar.gz
% tar xvf GRIMM-2.01.tar
% cd GRIMM-2.01
% make
```

For MGR:

```
% gzip -d MGR-2.01.tar.gz
% tar xvf MGR-2.01.tar
% cd MGR-2.01
% make
```

The executable for GRIMM-Synteny is grimm_synt; the executable for GRIMM is grimm; the executable for MGR is MGR. Either copy the executable files to your bin directory or add the directories to your PATH variable. For updated installation details, please check the web site and read the README file in each download.

2.3. Your Data

Various types of input datasets can be used to study large-scale rearrangements. Typically, every multi-genome comparative analysis project provides either sequence-based alignments or sets of homologous genes. The challenge is that the actual source and format differ from one study to the next. For this reason, a simple file format for GRIMM-Synteny, and another for GRIMM and MGR, was created containing only the information the respective programs needed. For each analysis project, it should be relatively straightforward to write custom conversion scripts (e.g., using PERL) to extract the information required from the original source dataset and output it in the standardized format required for the GRIMM/MGR suite. A description of the type of information needed is in the following subsections, and a detailed description of the formats is in **Section 3**.

It is also possible to acquire datasets with homologous genes or aligned regions from public databases, such as Ensembl, the National Center for Biotechnology Information's (NCBI) HomoloGene, and the University of California, Santa Cruz Genome Bioinformatics web site (*see* web site references). The web site interfaces and database formats tend to change on a frequent basis. It may be necessary to combine multiple tables or do computations to get the required information.

2.3.1. Data for Input to GRIMM-Synteny

The inputs to GRIMM-Synteny describe the coordinates of multi-way orthologous sequence-based alignments or multi-way orthologous genes. Either one of these is called an orthologous marker. There are several output files; principally, it outputs large-scale syntenic blocks (similar to conserved segments but allowing for micro-rearrangements) comprised of many orthologous elements that are close together and consecutive or slightly shuffled in order. The specific details are given in **Section 3.1**. The information that you must have for every orthologous element is its coordinates in *every* species. The coordinates include the chromosome, starting nucleotide, length in nucleotides, and strand (or relative orientation). Optionally, you may also assign an ID number to each orthologous element (*see* **Fig. 20.2**).

If you do not have the coordinates in nucleotides, but do know the order and strand of the elements across all the genomes, you may specify fake coordinates that put them into the correct order, and tell GRIMM-Synteny to use the "permutation metric" that only considers their order.

It is possible to deal with data in which the strand of the orthologous markers is unknown (e.g., if the source data come from Radiation-Hybrid mapping), but it is beyond the scope of GRIMM-Synteny (*see* **Note 1**).

2.3.2. Data for Input to GRIMM and MGR

GRIMM can be used to compare the order of elements within orthologous regions in two genomes, or the orders of syntenic blocks between two genomes on a whole genome scale. MGR can be used for these purposes with three or more genomes.

GRIMM-Synteny produces a file called mgr_macro.txt suitable for input to GRIMM or MGR. If you are not using GRIMM-Synteny, you have to number your orthologous regions or syntenic blocks and create a file in a certain format that specifies the orders and signs (orientations or strands) of these for each species. The format is described in **Section 3** (*see* **Fig. 20.3**).

The signs are very important and the quality of the results deteriorates if you do not know them. They should be available with alignments or genes obtained from current whole genome

```
(A) Excerpted lines from hmrc_genes_coords.txt
# Homologous gene coordinate inputs for GRIMM-Anchors
# genome1: Human
# genome2: Mouse
# genome3: Rat
# genome4: Chicken
0 2 69684252 160454 −1 6 87353142 137355 1 4 121355398 32464 1 22 23066 54613 1
0 2 69597801 41340 −1 6 87513091 18625 1 4 121416392 14848 1 22 89145 11384 1
0 X 127379875 52276 1 X 34167557 18067 1 X 134345833 48106 1 4 129940 18603 1
0 5 126929555 37439 1 18 57731473 26125 1 18 53194208 22005 1 W 4319028 17980 1
```

```
(B) Excerpted lines from hmrc_align_coords.txt
# 4-way alignment coordinate inputs for GRIMM-Anchors
# genome1: Human
# genome2: Mouse
# genome3: Rat
# genome4: Chicken
0 1 2041 54 + 17 64593043 54 + 4 158208520 52 − 1 56599921 54 +
0 1 2459 105 + 6 122223200 100 − 4 158202900 102 − 1 56600370 109 +
0 1 69708707 115 + 3 158985947 122 − 2 256411110 117 − W 2749741 117 −
0 2 19353347 207 + X 69783233 211 + X 71631432 211 + 1 108993546 211 +
0 X 153118976 182 + X 57378479 182 − X 159163656 182 − 4 1950771 182 −
0 _ 0 1 + _ 0 1 + _ 0 1 + Z 32174744 1081 +
```

Fig. 20.2. (**A**) Sample lines from gene coordinate file hmrc_genes_coords.txt used for GRIMM-Synteny. The first field (ID number) is set to 0 since it is not useful in this example. After that, each species has four fields: chromosome, start, length, sign. (**B**) Sample lines from alignment coordinate file hmrc_align_coords.txt. Notice the fourth alignment shown has human chromosome 2 aligned to mouse and rat X. The sixth "alignment" shown uses a fictitious chromosome "_" as a means to filter out a segmental duplication involving chicken chromosome Z.

assemblies. However, if your source data really do not have them (e.g., gene orders obtained in a radiation hybrid map), GRIMM has procedures to guess relative signs (for two or more species). These are described in **Sections 3.2.3** and **4**.

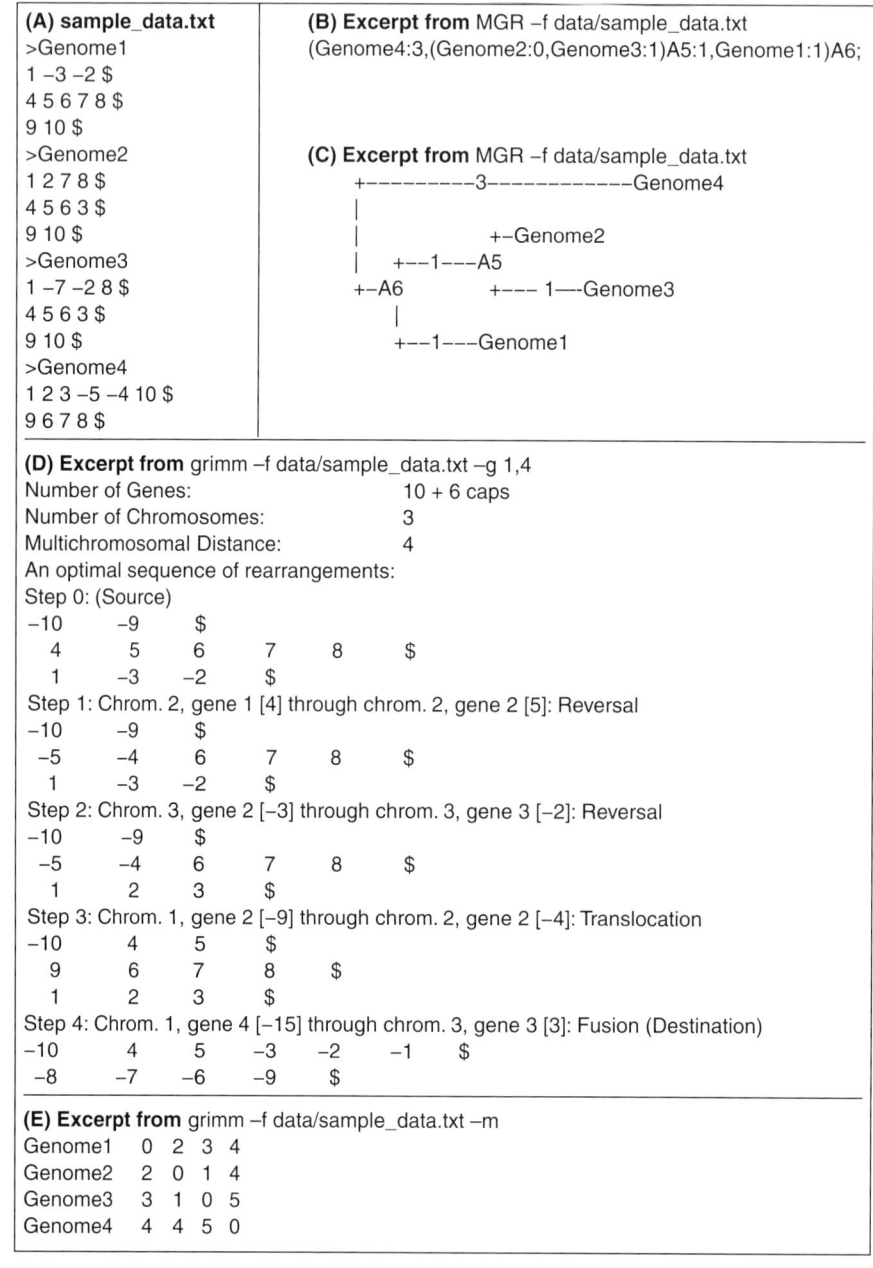

Fig. 20.3. (**A**) Sample input file for GRIMM or MGR. Part of MGR's output: (**B**) Newick representation of reconstructed phylogeny and (**C**) ASCII graphical tree representation of the same phylogeny. (**D**) Part of GRIMM's output: an optimal sequence of rearrangements from Genome1 to Genome4. (**E**) GRIMM's 4 × 4 pairwise distance matrix on the input genomes. MGR also produces a 6 × 6 matrix for the input genomes plus ancestral genomes.

3. Methods

3.1. GRIMM-Synteny: Identifying Homologous Synteny Blocks Across Multiple Genomes

There are two main uses of GRIMM-Synteny: *(1)* GRIMM-Anchors to filter out non-unique alignments (**Section 3.1.2**), and *(2)* forming synteny blocks from anchors (**Sections 3.1.3** and **3.1.4**). Both require the same input format, which is covered first.

3.1.1. Preparing the Input for GRIMM-Synteny

We work with the human/mouse/rat/chicken orthologous alignments as a starting point (computed by Angie Hinrichs) (10, 11). The discussion for genes would be similar. This example has k = 4 species. GRIMM-Synteny uses the coordinates of k-way alignments. The input file consists of many lines with the following format ("k-way coordinate format"), in which each line represents coordinates of a k-way alignment (but does not include the base-by-base details of the alignment). The same format is used in the output file blocks.txt that lists the coordinates of the synteny blocks. *See* **Fig. 20.2** for excerpts from a four-way coordinate file, and **Fig. 20.4A** for an illustration of the coordinate system (in two-way data). The k-way coordinate format is as follows:

```
ID chr1 start1 length1 sign1 … chrk startk
   lengthk signk
```

1. ID is a number, which can be used to number the alignments. If you do not care to do this, set it to 0. GRIMM-Synteny does not use the value you put there on input. On output, the same format is used for the file blocks.txt and the ID is used to number the blocks.

2. Species numbering: In the human/mouse/rat/chicken data, chr1, start1, length1, sign1 refer to coordinates in human. Species 2 is mouse. Species 3 is rat. Species 4 is chicken. For your own data, choose your own species numbers and use them consistently. Please note that the coordinates were for particular assembly versions; they will change as newer assemblies are produced, and care must be taken in comparing results produced from different assembly versions.

3. The four coordinate fields per species are as follows:
   a. chrN: Chromosome name, e.g., "1," "X," "A1."
   b. startN: Starting nucleotide on the positive strand. It does not matter if you used 0- or 1-based coordinates, so long as you are consistent.
   c. lengthN: Length in nucleotides. Combined with startN, this gives a half-open interval [startN, startN+lengthN) on species N. If your source data has start and end coordinates of a closed interval [start,end] then the length is end−start+1, whereas if it is a half-open interval [start,end), then the length is end−start.

d. signN: Strand (+ or 1 for positive, – or –1 for negative). Negative means that the aligned nucleotides are the ones paired to the positive strand nucleotides on the interval just specified. Be careful to check that you use the same coordinate for both members of a base pair, since it is also common to have complementary coordinates on the two strands.

The input file may also include comment lines, which begin with "#." Comments in a special format may be included to give the species names, as shown in **Fig. 20.2**.

Your multi-way alignment or multi-way ortholog procedure may produce partial alignments involving fewer than all k species; you must discard those. Your procedure may produce multiple hits involving the same coordinates. GRIMM-Synteny has a procedure GRIMM-Anchors *(9)* to assist in filtering out alignments with conflicting coordinates. Since all alignments with conflicting coordinates are discarded, we recommend that you first determine if your source data have information (e.g., scoring information) that you could use to choose a unique best hit and discard the others. This procedure is described next, followed by the main procedure (GRIMM-Synteny).

3.1.2. GRIMM-Anchors: Filtering Out Alignments with Non-Unique Coordinates

1. Create a file (e.g., align_coords.txt) with the alignment coordinates in the k-way coordinate format described in **Section 3.1.1**.
2. Create a directory (e.g., anchors) in which to place the output files. The current directory will be used by default.
3. Run the GRIMM-Anchors algorithm to filter out repeats and keep only the anchors. The basic syntax is:

```
% grimm_synt -A -f align_coords.txt -d anchors
```

You should replace align_coords.txt by the name of your alignment coordinates file, and anchors by the name of your output directory. The switch –A says to run GRIMM-Anchors.

4. Three output files are created in the directory anchors:
   a. report_ga.txt: This is a log file with information on the number of conflicting alignments, the number of repeat families detected and filtered out (by merging together collections of conflicting alignments), and the number of anchors (unique alignments) remaining. In some cases, such as directed tandem repeats, overlapping alignments are still uniquely ordered with respect to all other alignments, so they are merged into a larger "anchor" instead of being discarded. This is useful for the purpose of determining larger synteny blocks, even if it might not be crucial for most downstream analyses.
   b. unique_coords.txt: This has the same format as described in **Section 3.1.1**, but all conflicting alignments have been removed.

c. repeat_coords.txt: This lists the conflicting alignments that were merged or discarded, organized into repeat families.

3.1.3. Forming Synteny Blocks from Sequence-Based Local Alignments

A chromosome window is a specification of chromosomes $(c_1,...,c_k)$ over the k species. Synteny blocks are formed by grouping together nearby anchors in each chromosome window, as shown in **Fig. 20.4**. This is controlled by two sets of parameters: parameters that control the maximum allowable gap between anchors, and parameters that control the minimum size of a block.

Let $x = (x_1,...,x_k)$ and $y = (y_1,...,y_k)$ be two points in the same chromosome window, with coordinates expressed in nucleotides. The distance between them in species N is $d_N(x,y) = |x_N − y_N|$ and the total distance between them is the Euclidean distance $d(x,y) = |x_1 − y_1| + ... + |x_k − y_k|$.

Each anchor A can be represented as a diagonal line segment in k dimensions between two points within a chromosome window: $a = (a_1,...,a_k)$ and $a' = (a_1',...,a_k')$ (*see* **Fig. 20.4A**). These are the two terminals of A. They are determined by the start coordinates, lengths, and orientations in the k-way coordinate format. If the orientation in species N is positive, then a_N = start$_N$ and a_N' = start$_N$+ length$_N$− 1, whereas if the orientation in species N is negative then the definitions of a_N and a_N' are reversed.

Let A and B be two anchors in the same chromosome window. The total distance between A and B is the total distance between their closest terminals. Once the closest terminals have been determined, the distance between A and B in species N is the distance between those closest terminals in species N (*see* **Fig. 20.4B,C**). The per-species distances are shown as d_1 and d_2, whereas the total distance is $d_1 + d_2$. Had other combinations of anchor terminals been used (shown in **Fig. 20.4B,C** with dotted or dashed lines), the total distance would have been larger.

Anchors A and B in the same chromosome window are connected together if their distance in each species is less than a *per-species gap threshold* specified for that species (see the –g option). Groups of connected anchors form potential synteny blocks. Alternately, anchors may be joined together if the sum of the distances across all species is less than a specified *total gap threshold* (see the –G option). This was the approach Pevzner and Tesler used in their human–mouse comparison *(12, 13)*. The per-species threshold was added as additional species were considered, and seems to work better than the total gap threshold.

To distinguish between noise vs. real blocks, several measurements of the minimum size of a block are available: the span, the support, and the number of anchors. (*See* **Fig. 20.4B** and the –m, –M, and –n options in the next section.) Any potential block that is smaller than these minimums is discarded.

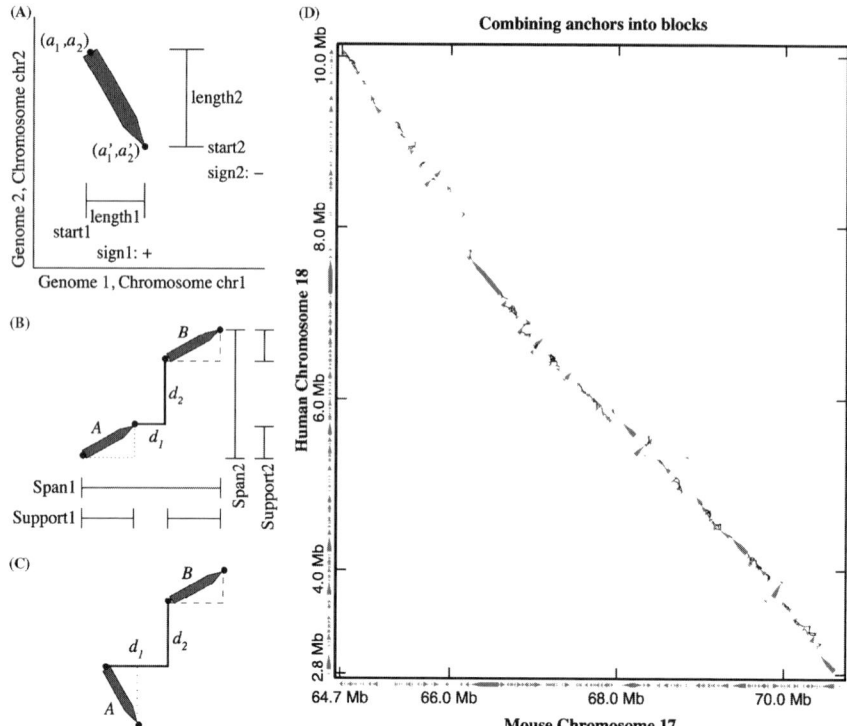

Fig. 20.4. Forming blocks from anchors in GRIMM-Synteny. (**A**) Anchor coordinates. Coordinates are given genome-by-genome, either as chromosome, start (minimum coordinate), length, sign (strand), or as chromosome window and the Cartesian coordinates of the two ends. (**B, C**). The total distance between two anchors is the Manhattan distance between their closest terminals *(thick solid line)*. Distances between other terminals *(dotted or dashed extensions)* increase the distance. The per-genome distance components d_1 and d_2 are indicated, and add up to the total $d = d_1 + d_2$. A block consisting of these two anchors has per-species measurements of the span (total size including gaps) and support (total size not including gaps), as well as the total number of anchors (2). (**D**) Several blocks between human and mouse from (12), with lines showing how the anchors are joined.

Determining the relative orientations of blocks in each species can be subtle if there are micro-rearrangements. GRIMM-Synteny uses a set of criteria that is geared toward signed permutations *(13)*.

If you only want to consider the order of the anchors (which is common in gene-order–based studies), prepare your data using either real coordinates or sham coordinates that put it into the correct order, and then use the *permutation metric* by specifying the option –p. This treats all anchors as length 2 nucleotides in every species (so that the two orientations of the anchor are distinguishable in each species) and no gaps between consecutive anchors in each species. This is achieved by keeping their chromosomes, orders, and signs, but recomputing their starting coordinates and lengths.

The process of joining anchors can result in blocks that are overlapping in some species or contained in another block in some species. The minimum size parameters filter out small blocks that are contained in large ones, but do not prevent overlaps among large blocks. This is in part due to different rearrangements of the anchors in each species, and also to the use of parameters that are constant across each species instead of somehow adapted to each region within a species. GRIMM-Synteny has a phase to detect and repair block overlaps and containments. When these are detected, the blocks are recursively broken into several segments. Any segments smaller than the size minimums will be filtered out. This phase is run by default, and can be prevented using the switch –O (letter "oh"). One reason to prevent it would be to study rearrangements in anchors at the ends of block, in which two large blocks were brought together in some species and then their ends were mixed through further rearrangements.

Strips are sequences of one or more consecutive blocks (with no interruption from other blocks) in the exact same order with the same signs, or exact reversed order with inverted signs, across all species. In **Fig. 20.3A**, genes 4 5 forms a strip of length 2 (it appears as 4 5 or –5 –4, depending on the species), and all other blocks form singleton strips (x or $-x$, depending on the species). A most parsimonious scenario can be found that does not split up the strip 4 5 (see (6)), so it would reduce the size of the permutations to recode it with blocks 4 and 5 combined into a single block. The option –c does this recoding by condensing each strip of blocks into a single block. (However, in applications such as studying the sequences of breakpoint regions, the separate boundaries of blocks 4 and 5 could be of interest, so it would not be appropriate to use the –c option.) If GRIMM-Synteny had produced a strip such as 4 5, it would either be because the two blocks are farther apart than the gap threshold, or because the overlap/containment repair phase split up a block and then its smaller pieces were deleted in such a way that what remained formed a strip of separate blocks.

3.1.4. GRIMM-Synteny: Usage and Options for Forming Synteny Blocks

The basic syntax of GRIMM-Synteny when used for forming synteny blocks is

```
% grimm_synt -f anchor_file -d output_dir [other options]
```

Input/output parameters:

- –f Input_file_name: This is required and should contain the path to the file with the non-conflicting anchor coordinates.
- –d Output_directory_name: This is required. It specifies a directory into which GRIMM-Synteny will write several output files.

Gap threshold:

- –g N
  –g N1,N2,N3,…

These specify the per-species gap threshold in nucleotides, either the same value in all species or a comma-separated list of values for the different species. Anchors are joined if in every species, their gap is below the per-species gap threshold for that species.

- –G N: This specifies the total gap threshold. Anchors are joined if the total gap (sum of the per-species gaps) is below this threshold.

Minimum block size:
- –m N
  –m N1,N2,N3,...

These specify the per-species minimum block spans, either the same value in all species or a comma-separated list of values for the different species. A block on chromosome c, with smallest coordinate x and largest coordinate y, has *span* $y - x + 1$, so anchors and gaps both contribute to the span (*see* **Fig. 20.4B**). Blocks are deleted if their span falls below this threshold in any species.

- –M N
  –M N1,N2,N3,...

These specify the per-species minimum block supports. The *support* of a block is the sum of its anchor lengths in nucleotides, which ignores gaps (*see* **Fig. 20.4B**). Blocks are deleted if their support falls below this threshold in any species.

- –n N: The minimum number of anchors per block. Blocks with fewer anchors are deleted. Since we do not consider deletions, this parameter consists of just one number, not a different number for each species.

Other settings:
- –c: If specified, strips of more than two blocks are condensed into single blocks.

- –O: If specified, the block overlap/containment repair phase is skipped.

- –p: If specified, use the "permutation metric." The anchor order and signs in each species are retained, but the coordinates are changed so that each anchor has length 2 and there is no gap between consecutive anchors (within each species, the ith anchor on a chromosome is regarded as having start coordinate $2i$ and length 2).

Output files: GRIMM-Synteny produces five files in the output directory specified by –d:

- report.txt: This is a detailed log file describing the computations and phases that GRIMM-Synteny performed. It includes measurements for each block, such as number of anchors, per-species support and span of anchors, and micro-rearrangement distance matrix. It also includes information about macro-rearrangements of all the blocks.

- blocks.txt: This has the coordinates of the blocks in the same format as described in **Section 3.1.1** for anchor files.

- mgr_macro.txt: This file gives signed block orders on each chromosome in each genome in the GRIMM/MGR input file format described in **Section 3.2.1** (*see* **Fig. 20.3A**). Coordinates and lengths in nucleotides are not included; to determine these, look up the blocks by ID number in blocks.txt.

- mgr_micro.txt: This file lists the anchors contained in each block. For each block, the k-way coordinates of all its anchors are given, and the GRIMM/MGR permutation of the anchors is given. Since blocks are in just one chromosome per species, and since blocks have a definite sign, these permutations should be regarded as directed linear chromosomes (–L option in GRIMM/MGR). Also, in the permutations, strips of anchors have been compressed.
- mgr_micro_equiv.txt: This lists which blocks have the same compressed anchor permutations in mgr_micro.txt. In other words, it identifies blocks whose anchors underwent similar micro-rearrangements.

3.1.5. Sample Run 1: Human-Mouse-Rat-Chicken Gene-Based Dataset

We discuss the data in the sample data directory hmrc_gene_data. Change to that directory. The four-way homologous genes were computed by Evgeny Zdobnov and Peer Bork *(10, 11)* and identified by their Ensembl IDs. We combined their data with Ensembl coordinates of those genes for the specific Ensembl builds they used. The file hmrc_genes_ensembl.txt is for informational purposes only, and shows the Ensembl IDs combined with the gene coordinates. The file hmrc_genes_coords.txt is in the GRIMM-Synteny four-way coordinate format.

First we filter out conflicting homologues:

```
% mkdir anchors
% grimm_synt -A -f hmrc_genes_coords.
  txt -d anchors
```

This produces a log file anchors/report_ga.txt and a coordinate file anchors/unique_coords.txt. The log file indicates there were 8,095 homologous gene quadruplets, but a number of them had conflicting coordinates, resulting in 6,447 anchors. Details about the conflicts are given in a third output file, anchors/repeat_coords.txt. In the sample data, we named the directory anchors_example instead of anchors so that you can run these examples without overwriting it. In the steps that follow, we also did this with directories gene7_example instead of gene7 and 300K_example instead of 300K.

Next we form synteny blocks. The main run analyzed in the paper, gene7, was produced as follows:

```
% mkdir gene7
% grimm_synt -f anchors/unique_coords.txt -d
  gene7 -c -p -m 6 -g 7
```

We used the permutation metric, so each gene is considered to have length 2 units. We required a minimum length of six units (i.e., size of 3 genes at length 2) in each species. Using –n 3 (minimum of three anchors, regardless of size) instead of –m 6 produced identical results, but at larger values, –n x and –m $2x$ would not be the same, since –m $2x$ would allow for gaps.

Finally, –g 7 is the per-species gap-threshold, which motivated naming this run gene7; we performed similar runs with thresholds from 1 through 20. We also varied other parameters. At smaller values of the gap-threshold, there is little tolerance for micro-rearrangements within blocks, so many small blocks are formed (but many of them are deleted for being below 6 units, from –m 6). Setting it too high would keep a high number of anchors but low number of blocks by merging too many blocks together. The selection –g 7 retained a relatively high number of anchors and high number of blocks. In addition to this, for each combination of parameter settings, we also examined plots of the microrearrangements in the blocks (similar to **Fig. 20.4D**), ran the blocks through MGR, and did other tests. Unfortunately, optimal parameter selection is still somewhat an art. (Tools to produce such plots are highly data dependent and are not provided with the current release of GRIMM-Synteny. The plots are based on the anchor and block coordinates in mgr_micro.txt.)

3.1.6. Sample Run 2: Human-Mouse-Rat-Chicken Alignment-Based Dataset

We discuss the data in the sample data directory hmrc_align_data. Change to that directory. In *(10)*, alignments between human and one to three of mouse, rat, and chicken, were computed by Angie Hinrichs and others at the UCSC Genome Bioinformatics group. The alignment files were several hundred megabytes per chromosome because they included the coordinates of the alignments as well as base-by-base annotations. We extracted the coordinates of four-way alignments in this data. One-, two-, and three-way alignments were discarded. This is further described in *(11)*. The file hmrc_align_coords.txt contains the four-way coordinates of the alignments.

The UCSC protocol included several ways of masking out repeats. However, there were still a number of repeats left in the data, particularly in chicken, in which repeat libraries were not so thoroughly developed at that time. Evan Eichler provided us with coordinates of segmental duplications in chicken *(10)*. We used GRIMM-Anchors to filter out alignments conflicting with the duplications, by adding the coordinates of the duplications into hmrc_align_coords.txt as shown in **Fig. 20.2**. We made a new chromosome "_" in human, mouse, and rat, and coded all the segmental duplications into four-way alignments at coordinate 0 on "_" in human, mouse, and rat, and their true coordinate in chicken. This way, all the segmental duplications conflicted with each other in human, mouse, and rat (so that GRIMM-Anchors would filter them out) and conflicted with any real alignments at those coordinates in chicken (so that GRIMM-Anchors would filter those alignments out).

We filter out conflicting homologues:

```
% mkdir anchors
% grimm_synt -A -f hmrc_align_coords.txt -d
  anchors
```

This produces a log file anchors/report_ga.txt and a coordinate file anchors/unique_coords.txt. Next, the main alignment-based run considered in the paper was produced as follows:

```
% mkdir 300K
% grimm_synt -f anchors/unique_coords.txt -d
  300K -c -m 300000 -g 300000
```

We used –c to condense strips of blocks into single blocks. We used –m 300000 to set a minimum span of 300000 nucleotides per species. We used –g 300000 to set a maximum gap size of 300000 per species.

We also produced blocks with other combinations of parameters and considered similar factors as for the gene7 run in determining to focus on the "300K" blocks. In **Fig. 20.2B**, notice that one of the alignments involves human chromosome 2 and mouse and rat chromosome X. Another sanity check we did on the output for each choice of parameters was to see if any blocks were formed between the X chromosome on one mammal and a different chromosome on another mammal, since such large-scale blocks would violate Ohno's Law *(14)*.

3.2. GRIMM: Identifying Rearrangements Between Two Genomes

GRIMM implements several algorithms for studying rearrangements between two genomes in terms of signed permutations of the order of orthologous elements. Most of the literature refers to this as *gene orders*, although we also apply it to the order of syntenic blocks, such as those produced by GRIMM-Synteny.

Hannenhalli and Pevzner showed how to compute the minimum number of reversals possible between two uni-chromosomal genomes in polynomial time *(2)*, and Bader, Moret, and Yan improved this to linear time and implemented it in their GRAPPA software (*see* web site references and *(5)*). GRIMM is adapted from the part of GRAPPA that implements this.

Hannenhalli and Pevzner went on to show how to compute the minimum number of rearrangements (reversals, translocations, fissions, and fusions) between two multi-chromosomal genomes in polynomial time *(3)*. Tesler fixed some problems in the algorithm and adapted the Bader-Moret-Yan algorithm to solve this problem in linear time *(6)*. Ozery-Flato and Shamir found an additional problem in the Hannenhalli-Pevzner algorithm *(7)*. GRIMM implements all of these for multi-chromosomal rearrangements.

Hannenhalli and Pevzner described an algorithm for studying rearrangements in genomes when the orientations of genes are not known *(4)*. This algorithm is only practical when the number of singleton genes is small. GRIMM implements this algorithm, a generalization of it for the multi-chromosomal case, and a fast approximation algorithm.

3.2.1. Input Format

The input for GRIMM and MGR is a file that gives the permutation of orthologous regions in each genome, split into chromosomes. *See* **Fig. 20.3A** for a sample file with four genomes with up to three chromosomes in each. All four genomes consist of the same 10 regions but in different orders.

Each genome specification begins with a line consisting of the greater-than symbol followed by the genome name. Next, the order of the orthologous regions 1, 2, … *n* is given, with dollar sign symbol "$" at the end of each chromosome. The numbers are separated by any kind of white space (spaces, tabs, and new lines). Chromosomes are delimited by "$," the start of the next genome, or the end of the file. Comments may be inserted in the file using the "#" symbol. The rest of the line is ignored.

3.2.2. Output

The main usage of GRIMM is to compute the most parsimonious distance between two genomes and give an example of one rearrangement scenario (out of the many possible) that achieves that distance. An excerpt of GRIMM's output for this usage is shown in **Fig. 20.3D**. There are other usages too, which have different outputs.

3.2.3. Usage and Options

There are several usages of GRIMM: *(1)* compute the most parsimonious distance between two genomes (along with other statistics about the breakpoint graph), *(2)* exhibit a most parsimonious rearrangement scenario between two genomes, *(3)* compute matrices of pairwise distances and pairwise statistics for any number of genomes, and *(4)* compute or estimate signs of orthologous regions to give a most parsimonious scenario. The command-line syntax is as follows:

```
% grimm -f filename [other options]
```

Input/output:
- –f Input_file_name: This field is required and should contain the path to the file with the starting permutations (e.g., data/sample_data.txt or a file mgr_macro.txt generated using GRIMM-Synteny).
- –o Output_file_name: The output is sent to this file. If –o is not specified, the output is sent to STDOUT.
- –v Verbose output: For distance computations, this gives information on the breakpoint graph statistics from the Hannenhalli-Pevzner theory. For other computations, this gives additional information.

Genome type:
- –C, –L, or neither: –C is uni-chromosomal circular distance and –L is uni-chromosomal directed linear reversal distance. If neither –C or –L is selected, the genomes have multi-chromosomal *undirected* linear chromosomes. (Undirected means flipping the whole chromosome does not count as a reversal, whereas *directed* means it does count. On single chromosome genomes, –L vs. multi-chromosomal are different in this regard.)

Genome selection: GRIMM is primarily used for pairs of genomes, but can also display matrices to show comparisons between all pairs. Input files with two genomes default to pairwise comparisons and files with more than two genomes default to matrix output, unless the following options are used:

- –g i,j: Compare genome i and genome j. Genomes in the input file are numbered starting at 1. With the –s option, a rearrangement scenario is computed that transforms genome i into genome j. **Fig. 20.3D** uses –g 1,4. For a file with two genomes, this option is not necessary, unless you want to compare them in the reverse order (–g 2,1).

- –m: Matrix format (default when there are more than two genomes if –g is not used). A matrix of the pairwise distances between the genomes is computed, as shown in **Fig. 20.3E**. When used in combination with the –v option, matrices are computed for breakpoint graph parameters between all pairs of genomes.

Pairwise comparison functions (not matrix mode): Defaults to –d –c –s for multi-chromosomal genomes and –d –s for uni-chromosomal genomes.

- –d: Compute distance between genomes (minimum number of rearrangement steps combinatorially possible).

- –s: Display a most parsimonious rearrangement scenario between two genomes. Not available in matrix mode.

- –c, –z: In multi-chromosomal genomes, a pair of additional markers ("caps") are added to the ends of each chromosome in each genome, and the chromosomes are concatenated together into a single ordinary signed permutation (without "$" chromosome breaks). The details are quite technical; *see* *(3, 6, 7)*. These options display the genomes with added caps in two different formats: –c displays the concatenation as an ordinary signed permutation (not broken up at chromosomes) suitable as input to GRIMM with the –L option, whereas –z breaks it up by chromosome.

Unsigned genomes:

- –U n: A fast approximation algorithm for determining the signs in unsigned genomes via hill-climbing with n random trials. This works for any number of genomes, not just two. It works for some signs known and some unknown, or for all signs unknown. This was used in Murphy et al. to determine signs of blocks with only one gene *(15)*. A paper about the technical details is in preparation (*see* **Note 1 and Fig. 20.6**).

- –u: Exact computation of the rearrangement distance between two unsigned genomes (all signs unknown). This also computes an assignment of signs that would achieve this distance if the genomes were regarded as signed. For uni-chromosomal genomes, this uses the algorithm by Hannenhalli and Pevzner

(4) and for multi-chromosomal genomes, this uses a generalization of that by Glenn Tesler (in preparation). The complexity is exponential in the number of singletons, so this option is only practical when the number of singletons is small (*see* **Fig. 20.6**).

3.2.4 Sample Run: Toy Example

We use the file data/sample_data.txt shown in **Fig. 20.3A**.

The command line to compute the scenario shown in **Fig. 20.3D** is:

```
% grimm -f data/sample_data.txt -g 1,4
```

The command line to compute the matrix shown in **Fig. 20.3E** is:

```
% grimm -f data/sample_data.txt -m
```

(Note: –m is optional; since there are more than two genomes, it is assumed unless –g is used.)

Additional details about the breakpoint graphs can be shown in either of these by adding the option –v.

```
% grimm -f data/sample_data.txt -g 1,4 -v
% grimm -f data/sample_data.txt -m -v
```

3.2.5. Sample Run 2: Human-Mouse-Rat-Chicken Dataset

GRIMM can also be run on the files mgr_macro.txt output by GRIMM-Synteny using similar command lines but changing the filename.

Of greater interest, however, would be to run GRIMM after MGR has computed the topology of a phylogenetic tree and possible gene/block orders at its ancestral nodes. GRIMM would then be appropriate to study the breakpoint graph or possible scenarios on a branch of the tree.

3.3. MGR: Reconstructing the Rearrangement Scenario of Multiple Genomes

The Multiple Genome Rearrangement Problem is to find a phylogenetic tree describing the most "plausible" rearrangement scenario for multiple species. Although the rearrangement distance for a pair of genomes can be computed in polynomial time, its use in studies of multiple genome rearrangements has been somewhat limited since it was not clear how to efficiently combine pairwise rearrangement scenarios into a multiple rearrangement scenario. In particular, Caprara demonstrated that even the simplest version of the Multiple Genome Rearrangement Problem, the Median Problem with reversals only, is NP-hard *(16)*.

MGR implements an algorithm that, given a set of genomes, seeks a tree such that the sum of the rearrangements is minimized over all the edges of the tree. It can be used for the inference of both phylogeny and ancestral gene orders *(8)*. MGR outputs trees in two different formats described in the following: *(1)* Newick format and *(2)* ASCII representation. The algorithm makes extensive use of the pairwise distance engine GRIMM. This section first provides a detailed description of the input and output format. Next, it describes two typical standard runs, one a toy example and one of the human-mouse-rat-chicken dataset.

3.3.1. Input Format

The gene order input format for MGR is the same as for GRIMM. For an example with four genomes with two or three chromosomes each *see* **Fig. 20.3A**. This small sample input file can also be found in the MGR package in the subdirectory data as sample_data.txt.

3.3.2. Output: Newick Format and ASCII Representation

The Newick Format uses nested parenthesis for representing trees. It allows the labeling of the leaves and internal nodes. Branch lengths corresponding to the number of rearrangements can also be incorporated using a colon. For instance, the example shown in **Fig. 20.3A** would produce a tree in Newick Format shown in **Fig. 20.3B**. Note that the internal nodes correspond to ancestral nodes and are labeled using the letter A followed by a number (e.g., A4). Also note that the Newick Format specifies a rooted tree with ordered branches, but MGR determines an unrooted tree, so MGR chooses an arbitrary location for the root. Your additional knowledge of the timeline should be used to relocate the root and order the branches.

The ASCII graphical representation of the tree is generated by a modified version of the RETREE program available in the PHYLIP package by Joe Felsenstein (*see* web site references). The number of rearrangements that occurred on each edge is shown and (unless the –F switch is selected, see the following) the edges are drawn proportionally to their length. When no number is shown on an edge it means that no rearrangement occurred on that edge. *See* **Fig. 20.3C** for the tree associated with the example from **Fig. 20.3A**.

3.3.3. Usage and Options

There are three main usages of MGR: *(1)* with data from a file, *(2)* with simulated data, and *(3)* to display previous results. The current description focuses on the first usage, which represents the most common application. The command-line syntax is as follows:

```
% MGR -f filename [other options]
```

Input/output:
- –f Input_file_name: Same as in GRIMM.
- –o Output_file_name: Same as in GRIMM.
- –v: Verbose output. This is very important to visualize and record the progress of MGR, especially for large datasets. Using this option, the initial input genomes are reported along with their pairwise distances. Following that, each rearrangement identified in the procedure and the intermediate genomes are reported. The program terminates once the total distance between the intermediate genomes of the various triplets has converged to zero.
- –w: Web output (html). Should not be combined with the –v option but allows for a more elaborate html report. This option can also be used to redisplay in html format a previous result (if used in combination with the –N option, see README).

- –W: Width (in characters) of the tree displayed (default is 80). Only affects the way the ASCII representation of the tree is displayed.
- –F: fixed size edges in the tree displayed. Displays fixed size edges in the ASCII representation instead of edges proportional to their length.

Genome type:
- –C, –L, or neither: Uni-chromosomal circular genomes, uni-chromosomal directed linear genomes, or multi-chromosomal undirected linear genomes. These are the same as in GRIMM.

Other options:
- –H: Heuristic to speed up triplet resolution:
  –H 1: only look at reversals initially, and pick the first good one.
  –H 2: only look at reversals initially, and take the shortest one.

  Especially for large instances of the problem (e.g., more than 100 homologous blocks, or more than five genomes), these options can greatly speed up the algorithm by restricting the search and the selection to specific categories of good rearrangements (*see* **Note 2**).

- –c: Condense strips for efficiency. This combines strips of two or more homologous blocks that are in the exact same order in all k genomes being considered. If this option is selected, the condensing procedure is called recursively but the whole process is seamless as the strips are uncondensed before the output is generated. This option can greatly speed up MGR, especially if the starting genomes are highly similar. This is related to GRIMM-Synteny's –c option; the difference is that GRIMM-Synteny's –c option changes the blocks that are output, whereas MGR's –c option affects internal computations but uncondensed block numbers are used on output.

- –t: Generate a tree compatible with the topology suggested in the file. Forces MGR to look for an optimal rearrangement scenario only on the tree topology provided by the user (*see* **Note 3**).

3.3.4. Sample Run 1: Toy Example

To run MGR on the example displayed in **Fig. 20.3A**, do:

```
% MGR -f data/sample_data.txt
```

The output should be similar to the output displayed in **Fig. 20.3B,C** with some additional information on the parameters used and on the permutations associated with the input genomes and the ancestors recovered. To view the same result but in html format, try:

```
% MGR -f data/sample_data.txt -w -o
  sample_out.html
```

3.3.5. Sample Run 2: Human-Mouse-Rat-Chicken Dataset

The file data/hmrc_gene_perm.txt is identical to the file gene7/mgr_macro.txt generated in **Section 3.1.5** based on orthologous genes. It contains the human, mouse, rat, and chicken genomes represented by four signed permutations of 586 homologous blocks. Run MGR on this example as follows:

```
% MGR -f data/hmrc_gene_perm.txt -H2 -c -o
  hmrc_gene_perm_out.txt
```

Even using the –H2 and –c switches to speed up computations, this is a challenging instance of the Multiple Genome Rearrangement problem that will probably take a few hours to complete on most computers. The final output (hmrc_gene_perm_out.txt) should be identical to the file data/hmrc_gene_perm_out.txt. To get a better sense of how quickly (or slowly) the program is converging, you can use the –v switch:

```
% MGR -f data/hmrc_gene_perm.txt -H2 -c -v -o
  hmrc_gene_perm_out1.txt
```

but of course, this will also generate a much larger output file. An actual rearrangement scenario between one of the initial genomes and one of the recovered ancestors can be obtained by extracting the permutations from the bottom of the output file, creating a new input file (e.g., hmrc_result.txt) and running:

```
% grimm -f hmrc_result.txt -g 1,5
```

To facilitate the comparison of the initial, modern day, genomes with the recovered ancestral genomes, it is also possible to plot the various permutations (*see* **Fig. 20.5**) *(15)*. However, the tools to produce such plots are highly data dependent and are not provided with the current release of MGR. For challenging examples, when the initial pairwise distances are significant as compared with the number of homologous blocks (e.g., in the current example), it is possible to find alternative ancestors satisfying the same overall scenario score. This in turn can lead to the identification of *weak* and *strong* adjacencies in the ancestors (*see* **Note 4**).

4. Notes

1. Radiation-hybrid maps and missing signs. In *(15)*, an eight-way comparison was done among three sequenced species (human, mouse, and rat) and five species mapped using a radiation-hybrid approach (cat, cow, dog, pig, and on some chromosomes, horse). GRIMM-Synteny is not appropriate to use due to the RH-mapped data. A method is described in that paper to construct syntenic blocks that take into account the mixed coordinate system and types of errors that occur with RH maps. In blocks with two or more genes, an inference about the orientation of the block in each species was easy to make. However, singleton blocks (supported by a single gene) had known orientations in the sequenced species

Fig. 20.5. Graphical visualization of the permutations associated with two modern genomes (Human and Mouse) and an ancestral permutation (Mammalian Ancestor) as recovered by MGR (15).

(human, mouse, and rat), and unknown orientations in the other species. GRIMM was used to guess the signs in the other species with the –U option.

2. MGR heuristics. The –H heuristics rely on the simple assumption that reversals, and specifically short reversals in the case of –H2, represent a more common evolutionary event as compared with translocation, fusions, and fissions. These heuristics also allow a more robust analysis of noisy datasets that may contain sign errors or local misordering.

3. Fixed topology. MGR can be invoked using the –t option to reconstruct a rearrangement scenario for a specific tree topology. There could be various reasons to use this option: to compare the score of two alternative topologies, accelerate computations, etc. The desired topology needs to be specified in a separate file using the Newick format without edge lengths. In such topology files, genomes are referenced using identifiers from 1 to k, where 1 is the genome that

```
(A) File data/unsigned1.txt

>genome1
1 2 3 4 5
>genome2
1 4 3 2 5
>genome3
2 1 3 4 5
```

```
(B) Excerpt from / grimm -f data/unsigned1.txt -L -u

Distance Matrix:
genome1            0 1 1
genome2            1 0 2
genome3            1 2 0
```

```
(C) Excerpt from / grimm -f data/unsigned1.txt -L -u -g 1,2

An optimal sequence of reversals:
Step 0: (Source)
    1   2   3   4   5
Step 1: Reversal (Destination)
    1  -4  -3  -2   5
```

```
(D) Excerpt from / grimm -f data/unsigned1.txt -L -U 100

Best score: 4

A best scoring solution:
>genome1
1 2 3 4 5
>genome2
1 -4 -3 -2 5
>genome3
-2 -1 3 4 5

Distance matrix for that specific solution:
genome1            0 1 1
genome2            1 0 2
genome3            1 2 0
```

Fig. 20.6. Unsigned data. (**A**) Input file. Each genome has one chromosome, directed linear, so the –L option is used on all commands. (**B–D**). Excerpts from runs. (**B, C**) The –u option does an exact computation for each pair of genomes. (**D**) One hundred trials of an approximation algorithm are performed that seeks a best global assignment of signs.

appears first in the main gene order file, 2 is the genome that appears second, etc. See the file data/sample_tree.txt for an example associated with data/sample_data.txt. The command line to run this example would be:

```
% MGR -f data/sample_data.txt -t data/sample_tree.txt
```

Note that the scenario recovered is slightly worse than the scenario shown in **Fig. 20.3C** with seven rearrangements instead of six, but the tree topology matches the tree topology in data/sample_tree.txt. The details of the algorithm used for this are in *(15)*.

4. Alternative ancestors. For a given ancestor, when the ratio between the total number of rearrangements of the three incident edges and the number of common blocks is high, it is often possible to find alternative ancestors also minimizing the total number of rearrangement events on the evolutionary tree. By exploring a wide range of such alternative ancestors, it is possible to distinguish between weak and strong areas of the ancestral reconstructions. Specifically, adjacencies that are present in all of the observed alternative ancestors are called *strong* adjacencies, whereas adjacencies that are not conserved in at least one of the alternative ancestors are called *weak* adjacencies (*see (15)* for more details). The number of weak adjacencies identified in this manner is actually a lower bound for the true number of weak adjacencies since only a subset of all the alternative solutions can be explored. This search for alternative ancestors is available in MGR for k = 3 using the –A switch but is not described further in this chapter.

Acknowledgments

G.B. is supported by funds from the Biomedical Research Council of Singapore. G.T. is supported by a Sloan Foundation Fellowship.

References

1. Tesler, G. (2002a) GRIMM: genome rearrangements web server. *Bioinformatics* 18, 492–493.
2. Hannenhalli, S., Pevzner, P. A. (1995a) Transforming cabbage into turnip (polynomial algorithm for sorting signed permutations by reversals). In Proceedings of the 27th Annual ACM Symposium on the Theorey of Computing. (Full journal version with same title appeared in (1999) *JACM* 46, 1–27.
3. Hannenhalli, S., Pevzner, P. A. (1995b) Transforming men into mice (polynomial algorithm for genomic distance problem). In *36th Annual Symposium on Foundations of Computer Science, Milwaukee, WI.*
4. Hannenhalli, S., Pevzner, P. A. (1996) To cut … or not to cut (applications of comparative physical maps in molecular evolution). In *Proceedings of the Seventh Annual ACM-SIAM Symposium on Discrete Algorithms, Atlanta, GA.*
5. Bader, D., Moret, B., Yan, M. (2001) A linear-time algorithm for computing inversion distances between signed permutations with an experimental study. *J Comput Biol* 8, 483–491.
6. Tesler, G. (2002b) Efficient algorithms for multi-chromosomal genome rearrangements. *J Comp Sys Sci* 65, 587–609.
7. Ozery-Flato, M., Shamir, R. (2003) Two notes on genome rearrangement. *J Bioinform Comput Biol* 1, 71–94.
8. Bourque, G., Pevzner, P. A. (2002) Genome-scale evolution: reconstructing gene orders in the ancestral species. *Genome Res* 12, 26–36.
9. Bourque, G., Pevzner, P. A., Tesler, G. (2004) Reconstructing the genomic architecture of ancestral mammals: lessons from human, mouse, and rat genomes. *Genome Res* 14, 507–516.
10. Hillier, L., Miller, W., Birney, E. et al.(2004) Sequence and comparative analysis of the chicken genome provide unique perspectives on vertebrate evolution. *Nature* 432, 695–716.

11. Bourque, G., Zdobnov, E., Bork, P., et al. (2005) Genome rearrangements in human, mouse, rat and chicken. *Genome Res* 15, 98–110.
12. Pevzner, P. A., Tesler, G. (2003a) Genome rearrangements in mammalian evolution: lessons from human and mouse genomes. *Genome Res* 13, 37–45.
13. Pevzner, P. A., Tesler, G. (2003b) Transforming men into mice: the Nadeau-Taylor chromosomal breakage model revisited. Proceedings of the 7th Annual International Conference on Research in Computational Molecular Biology (*RECOMB 2003*). ACM Press, New York, 247–256.
14. Ohno, S. (1967) *Sex Chromosomes and Sex Linked Genes*. Springer-Verlag, Berlin.
15. Murphy, W. J., Larkin, D. M., Everts-van der Wind, A., et al. (2005) Dynamics of mammalian chromosome evolution inferred from multispecies comparative maps. *Science* 309, 613–617.
16. Caprara, A. (1999) Formulations and complexity of multiple sorting by reversals, in (Istrail, S., Pevzner, P., Waterman, M. eds.), *Proceedings of the Third Annual International Conference on Computational Molecular Biology (RECOMB-99)*. ACM Press, Lyon, France.

Web Site References

Ensembl http://www.ensembl.org

Felsenstein, J., *PHYLIP*. http://evolution.genetics.washington.edu/phylip.html

Homolo Gene http://www.ncbi.nlm.nih.gov/Homolo Gene

Moret, B. et al. (2000) *GRAPPA*. http://www.cs.unm.edu/~moret/GRAPPA

UCSC Genome Bioinformatics web site http://www.genome.ucsc.edu

Chapter 21

Detecting Lateral Genetic Transfer

A Phylogenetic Approach

Robert G. Beiko and Mark A. Ragan

Abstract

Nucleotide sequences of microbial genomes provide evidence that genes have been shared among organisms, a phenomenon known as lateral genetic transfer (LGT). Hypotheses about the importance of LGT in the evolution and diversification of microbes can be tested by analyzing the extensive quantities of sequence data now available. Some analysis methods identify genes with sequence features that differ from those of the surrounding genome, whereas other methods are based on inference and comparison of phylogenetic trees. A large-scale search for LGT in 144 genomes using phylogenetic methods has revealed that although parent-to-offspring ("vertical") inheritance has been the dominant mode of gene transmission, LGT has nonetheless been frequent, especially among organisms that are closely related or share the same habitat. This chapter outlines how bioinformatic and phylogenetic analyses can be built into a workflow to identify LGT among microbial genomes.

Key words: Lateral genetic transfer, phylogenetic analysis, multiple sequence alignment, orthology, edit paths.

1. Introduction

Lateral genetic transfer among prokaryotes has been recognized as an evolutionary mechanism for some time (1–3), but quantifying its extent and role could be undertaken only after complete genome sequences began to become available. Many major bacterial lineages are now represented by one or more sequenced genomes, and several genera (*Chlamydia*, *Escherichia*, and *Streptococcus*) are represented by five or more isolates in the set of publicly available genome sequences.

This depth and breadth of taxonomic sampling allowed us to investigate LGT at "short" (e.g., within-genus) and "long" (e.g., intra-phylum or -domain) distances among 144 sequenced prokaryotic genomes *(4)*. The results from this analysis, the most extensive application so far of rigorous phylogenetic methods to this question, confirmed many previous conjectures about LGT, including the tendency of genes encoding information-processing proteins to be shared less frequently than genes encoding metabolic enzymes *(5, 6)* and, based on the hybrid nature of the *Aquifex aeolicus* and *Thermoplasma acidophilum* genomes, the apparent role of LGT in habitat invasion *(7, 8)*.

The best phylogenetic methods are explicitly based on models of the evolutionary process. Model-based methods provide a more robust framework for evaluating instances of LGT than do *surrogate* or *parametric* methods, which instead identify anomalous compositional features, patterns, or distributions that are not obviously consistent with vertical genetic transmission. When applied to the genome of *Escherichia coli* K12, four surrogate methods agree in their identification of anomalous genes less often than would be expected by chance *(9, 10)*. However, phylogenetically based approaches too are not without their complications. For example, they depend on accurate delineation of sets of orthologous genes and on appropriate alignment of multiple sequences, two problem areas that remain the subject of intensive research. Simulations also show that some phylogenetic methods are not robust to violations of underlying evolutionary assumptions *(11, 12)*, and all methods are of little use where phylogenetic signal has decayed completely.

This chapter outlines how procedures for recognizing putatively orthologous groups of genes, aligning multiple sequences, and inferring and comparing phylogenetic trees can be integrated into a high-throughput workflow to identify instances of LGT among microbial genomes. Alternative procedures, and opportunities for further refinement, are also presented.

2. Systems, Software, and Databases

2.1. Data Sources

1. Sequenced microbial genomes are available from the FTP site of the National Center for Biotechnology Information (ftp://ftp.ncbi.nlm.nih.gov/genomes/Bacteria/), with specific genomes available in subdirectories *Genus_species_strain*. Each molecule of DNA (e.g., chromosomes, plasmids) from the genome of a given organism is represented by a suite of files (**Note 1**). The file with the annotations needed to construct an LGT workflow is the ".gbk" file, which contains the complete DNA sequence as well as information on each

predicted protein. The "CDS" tag for a given protein identifies the genomic coordinates of its corresponding gene, and whether the coding sequence can be read directly from the displayed DNA sequence, or must instead be reverse complemented to obtain the correct amino acid translation. The .gbk file also contains the amino acid translation, the annotated function (if any), and database references that can be useful when working with the protein sequence in GenBank.

2. Functional annotations of gene products and organisms can be obtained from many sources, but a few web sites offer particularly comprehensive annotations of primary data. The NCBI Taxonomy database (http://www.ncbi.nlm.nih.gov/Taxonomy/) provides a useful although non-canonical reference for classification of microbial organisms. Where no consensus exists on microbial classification, the NCBI Taxonomy database may offer an adequate starting point. The Clusters of Orthologous Groups database *(13)* contains useful functional annotations of proteins. Although we used a different method (described in **Section 3.1**) to infer orthologs, we retained the functional categories proposed in the CoG database to examine the functional annotations of putatively transferred genes. The Institute for Genomic Research (TIGR) also provides a list of "role categories" that classify protein functions in detail, and a database of genome properties with information about the basic sequence properties of every genome in the database, and about the lifestyle and metabolic features of the corresponding organisms *(14)*.

2.2. Program Availability

All of the analytical tools described in the following are freely available from the World Wide Web or from the authors, and with two exceptions (TEIRESIAS and GBLOCKS) the source code is available as well.

MCL *(15)*: http://micans.org/mcl/

TEIRESIAS *(16)*: http://cbcsrv.watson.ibm.com/Tspd.html

WOOF *(17)*: http://bioinformatics.org.au/woof

CLUSTALW *(18)*: http://www.ebi.ac.uk/clustalw/

T-COFFEE *(19)*: http://www.tcofee.org/Projects_home_page/t_coffee_home_page.html

Poa *(20)*: http://www.bioinformatics.ucla.edu/poa/

Prrn *(21)*: http://www.cbrc.jp/~gotoh/softdata.html

MAFFT *(22)*: http://align.bmr.kyushuu.ac.jp/mafft/software/

MrBayes *(23)*: http://mrbayes.csit.fsu.edu/index.php

GBlocks *(24)*: http://molevol.ibmb.csic.es/Gblocks.html

CLANN *(25)*: http://bioinf.may.ie/software/clann/

EEEP *(26)*: http://bioinformatics.org.au/eeep

3. Methods

What follows is a general description of the methods we used in our analysis of 144 genomes *(4)*, and the parameter settings reported here correspond to those used in that analysis. More details on the use of these methods can be found in the Supporting Information to *(4)* and in the original papers cited there.

3.1. Clustering of Proteins into Sets of Putative Homologs and Orthologs

1. All-versus-all BLASTP *(27)* is performed on the complete set of predicted proteins from all genomes in the analysis. Every protein-protein BLASTP match with an associated expectation score $e \leq 1.0 \times 10^{-3}$ is kept and is normalized by dividing by its self-score (obtained from a BLASTP comparison of the protein against its own sequence).

2. The matches between proteins implied by normalized BLASTP similarity are used as the basis for Markov clustering using MCL. Validation of MCL on the Protein Data Bank *(28)* suggested that a relatively low inflation parameter of 1.1 was appropriate *(29)*. The Markov clusters are interpreted as sets of homologous proteins, from which orthologous relationships can be extracted.

3. Each Markov cluster is subjected to single-linkage clustering *(30)* to determine putatively orthologous groups (**Fig. 21.1**). Maximally representative clusters (MRCs) contain no more than one representative protein from any given genome, and are maximal in that either the next protein(s) added by lowering the normalized BLASTP threshold would duplicate genomes already in the MRC, or the MRC represents an entire Markov cluster, and no further proteins can be added by decreasing the normalized BLASTP threshold. MRCs are interpreted as putative ortholog families, and MRCs with ≥ 4 proteins (the minimum size that can yield meaningful unrooted phylogenetic trees) are retained.

3.2. Sequence Alignment, Validation, and Trimming

1. Alternative alignments are generated for the protein sequences in each MRC using different alignment algorithms (e.g., T-COFFEE, CLUSTALW, Poa, Prrn, and MAFFT) and, where appropriate, a range of gap-opening and gap-extension penalty settings.

2. In parallel with the preceding, conserved patterns are extracted from each pair of proteins in each unaligned MRC using the TEIRESIAS pattern-detection algorithm. To be considered as a pattern here, a pair of substrings must have at least 3 literal matches (L) within a window of total length (W) no greater than 15. Each pattern is assigned a weight

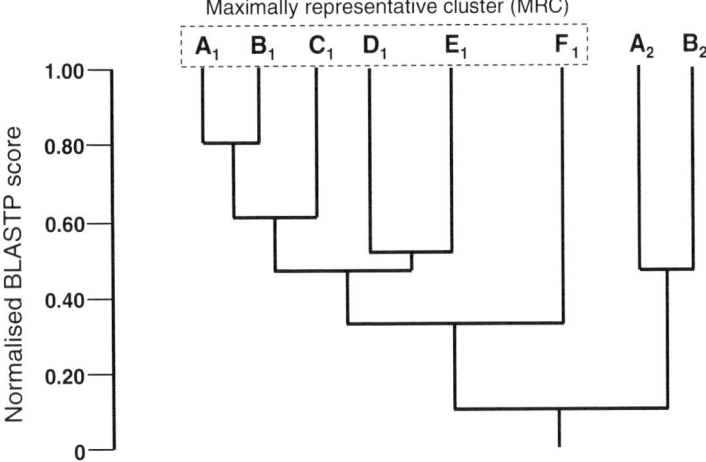

Fig. 21.1. Schematic for hybrid protein clustering strategy. Each protein in the hypothetical Markov cluster shown below is identified with a letter that corresponds to its host genome, with genomes A and B represented by two (putatively paralogous) proteins. Vertical and horizontal lines indicate clustering relationships within the set of proteins, with a horizontal line indicating the *maximum* normalized BLASTP threshold at which a given cluster exists. Every cluster that exists here within the group (A_1, B_1, C_1, D_1, E_1, F_1) is a *representative cluster* because no genome is represented in it more than once, but the cluster containing all six of these proteins is *maximally representative* because, in this example, any further extension of the cluster would lead to duplication of genomes A and B.

based on its probability of occurring by chance, and by its positional conservation relative to other extracted patterns.

3. The extent to which these conserved patterns are recovered intact and precisely aligned with each other in each alternative alignment is assessed using the word-oriented alignment function WOOF. The alignment yielding the highest score is retained for subsequent analysis.

4. Ambiguously aligned regions (e.g., ragged ends, sparsely populated columns and large gaps) are "trimmed" from each winning alignment of n sequences using GBLOCKS with relatively conservative settings (*see* **Note 2**).

Settings for GBLOCKS:

Minimum number of sequences for a conserved position: $n \times 0.5 + 1$

Minimum number of sequences for a flank position: $n \times 0.5 + 1$

Maximum number of contiguous non-conserved positions: 50

Minimum length of a block: 5

Allowed gap positions: All

3.3. Phylogenetic Inference and Supertree Construction

1. Trimmed alignments are converted into Nexus format by adding the appropriate header information *(31)*, and MrBayes command blocks *(23)* describing run parameters are added to the Nexus files. The following settings were selected based on extensive calibration and sensitivity testing of a subset of MRCs *(32)*:

   a. Number of samples (*see* **Note 3**): for data sets with <30 protein sequences, a single MCMC run of between 250,000 and 550,000 generations is performed. For data sets of ≥30 sequences, replicate MCMC runs are carried out, each of length 10^6. The number of replicates is equal to the number of sequences in the alignment set, divided by ten and rounded down: thus three replicate runs are performed for sets with 30–39 proteins, four runs for sets with 40–49 proteins, and so on. Each MCMC run involves four Markov chains, with the heating parameter set to 0.5.

   b. Model choice: Because it is not feasible to fit all of the free parameters in an amino acid general time-reversible (GTR) model, empirical models of sequence substitution are typically used. MrBayes 3.04β supports five substitution models potentially relevant to microbial data sets: PAM, WAG, VT, Blosum, and JTT *(23)*. Each is assigned an equal prior probability (0.2 in the case of five models), and the Markov chains are allowed to swap among them. Among-site rate variation (ASRV) is modeled by a four-category discrete approximation to the (continuous) gamma distribution, with uniform distribution over the interval [0.10, 50.00] and automatic estimation of the shape parameter α.

   c. Priors on branch lengths and trees: Each tree is assigned an equal prior probability, and edge-lengths are assigned a uniform prior on the interval [0.0, 10.0].

2. The burn-in phase is identified from the likelihood progress of a run, as follows: first, the mean of all log-likelihood scores from the final 100,000 iterations of a given run is determined. The end of the burn-in phase is then defined to be the first cold chain sample in the MCMC run that has a log-likelihood score exceeding this mean value (*see* **Note 4**). If the burn-in point has occurred too late in the run to yield the target number of post-burn-in samples, the entire run is discarded and performed again.

3. Where replicate runs have been carried out, the post-burn-in samples from the replicates can be combined into a single "metachain" prior to being summarized. However, this should be done only if all the chains have converged on the target distribution. Convergence can be evaluated by

comparing the range of log-likelihood values associated with each replicate run, but topological convergence should also be addressed using a criterion such as δ *(32)*, which measures the variation in bipartition posterior probabilities between replicates. A lack of convergence across short replicate runs strongly suggests that much longer runs are needed, either to permit convergence on the target distribution or to yield an adequate number of samples once the target distribution has been reached.

4. Sampled post-burn-in trees are summarized to yield posterior probabilities on trees and bipartitions by running the "sumt" command in MrBayes, specifying the appropriate burn-in point.

5. A reference hypothesis of organismal or genomic descent is required (*see* **Note 5**), for example a supertree. From the set of bipartitions with posterior probabilities ≥0.95 (*see* **Note 6**), a supertree is constructed using software such as CLANN. Several supertree methods are available in CLANN, but we favor the matrix representation with parsimony (MRP) method *(33, 34)* because of its wide use and applicability to very large data sets.

3.4. Inference of Lateral Transfer Events via Topological Comparisons of Trees

1. Given a reference tree that serves as a null hypothesis of organismal or genomic descent, its topology is compared with the topology of each protein tree in turn, using the Efficient Evaluation of Edit Paths (EEEP) method (*see* **Fig. 21.2** and **Note 7**). A subtree prune-and-regraft operation on a tree is an *edit*. A hypothesis of LGT can be represented graphically as coordinated set of such edits, or *edit path*. The number of edits in an edit path is the *edit path length* or *edit distance*. The goal is to find all shortest edit paths that can reconcile each observed protein tree with the reference tree.

2. There exists a single minimal edit distance of zero or greater between each protein tree and the reference tree, but there may be multiple non-identical edit paths of this length. An edit operation that appears in every alternative path is considered *obligate*, and the transfer that it implies must have occurred in the evolution of that MRC, given certain assumptions (*see* **Note 8**). Edit operations that occur in at least one, but not necessarily all, alternative paths are termed *possible* edits. They offer potential, but not definitive, explanations for the observed history of a given MRC. Edits can be of determinate or indeterminate direction, depending on whether or not at least one of the lineages involved in a given LGT operation can be identified as the donor or the recipient (*see* **Note 9**).

3. Obligate and possible edits can be pooled to test hypotheses about gene sharing within and between different groups of

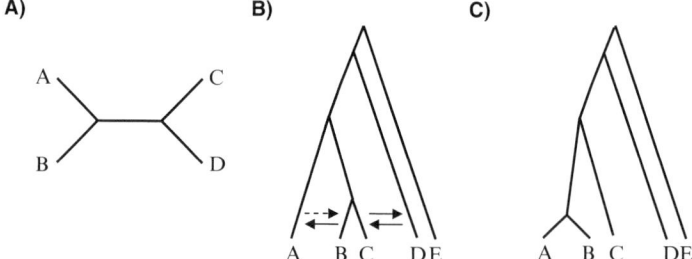

Fig. 21.2. Reconciling an unrooted protein tree (**A**) with a rooted reference tree (**B**) using EEEP. One or more subtree pruning and regrafting (SPR) operations (edits) are performed on the reference tree until it becomes topologically consistent with the protein tree. Ideally all minimum-length paths to all consistent topologies would be discovered, although when very large trees are compared, computational feasibility may require that the search space be limited in a manner that does not guarantee a complete set of edit paths. In the example shown, four alternative edit paths of length 1, each represented by a unidirectional arrow, can convert tree (**A**) into a tree that is topologically congruent with tree (**A**). The edit operation represented by the dashed arrow in (**B**) implies a donation of genetic material from the ancestor of genome A to the ancestor of genome B, yielding tree (**C**), which is consistent with the inferred protein tree. In this simplified example, no edit path is obligate; the donor/recipient pair cannot be uniquely identified because the implied lateral transfer implicates either A and B, or C and D.

organisms. For instance, although in our analysis the *Aquifex aeolicus* genome did not show a strong affinity for any particular branch among Archaea, collectively many transfer operations are implied between the *Aquifex* and archaeal lineages.

4. Notes

1. Genome sequence data can also be obtained from other locations. Open non-commercial sites include The Institute for Genomic Research (http://www.tigr.org) and the U.S. Department of Energy Joint Genome Institute (http://jgi.doe.gov).

2. The choice of parameter settings in GBLOCKS can have a dramatic effect on the amount of sequence that is retained within a given alignment. In principle, it is essential to eliminate all columns that contain non-homologous residues, and desirable as well to remove columns that have a high probability of being misaligned. However, overly aggressive settings remove many columns with data from rapidly evolving sites, useful for resolving relationships among closely related organisms. Since we used WOOF to score

sets of alignments of each data set, and since we adopted a global alignment approach to the analysis of protein sets, we chose GBLOCKS settings that were considerably more conservative (thus retaining more columns in the alignment of n sequences) than the default parameter settings shown in the following.

Default settings for GBLOCKS:

Minimum number of sequences for a conserved position: $n \times 0.5 + 1$

Minimum number of sequences for a flank position: $n \times 0.85$

Maximum number of contiguous non-conserved positions: 8

Minimum length of a block: 10

Allowed gap positions: None

3. There is extensive literature on whether it is preferable to run a single long or several short Markov chains. Certain theoretical distributions such as the "Witch's Hat" can produce highly misleading results if multiple short chains are used *(35)*, but in separate work we have observed substantial evidence for multi-modality of some protein data sets. Given the practical limitations on the length of MCMC chains, the pragmatic approach of multiple short runs is more likely to identify multi-modality than a single long run *(32)*. Also, although our calibration runs showed that the chain lengths stated in **Section 3.3.1** were sufficient for our sub-sampled data sets, the question of chain length is by no means closed, and we recommend longer runs (e.g., of at least 10^6 iterations for data sets of size ≤10, and more for larger data sets) and more replicates where computationally feasible.

4. Our chosen convergence diagnostic relies on the stabilization of likelihood values from a given Markov chain, and is a simple but consistent adaptation of the graphical inspection of serial log-likelihood values commonly used in phylogenetics. Other, more-sophisticated convergence diagnostics have been proposed *(36)*, but it is not clear which among these are appropriate to phylogenetic inference.

5. Reference hypotheses of organismal or genomic relationships can be derived by other approaches. For example, trees could be inferred from sequences of trusted single genes such as that encoding small-subunit (16S) ribosomal RNA, from weighted or unweighted proportions of genes held in common ("genome trees"), or from shared physiological or morphological characters. Each of these alternatives presents difficulties. For example, 16S rDNA exhibits insufficient

variability to resolve many closely related strains, whereas the interpretation of genome trees remains controversial *(6)*, and use of phenetic characters in phylogenetic inference has a highly problematic history. For recent comparisons of supertree methods see *(37)* and *(38)*.

6. The choice and validity of posterior probability thresholds for phylogenetic MCMC remains controversial. Several groups *(39)* have presented cases in which bipartition posterior probabilities tend to exaggerate the support for topological features with higher posterior probability. Such observations, and the manner in which competing hypotheses are evaluated using, e.g., Bayes factors *(40)*, have led to the use of thresholds of 0.90 or greater in many published studies. In *(4)* we chose 0.95 because: *(1)* this threshold requires that a given feature be strongly supported in all replicate runs; *(2)* our large data sets allowed the use of a stringent threshold while still retaining over 90,000 bipartitions; and *(3)* it corresponds to a minimum Bayes factor (expressed as the posterior probability ratio of two competing hypotheses) of $0.95/0.05 = 19$, which corresponds to "very strong" support for a given hypothesis *(40)*. The supertree we obtained was robust to changes in the PP threshold, with no more than 5 of 141 internal nodes differing between thresholds of 0.51 and 1.00. The proportion of resolved concordant bipartitions was more sensitive to threshold, ranging from about 77% at 0.51, to 92% at the maximum threshold 1.00.

7. EEEP allows unrooted test (in this case, protein) trees to be compared with a rooted reference tree (in this case, the supertree). This was the most appropriate comparison, since the reference tree could be rooted in accordance with previous molecular work *(41)*, but the test trees could not be rooted without assuming a molecular clock. Two other programs, LatTrans *(42)* and HorizStory *(43)*, require both reference and test trees to be rooted. LatTrans is extremely fast compared to EEEP and HorizStory *(26)*, but currently requires both trees to be completely resolved, which was not appropriate to our analysis.

8. Three fundamental assumptions underlie our inference of LGT: *(1)* that the evolution of these putatively orthologous sets of proteins can be described with a tree, *(2)* that we have recovered the correct tree, and *(3)* that the "true" donor taxon was a reasonably close relative of the one implied in our analysis. The first of these assumptions can be violated if some of the sequences in an MRC are not orthologous, due either to unrecognized (cryptic) paralogy, or to inter- or intra-species recombination. Many reasons have been documented by which orthologous sequences can yield

incorrect trees, including but not limited to inappropriateness of reconstruction method, rapid or highly variable rates of sequence substitution, and violation of sequence stationarity, rate homogeneity, and/or substitution reversibility *(12, 44–47)*. We carried out a battery of statistical tests in *(4)* to assess the impact of some of these issues, but many of these potential problems remain open issues. The third assumption relates to density of taxonomic sampling; better sampling will clarify this question for some relationships, but other donor lineages may have no extant representatives.

9. Although some of the limitations and pitfalls of surrogate methods have been documented *(9, 10, 48)*, in combination with a phylogenetic approach such as the one outlined herein, these approaches can be valuable in identifying donor and recipient lineages. Genes that have been acquired vertically retain, for some time, sequence biases characteristic of their donor lineage, and these biases can in favorable cases be used to distinguish donor from recipient. Similarly, a lineage within which most or all extant genomes possess a given gene or sequence feature is more likely to be the source than is a lineage in which that gene or features is rare.

Acknowledgments

Cheong Xin Chan, Nicholas Hamilton, Tim Harlow, and Jonathan Keith provided vital assistance in developing and executing the phylogenetic pipeline described in this chapter. We acknowledge the Australian Research Council (CE0348221) and the Australian Partnership for Advanced Computing for support.

References

1. Gurney-Dixon, S. (1919) *The Transmutation of Bacteria*. Cambridge University Press, Cambridge, UK.
2. Jones, D., Sneath, P. H. A. (1970) Genetic transfer and bacterial taxonomy. *Bacteriological Rev* 34, 40–81.
3. Medigue, C., Rouxel, T., Vigier, P., et al. (1991) Evidence for horizontal transfer in *Escherichia coli* speciation. *J Mol Biol* 222, 851–856.
4. Beiko, R. G., Harlow, T. J., Ragan, M. A. (2005) Highways of gene sharing in prokaryotes. *Proc Natl Acad Sci U S A* 102, 14332–14337.
5. Jain, R., Rivera, M. C., Lake, J. A. (1999) Horizontal gene transfer among genomes: the complexity hypothesis. *Proc Natl Acad Sci U S A* 96, 3801–3806.
6. Charlebois, R. L., Beiko, R. G., Ragan, M. A. (2004) Genome phylogenies, in (Hirt, R. P., Horne, D. S., eds.), *Organelles, Genomes and Eukaryote Phylogeny: An Evolutionary Synthesis in the Age of Genomics*. CRC Press, Boca Raton, FL.
7. Deckert, G., Warren, P. V., Gaasterland, T., et al. (1998) The complete genome of the hyperthermophilic bacterium *Aquifex aeolicus*. *Nature* 392, 353–358.

8. Nelson, K. E., Clayton, R. A., Gill, S. R., et al. (1999) Evidence for lateral gene transfer between Archaea and bacteria from genome sequence of *Thermotoga maritima*. *Nature* 399, 323–329.

9. Ragan, M. A. (2001) On surrogate methods for detecting lateral gene transfer. *FEMS Microbiol Lett* 201, 187–191.

10. Ragan, M. A., Harlow, T. J., Beiko, R. G. (2006) Do different surrogate methods detect lateral genetic transfer events of different relative ages? *Trends Microbiol* 14, 4–8.

11. Ho, S. Y., Jermiin, L. (2004) Tracing the decay of the historical signal in biological sequence data. *Syst Biol* 53, 623–637.

12. Jermiin, L., Ho, S. Y., Ababneh, F., et al. (2004) The biasing effect of compositional heterogeneity on phylogenetic estimates may be underestimated. *Syst Biol* 53, 638–643.

13. Tatusov, R. L., Fedorova, N. D., Jackson, J. D., et al. (2003) The COG database: an updated version includes eukaryotes. *BMC Bioinformatics* 4, 41.

14. Peterson, J. D., Umayam, L. A., Dickinson, T., et al. (2001) The comprehensive microbial resource. *Nucleic Acids Res* 29, 123–125.

15. Van Dongen, S. (2000) Graph clustering by flow simulation. Ph.D. Thesis: University of Utrecht, Utrecht.

16. Rigoutsos, I., Floratos, A. (1998) Combinatorial pattern discovery in biological sequences: the TEIRESIAS algorithm. *Bioinformatics* 14, 55–67.

17. Beiko, R. G., Chan, C. X., Ragan, M. A. (2005) A word-oriented approach to alignment validation. *Bioinformatics* 21, 2230–2239.

18. Thompson, J. D., Higgins, D. G., Gibson, T. J. (1994) CLUSTAL W: improving the sensitivity of progressive multiple sequence alignment through sequence weighting, position-specific gap penalties and weight matrix choice. *Nucleic Acids Res* 22, 4673–4680.

19. Notredame, C., Higgins, D. G., Heringa, J. (2000) T-Coffee: A novel method for fast and accurate multiple sequence alignment. *J Mol Biol* 302, 205–217.

20. Lee, C., Grasso, C., Sharlow, M. F. (2002) Multiple sequence alignment using partial order graphs. *Bioinformatics* 18, 452–464.

21. Gotoh, O. (1996) Significant improvement in accuracy of multiple protein sequence alignments by iterative refinement as assessed by reference to structural alignments. *J Mol Biol* 264, 823–838.

22. Katoh, K., Misawa, K., Kuma, K., et al. (2002) MAFFT: a novel method for rapid multiple sequence alignment based on fast Fourier transform. *Nucleic Acids Res* 30, 3059–3066.

23. Huelsenbeck, J. P., Ronquist, F. (2001) MRBAYES: Bayesian inference of phylogenetic trees. *Bioinformatics* 17, 754–755.

24. Castresana, J. (2000) Selection of conserved blocks from multiple alignments for their use in phylogenetic analysis. *Mol Biol Evol* 17, 540–552.

25. Creevey, C. J., McInerney, J. O. (2005) Clann: investigating phylogenetic information through supertree analyses. *Bioinformatics* 21, 390–392.

26. Beiko, R. G., Hamilton, N. H. (2006) Phylogenetic identification of lateral genetic transfer events. *BMC Evol Biol* 6, 15.

27. Altschul, S. F., Madden, T. L., Schaffer, A. A., et al. (1997) Gapped BLAST and PSI-BLAST: a new generation of protein database search programs. *Nucleic Acids Res* 25, 3389–3402.

28. Berman, H. M., Westbrook, J., Feng, Z., et al. (2000) The Protein Data Bank. *Nucleic Acids Res* 28, 235–242.

29. Harlow, T. J., Gogarten, J. P., Ragan, M. A. (2004) A hybrid clustering approach to recognition of protein families in 114 microbial genomes. *BMC Bioinformatics* 5, 45.

30. Sokal, R. R., Sneath, P. H. A. (1963) *Principles of Numerical Taxonomy*, W.H. Freeman & Co, London.

31. Maddison, D. R., Swofford, D. L., Maddison, W. P. (1997) NEXUS: an extensible file format for systematic information. *Syst Biol* 46, 590–621.

32. Beiko, R. G., Keith, J. M., Harlow, T. J., Ragan, M.A. (2006) Searching for convergence in Markov chain Monte Carlo. *Syst Biol.* 55, 553–565.

33. Baum, B. R. (1992) Combining trees as a way of combining data sets for phylogenetic inference, and the desirability of combining gene trees. *Taxon* 41, 3–10.

34. Ragan, M. A. (1992) Phylogenetic inference based on matrix representation of trees. *Mol Phylogenet Evol* 1, 53–58.

35. Geyer, C. J. (1992) Practical Markov chain Monte Carlo. *Stat Sci* 7, 473–483.

36. Cowles, M. K., Carlin, B. P. (1996) Markov chain Monte Carlo convergence diagnostics: a comparative review. *J Amer Statist Assoc* 91, 883–904.

37. Bininda-Emonds, O. R. P. (2004) *Phylogenetic Supertrees: Combining Information to Yield the Tree of Life*. Kluwer, Dordrecht.
38. Wilkinson, M., Cotton, J. A., Creevey, C., et al. (2005) The shape of supertrees to come: tree shape related properties of fourteen supertree methods. *Syst Biol* 54, 419–431.
39. Suzuki, Y., Glazko, G. V., Nei, M. (2002) Overcredibility of molecular phylogenies obtained by Bayesian phylogenetics. *Proc Natl Acad Sci U S A* 99, 16138–16143.
40. Kass, R., Raftery, A. E. (1995) Bayes factors. *J Amer Statist Assoc* 90, 773–795.
41. Linkkila, T. P., Gogarten, J. P. (1991) Tracing origins with molecular sequences: rooting the universal tree of life. *Trends Biochem Sci* 16, 287–288.
42. Addario-Berry, L., Chor, B., Hallett, M., et al. (2003) Ancestral maximum likelihood of evolutionary trees is hard. *Algorithms Bioinformat Proc* 2812, 202–215.
43. MacLeod, D., Charlebois, R. L., Doolittle, F., et al. (2005) Deduction of probable events of lateral gene transfer through comparison of phylogenetic trees by recursive consolidation and rearrangement. *BMC Evol Biol* 5, 27.
44. Kuhner, M. K., Felsenstein, J. (1994) A simulation comparison of phylogeny algorithms under equal and unequal evolutionary rates. *Mol Biol Evol* 11, 459–468.
45. Huelsenbeck, J. P. (1995) Performance of phylogenetic methods in simulation. *Syst Biol* 44, 17–48.
46. Waddell, P. J., Steel, M. A. (1997) General time-reversible distances with unequal rates across sites: mixing gamma and inverse Gaussian distributions with invariant sites. *Mol Phylogenet Evol* 8, 398–414.
47. Bruno, W. J., Halpern, A. L. (1999) Topological bias and inconsistency of maximum likelihood using wrong models. *Mol Biol Evol* 16, 564–566.
48. Lawrence, J. G., Hendrickson, H. (2003) Lateral gene transfer: when will adolescence end? *Mol Microbiol* 50, 739–749.

Chapter 22

Detecting Genetic Recombination

Georg F. Weiller

Abstract

Recombination is the major motor of evolution. While mutations result in gradual changes, recombination reshuffles entire functional modules and thus progresses evolution in leaps and bounds. We need to identify recombination breakpoints in sequences to understand the evolutionary process, the impact of recombination, and to reconstruct the phylogenetic history of genes and genomes. This chapter provides a step by step guide for detecting recombination even in large and complex sequence alignments.

Key Words: Recombination; PhylPro; multiple sequence alignment; phylogenetic profile; phylogenetic correlation; HIV-1; GAG.

1. Introduction

Genetic recombination is the formation of a new DNA sequence from two donor sequences, so that one region of the sequence corresponds to one donor and another region to the other donor. There are many different mechanisms of genetic recombination, including homologous recombination, gene conversion, transposition, transduction and intron-homing. In some forms only the information of a donor is copied, whereas others involve physical breakage and rejoining of the DNA strand. Recombination between the homologous chromosomes is reciprocal, whereas most forms of non-homologous recombination are unidirectional. Horizontal or lateral gene transfer is a recombination of genetic material from two different species *(1)*, and in many cases is mediated by mobile genetic elements such as viruses. Recombination of genetic material from sequences as distantly related as plants and vertebrates has been reported *(2)*.

Recombination can combine sequences or genes that were independently shaped and tested by evolution and therefore forms the major motor for evolution. Although evolution through mutation can only adjust individual nucleotides and results in gradual changes, evolution through recombination deals with entire functional modules and so progresses in leaps and bounds.

The depiction of the evolutionary process as a phylogenetic tree has considerable shortcomings as it ignores genetic recombination. Recombination hampers current methods of phylogenetic reconstruction. Therefore, before a phylogenetic reconstruction based on sequences is attempted, the possibility that individual sequences are recombinant must be assessed.

In recent years, many methods have been developed to find genetic recombination in sequences and a comprehensive list of programs for detecting and analyzing recombinant sequences has been maintained for many years by David Robertson at http://bioinf.man.ac.uk/recombination/programs.shtml. The site is frequently updated and currently lists 43 programs. Although almost all methods require a multiple sequence alignment to find recombination breakpoints, some require additional information, such as the knowledge of non-recombinant prototype sequences, the "true" phylogenetic tree or which sequence is recombinant. Rigorous evaluations of many of the major methods have been published using simulated *(3)* or empirical *(4)* sequence examples. Some of these programs are easy to use as they require few parameters. Other programs are more flexible but require a careful choice of parameters. The phylogenetic profile (PhylPro) method described below, if correctly used, is extremely sensitive and makes no assumptions as to which, if any, sequence is recombinant, or where to expect recombination breakpoints. However, the method requires much user intervention as it provides a platform for data exploration. PhylPro is widely used in recombination analyses. This chapter explains how the program can be used, even with a large and complex dataset, to detect recombinant sequences and recombinant breakpoints.

Details of the phylogenetic profile method are described elsewhere *(5, 6)*. Briefly, the method introduces the "phylogenetic correlation" measure that quantifies the coherence of the sequence interrelationships in two different regions of a multiple alignment. Positions in which sequence relationships in the upstream region clearly differ from their downstream counterpart exhibit low phylogenetic correlations and are likely recombination sites. For each individual sequence in the alignment, the phylogenetic correlations are computed at every position using a sliding window technique. The plot of the phylogenetic correlations against the sequence positions is termed a "phylogenetic profile," and the profiles of all individual sequences are typically superimposed in a single diagram. Such profiles support the identification of individual recombinant sequences as well as recombination hot spots.

In the following example, the *gag* region of HIV-1 has been chosen to provide a step-by-step guide to the use of PhylPro. It represents a challenging example, since recombinations in HIV-1 sequences are much more widespread than previously thought *(4, 6)*. The dataset contains over 600 sequences, all or most of which are likely to be the result of many recombinant events.

2. Materials

1. Program availability: The PhylPro program and accompanying manual can be downloaded free of charge from http://bioinfoserver.rsbs.anu.edu.au/. PhylPro is a MS Windows application and requires Windows 95 or above, or a Windows emulation software, to run. Install the program according to the instructions in the manual.

2. Data format: PhylPro requires a multiple alignment of nucleotide or protein sequences (*see* **Note 1**) and accepts NBRF-PIR, FastA, GDE, and GenBank formats.

3. Data source: For this example an alignment of the *gag* region of HIV-1 is used. Readers that wish to follow the example described below can download the alignment from the HIV Sequence Database at http://hiv-web.lanl.gov via the "Sequence Database" and "Alignments" links to the 2004 HIV and SIV alignments. Choose the alignment format "FastA" and select "HIV-1/SIVcpz" gag DNA and then click "Get alignment." Download the alignment to a folder of your choice and name the file "HIV1GAGDNA.fa" (*see* **Note 2**).

3. Methods

This section describes, step by step, how recombinant sequences and recombinant breakpoints can be found using PhylPro. It assumes that PhylPro is installed and that the HIV1GAG alignment has been downloaded. In the following description, menu options are displayed in italics, specifically *File:New* means select the "New" option of the "File" menu. Dialog box names are given with the first letter of each word in upper case and their options enclosed in single quotes.

3.1. Creating a PhylPro Project

To analyze sequences in PhylPro the sequences must be first imported into the PhylPro database system called VOP (virtual object pool).

1. Start the PhylPro program. Choose the menu option *File: New* and select *Project*. An empty VOP window will appear.

2. To import the alignment choose *File:Import sequence*. In the dialog box that displays, choose 'File type: FastA (Nuc)' and navigate to the folder containing the file HIV1GAGDNA.fa and open it (*see* **Note 3**). An information window displays that the file contains 614 nucleotide sequences (*see* **Note 4**). Click 'Import now.' Some information about the imported sequences appears in the project window. Name the project by saving the VOP, select *File:Save VOP as*, and type the file name 'HIV1gag' in the 'Save As' dialog box. The VOP window is no longer required and can be closed (by clicking the button on the top right of the window).

3.2. Creating a Dataset Containing All Sequences

PhylPro is able to manage many different datasets, differing in the sequences it contains, analysis parameters or display options. A given sequence can be a member of many datasets without physical duplication. Before a phylogenetic profile can be created, a dataset containing the sequences must be created.

1. Use the *Dataset:New* command to create a new dataset. In the Compose Set of Nucleotide Sequences dialog box, choose 'Select...all sequences' from the 'Sequence Pool' and click 'To set' to move them to the 'Sequence Data-Set.' Click 'OK' to confirm the creation of a new dataset containing all aligned sequences. The sequences are displayed in the Sequence View window.

2. Before exploring the alignment, ensure that the following Sequence View Preferences are set: Select *View:Preference* (or via the right mouse button) and select 'Windows: React to others,' 'Show features' and 'Show column usage', 'Columns: raw' and 'Nuc-colors: by Nucleotide.' Ensure that 'Display in blocks of 10' and 'Gray analysis region' are not selected. A different font can also be used if desired. Click 'OK' to close the preferences dialog.

3. Individual regions of the alignment can be marked. Here the different *gag* proteins are marked. Select *View:Feature* to open the 'FeatureView.' An empty window shows that there are no features defined. Select *Edit:Add Feature* and enter a range 'From' 1 'to' 558, with a 'Text' of P17 to indicate that this region encodes the P17 protein of *gag*. Click 'OK' to confirm. Repeat this process for P24 (559-1278), P7 (1366-1551), and P6 (1555-1947), choosing different colors for each (*see* **Note 5**). Close the 'FeatureView.' The

features are now displayed above the alignment at the top of the Sequence View. Use both scroll bars to examine the entire alignment.

4. The downloaded alignment is a subset of a bigger, multiple sequence alignment and contains columns that consist only of gap characters. Highly gapped columns are not helpful for further analysis and will be excluded. The 'Columns' row above the alignment indicates the alignment positions used for analysis. Individual columns can be chosen, in several ways, and excluded or included in the analysis. Select *Columns:Select special:Missing chars* and enter 'Minimum occurrence: 50' to select all columns in which more than 50 sequences have a gap (*see* **Note 6**). The selected columns are now grayed in the Columns row. Then select *Columns:Exclude selection*. The excluded columns will now have a minus sign in the 'Columns' row above the aligned sequences, and will not be used in further analyses. To also exclude them from the Sequence View, open the Sequence View Preferences (*View:Preferences*) and select *Columns:Included*.

3.3. Creating the First Phylogenetic Profile

1. Before generating the first phylogenetic profile ensure that the default Profile Parameters are set as follows: select *DataSet:Prof Parameter* and in the dialog set 'Distance scores: Score nucleic acids,' 'Analysis column: Variable,' 'Limit sliding window by: Comparisons 40,' 'Correlation measure: Correlation,' and click 'Save graph between sessions' to avoid having to recalculate the phylogenetic profile when viewing this dataset at a later stage. Click 'OK' to generate the profile.

2. Before exploring the profile (**Fig. 22.1A**) ensure that the following Profile View Preferences are set: select *View: Preferences*. In the Profile View Preferences dialog select 'Window: React to others,' 'Window: Show features,' 'Graph: Show inform. columns only,' 'Smooth: 1,' 'Show location: Cross hair,' 'Snap: to minimum,' and 'Summary line: Show.' Click 'OK.' Save the dataset by selecting *Dataset: Save As* and specify 'All40C' in the 'Object:' field of the Save Object dialog. The name indicates that this dataset contains all sequences and a phylogenetic profile using a window of 40 comparisons.

3.4. Interpretation of the Profile

This first profile is not ideal for finding recombinants, but gives a first view of the data variability. The graph is overwhelming, showing hundreds of low phylogenetic correlation minima. The graph is not uniform over the entire alignment and some areas, for example the junction of P17 and P24, and a region in the last half of P24, have particularly low phylogenetic correlation as can be seen

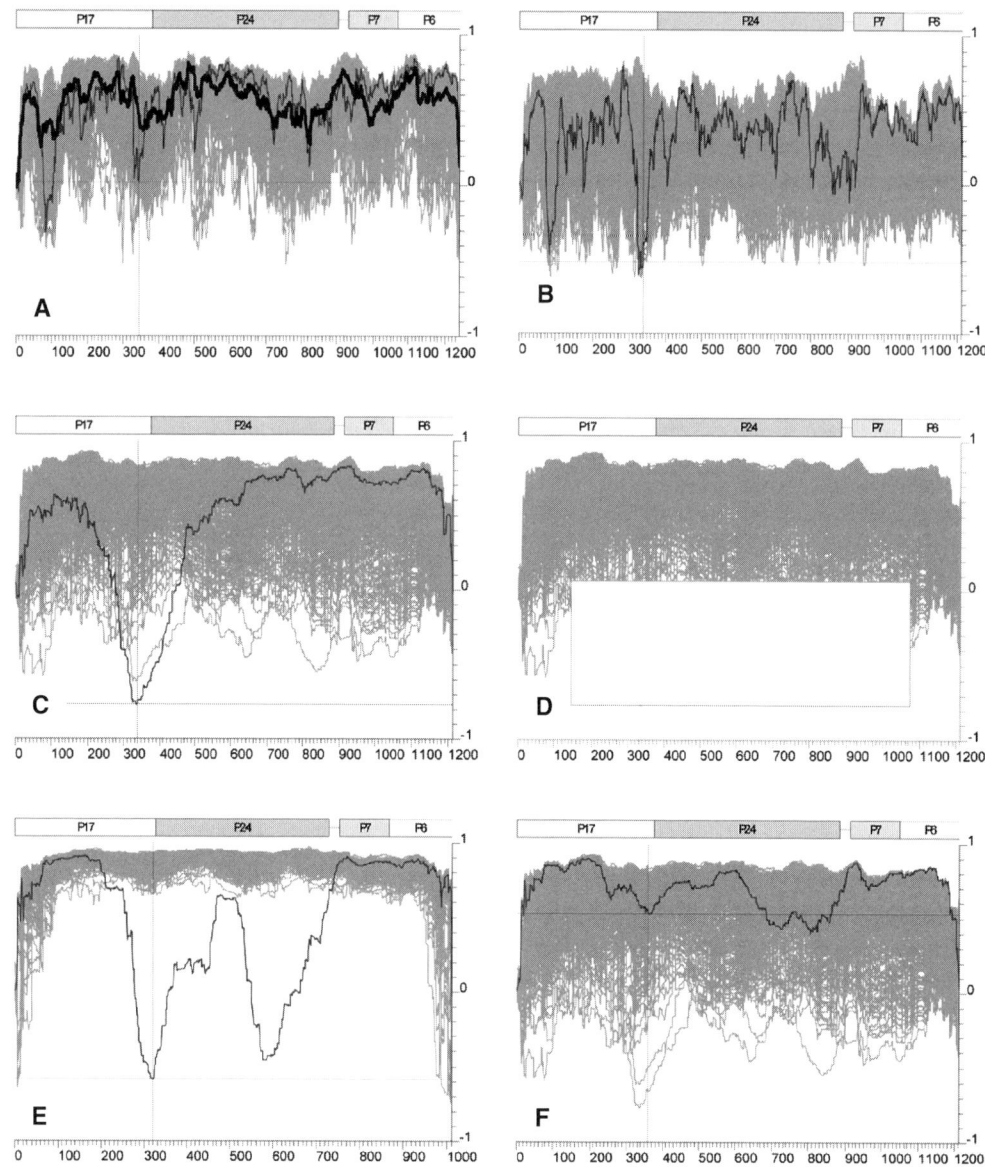

Fig. 22.1. Phylogenetic Profiles of the HIV-1 GAG region. All graphs are produced using PhylPro. (**A**) Dataset All40C. The bold dark line gives the average phylogenetic correlation of all sequences. Sequence A1C.TZ.97.97 is highlighted with a thin dark line. (**B**) Dataset M40C. Sequence A1C.TZ.97.97 is highlighted. (**C**) Dataset M30D. Sequence A1C.TZ.97.97 is highlighted. (**D**) Dataset M30D with a region of the profile selected. (**E**) Dataset M275+1. Sequence 08_BC.CN.98 is highlighted. (**F**) Dataset M30D. Sequence 08_BC.CN.98 is highlighted. See text for further details.

on the purple summary line (bold, dark line in **Fig. 22.1A**), which displays the average phylogenetic correlation of all sequences. The low phylogenetic correlations show that sequence relationships change very frequently in the chosen window size, indicating that recombinations are very frequent in the dataset. By double clicking on various areas in the graph a position in a sequence is selected and its phylogenetic profile is displayed in red (thin, dark line in **Fig. 22.1**), but rather than exploring this profile further we will create a new optimized profile. To improve the recombination signal, distantly related sequences will be removed and the analysis window enlarged for profile calculation.

3.5. Refinement of the Sequence Set

If a sequence alignment contains a small group of sequences that is more distantly related than the in-group, then recombinations between in-group sequences do not show up clearly since the relationship between in-group sequences will remain relatively close regardless of the recombination event. To enhance the signal for in-group sequences the out-group has to be removed.

1. To identify the more distantly related sequences, go to the Sequence View window and select *Sequences:Compare all*. In the dialog, select only 'Untypical sequences.' Click 'OK.' This calculates the distances between all sequences and shows the sum of distances for every sequence. The top 29 sequences in the display window are clearly more divergent, with the sums of differences ranging from c. 300000 to c. 230000. The next highest sum of differences is much lower (c. 130000, sequence D.ZA.85.R214) (*see* **Note 7**).

2. Create a new dataset that does not include the 29 out-group sequences. A convenient way to create a new dataset based on the initial dataset, and so keep the column and feature definitions, is to duplicate the existing set. Select *Dataset:Duplicate* and then change the sequence composition of the newly created dataset by selecting *Sequences:Add/Compose*. Highlight the last 29 sequences of the 'Sequence Dataset' starting with sequence N.CM.97.YBF1 and click 'To pool.' Click 'OK' to modify the dataset (*see* **Note 8**). The text window displaying the out-group sequences is no longer required and can be closed.

3. Create a new profile using the same parameters as in **Section 3.3.1**. Save the new dataset, which also includes the profile (*see* **Section 3.3.2**). Here the name M40C is chosen to indicate that this dataset contains the group M sequences using a window of 40 comparisons.

3.6. Refinement of the Profile Parameters

Before examining the new profile (**Fig. 22.1B**), create a third profile by changing the window size (*see* **Note 9**).

1. Duplicate the M40C dataset (*see* **Section 3.5.2**).

2. Change the profile parameters of the new dataset. Select *Dataset:Profile parameter*, and set the parameter as in **Section 3.3.1** except for 'Limit sliding window by: Differences 30.' Click 'OK.' A new profile is generated (**Fig. 22.1C**).

3. Save the dataset (*see* **Section 3.3.1**) with the name M30D, indicating the analysis window is 30 differences wide.

3.7. Comparing Profiles

To see the effect that sequence removal and a change in the window size have on the profile, compare the three profiles.

1. Ensure that all three datasets are open with only the Profile View of each displayed. Minimise or close the Sequence Views of the datasets as well as any other open windows.

2. Arrange the windows by selecting *Window:Tile horizontal*.

3. Ensure the following View Parameters in each of the Profile Views (right click in the profile and select Preferences) are selected: 'Window: React to others,' 'Window: Show features,' 'Graph: Show inform. columns only,' 'Smooth: 1,' 'Show location: Cross hair,' and 'Snap: to minimum.' Make sure 'Summary line: Show' is not checked. The three profiles are now easily compared.

4. By double clicking close to a prominent trough in one profile, the profile for this sequence will be highlighted, a crosshair will be displayed at the profile minimum, and the corresponding sequence and position will be shown in all windows containing the sequence. For example, if the lowest point in profile M30D is selected (i.e., the most prominent recombination site close to informative position 335), the recombination is also apparent in the profile for M40C, but is not prominent in the All40C profile (**Fig. 22.1A–C**).

3.8. Determining Recombinant Sequences

The M30D profile will be used to identify the most prominent recombination breakpoints.

1. Close all windows except the M30D Sequence View and Profile View windows. Select *View:Alignment* if it is not currently displayed. Arrange the windows by selecting *Window: Tile horizontal*.

2. In the Sequence View right click, select Preferences and then 'Columns: informative' while keeping the other settings set as in **Section 3.2.2** (*see* **Note 10**). Click 'OK.'

3. In the Profile View, double clicking on any of the troughs will select the sequence. On double clicking the lowest trough (c. position 335), the profile for this sequence will be highlighted and a crosshair will be displayed at the profile minimum. In the Sequence View the selected sequence becomes visible and its acronym A1C.TZ.97.97 is grayed, the sequence position corresponding to the crosshair position of the Profile View is highlighted with a yellow background and the selected column is indicated by a grayed '+' sign.

4. By double clicking the four most predominant profile troughs, the following recombinant sequences and their approximate recombination site (informative) can be identified: A1C.TZ.97.97 c. position 334, A1C.SE.96.SE position 334, A1C.RW.92.92 position 836, and 02C.BE.93.VI position 642.

The similarity of the profiles for the sequences A1C.TZ.97.97 and A1C.SE.96.SE suggests that both are descended from a common recombinant sequence.

5. To further examine any of the selected (crosshair) recombination breakpoints, select *View:Relationships*. A text window will now describe the selected sequence and position, and the sequences most closely related are given for the left and right analysis window. This is a convenient way to find the potential recombination donor sequences.

6. To see the area of analysis (i.e., the window size) in the Profile View select *View:Preferences* and select 'Show location: Analysis region.' Click 'OK.' Double clicking anywhere in the profile will highlight the extent to which any of the other sequences were involved in the calculation of the sequence location indicated by the crosshair. Note that the extent of the analysis region is different for every sequence because this dataset uses an analysis window that is limited by sequence differences. To see the area of analysis in the Sequence View select 'Gray analysis region' in the Sequence View Preferences dialog box. Double clicking in the profile will now not only mark the corresponding point in the Sequence View (yellow highlight) but will also mark the portion of the sequences used for analysis (i.e., the analysis window) with a gray background (**Fig. 22.2**, grayscale only).

3.9. Systematic Removal of Recombinant Sequences from the Dataset

After the most predominant recombinants have been identified it is useful to remove these sequences from the dataset as the detection of recombination sites in other sequences is impeded by the presence of strong recombinants in the background sequence set.

1. To select sequences for removal, single sequences can be highlighted in the Sequence View or Profile View, or groups of sequences can be highlighted in the acronym column of the Sequence View, or more conveniently a region of the Profile View can be selected using 'click' and 'drag.' With this latter method, a whole group of sequences with low phylogenetic correlation can be selected at once (**Fig. 22.1D**). In Sequence View, the number of selected sequences will be displayed in the lower left corner and the sequences acronyms are grayed. Select *Sequences:Remove* and generate a new profile.

2. It is often possible through consecutive removal of sequences with low phylogenetic correlation to arrive at a dataset of sequences with a consistently high phylogenetic profile. When previously removed sequences are reintroduced into the dataset, their recombination sites become apparent (*see* **Note 11**). Fig. 22.1E shows a profile consisting only of the 275 sequences with a high phylogenetic correlation

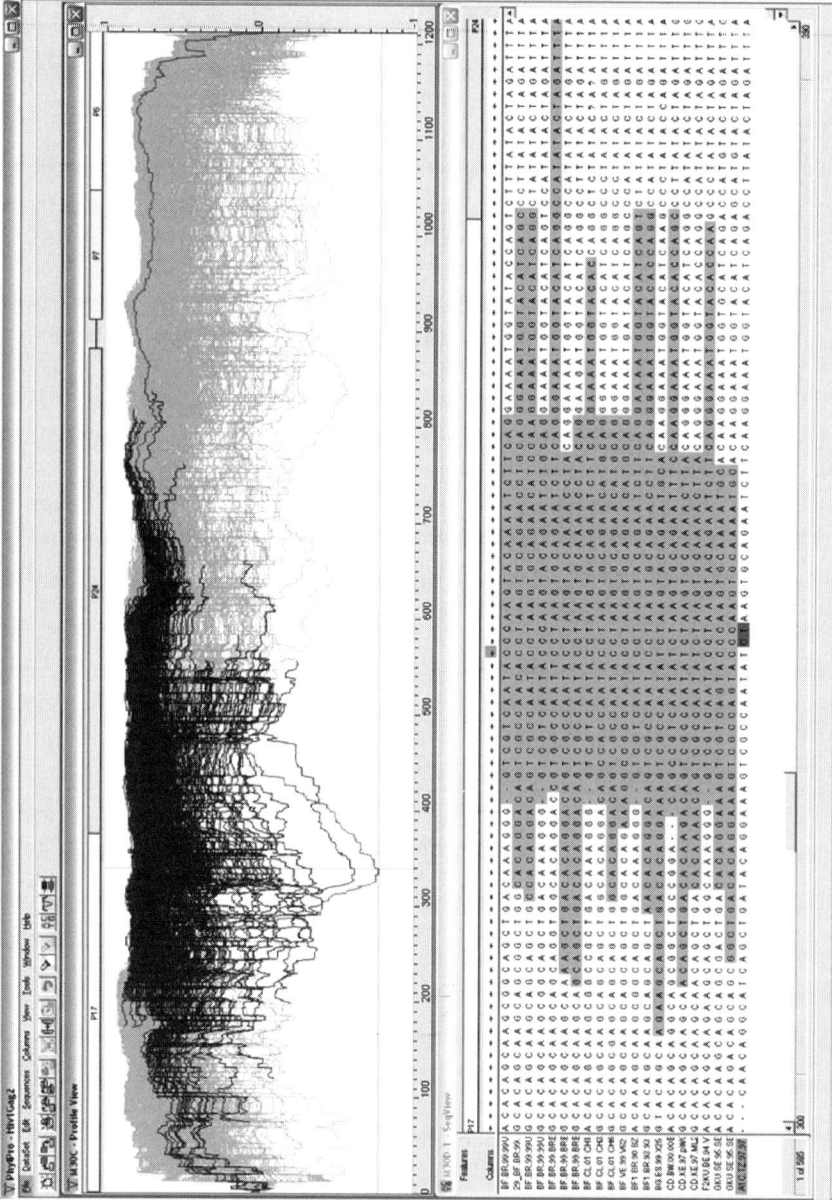

Fig. 22.2. PhylPro application window. The Profile View displays the M30D dataset of 585 sequences with sequence A1C.TZ.97.97 highlighted as in **Fig. 22.1C**. The analysis region in the profile is shown by the darker profile lines (window limit = 30 differences). The Sequence View gives the corresponding region; however, a window limit of four differences is chosen to illustrate the rugged window edges in the available space.

plus sequence 08_BC.CN.98. The two recombination junctions of this sequence become apparent in this profile, but could not have been detected in profile M30D (**Fig. 22.1F**).

3. Save the Project (*File:Save VOP*) and put the kettle on.

4. Notes

1. To produce a multiple alignment of sequences with extended recombination is potentially complicated as some alignment methods, such as CLUSTALW *(7)*, require a guide-tree. However, it may be difficult to get a reliable guide-tree from recombinant sequences. Other alignment methods such as DIALIGN *(8)* or methods that use hidden Markov models *(9, 10)* may be more advantageous. As some recombinant sequences may have regions that are not homologous to other sequences, some manual editing of the alignments may be necessary.

2. PhylPro expects text files to adhere to the Windows convention, which terminates lines with a CR (ascii 13) + LF (ascii 10) combination. Occasionally, downloaded sequences use either the Macintosh or Linux convention, which terminates lines with a CR or LF only. In these cases the file needs to be converted. Convert the file by, for example, opening it in MS WordPad or Word and saving it as a text file under the same name.

3. PhylPro assumes that FastA files have an extension .fa and displays only these files. To also display files that have a different naming convention write *.* in the file name field and press 'return.'

4. PhylPro has analyzed the file and found 614 nucleotide sequences. If the expected numbers of sequences have not been found, a problem with the file format is most likely the trouble (*see* **Note 2**).

5. Positions provided are correct for the current alignment (i.e., 2004 HIV and SIV alignments). Positions in subsequent alignments may change as additional sequences may require the introduction of additional gaps.

6. The number entered is not crucial in this example as most alignment columns have either a very high or a very low number of gaps; however, extensive gap regions will hamper the interpretation of phylogenetic profiles. The Sequence View should be used as a guide as to which positions or sequences are excluded from the analysis. For very large

alignments it may be helpful to temporarily set the font size to 1 point.

7. Readers familiar with HIV nomenclature will know that HIV-1 lineages were originally put into the groups M for main, O for outlier and N for non-M and non-O. For further information on HIV nomenclature see *(11, 12)*. The top 29 sequences are in the O and N groups, whereas the remaining 585 sequences belong to group M. The removal of sequences other than M will enhance the resolution of recombination involving group M subtype sequences. Out-group sequences cannot be removed if recombination involving out-group lineages is to be examined.

8. It is fortuitous that the 29 out-group sequences are grouped on the bottom of the alignment, otherwise the sequence information presented in the Compare All window can be used to find and remove the sequences individually.

9. The choice of window size is not easy. If few recombination sites are expected, the windows can be made sufficiently large. If more than one recombination site is within a window, the phylogenetic correlation calculation will be impeded and smaller windows should be chosen. However, the window size must be big enough to gather enough sequence differences between any two sequences if a reliable measure of sequence similarity is to be obtained. As one sequence is compared with many other sequences, the ideal window size may differ for each sequence with which the test sequence is compared. Therefore, PhylPro can use a different window size for every sequence. When the option 'Limit sliding window by: Differences' is chosen, the window for every sequence will be set to the size required to find the specified differences. We have done this in this example where the window was set to 30 differences. The remaining options are to set the window size to a fixed number of informative positions (by comparison) or to use a maximum window size (unlimited) whereby the window encompasses the whole alignment. It is important to note that the phylogenetic correlations near the alignment boundaries are typically low, as one of the analysis windows becomes too small. The low correlation toward the end of the sequence must therefore be ignored.

10. PhylPro has three ways to refer to sequence positions. *Raw positions* refer to sequence positions in the original imported alignment file. *Included positions* ignore alignment columns that have been excluded from the analysis. *Informative positions* are the positions that are useful in calculating phylogenetic profiles. These are either variable positions or parsimoniously informative positions as set in the Profile

Parameter dialog box. Parsimoniously informative positions are those with at least two of each of two different nucleotides. The Profile View can display raw or informative positions, whereas the Sequence View can display raw, included, or informative positions.

11. Clear-cut recombinations will show up by deep v-shaped troughs in an otherwise relatively high phylogenetic correlation. However, there will always be borderline cases. If a dataset would consist only of one or a small number of recombinant sequences, together with a set of non-recombinant prototype sequences, determination of recombination sites would be easy. However, such circumstances are rare. In reality, all sequences are the product of a plethora of recombinations involving donors, which are themselves recombinant. Sequence comparisons such as done with PhylPro will only be able to find sufficiently recent recombination sites involving sufficiently different donor sequences. PhylPro makes no attempt to assess the statistical significance of a recombination event. Rather, it tries to provide the user with the available sequence information.

References

1. Beiko, R. G., Ragan, M. A. (2007) Detecting lateral gene transfer: a phylogenetic approach, chapter 21, this volume.
2. Gibb, M. J., Weiller, G. F. (1999) Evidence that a plant virus switched hosts to infect a vertebrate and then recombined with a vertebrate-infecting virus. *PNAS* 96, 8022–8027.
3. Posada, D., Crandall, K. A. (2001) Evaluation of methods for detecting recombination from DNA sequences: computer simulations. *PNAS* 98, 13757–13762.
4. Posada, D. (2002) Evaluation of methods for detecting recombination from DNA sequences: empirical data. *Mol Biol Evol* 19, 708–717.
5. Weiller, G. F. (1998) PhylPro: a graphical method for detecting genetic recombinations in homologous sequences. *Mol Biol Evol* 15, 326–335.
6. Weiller, G. F. (2000) Graphical methods for exploring sequence relationships, in (Rodrigo, A.G., Learn, G.H., eds.), *Computational and Evolutionary Analysis of HIV Molecular Sequences*. Kluwer, Boston.
7. Thompson, J. D., Higgins, D. G., Gibson, T. J. (1994) CLUSTAL W: improving the sensitivity of progressive multiple sequence alignment through sequence weighting, position-specific gap penalties and weight matrix choice. *Nucleic Acids Res.* 22, 4673–4680.
8. Morgenstern, B., Frech, K., Dress, A., et al. (1998) DIALIGN: Finding local similarities by multiple sequence alignment. *Bioinformatics* 14, 290–294.
9. Hughey, R., Krogh, A. (January 1995) SAM: Sequence alignment and modeling software system. Technical Report UCSC-CRL-95-7, University of California, Santa Cruz, CA. http://www.cse.ucsc.edu/research/compbio/sam.html.
10. (Accessed July 2007) HMMER: profile HMMs for protein sequence analysis. Washington University in St Louis, http://selab.janelia.org/.
11. (Accessed July 2006) *HIV and SIV nomenclature*, HIV Sequence Database, Los Alamos National Laboratory, http://hiv-web.lanl.gov/content/hiv-db/HelpDocs/subtypesmore.html.
12. Robertson, D. L., Anderson, J. P., Bradac, J. A., et al. (Accessed July 2006) *HIV-1 Nomenclature Proposal, A Reference Guide to HIV-1 Classification*, HIV Sequence Database, Los Alamos National Laboratory, http://hiv-web.lanl.gov/content/hiv-db/REVIEWS/nomenclature/Nomen.html.

Chapter 23

Inferring Patterns of Migration

Paul M.E. Bunje and Thierry Wirth

Abstract

The historical movement of organisms, whether recent or in the distant past, forms a central aspect of evolutionary studies. Inferring patterns of migration can be difficult and requires reliance on a large suite of bioinformatic tools. As it is primarily the movement of groups of related individuals or populations that are of interest, population genetic and phylogeographic methods form the core of tools used to decipher migration patterns. Following a description of these tools, we discuss the most critical—and potentially most difficult—aspect of these studies: the inference process used. Designing a study, determining which data to collect, how to analyze the data, and how to coordinate these results into a coherent narrative are all a part of this inference process. Furthermore, using different types of data (e.g., genotypic and DNA sequence) from different types of sources (direct, or from the organisms of interest; and indirect, from symbiotic organisms) produces a powerful suite of techniques that are used to infer patterns of migration.

Key words: Migration, inference process, population genetics, indirect evidence, genotypic data, DNA sequence analysis, molecular evolution.

1. Introduction

Determining the migratory history of organisms and ongoing patterns of migration requires a complex inference process that employs several different analytical techniques. Most contemporary studies of migration use genetic data to infer patterns since direct observation of migration (e.g., tagging and tracking) is often difficult or impossible. When beginning a study to elucidate patterns of migration, it is critical that an appropriate strategy of data collection, data analysis, and interpretation is followed. This necessitates that any investigation begin with a detailed plan:

1. Identify the hypothesis being tested.
2. Determine the geographic scope of the population(s) or species being investigated.
3. Identify the type of genetic data to be used.
4. Conduct a preliminary sample across the geographic range using several genetic markers.
5. Use the pilot study to determine the appropriate genetic marker(s) and density of geographic sampling required to evaluate the hypothesis.
6. Collect samples for analysis.
7. Collect genetic data from samples.
8. Analyze the data using appropriate statistical techniques.
9. Combine the various statistical analyses and data types into a logical interpretive framework. At this point, ancillary data and/or information can be used to improve the inference of migration pattern.

Although we discuss how these methods are used for inferring patterns of migration among related populations, this process can be easily modified to answer many other questions of intra-specific geographic distribution.

2. Data and Materials

2.1. Organism(s) Under Investigation

A thorough understanding of the life history and ecology of the organism(s) under investigation is critical to an appropriately designed study. Some aspects of an organism's biology that are particularly germane to patterns of migration are vagility (*see* **Note 1**), fecundity, territoriality, breeding behavior, dispersal ability, and type of propagule/offspring. This knowledge is critical in determining an appropriate hypothesis for migration pattern. For example, an organism with wind-borne spore dispersal can be hypothesized to have migratory patterns that coincide with dominant wind patterns. If using a symbiont to trace migration patterns of its host, the mode of transmission, rate of recombination, and mutation rate are also critical. Furthermore, knowledge of the demographic structure derived from previous studies of demography or inferred from other biological characteristics is important for designing the preliminary sampling scheme.

2.2. Sampling Scheme

An appropriately designed sampling scheme is critical to obtaining a sound estimate of migration patterns. While designing the sampling scheme, continually analyze preliminary data to determine both where and how intensively to sample, as well as to identify appropriate genetic markers (see the following).

1. Determine the geographic range of interest and identify all known or putative populations of study organism within that area.

2. Determine the scale at which migration patterns are to be determined (i.e., at how finely resolved a geographic scale the hypothesized movements are to be observed).

3. Collect organisms for a preliminary analysis (pilot study). To do this, collect 5–10 samples/individuals from ~10% of the total populations expected in the full study. Collect from populations in several clumps distributed around the entire range. In other words, collect several populations that are at the smallest possible distance from each other. Collect these groups of populations from all areas of the study area, including the center (**Fig. 23.1**).

4. Perform preliminary analyses on the genetic markers of choice (see the following). Using these preliminary analyses, it should be possible to predict how many populations are

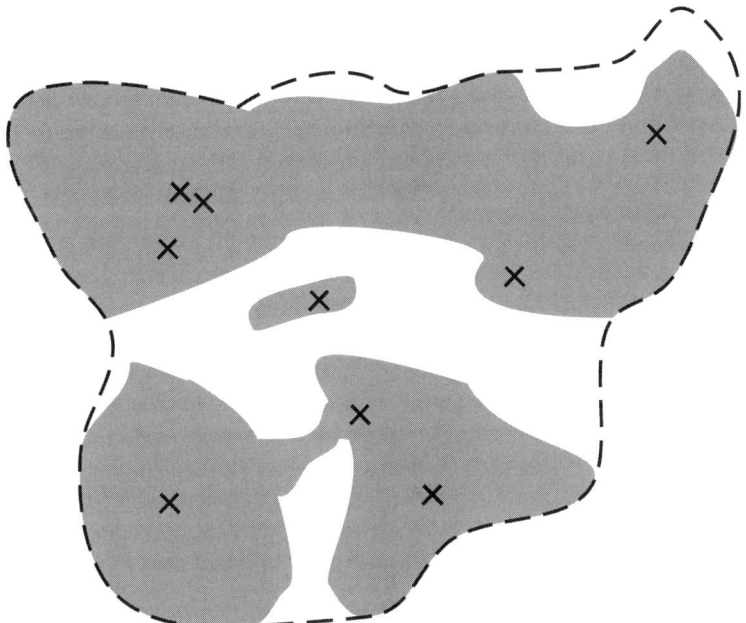

Fig. 23.1. Hypothetical preliminary sampling scheme. Suppose the organism of interest inhabits a range delimited by the dashed line. Populations can be found throughout the gray shaded area of this range. In the full study, it is expected that ~80 populations will be sampled. Sampling 8–10 of these populations for a preliminary study, demarcated by the crosses, in a pattern such as that shown will help to identify regions that need to be sampled more or less densely, as well as the distance between populations that should be sampled in the full study.

needed to provide a statistically significant estimation of migration patterns. This stage is also where the appropriate level of genetic variation and/or the appropriate genetic marker can be identified, based on how the preliminary genetic variation appears to be distributed in space.

5. Using these preliminary data, perform a power analysis to determine how many samples per population are necessary to confidently assess the amount of genetic diversity present *(1, 2)*. This can be done by performing an ANOVA on the genetic variation present in each population and estimating β to determine how many samples per population are necessary to define each population as statistically distinct. As neighboring populations are expected to be similar or identical genetically, it is appropriate to perform this power analysis between populations that are approximately the average distance apart of all populations. Importantly, this step is not meant to determine how populations are related to each other, simply to determine how many samples are necessary to capture most of the existing genetic diversity.

6. Collect the determined number of samples from populations that are distributed across the range, being sure to collect from all possible areas but collecting more populations in areas that are hypothesized to be especially interesting. Coalescent theory predicts that between 20 and 50 individuals per population will capture >95% of haplotype diversity for most organisms *(1)*. Additionally, distribute the sampling effort across the study area so as to sample as much genetic diversity as possible, avoiding sampling populations that are too close together.

7. Continually analyze the data as it is collected, thus identifying geographic areas that have more genetic variation than average. Collect more populations and/or samples from these areas as they are identified.

2.3. Types of Data

There are two general types of data, those representing haplotypic sequence of particular genetic loci (in haploid or diploid organisms/genomes) or those representing genotypic data describing allelic variation in one or more loci (which may be either haploid or diploid). We first discuss DNA sequence data and then the several most common types of genotypic data. These two types of data require different, though often analogous, analytical methods relating to their biological properties.

Common to both types of data, it is critical to have data that exhibit an appropriate amount of genetic variation. This means that the DNA sequence or genotypes must have enough variability (either in haplotype or allelic diversity) to provide a statistically significant result and must not be so variable as to introduce either excess homoplasy (*see* **Note 1**), in the case of sequence data, or too little common allelic variation, in the case of genotypes. The amount of variability can only be determined through preliminary analysis of the study organism, the geographic study area, and several potentially useful loci (*see* **Notes 2–5** for caveats to be considered when identifying appropriate genetic loci).

1. Plastid DNA. DNA sequences from mitochondrial and chloroplast genomes are commonly used to study migration patterns. This is due to the relatively fast rate of evolution of mitochondria in animals and fungi and chloroplast genomes in plants (plant mitogenomes tend to evolve slowly), the uniparental inheritance mode (obviating allelic variation), and the intracellular abundance of these organelles. To obtain these data, whole genomic DNA must be isolated from the organisms of interest using one of several available methods, such as phenol-chloroform and salt-based extractions, or commercially available kits. Amplification by PCR using primers specific to the target sequence then proceeds. Several universal primers have been developed that target highly variable loci, including the mitochondrial control region, cytochrome b, and cytochrome c oxidase subunit I in animals and plants. Alternatively, it may be necessary to develop new primers for a particular group or for a locus that demonstrates the appropriate amount of genetic variation. Finally, direct sequencing of purified PCR product results in haplotypes ready for alignment and analysis. These haplotypes, like those from nuclear DNA, can be used either as DNA sequence data or as haploid genotypes in most methods described in the following.

2. Nuclear DNA. Using nuclear DNA sequences for estimating migration history is hindered by the generally slow rate of mutation in the nuclear genome. Nonetheless, some loci with relatively fast rates of nucleotide substitution have been used successfully, notable the internal transcribed spacer regions of the ribosomal DNA locus and intronic regions of some genes. The procedure for obtaining haplotype sequence of these genes (presuming negligible allelic variation) is the same as that in the preceding, although universal primers are less likely to be available.

3. Single Nucleotide Polymorphisms. Single base-pair differences between haplotypes (e.g., from different chromosomes), called single nucleotide polymorphisms (SNPs), represent the most common form of genetic diversity in model organisms whose whole genomes have been sequenced, and tend to have a moderate mutation rate of 10^{-8}–10^{-9}/my *(3)*. For non-model organisms, SNPs can be identified by either comparing homologous loci from genetic databases or by sequencing anonymous loci from many individuals. SNPs can be identified in sequence chromatograms using programs (**Table 23.1**) such as Phred

Table 23.1
List of commonly used applications (in alphabetical order)

	URL	Data types	Analyses performed	Platforms supported
Ape	http://pbil.univ-lyon1.fr/R/ape/	DNA sequence	Computing distances, Minimum spanning trees, Generalized skylines	OS X Windows Linux
Arlequin	http://anthro.unige.ch/arlequin/	DNA sequence	Population genetics, Fst Miscellaneous	Macintosh Windows Unix
BayesAss+	http://www.rannala.org/labpages/software.html	Multi locus genotypes	Bayesian estimation of recent migration rates	OS X Windows Linux
BEAST	http://evolve.zoo.ox.ac.uk/beast/	DNA sequence	Clock and divergence estimates, Coalescent models of population parameters	OS X Windows Linux
DMLE+	http://dmle.org/	Genotypes, Haplotypes	Linkage disequilibrium mapping	OS X Windows Linux
DnaSP	http://www.ub.es/dnasp	DNA sequence	Linkage disequilibrium, Recombination miscellaneous	Windows
FSTAT	http://www2.unil.ch/popgen/softwares/fstat.htm	Genotypes	Population genetics, Fst	Windows
GDA	http://hydrodictyon.eeb.uconn.edu/people/plewis/software.php	DNA sequences, Genotypes	Linkage disequilibrium, Fst	Macintosh Windows
GenePop	http://genepop.curtin.edu.au/	Genotypes	Population genetics, Fst	Windows
GENETREE	http://www.stats.ox.ac.uk/~griff/software.html	DNA sequence	Mutation, Migration, Growth rates	Windows
GENIE	http://evolve.zoo.ox.ac.uk/software.html?id=genie	Molecular phylogenies	Inference of demographic history	OS X Windows Linux
GeoDis	http://darwin.uvigo.es/software/geodis.html	DNA sequence	Nested clade analysis	OS X Windows Linux
Hickory	http://darwin.eeb.uconn.edu/hickory/software.html	DNA sequence, Microsatellites	Bayesian methods for F-statistics	Windows Linux
LAMARC	http://evolution.gs.washington.edu/lamarc/	DNA sequence, SNP, Microsatellite	Population size, growth recombination, migration rates	OS X Windows Linux
LDhat	http://www.stats.ox.ac.uk/~mcvean/LDhat/LDhat1.0/LDhat1.0.html	DNA sequence	Linkage disequilibrium, Recombination rates	Windows Unix Linux

Inferring Migrations 491

Name	URL	Data	Description	OS
MDIV	http://www.binf.ku.dk/~rasmus/webpage/programs.html#MDIV	DNA sequence	Divergence time, Migration rates	Windows
ModelTest	http://darwin.uvigo.es/software/modeltest.html	DNA Sequence	Tests models of sequence evolution	OS X Unix Windows
MrBayes	http://mrbayes.csit.fsu.edu/index.php	DNA sequence	Phylogenetic analyses, Bayesian inference	Macintosh Windows Unix
MrModelTest	http://darwin.uvigo.es/software/modeltest.html	DNA Sequence	Tests models of sequence evolution	OS X Unix Windows
MultiDivTime	http://statgen.ncsu.edu/thorne/multidivtime.html	DNA sequence	Molecular rates, Divergence times	OS X Linux
PAL	http://www.cebl.auckland.ac.nz/pal-project/	DNA sequence	Phylogenetic analyses, Coalescence, Demographic parameters	OS X Unix MSDOS
PAML	http://abacus.gene.ucl.ac.uk/software/paml.html	DNA sequence, Proteins	Phylogenetic inferences, Test for positive selection (ML)	OS X Unix Linux
PAUP*	http://paup.csit.fsu.edu/	DNA sequence, Morphological data	Phylogenetic inference	Macintosh Unix Windows
PHASE	http://www.stat.washington.edu/stephens/	DNA sequence, Haplotypes	Haplotype reconstruction, Recombination rate estimation	Windows Linux
Phrap	http://www.phrap.com/	DNA sequence	Sequence comparisons and assembly	OS X Windows
Phred	http://www.phrap.com/	DNA sequence	Base calling	OS X Windows
PHYLIP	http://evolution.genetics.washington.edu/phylip.html	DNA sequence, Protein, Restriction sites	Phylogeny inference	OS X Windows Linux
PHYML	http://atgc.lirmm.fr/phyml/	DNA sequences	Phylogenetic inferences, Maximum Likelihood	Macintosh Windows Unix
R8S	http://loco.biosci.arizona.edu/r8s/	DNA sequence	Molecular rates, Divergence times	OS X Linux
SPAGEDI	http://www.ulb.ac.be/sciences/ecoevol/spagedi.html	DNA sequence	Population genetics, Fst Miscellaneous	Windows
Structure	http://pritch.bsd.uchicago.edu/structure.html	Multi locus genotype	Population structure, Bayesian inference	OS X Windows Unix
Strucurama	http://www.structurama.org/	Multi locus genotype	Population structure, Bayesian inference	OS X Windows
TCS	http://darwin.uvigo.es/software/tcs.html	DNA Sequence	Network construction using statistical parsimony	OS X Windows Linux

and Phrap (4). SNPs can then be analyzed either by resolution into haplotypes using Bayesian statistics in PHASE (5) or Markov chain Monte Carlo analysis (6) or through population genetic and coalescent methods. Importantly, when performing these analyses, the SNP data must be corrected for ascertainment bias (6), that is:

   a. Define the minimum frequency of variability among samples that constitutes a SNP.

   b. Sequence all loci in either some individuals or all individuals to identify variable sites.

   c. Correct for ascertainment strategy. Specifically, if a limited subset of individuals (a panel) is used to identify SNPs, a correction factor must be used that accounts for the missing rare SNPs, many of which may be represented in recently migrated individuals (3). LAMARC can perform such analyses while correcting for ascertainment bias.

4. Microsatellites. Microsatellites are codominant markers (see **Note 1**) consisting of short repeat elements (generally tandem or triplet repeats). Due to their high mutation rates, ca. 10^{-4}/my (7), frequency in the genome, and ease of identification, they have become especially popular in studies of migration. Despite their popularity, little is known about the evolutionary mechanisms that govern their diversity, making it difficult to construct genealogical inferences from them (7). Ideally, >5 unlinked microsatellite loci should be used for intra-specific evolutionary research (see **Note 3**). If appropriate microsatellite primers are available for a given study organism, then genotyping can begin with step i. Otherwise, they can be obtained as follows:

   a. Digest genomic DNA (e.g., using MboI or MseI) and size fraction the product on an agarose gel, purifying out 500–1,000 bp fragments.

   b. Ligate linker fragments of known sequence to these products and test quality via PCR.

   c. Enrich the concentration of fragments with linkers via PCR.

   d. Using biotinilated probes of di- or tri-nucleotide repeats (i.e., CA, CT, etc. in oligonucleotides containing ~10 repeats), hybridize the enriched fragments from (c) to the probes and isolate only those fragments that hybridize to the probes (e.g., using magnetic beads).

   e. Test for the presence of microsatellite repeats in the fragments from (d) via PCR.

   f. Subclone the PCR products and identify sequences via PCR and cycle sequencing of the cloned products.

g. Design primers for loci that have (close to) perfect repeat stretches of >10 repeats.

h. Screen the microsatellite loci designed in (g) from several populations in the study organism to identify microsatellites with variation.

i. PCR several good microsatellites from all samples using the primers identified in (8), one of which is fluorescently labeled.

j. Perform gel electrophoresis and scoring on an automatic genotyping detection system.

5. Amplified Fragment Length Polymorphisms. Amplified fragment length polymorphisms (AFLPs) are dominant markers derived from two older methods (restriction fragment length polymorphisms and randomly amplified polymorphic DNA). This type of data is relatively inexpensive to obtain and quickly results in large numbers of analyzed loci (hundreds to thousands) whose mutation rates depend on the restriction enzymes chosen and so can be tailored to a given question and organism (8). They can be obtained as follows:

    a. Digest genomic DNA using two different restriction enzymes (e.g., *EcoRI* and *MseI*).

    b. Ligate adaptors of known sequence to the digested DNA.

    c. Preamplify the fragments via PCR using primers that complement the sequence made by the adaptor plus the enzyme recognition sequence and a single additional base.

    d. Selectively amplify these fragments via PCR using primers that correspond to those used in (c) plus two additional bases. One of these two primers (usually the one containing the longer restriction enzyme recognition sequence) should be fluorescently labeled.

    e. Perform gel electrophoresis and scoring on an automatic genotyping detection system.

2.3.6. Other Data Types

There are several other types of genotypic data that are still in use, although their popularity is waning. These include restriction fragment length polymorphisms (RFLP), randomly amplified polymorphic DNA (RAPD), single-stranded conformation polymorphism (SSCP), and allozymes (7). All of these data can be analyzed using the methods outlined below in a fashion analogous to microsatellites or AFLPs.

3. Methods

3.1. Classical Population Genetics

3.1.1. Fixation Indices

The F-statistics of Wright (1965) describe the proportion of genetic variation in a sample that can be explained by defined sub-samples *(9)*. The most common index, F_{st}, estimates how much genetic variation is accounted for by differences within populations versus differences between populations. This operates as an estimate of the amount of divergence (or alternatively cohesion) among populations. It can be estimated from genotypic and haplotypic data (often referred to as Φ_{st} for haplotypic data such as mtDNA sequences). To estimate F_{st}, genetic data from individuals are assigned to one or more populations and then the variance in allele frequency among the various sub-populations is estimated using one of several models *(10)*. These models are generally equivalent and can be implemented in most population genetics software. R_{st} is a related statistic that explicitly employs a stepwise mutation model and is appropriate for analyzing microsatellite evolution. Under certain assumptions, F_{st} can be used to infer rates of symmetrical migration between equally sized populations *(11)*.

3.1.2. Nucleotide and Haplotype Diversity

These two estimates of genetic diversity for DNA sequence were developed by Nei *(12)*. They provide a general estimate of genetic diversity at both the nucleotide level and the gene-locus level and can be estimated by most population genetics software packages. Nucleotide diversity (π) is the probability that two randomly chosen homologous nucleotides from a pool of haplotypes are different. Haplotype diversity (H) is the probability that two randomly chosen haplotypes from a pool are different, and is analogous to expected heterozygosity for genotypic data. Low values of π and H in a population could be the result of recent colonization or other bottleneck-like events.

3.1.3. Heterozygosity

Heterozygosity, derived from the Hardy-Weinberg (H-W) equation, can be calculated for any genotypic data, including nuclear sequences in which both alleles are known for given individuals. By calculating expected and observed heterozygosity (H_E and H_O, respectively), which can be done by most population genetics software, deviations from H-W equilibrium can be identified statistically. One of the most commonly observed causes for deviation from H-W equilibrium within a population is migration. Another case of deviation is the so-called Wahlund effect in which a deficit of heterozygotes is observed as the result of a mixture of populations within a sample.

3.1.4. Allelic Richness

For genotypic data, a simple count of the number of alleles per locus in a sample provides an estimate of the pool of genetic diversity present. This estimate should be statistically verified by bootstrapping the total samples from each population to a common number of samples *(13)* or via rarefaction *(14)*. Populations that resulted from a recent colonization event generally have much lower allelic richness than the source population(s). Ongoing migration between established populations can increase allelic richness by exchanging novel alleles that have evolved in different localities. As a general rule, the more alleles shared between populations, the more gene flow (viz. migration) is occurring between them.

3.1.5. Theta

Theta is a classic population parameter than can be effectively used to estimate migration rates between populations. Theta is calculated as $\Theta = 4N_e\mu$ for diploid populations and $\Theta = 2N_e\mu$ for haploid populations, where N_e is the effective population size and μ is the mutation rate of the locus being studied. There are several methods for estimating Θ based on either homozygosity, variance in the number of segregating sites, the relationship between number of alleles and sample size, and the estimated number of pairwise differences in a population *(15)*. This quantity is particularly useful since the value of $4N_em$, where m is the migration rate between population pairs, can be estimated using a coalescent approach (see the following). Therefore, if an estimate for μ is available, migration rate between diploid populations can be estimated simply as $m = (4N_em)\mu/\Theta$.

3.1.6. AMOVA

Analysis of molecular variance (AMOVA) is an analysis of variance method for describing the population structure of genetic diversity *(16)*. By first defining populations, then grouping those populations into a reasonable structure (i.e., to test geographic regions and relationships between them), AMOVA partitions the genetic variance within populations, within groups of populations, and between groups of populations. The covariance components of these partitions are then used to calculate the fixation indices described in the preceding. An AMOVA can be performed in Arlequin *(15)*. It is useful for inferring migration patterns by generating estimates of which groups of populations (or which geographic areas) are more or less integrated genetically. Low variance components between populations or areas indicate a lack of gene flow between them.

3.2. Coalescent-Based Population Genetics

Coalescent theory allows an estimate of the statistical distribution of branch lengths in genealogies described by heritable units such as chromosomes or genes (but *see* **Note 5**). Taking into account variation in life history parameters, the temporal and demographic pattern of coalescence of extant genes into their hierarchically inclusive ancestors can be estimated, thus providing a picture of population changes through time.

3.2.1. Estimation of $4N_em$

Using coalescent theory, Beerli and Felsenstein (2001) have developed a model that estimates, via Metropolis-Hastings Monte Carlo simulation, the maximum likelihood value of the quantity $4N_em$, where N_e is the effective population size and m is the migration rate *(11)*. This method assumes a known mutation rate at each locus, which is what relates this method to estimates of theta (see the preceding). As a result, with an estimate of mutation rate (π) and either N_e or Θ, asymmetric (or symmetric) migration rates between multiple populations of any size can be estimated. This method is implemented in the program MIGRATE (also LAMARC). Similar methods for estimating $4N_e m$ and $2m$ are implemented in GENETREE and MDIV.

3.2.2. Mismatch Distribution

Using the pairwise mismatch distribution suggested by Rogers and Harpending *(17)*, one can identify whether a group of related haplotypes is the likely result of recent population expansion *(18)*, as may result from range expansion, colonization, or extremely asymmetric migration between populations. Slatkin and Hudson *(18)* have shown that the distribution of pairwise differences between haplotypes in a sample will be a unimodal Poisson-like distribution if the haplotypes are the result of exponential population growth, whereas a multimodal distribution is typical of stable population sizes, which may result from stable or no migration between populations. Coalescent theory allows one to calculate the value $\tau = m - (v-m)^{1/2}$, where m is the mean of the normal mismatch distribution and v is the variance *(15)*. Since $\tau = 2\mu t$, where u is mutation rate, and t is time, the onset of a rapid expansion can be estimated. This method is implemented in Arlequin and DnaSP.

3.2.3. Lineage Through Time Plots

Coalescent theory allows for the estimation of when in the past various haplotypes (or haplotypes inferred from genotypic data) last shared a common ancestor. Lineage through time (LTT) plots display the relative age of these coalescent events as they accumulate in a species, clade, or population *(19)*. LTT plots can be used on individual populations to determine when in the past a notable change in the rate of lineage accumulation occurred. Assuming a constant mutation rate, these changes are likely to represent the time of changes in migration rate or other related demographic changes *(20)*. LTT plots can be created using Ape and GENIE *(21)*.

3.2.4. Skyline Plots

The LTT plots described above can be used to explicitly estimate changes in demographic parameters through time by the use of the derived skyline plot *(22)*. Taking into account error in phylogeny and mutation rates as well as the possibility of multiple coalescent events occurring simultaneously, the generalized skyline plot can estimate changes in population size through time by accounting for variance in the rate of coalescent events for a given set of (inferred) haplotypes *(23)*. As with LTT plots, these parameters are particularly useful when inferring patterns and timing of migration events. Generalized skyline plots can be made using GENIE, PAL *(24)*, and Ape.

3.3. Bayesian-Based Population Genetics

The recent interest in applying Bayesian inference to biological statistics has also resulted in numerous Bayesian methods for estimating population genetic parameters such as those described in the preceding. Software that can estimate such parameters in a Bayesian framework includes Hickory *(25)*, BEAST *(26)*, BayesAss *(27)*, and DMLE+ *(28)*.

3.3.1. Structure

One of the most commonly used Bayesian applications is the program Structure *(29)*. Structure allows assignment of individuals to populations based on multiple loci of genotypic data, though haplotypic data such as those from asexual organisms can also be used in a modified form *(30)*. Most relevant for migration inference, Structure assigns individuals and loci to source populations in a probabilistic way such that introgressed individuals that result from immigration are readily apparent (**Fig. 23.2**).

3.4. Multivariate Statistical Analysis

In addition to the methods designed expressly for genetic data outlined in the preceding, traditional statistical techniques can be applied when the appropriate precautions and assumptions are taken into account. Because of the multivariate nature of genetic (particularly genotypic) data, with gene frequencies of multiple loci distributed in multiple populations, multivariate statistical techniques such as principal components analysis, multidimensional scaling, and factor analysis can effectively identify relevant migratory patterns in genetic data *(44)*. In particular, migration is predicted to have a linear effect on gene frequencies, making principal components analysis, which can identify the genetic variation (i.e., geographic gene frequencies) that is most significant to realized genetic diversity in related populations, especially useful (**Fig. 23.3**).

3.5. Phylogenetic Methods

The details of reconstructing a phylogenetic tree are described in **Chapters 13–16**. In short, the process involves: *(1)* sequence alignment, *(2)* determination/estimation of model parameters, *(3)* algorithmic determination of optimal tree(s), and *(4)* statistical assessment of branch support. By tracing the geographic data of the tips of a reconstructed phylogeny (gene genealogy in the case of intraspecific migrations; *see* **Note 6**), the direction and timing (if estimates of branching events are available from, e.g., a molecular clock) of migration events can be determined.

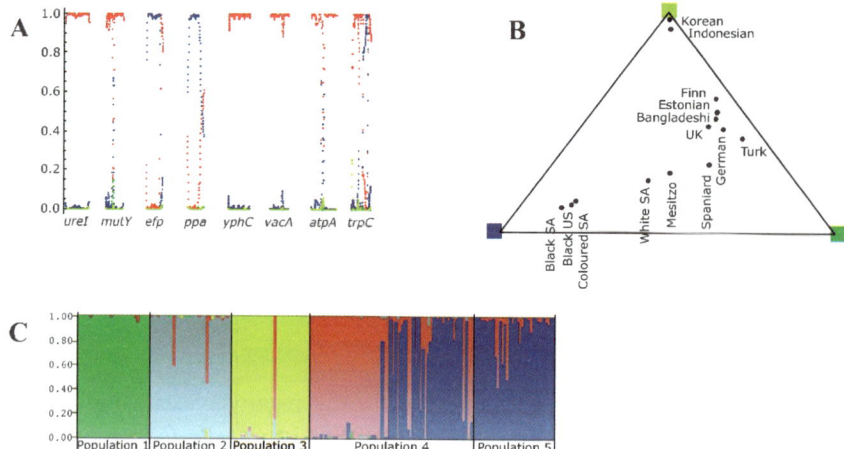

Fig. 23.2. Examples of Structure output demonstrating how patterns of locus and individual assignment can inform locations of origin and patterns of migration. (**A**) Ancestry estimates for a strain of the bacterial parasite *Helicobacter pylori* isolated from a black South African patient. Every single point corresponds to a polymorphic nucleotide, which is assigned to each population with a given probability. This isolate reveals that six genes segments are clearly from one (red) origin, with high probabilities whereas two other genes seem to be imported from another region (blue). Mosaic patterns observed in *H. pylori* are a signature of direct contact and homologous recombination between different microbial populations, reproduced from *(30)*. (**B**) Assignment data of individual *H. pylori* isolates from different locations. The proportion of ancestry from each of three ancestral sources (blue, Africa; yellow, East Asia; and green, Indo-Europe) is represented by its proximity to the corresponding corner of the triangle, reproduced from *(30)*. (**C**) Typical Structure output showing the posterior probability that a particular individual comes from one of five different populations (represented by different colors). Individuals with mosaic patterns are indicative of historic or ongoing contact between populations as a result of migration, reproduced from unpublished data.

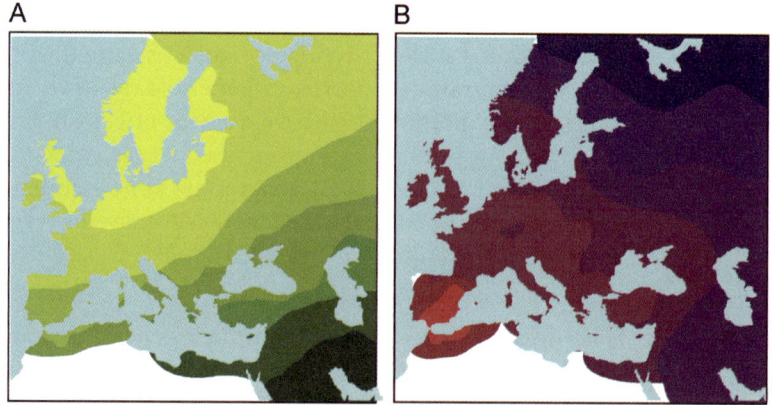

Fig. 23.3. Principal components of the major genetic variation in European humans, based on allele frequency data from multiple loci and modified from *(44)*. (**A**) The first principal component. (**B**) The second principal component. The gradients represent relative concentration of the loaded allele variants in each area. (**A**) Higher concentrations of the PC1-loaded alleles are shown in green and lower concentrations grade toward yellow. These patterns closely follow the migration of early agriculturalists from the Near East. (**B**) Higher concentration of PC2 loaded alleles are shown in purple and lower concentrations grade toward red. These patterns closely follow the migration of Finno-Ugric language speaking groups. Together these patterns correlate closely with known historical migrations and demonstrate how multi-locus genotypic data can be analyzed to understand recent migrations.

First, determine the geographic distribution of all tips. Then, following the rules of parsimony (*see* **Note 1**), trace the distribution as any other phenotypic character to ancestors. The geographic areas that are occupied by more basal branches are most likely ancestral ranges. Derived branches that occupy novel geographic areas are most likely to have resulted from migration (i.e., colonization or range expansion). Geographic areas inhabited by distantly related gene lineages are a signature of migration as well.

3.5.1. Maximum Parsimony

Maximum parsimony determines the most parsimonious relationships among an aligned set of DNA sequences. Parsimony is determined by the fewest number of base changes required along a reconstructed tree, which can result from either equally or unequally weighted character positions (i.e., base pair loci, usually considered as codon position in coding sequences). The most popular applications for reconstructing phylogenetic trees using parsimony are PAUP* *(32)* and PHYLIP *(33)*. Trees can be rooted using an outgroup, thus providing temporal polarity for related sequences. In the context of geographic data of haplotypes, this polarity can be used to determine the direction and timing of migration events (**Fig. 23.4A**).

3.5.2. Maximum Likelihood

These methods utilize a model of sequence evolution that is expected to best approximate the substitution pattern of a given set of aligned DNA haplotypes. An appropriate model of sequence evolution can be determined using either hierarchical likelihood ratio tests (hLRT) or the Akaike Information Criterion (AIC) in the programs ModelTest *(34)* and MrModelTest *(35)*. Using a given model, gene trees can be built using any number of applications, the most popular of which include PAUP*, PAML, and PHYML. As above, trees can be rooted using an outgroup, thus providing temporal polarity for related sequences.

3.5.3. Bayesian Phylogenetics

Bayesian inference can be used to assess the posterior probability of a given tree (or branch on a tree) under a particular model of sequence evolution (as in maximum likelihood) and with a given dataset. These methods typically use Metropolis-Coupled Markov Chain Monte Carlo analysis to reconstruct phylogenetic trees. The most popular implementation is MrBayes *(36)*.

3.6. Network Construction

Quite often, the relationships among genes within a species or population are poorly represented by bifurcating or multifurcating trees produced by phylogenetic analysis. Networks, which allow haplotypes to be connected simultaneously to multiple other haplotypes in ways that can also produce loops, can better

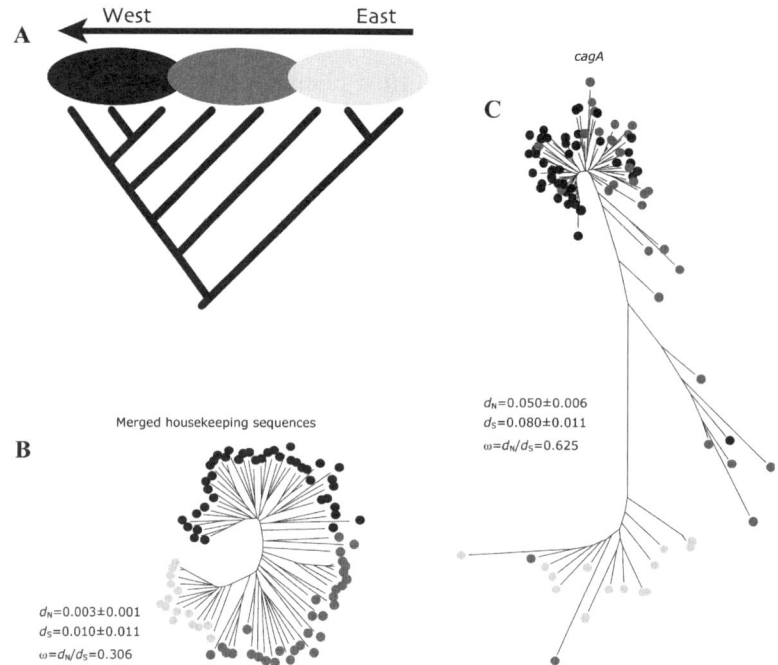

Fig. 23.4. The use of gene trees in inferring migrations. (**A**) In general, the polarity of ancestry for gene trees within species can inform the direction of movement. For instance, if most basal lineages are found in Eastern areas *(light gray)* and derived lineages are found in Western areas *(black)*, the inference of east-to-west migration of individuals carrying those genes, as in range expansion, is reasonable. More complex patterns often emerge, particularly when the genes/loci under investigation are inappropriate in either mutation rate or recombination pattern to resolve the migratory patterns of interest. (**B**) A tree inferred from concatenated housekeeping genes for 87 *H. pylori* isolates from Africa, East Asia and IndoEurope and (**C**) a tree for the same isolates derived from a single virulence gene, reproduced from *(30) (black: African strains; dark gray: Indo-European strains; light gray: East Asian strains)*. Notice that evidence for selection in (**C**) is much greater and the branch lengths are longer, together indicating that this gene has inappropriate evolutionary properties for inferring migrations. On the other hand, the housekeeping genes, which are evolving more slowly and under purifying selection, carry a strong signature of migration out of Africa.

represent the processes of reticulation (*see* **Note 1**), homoplasy, and recombination that occur within populations. As a result, reconstructing a loop in a network is a clue that some individuals possessing similar haplotypes may have interbred. When this pattern correlates with distantly distributed haplotypes, migration is a possible explanation (as is the independent evolution of identical haplotypes). Specifically, a loop in a network may indicate recombination, such as may be produced by individuals possessing distantly related haplotypes interbreeding following a migratory event. Furthermore, when multiple genes are used, conflict among haplotype networks is a common sign of interbreeding. If a population tree can be estimated from these multiple genes, these conflicts can pinpoint putative migration.

Many methods have been developed to reconstruct haplotype networks. Most of these methods are based on genetic distance between haplotypes and include the pyramids technique, split decomposition, median-joining, and statistical parsimony. Likelihood methods have also been developed to reconstruct haplotype networks.

3.6.1. Nested Clade Phylogeographical Analysis

The NCPA of Templeton *(37)* uses a haplotype network and a series of nested clades (each clade level another mutational step more inclusive) to estimate the geographic distribution of related haplotypes. It can indicate how haplotypes (and the individuals possessing them) came to be distributed in space, including through various migratory processes such as long-distance dispersal and contiguous range expansion. Although the beginning of the process employs statistical methods for determining the geographic distribution of haplotypes, the NCPA has been criticized for both the arbitrariness of the nesting procedure and the non-statistical nature of the inference procedure. There are four steps in this method:

1. Construction of a haplotype network using a statistical parsimony procedure (uses 95% confidence in a given parsimony-estimation for each branch, thus allowing for loops in the network) implemented in the program TCS v. 1.13 *(38)*.

2. The networked haplotypes are then nested into hierarchical clades following the nesting rules outlined in Templeton et al. *(39)* and extended in Templeton and Sing *(40)*. These rules group the most closely related haplotypes within clades and then group those clades into higher-level clades, at each step increasing the inclusion by one mutational difference.

3. The statistical test of geographic association among and between haplotypes and nested clades is then performed using GeoDis v. 2.0 *(41)*. This is done using two statistics that measure the distances within (D_c) and between (D_n) nested clades. Distances are calculated from the locality data of haplotypes, with these locations considered continuous variables in a geographic distance matrix. At this step, it is possible to assess the probability of geographic association or non-association with phylogenetic structure because the test uses randomized geographic distances to test the probability that the particular geographic distances sampled within a given clade (or haplotype) were sampled from a null distribution of geographic distances *(37, 42)*.

4. Finally, the updated inference key of Templeton *(43)* is used to infer particular historical processes for clades in which significant geographic association or dispersion was found.

3.7. Inference Structure

In inferring patterns of migration, it is critical to follow as logical and refutable a pattern of deduction/induction as possible. The complex nature of both ecological and evolutionary forces acting upon migrating organisms, in addition to the complex mechanisms of genetic evolution, make inferring patterns of migration a difficult process. In order to effectively understand past (and/or ongoing) migrations, special attention must be paid to the inference structure. Following the outline laid out in the introduction, this inference structure should include:

- Explicit description of hypotheses
- Accurate assessment of how the data can address aspects of the hypotheses
- A logical pattern of hypothesis support and rejection based on different data or different analyses that proceeds from less complex aspects of migration to more global patterns
- Identification of all ancillary information and indirect evidence for the hypotheses
- Appropriate description of how all ancillary information and indirect evidence relates to the hypotheses or to identified patterns of migration (i.e., degree of congruence)
- Identification of what data remains missing that could clearly support or reject the migration pattern inference

3.7.1. Direct Evidence

The most direct evidence available for observing migratory patterns are direct observations of the movements themselves. In some cases this is possible, although generally the patterns of interest are either too complex or have occurred in the past. Nonetheless, observations regarding vagility, diurnal or annual migration, and habitat use are extremely useful in clearly identifying historic migration patterns. We consider any data that is indivisibly tied to the organism's genealogical history to be direct evidence. Genetic data are the most commonly used, and useful type, because its geographic distribution is dependent explicitly, and predictably, on the distribution and genealogy of the organism.

3.7.2. Indirect Evidence

Indirect evidence can be an especially powerful tool when deciphering patterns of migration. Any evidence that is not intrinsic to an organism's genealogical (or kinship) history can be considered indirect. This includes, for humans, data regarding language distribution because the geographic patterns do not depend entirely upon genealogical relationships. Nevertheless, language has been one of the most powerful tools in both predicting and clarifying patterns of human migration *(45)*. It is also possible to deduce important aspects of the evolutionary history of organisms through the population parameters of symbionts, in particular parasitic and commensalist microbes that can reveal startling patterns because they have co-evolved with their host. Host microbes often evolve at much faster rates than their hosts, meaning that recent and short-lived migration events are preserved in their genealogy that may have been too fast or ephemeral for meaningful patterns of host genetic diversity to evolve *(30)*. The use of indirect evidence from symbionts has been mostly restricted to humans *(46, 47)*, although in principal any organism is amenable (*see* **Note 7**). For humans, useful parasites include the JC virus that infects renal tissue, human papillomavirus, the human T-cell lymphotropic virus, the hepatitis G virus, HIV, the eubacterium *Helicobacter pylori*, and the fungus *Coccidioides immitis* *(30)*. Indirect genetic evidence can be analyzed using the same tools as for direct evidence. However, in the inference process, the relationship between the indirect source and the organism(s) of interest must be considered and analyzed explicitly.

4. Notes

1. Definitions

    Vagility: The ability of an organism or a species to move about or disperse in a given environment. This can be measured by tagging or mark-recapture experiments.

    Homoplasy: Similarity in the characters of two taxa that are not the result of inheritance from a common ancestor. This contrasts with homology.

    Codominant: Equally expressed alleles of a gene (i.e., not dominant or recessive).

    Parsimonious character reconstruction: To trace a character with the fewest numbers of steps required to go from an ancestral state to a derived state.

    Reticulation: When independent branches of a phylogeny are intermingled, as in hybridization or species breakdown.

2. Mutation Rate: The rate at which genetic changes occur is critical to obtaining an appropriate marker for inferring migration patterns. For nucleotide substitution changes, this can be best estimated using maximum likelihood models (using r8s or PAUP*) or Bayesian inference (using MultiDivTime). For microsatellite changes, two models of evolution are assumed, the infinite alleles model and the stepwise mutation model, both of which can be incorporated into population genetic analyses.

3. Linkage Disequilibrium: Most statistical analyses of population genetic data assume that the loci being analyzed are unlinked, that is that they evolve as basically independent markers, and the results can be heavily biased if this assumption is violated. Linkage disequilibrium can be tested by evaluating the probability of genetic association between pairs of loci given the observed frequency and is implemented in most comprehensive population genetics software packages.

4. Selection: Selection can result in artificial groupings of populations with different histories and genealogies (**Fig. 23.4B**). Comparison of non-synonymous amino acid changes (d_N) and synonymous amino acid changes (d_S) in protein-coding genes is a common measure of selection pressure. The ratio of these two rates is measured by $\omega = d_N/d_S$ with $\omega = 1$, <1, and >1 indicating neutral evolution, purifying selection, and directional selection, respectively. Additionally, several evolutionary models exist that account for site-specific differences in adaptive selection at the protein level *(48)*, and are implemented in PAML. Classic estimates of selection on one or more loci of genotypic data such as Tajima's D and Fu's F are implemented in several population genetic software packages.

5. Recombination: The population recombination rate can be estimated by a composite likelihood method such as LD_{HAT} *(49)*. LD_{HAT} employs a parametric approach, based on the neutral coalescent, to estimate the scaled parameters $2N_e r$, where N_e is the effective population size and r is the rate at which recombination events separate adjacent nucleotides. An alternative solution is to use the homoplasy ratio *(50)*.

6. Multiple Gene Trees: A gene tree is not an organismal tree and therefore phylogenies should be based on multiple genes. By doing so, a "partition homogeneity test" *(51)* can be used to detect heterogeneity in phylogenetic signals among the individual genes. If the test reveals insignificant heterogeneity, the different genes can be concatenated for subsequent analyses. The use of multiple genes typically results in stronger statistical support and more accurate phylogenies. The Shimodaira-Hasegawa test *(52)*, Bayesian compatibility *(53)*, and quartet puzzling *(54)* can be applied in order to estimate the likelihood of different topologies.

7. Transmission Mode: Extensive knowledge of the transmission mode should be available before a microbe should be considered as a tracer of the history of its host. This knowledge can be acquired in many ways, including investigations of transmission from mother to child, parent to child, and within families over several generations *(30)*.

Acknowledgments

This work was supported by a Deutsche Forschungsgemeinschaft grant to Thierry Wirth (WI 2710/1) and the University of Konstanz.

References

1. Crandall, K. A., Templeton, A. R. (1993) Empirical tests of some predictions from coalescent theory with applications to intraspecific phylogeny reconstruction. *Genetics* 134, 959–969.
2. Nickerson, D. M., Brunell, A. (1998) Power analysis for detecting trends in the presence of concomitant variables. *Ecology* 79, 1442–1447.
3. Brumfield, R. T., Beerli, P., Nickerson, D. A., et al. (2003) The utility of single nucleotide polymorphisms in inferences of population history. *Trends Ecol Evol* 18, 249–256.
4. Ewing, B., Hillier, L., Wendl, M. C., et al. (1998) Base-calling of automated sequencer traces using phred. I. Accuracy assessment. *Genome Res* 8, 175–185.
5. Stephens, M., Smith, N., Donnelly, P. (2001) A new statistical method for haplotype reconstruction from population data. *Am J Hum Genet* 68, 978–989.
6. Nielsen, R., Signorovitch, J. (2003) Correcting for ascertainment biases when analyzing SNP data: applications to the estimation of linkage disequilibrium. *Theor Popul Biol* 63, 245–255.
7. Zhang, D.-X., Hewitt, G. M. (2003) Nuclear DNA analyses in genetic studies of populations: practice, problems and prospects. *Mol Ecol* 12, 563–584.
8. Bensch, S., Akesson, M. (2005) Ten years of AFLP in ecology and evolution: why so few animals? *Mol Ecol* 14, 2899–2914.
9. Wright, S. (1965) The interpretation of population structure by F-statistics with special regard to system of mating. *Evolution* 19, 395–420.
10. Balloux, F., Lugon-Moulin, N. (2002) The estimation of population differentiation with microsatellite markers. *Mol Ecol* 11, 155–165.
11. Beerli, P., Felsenstein, J. (2001) Maximum likelihood estimation of a migration matrix and effective population sizes in n subpopulations by using a coalescent approach. *Proc Natl Acad Sci U S A* 98, 4563–4568.
12. Nei, M. (1987) *Molecular Evolutionary Genetics*. Columbia University Press, New York.
13. Hauser, L., Adcock, G. J., Smith, P. J., et al. (2002) Loss of microsatellite diversity and low effective population size in an overexploited population of New Zealand snapper (*Pagrus auratus*). *Proc Natl Acad Sci U S A* 99, 11742–11747.
14. Goudet, J. (2002) *FSTAT: A Program to Estimate and Test Gene Diversities and Fixation Indices*. Institut d'Ecologie, Université de Lausanne, Switzerland.
15. Schneider, S., Roessli, D., Excoffier, L. (2000) *Arlequin. A Software for Population Genetics Data Analysis*. Genetics and Biometry Laboratory, University of Geneva.
16. Excoffier, L., Smouse, P. E., Quattro, J. M. (1992) Analysis of molecular variance inferred from metric distances among DNA haplotypes: application to human mitochondrial DNA restriction data. *Genetics* 131, 479–491.
17. Rogers, A.R., Harpending, H. (1992) Population growth makes waves in the distribution of pairwise genetic differences. *Mol Biol Evol* 9, 552–569.
18. Slatkin, M., Hudson, R. R. (1991) Pairwise comparisons of mitochondrial DNA sequences in stable and exponentially growing populations. *Genetics* 129, 555–562.
19. Nee, S., Holmes, E. C., Rambaut, A., et al. (1995) Inferring population history from molecular phylogenies. *Philos Trans Roy Soc B* 349, 25–31.
20. Pybus, O. G., Rambaut, A., Holmes, E. C., et al. (2002) New inferences from tree shape: numbers of missing taxa and population growth rates. *Syst Biol* 51, 881–888.
21. Pybus, O. G., Rambaut, A. (2002) GENIE: estimating demographic history from molecular phylogenies. *Bioinformatics* 18, 1404–1405.
22. Pybus, O. G., Rambaut, A., Harvey, P. H. (2000) An integrated framework for the inference of viral population history from reconstructed genealogies. *Genetics* 155, 1429–1437.
23. Strimmer, K., Pybus, O. G. (2001) Exploring the demographic history of DNA sequences using the generalized skyline plot. *Mol Biol Evol* 18, 2298–2305.
24. Drummond, A., Strimmer, K. (2001) PAL: an object-oriented programming library for molecular evolution and phylogenetics. *Bioinformatics* 17, 662–663.
25. Holsinger, K. E. (1999) Analysis of genetic diversity in geographically structured populations: a Bayesian perspective. *Hereditas* 130, 245–255.
26. Drummond, A. J., Nicholls, G. K., Rodrigo A. G., et al. (2002) Estimating mutation parameters, population history and genealogy simultaneously from temporally spaced sequence data. *Genetics* 161, 1307–1320.

27. Wilson, G. A., Rannala, B. (2003) Bayesian inference of recent migration rates using multilocus genotypes. *Genetics* 163, 1177–1191.
28. Reeve, J., Rannala, B. (2002) DMLE+: Bayesian linkage disequilibrium gene mapping. *Bioinformatics* 18, 894–895.
29. Pritchard, J. K., Stephens, M., Donnelly, P. (2000) Inference of population structure using multilocus genotype data. *Genetics* 155, 945–959.
30. Wirth, T., Meyer, A., Achtman, M. (2005) Deciphering host migrations and origins by means of their microbes. *Mol Ecol* 14, 3289–3306.
31. Swofford, D. L. (2003) PAUP*: Phylogenetic Analyses Using Parsimony and other methods, Version 4.0. Sinauer, Sunderland, MA.
32. Felsenstein, J. (2005) PHYLIP (Phylogeny Inference Package) version 3.6. Distributed by the author. Department of Genome Sciences, University of Washington, Seattle.
33. Posada, D., Crandall, K. A. (1998) Modeltest: testing the model of DNA substitution. *Bioinformatics* 14, 817–818.
34. Nylander, J. A. A. (2004) MrModeltest v2. Program distributed by the author. Evolutionary Biology Centre, Uppsala University.
35. Huelsenbeck, J. P., Ronquist, F. (2001) MRBAYES: Bayesian inference of phylogeny. *Bioinformatics* 17, 754–755.
36. Templeton, A. R. (1998) Nested clade analyses of phylogeographic data: testing hypotheses about gene flow and population history. *Mol Ecol* 7, 381–397.
37. Clement, M., Posada, D., Crandall, K. A. (2000) TCS: a computer program to estimate gene genealogies. *Mol Ecol* 9, 1657–1659.
38. Templeton, A. R., Boerwinkle, E., Sing, C. F. (1987) A cladistic analysis of phenotypic associations with haplotypes inferred from restriction endonuclease mapping. I. Basic theory and an analysis of Alcohol Dehydrogenase activity in *Drosophila*. *Genetics* 117, 343–351.
39. Templeton, A. R., Sing, C. F. (1993) A cladistic analysis of phenotypic associations with haplotypes inferred from restriction endonuclease mapping. IV. Nested analyses with cladogram uncertainty and recombination. *Genetics* 134, 659–669.
40. Posada, D., Crandall, K. A., Templeton, A. R. (2000) GeoDis: a program for the cladistic nested analysis of the geographical distribution of haplotypes. *Mol Ecol* 9, 487–488.
41. Knowles, L. L., Maddison, W. P. (2002) Statistical phylogeography. *Mol Ecol* 11, 2623–2635.
42. Templeton, A. R. (2004) Statistical phylogeography: methods of evaluating and minimizing inference errors. *Mol Ecol* 13, 789–809.
43. Cavalli-Sforza, L. L. (1998) The DNA revolution in population genetics. *Trends Genet* 14, 60–65.
44. Diamond, J., Bellwood, P. (2003) Farmers and their languages: the first expansion. *Science* 300, 597–603.
45. Wirth, T., Wang, X., Linz, B., et al. (2004) Distinguishing human ethnic groups by means of sequences from *Helicobacter pylori*: lessons from the Ladakh. *Proc Natl Acad Sci U S A* 101, 4746–4751.
46. Falush, D., Wirth, T., Linz, B., et al. (2003) Traces of human migration in Helicobacter pylori. *Science* 299, 1582–1585.
47. Yang, Z., Nielsen, R. (2002) Codon-substitution models for detecting molecular adaptation at individual sites along specific lineages. *Mol Biol Evol* 19, 908–917.
48. McVean, G., Awadalla, P., Fearnhead, P. (2002) A coalescent-based method for detecting and estimating recombination from gene sequences. *Genetics* 160, 1231–1241.
49. Maynard Smith, J. (1999) The detection and measurement of recombination from sequence data. *Genetics* 153, 1021–1027.
50. Farris, J. S., Kallersjo, M., Kluge, A. G., et al (1995) Testing significance of incongruence. *Cladistics* 10, 315–319.
51. Shimodaira, H., Hasegawa, M. (1999) Multiple comparisons of log-likelihoods with applications to phylogenetic inference. *Mol Biol Evol* 16, 1114–1116.
52. Buckley, T. R. (2002) Model misspecification and probabilistic tests of topology: evidence from empirical data sets. *Syst Biol* 51, 509–523.
53. Strimmer, K., von Haeseler, A. (1997) Likelihood-mapping: A simple method to visualize phylogenetic content of a sequence alignment. *Proc Natl Acad Sci U S A* 94, 6815–6819.

Chapter 24

Fixed-Parameter Algorithms in Phylogenetics

Jens Gramm, Arfst Nickelsen, and Till Tantau

Abstract

This chapter surveys the use of fixed-parameter algorithms in phylogenetics. A central computational problem in this field is the construction of a likely phylogeny (genealogical tree) for a set of species based on observed differences in the phenotype, differences in the genotype, or given partial phylogenies. Ideally, one would like to construct so-called perfect phylogenies, which arise from an elementary evolutionary model, but in practice one must often be content with phylogenies whose "distance from perfection" is as small as possible. The computation of phylogenies also has applications in seemingly unrelated areas such as genomic sequencing and finding and understanding genes. The numerous computational problems arising in phylogenetics are often NP-complete, but for many natural parametrizations they can be solved using fixed-parameter algorithms.

Key words: Phylogenetics, perfect phylogeny, fixed-parameter algorithms, fixed-parameter tractable.

1. Introduction

1.1. Phylogenetics

In phylogenetics, one studies how different species are evolutionarily related. Instead of species, the basic building blocks of biodiversity, one can also more generally consider *taxa*, which are arbitrary groupings of organisms and sometimes even single organisms. The basic paradigm is that species spawn new species, for example when part of a species' population adapts to a changing environment. Over time the set of extant species changes as new species emerge and other species become extinct. The ancestral relationship between the species can be depicted by arranging them in a tree, called a *phylogenetic tree* or *phylogeny*, in which the leaves are labeled with extant species and bifurcations correspond to events such as adaptations that lead to new species. Interior nodes are labeled with ancestral species or not at all when the ancestral species are unknown or not of interest.

Building phylogenies is not an easy task. The problems start with determining the set of taxa since it is not always clear where one should draw the line between, say, different species. However, suppose a set of taxa has been agreed on and the task is to arrange them in a phylogeny. Then one only has data from extant taxa, but not ancestral taxa. In rare, fortunate cases one might have access to fossils, but normally the evolution path taken is unknown.

Even if all taxa in an anticipated phylogeny are known, how they should be arranged often remains subject to debate. One way to solve this problem is to infer the phylogeny by looking at different *characters* of the taxa such as, say, the form of the skeleton. Taxa for which the form of the skeleton is similar should be in the same subtree of the phylogeny. The joint information from many characters often leaves us with a single phylogeny or at least few possible ones. In biology, principal sources of characters are the phenotype of a taxon, which is roughly "the way the organisms of the taxon look," but also genomic information such as which genes are present in the organisms of a taxon.

The construction and study of phylogenetic trees has many applications. First of all, a phylogeny allows a glimpse at how evolution works and can help in classifying organisms. Second, one can compare multiple phylogenies built for the same set of species based on, say, different parts of the genome. A third, rather intriguing application of phylogenies, is their use as measures in other applications. One such application is the *haplotype phase determination problem*, presented in detail in **Section 7.2**, in which the tricky part is not so much *finding* the biologically most likely solution, but *measuring* how likely is that solution. An elegant way of doing this is to declare as "good solutions" those that can be arranged in a phylogeny.

1.2. Computational Problems in Phylogenetics

The most fundamental problems in phylogenetics are not computational in nature. Deciding what exactly counts as a taxon, choosing a set of characters, or deciding which states a taxon is in are not computational problems, but require human experience and judgment. However, once these decisions have been made, many computational problems arise that cannot be solved "by hand" because large amounts of input data need to be processed, as is the case in phylogeny-based haplotyping, for instance. The following provides an overview of the computational problems addressed in this survey; good starting points for further reading on computational issues in phylogenetics are (1–4).

A fundamental problem in phylogenetics is the construction of a phylogeny for a set of taxa (detailed mathematical definitions

Table 24.1
Character-state matrix that admits a perfect phylogeny

Species	Hyaline margin	Marginal carina	Premarginal carina
Chelopistes guttatus	0	0	0
Osculotes macropoda	0	1	0
Oxylipeurus dentatus	1	1	1
Upupicola upupae	0	0	2
Perineus nigrolimbatus	2	0	2

It is a submatrix of a much larger character-state matrix, compiled by Smith (5), that contains entries for 56 lice species and 138 characters. The numbers in the matrix encode the different states of the characters. For example, for the character *marginal carina* the 0 entries mean that the adult marginal carina "forms a complete thickened band running anteriorly around the preantennal region of the head" and the 1 entries mean that it "forms a band which is interrupted laterally (partially or completely), medially (dorsally and/or ventrally) or both" (5).

are given later). One is given a set of taxa, a set of characters, and for each character and taxon the *state* of the taxon with respect to the character. State information for taxa and characters is typically arranged in matrices such as the one shown in **Table 24.1**.

The model of evolution that one chooses determines which phylogenies are considered good explanations of the observed character-state matrix. A basic model is the following: All taxa sharing a state for some character are descendants of the same taxon and mutations of a character to some state happen only once. One possible way of checking whether this condition is true for a phylogeny is to check whether for each character and each pair of states the path between any two taxa in the first state and the path between any two taxa in the second state do not intersect. Such a phylogeny is called *perfect*.

A second set of computational problems arises when it is not possible to arrange taxa in a perfect phylogeny. One then has several options: First, one can lower one's standards of what counts as a good phylogeny by allowing a small number of "backward mutations" in the tree. Second, one can still try to find a perfect phylogeny, but only for a subset of the taxa or for a subset of the characters. Third, one can claim that the data must be in error and try to find a way—as little disruptive as possible—to modify the data such that a perfect phylogeny can be obtained. Although this is not advisable in general (we cannot simply claim that elephants can fly, just to fit them into a perfect phylogeny), genomic data are often obtained through laboratory processes in which one cannot avoid a percentage of wrong entries.

Phylogenies need not always be constructed "from scratch" based on character state data. Rather, one often has access to *partial* phylogenies that are subtrees of the sought-for phylogeny. An example is the tree of life: One cannot construct the phylogenetic tree of all taxa on this planet based on an enormous character database for millions of taxa. Rather, the objective is to merge many different small phylogenies from the literature into one big phylogeny, appropriately called a *supertree*.

For a related problem, one also does not need to construct phylogenies from scratch, but several complete candidate phylogenies obtained through external means are given and the job is to compute biologically meaningful *distances* between them—a difficult problem all by itself.

1.2. Parametrization and Phylogenetics

Most computational problems in phylogenetics turn out to be NP-complete, forcing us to look for heuristics, approximation algorithms, or fixed-parameter algorithms. The fixed-parameter approach turns out to be especially successful. (Readers unfamiliar with NP-completeness or fixed-parameter algorithms are invited to have a look at **Sections 1.1** and **1.2** in **Chapter 21** of **Volume 2** of this book.)

The reason for the success of the fixed-parameter approach is that a number of problem parameters are small in realistic instances for phylogenetic problems, such as the number of states per character, the number of characters, or our tolerance for errors. For instance, the number of states per character is at most four (and in many cases even two) whenever single nucleotide polymorphisms in genomic data are involved and the running time of many algorithms is exponential in the number of states per character, but polynomial otherwise. The present chapter presents further examples of parameters that are small in practice, allowing us to construct efficient, exact algorithms for many computational problems arising in phylogenetics.

1.3. Goals and Overview

The central goal of this survey is to highlight selected fixed-parameter algorithms from the area of phylogenetics. The chosen examples are intended to illustrate the diversity of computational problems for which fixed-parameter algorithms have been developed within the area of phylogenetics. No detailed proofs are included in this survey, except for three short proofs of new easy results that are included for completeness.

Section 2 introduces one possible version of the formal problem of constructing a perfect phylogeny and studies how the parameters *number of taxa, number of characters*, and *number of states per character* influence the tractability of the problem. **Section 3** studies ways of measuring the deviation of a given phylogeny from "perfection." **Section 4** treats problems in which the task is to find a phylogeny that is near to perfection with respect to the introduced measures. **Section 5** looks at problems in which the task is to compute distances between phylogenies. **Section 6** treats the problem of merging several partial phylogenies. **Section 7** considers applications in which construction of a phylogeny is just a means to find regulatory genomic elements or determine haplotype phase. The Conclusion provides an outlook.

2. Construction of Perfect Phylogenies

This section studies how computationally difficult it is to construct a perfect phylogeny. First, it defines the problem PP (perfect phylogeny) formally, and discusses possible variations. Then it considers the complexity when one of the three central parameters: *number of taxa*, *number of characters*, or *number of states per character*, is fixed.

2.1. Formalization of the Perfect Phylogeny Problem

Fix a set C of *characters*, such as size or color, and for each character $c \in C$ fix a set Σ_c of states for this character, such as $\Sigma_{size} = \{small, medium, big\}$. Then the input for the perfect phylogeny problem is a set S of taxa together with one mapping for each taxon $s \in S$, each of which assigns an element of Σ_c to each character $c \in C$.

There are three natural parameters in such inputs:

- The number n of taxa
- The number m of characters
- The maximum number r of states a character can have

For computational issues, the names of the characters, states for each character, and taxa are not really important. Consequently, the notation can be made simpler by assuming that the set S of taxa is $\{1,\ldots,n\}$, the character set C is $\{1,\ldots,m\}$, and each state set is $\Sigma_i = \{0,\ldots,r-1\}$. It is customary to start the states with 0 so that if there are just two states, then they are 0 and 1. The states of a taxon can now be described by a vector from the set $\Sigma_1 \times \ldots \times \Sigma_m = \{0,\ldots,r-1\}^m$. Thus, the n input taxa are described by vectors of length m with entries from $\{0,\ldots,r-1\}$. Another way to think about the input is in terms of an $(n \times m)$-matrix with entries from $\{0,\ldots,r-1\}$. Be cautioned that in the biological literature these matrices are sometimes presented in transposed form.

Before defining *perfect* phylogenies, first define *phylogenies*. Readers familiar with phylogenetics may be surprised that internal nodes are allowed to be labeled. The reasons for this are explained in the following.

Definition 1 (phylogeny): Let A be a matrix describing n taxa. A *phylogeny for the matrix A* is a tree T whose node set V is labeled using a labeling function $l: V \to \{0,\ldots,r-1\}^m$ such that:

1. Every row of A (i.e., each taxon's state vector) is a label of some node in T.

2. The labels of the leaves of T are rows of A. (The labels of inner nodes correspond to ancestral taxa, which need not, but may, be part of the input.)

Be cautioned that phylogenetic *trees* may not always be the best way of describing evolutionary relationships, because they do not account for horizontal gene transfers in which a gene is transferred between unrelated taxa by a "mixing" of the genetic material. The present survey restricts attention to phylogenetic *trees*, nevertheless, since one should try to understand these before tackling the more difficult phylogenetic *networks*.

Recall that in the evolutionary model assumed in perfect phylogeny all taxa sharing a state for some character have the same ancestor and a mutation of a character to a state occurs only once. This can be formalized as follows:

Definition 2 (perfect phylogeny): A phylogeny for a matrix A is *perfect* if for every character $c \in C$ and every state $j \in \Sigma_c = \{0,\ldots,r-1\}$, the graph induced by the set of nodes labeled by a state vector $(S_1,\ldots,S_m)$ with $S_c = j$ is connected.

Definition 3 (perfect phylogeny problem): The input for the *perfect phylogeny problem* (abbreviated PP) is a character-state A for n taxa. The task is to decide whether there exists a perfect phylogeny for A.

Some remarks on the definition are in order both with respect to its biological relevance and to the chosen mathematical formalization.

Concerning biological relevance, one can object that real biological character-state matrices rarely admit a perfect phylogeny. For example, **Table 24.1** displays a real-life instance of PP, and a perfect phylogeny for this matrix is shown in **Fig. 24.1**. However, this table is just a small part of a much larger matrix compiled by Smith *(5)*, and the whole matrix does not admit a perfect phylogeny. Nevertheless, there are several reasons to study perfect phylogenies:

- PP is a basic computational problem in phylogeny construction and we would like to understand this problem well before we attack more complicated settings.

- Even if data cannot be arranged in a perfect phylogeny, we may still try to find a phylogeny that is "as perfect as possible," *see* **Section 4**.

- There are biological settings where the perfect phylogeny model works quite well.

We have chosen, for this survey, a rather broad formalization of perfect phylogenies. We allow an arbitrary tree topology, the input taxa can be found both at the leaves and at inner nodes,

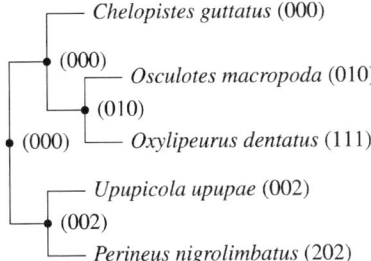

Fig. 24.1. One possible perfect phylogeny for the character-state matrix from Table 24.1. The labels assigned to the vertices of the phylogeny are shown in parentheses.

and the same label may be found at different nodes. We only insist that there are no superfluous leaves, although this condition is not necessary either. Other definitions in the literature impose more structure on perfect phylogenies:

- It is often required that the set of leaf labels equals the set of rows of the input matrix; that is, it is not allowed to place a taxon only at inner nodes. A perfect phylogeny in the sense of the second definition can be turned into a phylogeny with the input taxa at the leaves by adding a pending leaf to all inner nodes that harbor an input taxon.

- It is often convenient to have more control over the tree topology. It is particularly convenient to consider binary trees, which are trees in which every node either has degree one (leaves) or degree three (inner vertices). This can also be achieved easily: Replace all nodes of too high degree by small binary trees with all nodes labeled by the original node's label; remove all nodes of degree two and join their adjacent nodes.

- It is sometimes required that all labels are distinct. This can be accomplished by contracting subtrees whose nodes all have the same label (and, indeed, sets of nodes that are labeled identically must form a connected subtree in a perfect phylogeny). However, this contraction process may destroy the binary tree property and also the property that input taxa must label leaves.

In a perfect phylogeny there is no designated root node and, in general, it may be debatable which taxon should be considered the "root." If, for whatever reason, a root node has been chosen, the phylogeny is called *directed*.

Having defined (the decision version of) the perfect phylogeny problem, the natural question is, how difficult is this problem? Unfortunately, it is NP-complete.

Theorem 4 (6, 7): PP *is* NP-*complete*.

This result suggests that in order to tackle the problem we must look at restricted versions. We do so by fixing the various central parameters: number n of taxa, number m of characters, and the maximum number r of states per character.

2.2. Number of Taxa as the Parameter

The first restriction is to limit the number n of taxa in the input. Intuitively, if there are just, say, four taxa, it should not be particularly hard to find out whether we can arrange them in a perfect phylogeny; after all, there are only a fixed number of tree topologies for them.

Theorem 5: PP *can be solved in time* $O(2^n \, n! \cdot m)$.

Proof: For a fixed binary tree topology T and a one-to-one assignment of elements in S and leaves of T, it can be tested in time $O(nm)$, whether the inner nodes of T can be labeled in a way such that T is a perfect phylogeny for S. The number of possible binary trees with n distinctly labeled leaves is known to be $1 \cdot 3 \cdot 5 \cdot \ldots \cdot (2n - 5) \leq 2^n (n - 2)!$. Therefore, enumerating all binary trees for S and testing each for being a perfect phylogeny for S yields the stated running time.

Theorem 5 shows that PP is (more or less trivially) fixed-parameter tractable with respect to n. The algorithm simply enumerates all possible trees and tests whether they are a phylogeny for the data. It cannot handle a number n of taxa greater than perhaps 10, although in practical situations one typically has over 100 taxa. More clever exhaustive search algorithms in phylogenetics push the maximum number of taxa that can be handled to between 12 on desktop machines and about 15 on workstations; but what we really would like to find is a fixed-parameter algorithm for the parameter n based, ideally, on a kernelization algorithm followed by a search tree algorithm, yielding a running time such as the one stated in the open problem that follows. (For an introduction to kernelization and search tree algorithms *see* **Sections 2** and **3** in **Chapter 21** of **Volume 2**).

Open Problem 6: Is there a fixed-parameter algorithm for PP with respect to the parameter n with a running time in $O(c^n + (mr)^{O(1)})$ for some c close to 1?

2.3. Number of Characters as the Parameter

Returning to an arbitrary number of taxa, look at what happens when the number m of characters is fixed. This is justified in an important practical application. As argued by Gusfield in *(8)*, the perfect phylogeny model explains genomic variations well when crossing over effects are not present. This implies that for *short* genomic sequences, the perfect phylogeny model applies, and for longer sequences, we can try to partition the sequence into short intervals and derive perfect phylogenies for these small sets of characters.

Once more, the intuition is that it should be easy to find a perfect phylogeny if there are only, say, three characters and, indeed, Morris, Warnow, and Wimer present an algorithm with the following running time:

Theorem 7 (9): PP *can be solved in time* $O(r^{m+1} m^{m+1} + nm^2)$.

Using a different approach, Agarwala and Fernández-Baca arrive at the following running time:

Theorem 8 (10): PP *can be solved in time* $O((r - n/m)^m \cdot rnm)$.

For fixed m, both of the preceding time bounds are polynomial in n and r. However, neither algorithm shows that the problem is fixed-parameter tractable, as there are still m in the exponent and another input parameter in the base. Unfortunately, the work of Bodlaender et al. *(11)* shows that it is unlikely that this can be remedied, since it would have consequences (namely, certain complexity class collapses) that many theoreticians consider unlikely.

2.4. Number of States per Character as the Parameter

The third natural parameter for the perfect phylogeny problem is the number of states per character. Fixed-parameter results for this number are especially important since it is, indeed, small in many applications. For instance, there are only four genomic characters or, if one also takes alignment-induced gaps into account by adding a "no-data" or "gap" state, five. Even better, in applications such as the phylogeny-based haplotyping presented in **Section 7.2**, there are only two different states for each character.

The first fixed-parameter algorithm for the parameter r was proposed by Agarwala and Fernández-Baca. It has the following running time:

Theorem 9 (12): PP can be solved in time $O(2^{3r} \cdot (m^3 n + m^4))$.

This result was later improved by Kannan and Warnow.

Theorem 10 (13): PP can be solved in time $O(2^{2r} \cdot m^2 n)$.

An $O(m^2 n)$ algorithm for the special case $r=3$ had already been achieved by Dress and Steel *(14)*. Kannan and Warnow *(15)* give an $O(mn^2)$ algorithm for $r=4$.

For the special case $r=2$ one can make use of a simple but powerful characterization of matrices that admit a perfect phylogeny. The characterization is in terms of a forbidden induced submatrix and has been rediscovered independently by several authors *(16, 17)*.

Theorem 11: For $r=2$, a matrix A of taxa has a perfect phylogeny if and only if it does not contain the following induced submatrix:

$$\begin{pmatrix} 0 & 0 \\ 0 & 1 \\ 1 & 0 \\ 1 & 1 \end{pmatrix}.$$

Employing this characterization, which also plays a role in **Section 4.2**, Gusfield devised an algorithm running in linear time.

Theorem 12 (18): For $r=2$, a PP can be solved in time $O(mn)$.

The results of this section can be summed up as follows: PP with respect to either of the parameters n and r (number of taxa and number of states per character) is in FPT, but with respect to the parameter m (number of characters) it is unlikely to be fixed-parameter tractable.

3. Measures of Deviation from Perfection of Phylogenies

The previous section studied *perfect* phylogenies, but in practice one often has to deal with imperfect phylogenies. In this case one may look for a phylogeny that is at least "near" to being perfect.

For this, one needs to measure how strongly a phylogeny deviates from being perfect.

3.1. Measures Based on Relaxed Evolutionary Models

The basic assumption underlying perfect phylogenies is that mutations of a character to some state happen only once. We start with two measures that count, in different ways, how often this assumption is violated. Note that the input for these problems is a *phylogeny*, not a matrix. The closely related **Section 4.1** treats, for each measure, the question of finding *some* phylogeny for an input matrix that minimizes the distance to perfection.

The first measure is the *penalty* of a phylogeny, due to Fernández-Baca and Lagergren *(19)*.

Definition 13 (length and penalty of a phylogeny): For an edge of a phylogeny connecting nodes u and v, we define the *length* of the edge as the Hamming distance of u and v (the number of characters where the states differ). The *length of a phylogenetic tree T* is the sum of lengths taken over all edges of the tree. The *penalty of a phylogenetic tree T* is defined as:

$$penalty(T) = length(T) - \sum_{c \in C}(r_c - 1),$$

where r_c is the number of states of character c that are present in the phylogeny.

The idea behind this measure is the following: The length of an edge e connecting taxa u and v is the number of mutations that occurred between u and v. For a perfect phylogeny, a new state is introduced by a mutation only once; therefore, every character c contributes exactly $r_c - 1$ to the length of the tree. Hence, the penalty of a tree counts how often the assumption "each new state is introduced only once by a mutation" is violated. Perfect phylogenies have penalty 0.

The second measure is the *phylogenetic number*, due to Goldberg et al. *(20)*. For a state j and a character c let $T_{c,j}$ denote the subgraph of the phylogenetic tree T induced by the set of nodes whose labels are in state j for the character c. Then the phylogenetic number is defined as follows:

Definition 14 (phylogenetic number): The *phylogenetic number* of a phylogeny T is the maximum number of times that any given state arises in T, that is, the maximum number, taken over all characters c and all states j, of connected components in $T_{c,j}$. Phylogenies with phylogenetic number l are called l-phylogenies.

Perfect phylogenies are 1-phylogenies. Unlike penalty, which bounds the *total* number of violations of the basic evolutionary model, the parameter l does not restrict the total number of violations, but violations may not "concentrate" at a single state.

A third measure with a similar flavor is the *number of bad states*. It is due to Moran and Snir *(21)*, who study how to get rid of bad states by a minimal number of recolorings. (Compare Definition 21 in the next section.)

Definition 15 (number of bad states): Given a phylogeny T and a character c, the character's *number of bad states* is the number of states j for which $T_{c,j}$ is not connected. The number of bad states of a phylogeny T is the maximum of the numbers of bad states at any character.

Clearly, for a given phylogeny all of the preceding measures can be computed in polynomial time.

3.2. Measures Based on the Modification of Input Phylogenies

Measures are now introduced that are based on the idea that if the input data does not admit a perfect phylogeny, the data must be flawed. One then tries to modify or even remove the taxa of a given phylogeny until a perfect phylogeny is reached. Note, again, that the input is a *phylogeny*, not a matrix. The taxa are already arranged in a tree and we only wish to know how the particular input phylogeny needs to be modified to arrive at a perfect phylogeny. **Section 4.2** studies the related, but different, problem of modifying a character-state input matrix so that the resulting matrix admits a perfect phylogeny.

For the first measure of this type one tries to prune a phylogeny until it becomes perfect.

Definition 16 (tree perfection by taxa removal problem): The input for the *tree perfection by taxa removal problem* is a phylogeny T and a number k. The task is to decide whether one can turn the phylogeny T into a perfect phylogeny by repeatedly cutting away leaves such that at most k of the original leaves are removed.

It is not straightforward to minimize the number of taxa removals since there are many ways to prune a phylogeny; indeed, this problem is NP-complete already for $r = 2$:

Theorem 17: For every $r \geq 2$, the *tree perfection by taxa removal problem is* NP-*complete*.

Proof. The problem clearly is in NP. Hardness is shown by reducing the vertex cover problem to it. For a graph $G = (V, E)$ with $|V| = n$ and $|E| = m$ construct a star-shaped phylogeny T with one center node and n leaves, one for each vertex $v \in V$. The taxa have m characters c_e, one for each edge $e \in E$. Each character has two states 0 and 1. The center node is labeled 0^m. The leaf in T corresponding to vertex v in G is labeled with the character-state vector that has state 1 for character c_e if and only if v is an endpoint of e. Now, for each edge there are two taxa (leaves) in the phylogeny for which the state of c_e is 1. At least one of these taxa has to be removed to make the phylogeny perfect, because of the 0^m vector in the "center." Therefore, the vertex covers of G correspond exactly to sets of leaves whose removal lets T become perfect.

Open Problem 18: Is the tree perfection by taxa removal problem fixed-parameter tractable with respect to the parameter k (number of removed taxa)?

A second measure counts how many characters must be removed (disregarded) so that the phylogeny becomes perfect. This number is much easier to compute.

Definition 19 (tree perfection by character removal problem): The input for the *tree perfection by character removal problem* are a phylogeny T and a number k. The task is to decide whether the phylogeny T can be turned into a perfect phylogeny by disregarding at most k characters.

Theorem 20: The tree perfection by character removal problem can be solved in polynomial time.

Proof: A character is either "in error" (because there is a state such that the set of all taxa of this state for the character is not connected, which can be checked in polynomial time) or the character is "clean." Disregard all erroneous characters and this suffices.

A third measure, implicitly introduced by Moran and Snir (21), is based on a more fine-grained analysis of the erroneous characters. Instead of just disregarding those characters that violate the connectedness condition, we try to "fix them" by changing the states at a minimal number of places. Such a change of state may also be regarded as a *recoloring* since states correspond to colors in equivalent formulations of the perfect phylogeny problem.

Definition 21 (recoloring number): Given a phylogeny T, the *recoloring number* is the minimal number of state changes (the number of times one needs to change a state in some node label) needed to arrive at a perfect phylogeny.

Definition 22 (tree perfection by recoloring problem): The input for the *tree perfection by recoloring problem* consists of a phylogeny T and a number k. The task is to decide whether the recoloring number of T is at most k.

Finding an optimal recoloring for one character is not influenced by recolorings necessary for another character, so one can compute the recoloring number for each character separately. Hence, the problem reduces to the problem for a single character (called *convex recoloring of trees problem* by Moran and Snir), which Moran and Snir show to be NP-complete. Indeed, Moran and Snir show something even stronger.

Theorem 23 (21): The tree perfection by recoloring problem is NP-complete, even if one allows only instances in which the phylogeny forms a path and where there is only a single character.

On the other hand, Moran and Snir present an algorithm for computing the recoloring number. Recall that b is the number of bad states, see Definition 15, which are the states (or colors) for which some action needs to be taken.

Theorem 24 (21) The tree perfection by recoloring problem can be solved in time $O((b/\log b)^b \cdot bmn^4)$.

The preceding theorem shows that computing the recoloring number is fixed-parameter tractable with respect to the number of bad states.

Open Problem 25: With respect to which other parameters is the tree perfection by recoloring problem fixed-parameter tractable?

4. Construction of Good Phylogenies

This section studies algorithms that construct phylogenies that are "almost" or "nearly" perfect. To define what counts as a good phylogeny, we use the measures introduced in the previous section. Having fixed a measure, our objective is now to find a phylogeny of minimal measure for a given input matrix. Intuitively, this is a much more difficult problem than the ones studied in the previous section, which just computed the measure of a single phylogeny; and often already this seems difficult.

4.1. Minimizing Penalty, Phylogenetic Number, and Number of Bad States

Start with algorithms that find phylogenies, minimizing the penalty (number of excess mutations) from Definition 13, the phylogenetic number (one plus the maximum of the number of excess mutations per state) from Definition 14, or the number of bad states (number of states for which an excess mutation has occurred) from Definition 15.

Definition 26 (measure minimization problems): The input for the problems *phylogenetic penalty minimization*, *phylogenetic number minimization*, and *phylogenetic bad states minimization* is a matrix A of taxa and a number p. The task is decide whether there exists a phylogeny for A of penalty at most p, with a phylogenetic number of at most p, or with at most p bad states.

Fernández-Baca and Lagergren *(19)* call phylogenies that minimize the penalty "near-perfect," but this chapter uses *phylogenetic penalty minimization* for consistency.

All problems are generalizations of PP since when the penalty is 0 (or 1, for the phylogenetic number), the task is simply to decide whether a perfect phylogeny exists. This shows that one cannot hope for a fixed-parameter algorithm for any of these problems with respect to the parameter p alone. If we take the parameter r also into account, two theorems are known about minimizing the penalty.

Theorem 27 (19): The phylogenetic penalty minimization problem can be solved in time $O(m^{O(p)} 2^{O(p^2 r^2)} \cdot n)$.

Theorem 28 (22, 23): For $r = 2$, the *phylogenetic penalty minimization* problem can be solved in time $O(2^{O(p^2)} \cdot nm^2)$.

Theorem 28 tells us that for the particularly interesting case of only two states per character there is a fixed-parameter algorithm for finding a good phylogeny with respect to the parameter penalty.

Much less is known about minimizing the phylogenetic number or the number of bad states.

Open Problem 29: For which, if any, parameters or parameter pairs are the phylogenetic number minimization or phylogenetic bad states minimization problems fixed-parameter tractable?

4.2. Minimizing the Modification of Input Data

The next measures that we considered were based on the idea that one may modify the input to arrive at a perfect phylogeny. Trying to minimize these measures leads to the following problems:

Definition 30 (PP by taxa removal problem): The input for the *PP by taxa removal problem* is a matrix A of taxa and a number k. The task is to remove at most k taxa (rows) from A such that the resulting matrix admits a perfect phylogeny.

Definition 31 (PP by character removal problem): The input for the *PP by character removal problem* is a matrix A of taxa and a number k. The task is to remove at most k characters (columns) from A such that the resulting matrix admits a perfect phylogeny.

For the case $r=2$ there is a characterization of matrices that admit a perfect phylogeny by a forbidden submatrix (*see* Theorem 11). Combining the results proved in *(24)* on forbidden submatrices and results on the fixed-parameter tractability of the hitting set problem, one gets the following results.

Theorem 32: For every $r \geq 2$, both the *PP by taxa removal* and *PP by character removal* problems are NP-complete.

Theorem 33: For $r=2$, the *PP by taxa removal problem* can be solved in time $O(3.30^k + n^4)$ and also in time $O(2.18^k n + n^4)$.

Theorem 34: For $r=2$, the *PP by character removal problem* can be solved in time $O(1.29^k + m^2)$.

For larger r, where no characterization in terms of forbidden submatrices is known, the complexity of the removal problems is open.

Open Problem 35: Are the PP by taxa removal and PP by character removal problems with parameter k fixed-parameter tractable for all r?

Definition 36 (PP by recoloring problem): The input for the *PP by recoloring problem* is a matrix A of n taxa and a number k. The task is to decide whether A has a phylogeny with recoloring number of at most k.

Recall that computing the recoloring number of a *given* phylogeny is fixed-parameter tractable with respect to the parameter b (number of bad states), but nothing is known for the PP by recoloring problem.

Open Problem 37: How difficult is the PP by recoloring problem?

5. Computing Distances Between Phylogenies

This section studies how difficult it is to compute the distance between phylogenies, which need not be perfect. Computing such distances is important when several candidate phylogenies are given, obtained either computationally by different methods or compiled from different literature sources. We discuss three classical, well-known editing distance measures as well as a recently introduced distance measure based on planar embeddings of the involved trees.

One way to define a distance between phylogenies is to count the number of modifications necessary to transform one phylogeny into another. Possible modifications include deletion and insertion of taxa or the movement of a subtree of the phylogeny to another place. For different sets of allowed modifications, one gets different notions of distance. Three increasingly general modifications have been studied extensively in the literature (see the book by DasGupta et al. for an entry point) *(25)*. Usually these operations are considered only on *undirected, binary* phylogenies in which *taxa label only leaves*.

The first operation is the *nearest neighbor interchange*. In a binary phylogeny, every internal edge has four subtrees attached to it (two at the one end of the edge and two at the other end) and the nearest neighbor interchange exchanges two such subtrees. This means that the tree: $\begin{smallmatrix}A\\B\end{smallmatrix}\!\!\succ\!\!\bullet\!\!-\!\!\bullet\!\!\prec\!\!\begin{smallmatrix}C\\D\end{smallmatrix}$

can be changed into: $\begin{smallmatrix}A\\C\end{smallmatrix}\!\!\succ\!\!\bullet\!\!-\!\!\bullet\!\!\prec\!\!\begin{smallmatrix}B\\D\end{smallmatrix}$

or into: $\begin{smallmatrix}A\\D\end{smallmatrix}\!\!\succ\!\!\bullet\!\!-\!\!\bullet\!\!\prec\!\!\begin{smallmatrix}C\\B\end{smallmatrix}$

The second operation is the *subtree prune and regraft* operation. Here one is allowed to cut an edge anywhere in the phylogeny and to reattach (regraft) the root of the subtree that has been cut at some other place (**Fig. 24.2**). It is not too hard to see that nearest neighbor interchange is a special case of subtree prune and regraft. The subtree prune and regraft operation models a

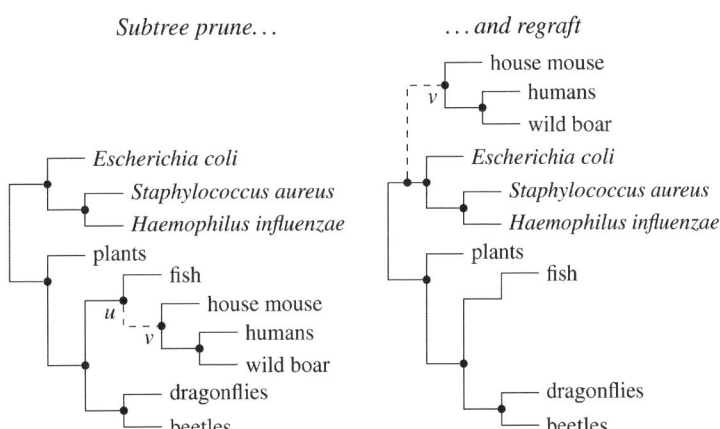

Fig. 24.2. Example of how the subtree prune and regraft operation works. In the left phylogeny the edge between u and v is cut and then the tree rooted at v is regrafted at a new position in the right phylogeny. The right phylogeny takes the presence or absence of the gene encoding *N*-acetylneuraminate lyase into account. This gene is present in vertebrates and bacteria but not in the other taxa, suggesting that a horizontal gene transfer took place. The announcement *(55)* in *Nature* that humans may have acquired over a hundred genes directly from bacteria made newspaper headlines, but the two *Science* articles *Microbial Genes in the Human Genome: Lateral Transfer or Gene Loss? (56)* and *Are There Bugs in Our Genome? (57)* quickly challenged the findings and suggest other explanations, at least for most genes.

horizontal gene transfer, in which a gene is transfered between unrelated taxa by a mixing of their genetic material. As argued for instance in *(26)*, such transfers must be taken into account when one intends to fully understand evolutionary processes.

The third operation is the *tree bisection and reconnection* operation. It is nearly the same as the subtree prune and regraft operation, only we now allow any node (rather than only the root) of the tree that has been cut away to be connected to the remaining tree. This operation is more general than the subtree prune and regraft operation, but one can simulate a tree bisection and reconnection operation by two subtree prune and regraft operations.

Definition 38 (distance problems): The input for the three problems *NNI distance*, *SPR distance*, and *TBR distance* are two binary, undirected phylogenies with input taxa labels only at the leaves and a distance d. The task is to decide whether at most d nearest neighbor interchanges, d subtree prune and regraft operations, or d tree bisection and reconnection operations suffice to transform the first phylogeny into the second, respectively.

Computing distances for phylogenies turns out to be a hard job. Computing the distance of two phylogenies with respect to either the nearest neighbor interchange operation or the tree bisection and reconnection operation is NP-hard and it is strongly suspected that the same is true for the subtree prune and regraft operation.

Theorem 39 (27): *The NNI distance problem is* NP-*complete*.

Open Problem 40: Is the SPR distance problem also NP-complete?

(An NP-completeness proof for *SPR distance* given in *(28)* turns out to be incorrect as argued by Allen and Steel *(29)*, but it might be possible to fix the proof.)

Theorem 41 (28, 29): *The TBR distance problem is* NP-*complete*.

Theorem 42 (29): *The TBR distance problem can be solved in time* $O(d^{O(d)} + n^4)$.

Open Problem 43: Are the NNI distance or SPR distance problems with parameter d (distance) also fixed-parameter tractable?

This section concludes with a distance measure that was introduced in *(30)*. It deviates from the preceding distance measures in that it is based on planar embeddings of the two trees involved. Given a leaf-labeled tree, a linear ordering on its leaves is called *suitable* if the tree can be embedded into the plane such that its leaves are mapped to a straight line in which the given order is maintained. Given two orderings on the same label set, their *crossing number* is the number of edge crossings when drawing the orderings onto two parallel layers and connecting the corresponding labels by edges, *see* **Fig. 24.3** for an example. We then obtain a definition of distance for trees as follows:

Definition 44 (crossing distance): Given two leaf-labeled trees T_1 and T_2 with the same leaf set, their *crossing distance* is the minimal

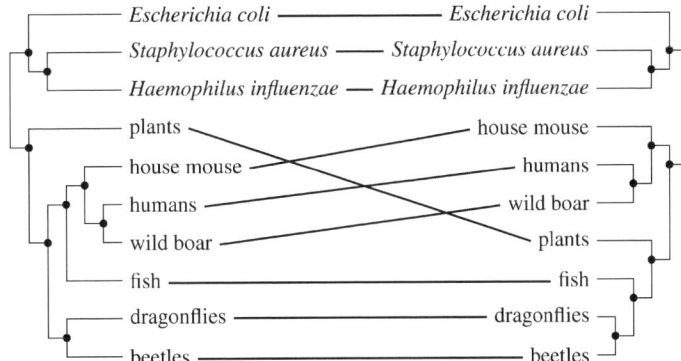

Fig. 24.3. Visualization of the crossing number computation for the two phylogenies from Fig. 24.2. The two phylogenies are drawn in such a way that the taxa lie on two parallel lines and the resulting number of crossings (three in the preceding example) is counted when identical taxa in the different phylogenies are connected.

crossing number between two suitable orderings, one with respect to T_1 and one with respect to T_2.

Note that under this definition trees with different topologies may have distance 0.

Definition 45 (crossing distance problem): The input for the *crossing distance problem* consists of two leaf-labeled trees T_1 and T_2 with the same n element leaf set and a distance d. The task is to check whether the crossing distance between T_1 and T_2 is at most d.

The problem is called *two-tree crossing minimization* by Fernau et al. (30). They show that it is NP-complete, but fixed-parameter tractable with respect to parameter d.

Theorem 46 (30): The crossing distance problem *is NP-complete*.

Theorem 47 (30): The crossing distance problem *can be solved in time* $O(2^{10d} \cdot n^{O(1)})$.

Unfortunately, due to its high running time, the preceding result merely classifies the problem as fixed-parameter tractable.

Open Problem 48: Give a practical fixed-parameter algorithm for computing the crossing distance.

6. Combining Phylogenies

This section studies approaches to combining several phylogenies into a single phylogeny. Suppose two researchers have accumulated character data for two partially overlapping sets of taxa and both have constructed phylogenies based on their data (*see* **Fig. 24.4** for an example). A natural question to ask is: How can one combine these two phylogenies into a single phylogeny?

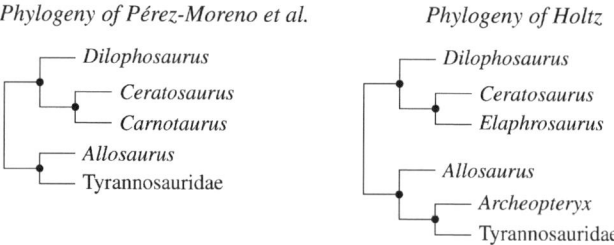

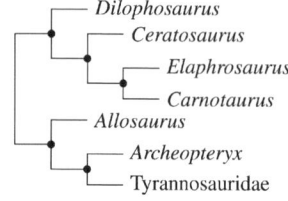

Fig. 24.4. (Parts of) two phylogenies for Dinosauria from two different publications (*(58)* and *(59)*) and a strict consensus supertree for them.

The first approach is to combine the character-state matrices into a *supermatrix* (as it is called in *(31)*) and build a phylogeny based on this combined *primary data* or *total evidence* (as it is called in *(32)*). Another approach, which has gained momentum only in recent years, is to ignore the primary data and build a phylogeny based only on the *topologies of the two phylogenies*. Phylogenies that are based on the topology of other phylogenies rather than on the underlying character-state data are called *supertrees*.

An obvious shortcoming of the supertree approach is that one expects phylogenies based on total evidence to be more exact than phylogenies based only on "second-hand, indirect data" like tree topologies. To make matters worse, the supertree approach can yield phylogenies that are outright contradictory to the primary data. Nevertheless, over the last few years numerous papers have presented supertrees, motivated by a number of arguments that count in favor of the supertree approach:

- The literature contains thousands of phylogenetic studies. When combining published phylogenetic trees to obtain larger trees, it is often hard or impossible to revisit the underlying methods or data; especially when publications date back decades.

- In order to increase efficiency, one can try to compute phylogenies in a two-phase process. In a first phase, one computes small trees based on a phylogenetic method of choice. Because the trees are small, one can use time-intensive methods. In a second phase, one combines these trees into one phylogeny.

- Phylogenetic trees can be computed based on different character sets and the task is to combine the resulting trees into a single supertree. Not all data may be available for all taxa of interest, for instance, genomic sequences may be available only for a small set of species, making it hard or impossible to construct a supermatrix for the primary character data.

The term "supertree" stems from the 1986 paper *Consensus Supertrees: The Synthesis of Rooted Trees Containing Overlapping Sets of Labeled Leaves*, by Allan Gordon *(33)*. However, the strict consensus supertrees of Gordon can only be built for conflict-free input phylogenies, which are only rarely available. Today, the term is also used for trees constructed using methods that handle conflicting input phylogenies more gracefully like the *matrix representation with parsimony* (MRP) method, which was proposed independently by Baum *(34)*, Doyle *(35)*, and Ragan *(36)*. For a more detailed discussion and critical appraisal of the different supertree methods, the reader is referred to the monograph edited by Bininda-Emonds *(37)*.

6.1. Combining Phylogenies Using Strict Consensus Supertrees

For every method, including the strict consensus supertree method, the most basic problem is to decide whether a supertree exists. For the following definitions, recall that in a binary phylogeny all nodes have degree one or three.

Definition 49 (strict consensus supertree): A phylogeny T induces a phylogeny T' if T' can be obtained from T by repeatedly deleting leaves and contracting edges. A phylogeny T is a *strict consensus supertree* of trees $T_1, ..., T_t$ if each T_i is induced by T.

Definition 50 (compatible undirected phylogenies problem): The input for the *compatible undirected phylogenies problem* (abbreviated CUP) are binary phylogenies $T_1, ..., T_t$. The task is to decide whether there is a binary strict consensus supertree for $T_1, ..., T_t$.

Already this basic problem turns out to be hard.

Theorem 51 (7): CUP *is* NP-*complete, even if all input trees have four leaves.*

The corresponding problem for *directed* trees is solvable in time $O(n^3)$ for n taxa using an algorithm of Aho et al. *(38)*. Steel *(39)* raised the question of whether the undirected version is fixed-parameter tractable with respect to the number t of input trees. This parametrization is reasonable since the combination of a small number of possibly large trees is a realistic scenario. Bryant and Lagergren have recently answered Steel's question positively.

Theorem 52 (40): CUP *can be solved in time* $O(f(t) \cdot n^{O(1)})$ *for some function f.*

Unfortunately, both theoretical results on which the fixed-parameter algorithm for CUP is based are, indeed, theoretical and do not have efficient, practical implementations. No one has yet determined an explicit upper bound on the function *f* mentioned in the preceding theorem.

> *Open Problem 53*: Give an efficient and practical parametrized algorithm with explicit running time bounds for CUP for the parameter *t* (number of input trees).

A parametrization of CUP with respect to the maximum size of the input trees does not even lead to a "theoretical" fixed-parameter algorithm by Theorem 51. On the other hand, the problem is fixed-parameter tractable with respect to the total number of *n* of input taxa since one can try all possible tree topologies over the taxa (*see* also Theorem 5 and Open Problem 6).

In practice, multiple phylogenies can only rarely be combined into a strict consensus supertree. Similar to the case of input matrices that do not permit a perfect phylogeny, one must now find ways of resolving the conflicts. Perhaps the simplest approach is to delete potentially erroneous input trees until a solution can be found. Here the number of deleted trees is a natural problem parameter.

> *Definition 54 (CUP by tree removal problem)*: The input for the *CUP by tree removal problem* is the same as for CUP plus a number *k*. The task is to remove at most *k* trees from the input such that the remaining trees are an instance of cup.

Theorem 51 implies that the above problem is NP-complete for *k* = 0 even for the extreme case that all input trees are *quartet* trees (binary trees with four leaves); so it is unlikely that one will make progress on the fixed-parameter tractability of the CUP by tree removal problem. However, in one particular case there is, at least, still hope:

> *Open Problem 55*: Is the CUP by tree removal problem with parameter *k* fixed-parameter tractable when one allows only quartets as input and all of them share a common taxon? (Note that a set of quartets that share a common taxon can be thought of as a set of directed triplets.)

The situation is more favorable when we turn towards the following "dense" version of the problem:

> *Definition 56 (minimum quartet inconsistency problem):* The input for the *minimum quartet inconsistency problem* is a set *S* of *n* taxa, a set *Q* containing a quartet tree for each four element subset of *S*, and a number *k*. The task is to remove *k* quartets from *Q* so that the remaining quartets have a binary supertree *T*.

> *Theorem 57 (41)*: The minimum quartet inconsistency problem *can be solved in time* $O(4^k \cdot n + n^4)$.

The running time in Theorem 57 is linear in the input size since there are $O(n^4)$ input quartets for *n* taxa. The algorithm described in *(41)* also exhibits how search tree algorithms can be complemented by heuristic strategies to prune the search space beyond the running time guarantee (for an introduction to search tree algorithms *see* **Section 3** in **Chapter 21** of **Volume 2**).

6.2. Combining Phylogenies Using Agreement Subtrees

Combining phylogenies using strict consensus supertrees is rarely possible in practice, but *always* bound to fail when one has to combine multiple phylogenies over *identical* leaf sets—a situation that arises in important applications. For example, common heuristic phylogeny reconstruction methods that optimize maximum parsimony or likelihood criteria usually produce multiple optimal or near-optimal trees. Choosing one of the near-optimal trees is arbitrary and a "consensus" of the trees may be preferable. Ad hoc methods for finding a consensus like the majority consensus tree method work in polynomial time—for instance, the randomized algorithm presented in *(42)* runs in linear time—but they may yield poorly resolved output trees. The following discusses a more sophisticated version of the consensus problem.

The rest of this section considers only *directed* phylogenies, which no longer need to be binary.

Definition 58 (maximum agreement subtree problem): The input for the *maximum agreement subtree problem* is a set S of n taxa, directed input trees $T_1, \ldots, T_t$ over S, and a number k. The task is to find a subset $S' \subseteq S$ of size $n - k$ such that there is a directed phylogeny T over S' such that each of $T_1, \ldots, T_t$ induces T.

Perhaps not surprisingly, this problem is NP-complete. The following theorem shows that the situation is even worse:

Theorem 59 (43): The maximum agreement subtree problem *is* NP-complete even for $t = 3$.

Concerning the maximum degree d of nodes in the trees, the following result is known, which is *not* a fixed-parameter result since the parameter d is in the exponent.

Theorem 60 (44): The maximum agreement subtree problem *can be solved in time* $O(n^d + tn^3)$.

For a more complete overview on agreement subtrees, refer to (45). Here, it is of particular interest that the maximum agreement subtree problem is fixed-parameter tractable with respect to parameter k (number of removed taxa):

Theorem 61 (45): The maximum agreement subtree problem *can be solved in time* $O(2.18^k + tn^3)$ *and also in time* $O(3^k \cdot tn)$.

The result can be extended to the closely related maximum compatibility tree problem *(45)*. For input trees with non-identical leaf sets, Berry and Nicolas show that the resulting maximum agreement supertree problem is unlikely to be fixed-parameter tractable as, once more, unlikely complexity class collapses would result.

7. Applications of Phylogenies

Two applications of phylogenetics are presented that are not related to taxonomy. In these applications one is not ultimately interested in finding a solution phylogeny. Rather, a phylogeny or the phylogenetic model is used to determine something seemingly unrelated. In the first application phylogenies help in the prediction of regulatory elements of the genome, in the second application perfect phylogenies are a measure of the quality of haplotype phase predictions.

7.1. Prediction of Regulatory Elements Using Phylogenetic Footprinting

Phylogenetic footprinting, first proposed by Tagle et al. (46), is a method for predicting which regions of the genome are regulatory (involved in the regulatory process). The basic idea relies on the following observation: Suppose one has identified a gene and it is expected that there are regulatory elements before and after the gene, but their exact location is unknown. Regulatory elements will not change (greatly) as mutations occur throughout the genome, because if a non-regulatory part mutates, this does not change the chances of survival, but when a mutation occurs inside a gene or a regulatory area, then the individual may not survive. Thus, a possible approach to predicting regulatory elements is to do a sequence alignment of multiple genomic data and to search for parts of the genome that stay (relatively) stable over evolutionary time spans.

In phylogenetic footprinting one attempts to improve the prediction by using a phylogenetic tree to judge how important a mutation is. If one sees only, say, three different sequences in a candidate regulatory region, but the sequences of closely related species vary strongly among the three sequences, it is less likely that the region is regulatory than if related species all share the same sequence inside the regulatory region.

The preceding ideas lead to a problem called *substring parsimony problem*. To state it formally we first define the *parsimony score*.

Definition 62 (parsimony score): Recall the notion of the length of a phylogenetic tree from Definition 13. Given a partially labeled phylogenetic tree T, the *parsimony score* of the tree is the minimal length of a label completion of T.

Definition 63 (substring parsimony problem): The input for the *substring parsimony problem* is a partially labeled phylogeny T in which only the leaves are labeled and two integers l and s. The task is to decide whether each leaf label can be replaced by a substring of length l such that the parsimony score of the resulting tree is at most s.

The substrings of length l that are chosen from each leaf are the predicted regulatory elements. Note that in the substring parsimony problem the phylogeny T is fixed and part of the input. The idea is that it is typically already available in the literature or can be computed using one of the methods presented in the previous sections.

Blanchette, Schwikowski, and Tompa prove the following theorem:

Theorem 64 (47): The substring parsimony problem *can be solved in time* $O((r^{2l} + m) \cdot ln)$.

The theorem shows that substring parsimony is fixed-parameter tractable with respect to the parameter pair (r,l). The parameter r is 4 in practice, but even for this low value the dominating part of the running time is $r^{2l} = 16^l$, which grows too quickly. Therefore, Blanchette et al. develop a number of improvements for the original algorithm and lower the dominating term first to r^l and even further for typical inputs.

7.2. Prediction of Haplotypes Using Perfect Phylogenies

The *haplotyping problem* arises when one searches for genetic variations of diploid organisms like humans. An example of important genetic variations are *single nucleotide polymorphisms* (SNPs), which are variations across the population of a single nucleotide in the genome. Knowing which nucleobase is present can be important for the prediction of drug response or susceptibility to diseases. The sequence of nucleobases at SNP positions on a chromosome is called a *haplotype*. Being diploid organisms, humans have two (possibly identical) haplotypes for every set of positions.

Existing methods that determine the state of a specific SNP for a person quickly and inexpensively (e.g., in a hospital during a study on drug response), are based on using two primers for each SNP, one for each of the two possible bases (SNP sites with more than two possible bases are very rare). By detecting which primer(s) react, one can determine whether an SNP site is heterozygous (there are different bases on the two copies of chromosomes) or homozygous (both chromosomes agree) and which base(s) are present. This information is called the *genotype* of the sites under consideration.

For heterozygous SNP sites the genotype does not tell us which base belongs on which chromosome. In other words, it lacks the often crucial information which base belongs to which *haplotype*. The *haplotyping problem* is the problem of computationally predicting this information based on the observed genotypes alone.

To make predictions, one must make assumptions about which haplotypes are more likely than others. For example, one could assume that haplotypes change only rarely (they certainly do not change within a few generations). Then if the genotypes of hundreds of persons of the same ethnic group are given, one can try to find a minimal set of haplotypes such that every observed genotype can be explained by assuming that the person has two haplotypes from the small set. Many statistical methods for haplotype phase determination are based on this parsimony assumption.

In a seminal paper, Gusfield *(8)* proposed a different idea. Gusfield argues that haplotypes evolve according to the evolutionary model underlying perfect phylogenies: Mutations occur only rarely and there are no back-mutations. Therefore, one should look for a set of haplotypes explaining the genotypes that forms a perfect phylogeny (the taxa being the haplotypes, the SNP sites being the characters, and the nucleobases being the states). The following definitions formalize the problem.

Definition 65 (haplotype, genotype): A *haplotype* is a state vector. The set Σ_i of permissible states at position i is typically (but need not be) a two element subset of {A,C,G,T}. A *genotype* is a sequence of sets, where the ith set is a subset of size one or two of Σ_i. Two haplotypes *explain* a genotype if the ith subset of the genotype contains exactly the two states of the ith positions of the two haplotypes.

Definition 66 (PP haplotyping problem): The input for the *PP haplotyping problem* is a set of genotypes. The task is to decide whether there exists a set of haplotypes forming a perfect phylogeny such that each genotype can be explained by two haplotypes in the set.

The PP haplotyping problem is at least as hard as the PP problem since we can reduce PP to the PP haplotyping problem by turning each taxon into a "genotype"' whose ith set contains only the ith state of the taxon. Then every set of "haplotypes" that explains the "genotypes" contains the original set of taxa. This shows that the PP haplotyping problem is NP-complete.

The question arises which fixed-parameter results on the PP problem carry over to the more general haplotyping problem. Not too much is known on this since research has almost entirely focused on the case $r = 2$. For this, the following remarkable result is known:

Theorem 67 (48): *For $r = 2$, the PP haplotyping problem can be solved in time $O(mn)$.*

Open Problem 68: How difficult is the PP haplotyping problem for $r > 2$?

In practice, the perfect phylogeny haplotyping problem is, unfortunately, not quite the problem that one wants to solve. Genotype data that is obtained via the laboratory process sketched earlier will always contain a certain amount of *missing data* caused by impurities or incorrect handling. Such missing data are commonly represented by question mark entries in the genotype input.

Definition 69 (incomplete PP haplotyping problem): The input for the *incomplete PP haplotyping problem* is a set of genotypes that may contain question marks for certain characters. The task is to decide whether the question mark entries can be completed in such a way that the resulting set of genotypes is an instance of the PP haplotyping problem.

The missing entries add yet another level of complexity. This new problem, which is of great practical interest, is (presumably) no longer fixed-parameter tractable with respect to the central parameter r. Indeed, the problem is difficult for all values of r as the following theorem shows.

Theorem 70 (49, 50): For every r >= 2, the incomplete PP haplotyping problem is NP-complete.

Because of the preceding result, already for $r = 2$ one has to look for some new parametrizations if one wishes to find a fixed-parameter haplotyping algorithm that can deal with missing data. An obvious parametrization is to consider the total number q of question mark entries in the data.

Theorem 71: For $r = 2$, the incomplete PP haplotyping problem can be solved in time $O(3^q \cdot mn)$.

Open Problem 72: How difficult is the incomplete PP haplotyping problem for $r > 2$?

Unfortunately, the total number q of question marks typically is *not* small in practice. Because of this, a different parameter was studied by Gramm et al. in *(49)*, namely the maximal number c of question mark entries *per character*. An analysis of publicly available genotype data shows that, typically, this parameter is reasonably small. The second key idea of the paper is to assume that phylogenies are directed and that they are *paths* (no branching occurs, except possibly at the root). At first sight it may seem strange to consider path phylogenies, but in the human genome for around 75% of the genomic loci one finds genotypes where all SNP sites are heterozygous *(51)*. The only phylogenies that explain such highly heterozygous genotypes are path phylogenies.

Theorem 73 (49): For $r = 2$, the incomplete PP path haplotyping problem can be solved in time $O(3^{O(c^2 \cdot 6c \cdot c!)} \cdot n^2 m^3)$.

Open Problem 74: How difficult is the incomplete PP path haplotyping problem for $r > 2$?

Open Problem 75: Find a fixed-parameter algorithm for the incomplete PP haplotyping problem for the parameter c (maximum number of question mark entries per column).

8. Conclusion

Fixed-parameter algorithms are a valuable tool in phylogenetics. Phylogenetics abounds in computational problems, many of which are NP-complete, and one cannot expect that efficient exact algorithms will be available for them in the near future, if ever. However, many of the computational problems *can* be solved efficiently and exactly if some of the natural input parameters are reasonably small.

In addition to the concrete open problems pointed out throughout this survey, we want to sketch two broader, less concrete prospective directions of future research.

8.1. Future Research Field: From Discrete to Stochastic Problems

The results presented in this survey refer to problem formulations for discrete input objects and discrete optimization criteria. In computational biology there is a general lack of and a need for fixed-parameter results addressing non-discrete computational problems arising in stochastic analyses. Examples include probabilistic sequence analysis *(52)* and maximum likelihood analysis.

A concrete stochastic computational problem is the following: The input for the maximum likelihood phylogeny problem is a character-state matrix and transition probabilities for the transitions between character states. The task is to find a phylogeny with the input taxa at the leaves that has a maximal "likelihood" among all such phylogenies. Intuitively, the likelihood of a phylogeny is the sum of the likelihoods that the character states at the leaves were generated given the labeling of the inner nodes. Computing this likelihood is a non-trivial task itself (see for instance *(52–54)*). Only recently it has been shown that the maximum likelihood phylogeny problem is NP-hard *(53, 54)*. It remains open to address this and related problems with appropriate fixed-parameter algorithms.

8.2. Future Research Field: From Phylogenetic Trees to Networks

The basic assumption made in this survey, namely that hypotheses on evolutionary history can be represented by trees, is often inappropriate. Phylogenetic trees cannot explain—among other biological effects—the *recombination* effect, in which a genomic sequence is combined from two source sequences by taking a prefix from the first and a suffix from the second sequence. The resulting evolutionary history can no longer be represented by a tree; rather, one must use *phylogenetic networks*.

Fixed-parameter algorithms might be particularly useful in the study of these networks since these are not arbitrary, but "tree-like." They deviate from trees only by a small amount and we propose this extent of deviation (however measured) as a natural problem parameter. It is known (*see* **Chapter 21** of **Volume 2** for an introduction) that fixed-parameter algorithms can often be obtained for such tree-like graphs. Many of the problems addressed in this survey can be extended to phylogenetic networks, but almost all of the resulting problems are open.

References

1. Felsenstein, J. (2004) *Inferring Phylogenies.* Sinauer Associates, Sunderland, MA.
2. Gusfield, D. (1997) *Algorithms on strings, trees, and sequences. Computer Science and Computational Biology.* Cambridge University Press, Cambridge, MA.
3. Page, R. D. M., Holmes, E. C., eds. (1998) *Molecular Evolution: A Phylogenetic Approach.* Blackwell Science, Ames, IA.
4. Semple, C., Steel, M. (2003) *Phylogenetics.* Oxford University Press, New York.
5. Smith, V. S. (2001) Avian louse phylogeny (phthiraptera: Ischnocera): a cladistic study based on morphology. *Zool J Linnean Soc* 132, 81–144.
6. Bodlaender, H. L., Fellows, M. R., Warnow, T. (1992) Two strikes against perfect phylogeny. In *Proceedings of the 19th Interna-*

tional Colloquium on Automata, Languages and Programming (ICALP), Springer-Verlag, New York.

7. Steel, M. (1992) The complexity of reconstructing trees from qualitative characters and subtrees. *J Class* 9, 91–116.

8. Gusfield, D. (2002) Haplotyping as perfect phylogeny: Conceptual framework and efficient solutions. In *Proceedings of the Sixth Annual International Conference on Computational Molecular Biology (RECOMB)*, ACM Press, New York.

9. McMorris, F. R., Warnow, T. J., Wimer T. (1993) Triangulating vertex colored graphs. In *Proceedings of the Fourth Symposium on Discrete Algorithms (SODA)*, SIAM Press, Austin, Texas.

10. Agarwala, R., Fernández-Baca, D. (1996) Simple algorithms for perfect phylogeny and triangulating colored graphs. *Int J Found Comput Sci* 7, 11–21.

11. Bodlaender, H. L., Fellows, M. R., Hallett, M. T., et al. (2000) The hardness of perfect phylogeny, feasible register assignment and other problems on thin colored graphs. *Theoret Comput Sci* 244, 167–188.

12. Agarwala, R., Fernández-Baca, D. (1994) A polynomial-time algorithm for the perfect phylogeny problem when the number of character states is fixed. *SIAM J Comput* 23, 1216–1224.

13. Kannan, S., Warnow, T. (1997) A fast algorithm for the computation and enumeration of perfect phylogenies. *SIAM J Comput* 26, 1749–1763.

14. Dress, A., Steel, M. (1992) Convex tree realizations of partitions. *Appl Math Lett* 5, 3–6.

15. Kannan, S., Warnow, T. (1994) Inferring evolutionary history from dna sequences. *SIAM J Comput.* 23, 713–737.

16. Estabrook, G. F., Johnson Jr., C. S., McMorris, F. R. (1975) An idealized concept of the true cladistic character. *Math Biosci* 23, 263–272.

17. Meacham, C. A. (1983) Theoretical and computational considerations of the compatibility of qualitative taxonomic characters. *Nato ASI series, volume G1 on Numercal Taxonomy.*

18. Gusfield, D. (1991) Efficient algorithms for inferring evolutionary trees. *Networks* 21, 19–28.

19. Fernández-Baca, D., Lagergren, J. (2003) A polynomial-time algorithm for near-perfect phylogeny. *SIAM J Comput* 32, 1115–1127.

20. Goldberg, L. A., Goldberg, P. W., Phillips, C., et al. (1996) Minimizing phylogenetic number to find good evolutionary trees. *Dis Appl Math* 71, 111–136.

21. Moran, S., Snir, S. (2005) Convex recolorings of phylogenetic trees: definitions, hardness results and algorithms. In *Proceedings of the Ninth Workshop on Algorithms and Data Structures (WADS)*, Springer-Verlag, New York.

22. Blelloch, G. E., Dhamdhere, K., Halperin, E., et al. (2006) Fixed-parameter tractability of binary near-perfect phylogenetic tree reconstruction. In *Proceedings of the 33rd International Colloquium on Automata, Languages and Programming*, Springer-Verlag, New York.

23. Sridhar, S., Dhamdhere, K., Blelloch, G. E., et al. (2005) FPT algorithms for binary near-perfect phylogenetic trees. Computer Science Department, Carnegie Mellon University. Technical Report CMU-CS-05-181.

24. Wernicke, S., Alber, J., Gramm, J., et al. (2004) Avoiding forbidden submatrices by row deletions. In *Proceedings of the 31st Annual Conference on Current Trends in Theory and Practice of Informatics (SOFSEM)*, Springer-Verlag, New York.

25. DasGupta, B., He, X., Jiang, T., et al. (1998) *Handbook of Combinatorial Optimization.* Kluwer Academic Publishers, Philadelphia.

26. Gogarten, J. P., Doolittle, W. F., Lawrence J. G. (2002) Prokaryotic evolution in light of gene transfer. *Mol Biol Evol* 19, 2226–2238.

27. DasGupta, B., He, X., Jiang, T., et al. (2000) On computing the nearest neighbor interchange distance. In *Discrete Mathematical Problems with Medical Applications, DIMACS Series in Discrete Mathematics and Theoretical Computer Science.* American Mathematical Society, Providence, RI.

28. Hein, J., Jiang, T., Wang, L., et al. (1996) On the complexity of comparing evolutionary trees. *Disc Appl Math* 71, 153–169.

29. Allen, B. L., Steel, M. (2001) Subtree transfer operations and their induced metrics on evolutionary trees. *Annals of Combinatorics.* 5, 1–13.

30. Fernau, H., Kaufmann, M., Poths, M. (2005) Comparing trees via crossing minimization. In *Proceedings of the 25th Conference on Foundations of Software Technology and Theoretical Computer Science (FSTTCS)*, Springer-Verlag, New York.

31. Sanderson, M. J., Purvis, A., Henze, C. (1998) Phylogenetic supertrees: Assembling the tree of life. *Trends Ecol Evol* 13, 105–109.

32. Kluge, A. G. (1989) A concern for evidence and a phylogenetic hypothesis of relationships among *epicrates* (boidæ, serpents). *Syst Zool* 38, 7–25.

33. Gordon, A. D. (1986) Consensus supertrees: the synthesis of rooted trees containing overlapping sets of labeled leaves. *J Class* 3, 31–39.

34. Baum, B. R. (1992) Combining trees as a way of combining data sets for phylogenetic inference, and the desirability of combining gene trees. *Taxon* 41, 3–10.

35. Doyle, J. (1992) Gene trees and species trees: molecular systematics as one-character taxonomy. *Syst Botany* 17, 144–163.

36. Ragan, M. (1992) Phylogenetic inference based on matrix representation of trees. *Mol Phylogenet Evol* 1, 53–58.

37. Bininda-Emonds, O., ed. (2004) *Phylogenetic Supertrees*. Kluwer Academic Publishers, Dordrecht.

38. Aho, A. V., Sagiv, Y., Szymansk, T. G., et al. (1981) Inferring a tree from lowest common ancestors with an application to the optimization of relational expressions. *SIAM J Comput* 10, 405–421.

39. Steel, M. (August 2001) Personal communication. Open question posed at the Dagstuhl workshop 03311 on fixed-parameter algorithms.

40. Bryant, D., Lagergren, J. (2006) Compatibility of unrooted trees is FPT. *Theoret Comput Sci* 351, 296–302.

41. Gramm, J., Niedermeier, R. (2003) A fixed-parameter algorithm for Minimum Quartet Inconsistency. *J Comp Syst Sci* 67, 723–741.

42. Amenta, N., Clarke, F., St. John, K. (2003) A linear-time majority tree algorithm. In *Proceedings of the Third Workshop on Algorithms in Bioinformatics (WABI)*, Springer-Verlag, New York.

43. Amir, A., Keselman, D. (1997) Maximum agreement subtree in a set of evolutionary trees: metrics and efficient algorithm. *SIAM J Comput* 26, 1656–1669.

44. Farach, M., Przytycka, T. M., Thorup, M. (1995) On the agreement of many trees. *Inf Proc Lett* 55, 297–301.

45. Berry, V., Nicolas, F. (2006) Improved parametrized complexity of maximum agreement subtree and maximum compatible tree problems. *IEEE/ACM Trans Comput Biol Bioinform* 3, 284–302.

46. Tagle, D. A., Koop, B. F., Goodman, M., et al. (1988). Embryonic ε and γ globin genes of a prosimian primate (Galago crassicaudatus) nucleotide and amino acid sequences, developmental regulation and phylogenetic footprints. *J Mol Biol* 203, 439–455.

47. Blanchette, M., Schwikowski, B., Tompa, M. (2002) Algorithms for phylogenetic footprinting. *J Comput Biol* 9, 211–223.

48. Ding, Z., Filkov, V., Gusfield, D. (2005) A linear-time algorithm for the perfect phylogeny haplotyping (PPH) problem. In *Proceedings of the Ninth Annual International Conference on Research in Computational Molecular Biology (RECOMB)*, Springer-Verlag, New York.

49. Gramm, J., Nierhoff, T., Tantau, T., et al. (2007) Haplotyping with missing data via perfect path phylogenies. *Discrete Appl Math* 155, 788–805.

50. Kimmel, G., Shamir, R. (2005) The incomplete perfect phylogeny haplotype problem. *J Bioinformatics and Comput Biol* 3, 359–384.

51. Zhang, J., Rowe, W. L., Clark, A. G., et al. (2003) Genomewide distribution of high-frequency, completely mismatching snp haplotype pairs observed to be common across human populations. *Amer J Hum Genet* 73, 1073–1081.

52. Durbin, R., Eddy, S. S., Krogh, A., et al. (1998) *Biological Sequence Analysis: Probabilistic Models of Proteins and Nucleic Acids.* Cambridge University Press, Cambridge, MA.

53. Chor, B., Tuller, T. (2005) Maximum likelihood of evolutionary trees: Hardness and approximation. *Proceedings of the 13th International Conference on Intelligent Systems for Molecular Biology (ISBM). Bioinformatics* 21, 97–106.

54. Chor, B., Tuller, T. (2005) Maximum likelihood of evolutionary trees: Hardness and approximation. In *Proceedings of the Ninth Annual International Conference on Research in Computational Molecular Biology (RECOMB)*, Springer-Verlag, New York.

55. Lander, E. S., Linton, M., Birren, B., et al. (2001) Initial sequencing and analysis of the human genome. *Nature* 409, 860–921.

56. Salzberg, S. L., White, O., Peterson, J., et al. (2001) Microbial genes in the human genome: lateral transfer or gene loss. *Science* 292, 1903–1906.

57. Andersson, J. O., Doolittle, W. F., Nesbø, C. L. (2001) Are there bugs in our genome? *Science* 292, 1848–1850.

58. Pérez-Moreno, B. P., Sanz, J. L., Sudre, J., et al. (1993). A theropod dinosaur from the lower cretaceous of southern France. *Revue de Paléobiologie* 7, 173–188.

59. Holtz T. R. Jr. (1994) The phylogenetic position of the *Tyrannosauridae*: implications for theropod systematics. *J Paleontol* 68, 1100–1117.

INDEX

A

Ababneh, F., ... 334, 338, 340, 342
ACCESSION field in GenBank flat file 13
Affy, bioconductor packages 96
affy_hg_u95av2 as filter 107–108
Affymetrix GeneChip™ chips 90
Agarwala, R., ...514, 515
Aho, A. V., ... 525
Akaike Information Criterion (AIC) 354, 355, 499
AlignACE algorithms 241
Alignment editors 158–159
Allen, B. L., ... 522
Alternative splicing 179–180
   methods
      alignment filtering 189
      and available public data 181, 184–186
      candidates for 189–190
      data mining 190–192
      pre-processing and 187–188
      transcript sequences alignment of 188
   and splice graph 201
   transcript, types of 180–181
AMBER module 50
Amino acid replacement models, comparison of 264
Among-site rate variation (ASRV) 462
AMoRe programs 72
Amplified Fragment Length Polymorphisms
      (AFLPs) 493
Analysis of molecular variance (AMOVA) 495
Anchor optimization 151–152
Ancient outgroup 372
annaffy and *annBuilder* annotation package 93
ANN-Spec motif discovery algorithms 242
ANOVA normalization methods 92
Aquifex aeolicus 458, 464
ArrayExpress microarray datasets 95
ASAP database 191
ATP citrate synthase activity (GO:0003878) 117

B

Bacteriophage (PHG) and bacterial (BCT) sequences,
      GenBank division 16
BAliBASE MSA benchmark database 147
BankIt, GenBank's web-based submission tool 11

Base-to-base H(C/N)H-type correlation
      experiments 46
Baum, B. R., ... 525
Bayes factors (BF) 354, 355
Bayesian Information Criterion *(BIC)* 354, 355
Bayesian phylogenetics 301, 499
Bayesian schema design 190
Bayesian tree inference 300–302
   estimation of 301–302
   MCMC algorithm and
      chain convergence and mixing 303–304
      sampling 302–303
      priors specification of 304–305
Beerli, P., ... 495
Benjamini, Y., ... 323
Bergeron, A., ... 399, 402, 414
Berry, V., ... 527
Bininda-Emonds, O., ... 525
Bioconductors
   packages 93
   software project for analysis and comprehension
      of genomic data 95
Bio-dictionary gene finder (BDGF) 170
BioMart
   generic data management system 93
   package 105
biomaRt annotation package 93
BioProspector algorithms 241
Blaisdell, J., ... 259
Blanchette, M., ... 529
BLAST
   all-*vs*-all BLASTP 460
   BLASTP threshold for cluster existance 461
   conversion output into GFF 135
   databases, for genome 374
   general platform and computing/
      programming tools 419
   graphical display VecScreen analysis 21
   -m9 option for a tab-delimited format 133–134
   output and GFF format for human genome 134
   protein BLAST bit score 422
   PSI-BLAST homology search engine 150
   Saccharomyces cerevisiae genome 418, 420
   searches 17
   tool 137

vs. SIM4 188
WU-BLAST program 374
BlockMaker algorithms 241
BLOCKS motif databases 232
BLOSUM
    amino acid exchange matrices 144
    matrix 154–155
    series of matrices 265, 266
Bodlaender, H. L., ... 514
Bootstrap
    procedure 172
    techniques, for phylogeny 320
    values 298
Bork, P., ... 443
Bourque, G., ... 433
Bowker matched-pairs test of symmetry 341
Bowker's, A. H., ... 341
Boxplot comparing intensities for
    arrays after normalization 98
Bragg, W. L., ... 65
Bryant, D., ... 334, 525
Buckley, T. R., ... 354

C

Cantor, C. R., ... 259, 353
Carrillo, H., ... 145
CCP4i package programs 72
CC-TOCSY transfer 44
Change point analysis 209–210
Changept 209, 210, 212, 213–215, 217, 219, 222, 226
Character-state matrix, for phylogeny .... 509, 513.
                  See also Perfect phylogenies
Charleston, M. A., ... 331
ChEBI, GO biological process ontology 119
Chemokine receptor genes 256
chloroplast *is_a* plastid 112
Chromosomes, ancestral and gene orders reconstruction
    hypotheses, DreLG3 and DreLG12 distribution 379
        pufferfish fusion hypothesis and
            zebrafish fission hypothesis 379–380
    linkage Group 3 (DreLG3) and
        Group 12 (DreLG12) 377
    Tni2
        regions of conserved gene order 381
        and Tni3 relation 377–379
        in zebrafish and pufferfish 379
Churchill, G. A., ... 261
CINEMA, Java interactive tool 159
Circular and linear chromosomes, DCJ on
    adjacency graph, applications of 399
        circular intermediates, restriction of 403
        distance formula 402–403
        DCJ distances, genome of
            Shigella and *E. coli* 402
        0 labels 400–401

        paths and cycles, odd and even paths 401–402
        procedures 403
        telomeres 399–400
    edge graph, applications of
        on capped lines 395–396
        caps and null chromosomes 395
        closure and distance 397, 399
        menu of operations 396–397
        paths and cycles 397
Clark, A. G., ... 274
Clone-based sequencing technology 8–9
ClustalW, ... 152, 159, 187, 382, 460, 481
    multiple alignment program 159
CNS programs 72
CNS/Xplor syntax 48–49
3D-Coffee protocol 152–153
Cold Spring Harbor Laboratory (CSHL) 93
COMMENT field in GenBank flat file 14
CompareProspector algorithms 241
CON division 9
Confidence in phylogenetic trees, bootstrapping 297
Confidence intervals on point estimates 296–297
    limitations of 300
    non-parametric bootstrap 297–298
    parametric bootstrap 298–299
Consensus motif discovery algorithms 242
Constructed entries (CON), GenBank division 16
Content sensors for genomic sequence 170–171
Contig sequences 9
Coo$_t$ programs 72
CRAN mirror website 95
Critica programs 174–175
Cross_Match packages 187

D

Danio rerio 370, 377
    Linkage Group 3 (DreLG3) 377
DasGupta, B., ... 521
Database of Interacting Proteins (DIP) 419
data.frame, ... 95
".data.gz" for cluster results and annotation 186
Dayhoff matrices, in database search methods 263–264
Dayhoff, M. O., ... 263, 265, 267, 306
DBTBS motif databases 232
D:/DATA directory 99
DDBJ's webbased submission tool SAKURA 11
Decision theory *(DT)* 354, 355
DEFINITION field in GenBank flat file 13
Degenerative mutations, in regulatory elements 273
Deleterious mutations 273
de novo gene-finding accuracy 173
de novo motif discovery 236–237
Denzo
    HKL2000 suite component 77
    macromolecular crystallography 72

Dermitzakis, E. T., ... 274
develops_from relationship ... 116
Dialign, ... 156, 157, 481
Directed acyclic graph (DAG) ... 201
Diversifying selection, location of ... 319–320
    likelihood ratio test and SLR approach ... 320
    posterior probability calculation ... 319
D. melanogaster and *D. simulans* in axt format ... 211–217
    attribution of function ... 226
    conservation levels for ... 220
    convergence assessing ... 217–218
    genomic regions, enrichment and depletion of ... 224–225
    number of groups and degree of pooling ... 219
    profile generation and ... 221–223
    segment generation ... 225–226
DNA
    microarrays (ChIP/chip) and sequencing paired-end di-tags (ChIP/PETs) ... 127
    repeats, ... 256
    repeats, in human genome ... 256
    template sequences for *in vitro* transcription of HIV-2 TAR RNA ... 35–36
    transposons ... 256
cDNA microarrays ... 92
Double cut and join (DCJ), circular chromosomes ... 386, 387–389
    adjacency graph ... 390–393
    block interchange ... 389
    on circular and linear chromosomes, ... 395–403
    on circular chromosomes ... 386–395
    distance
        on cycle ... 393
        formula ... 393
        removal of 1-cycle ... 395
        sorting of edge graph ... 394
    distance formula ... 393
    edge graph ... 389–390
    on genome ... 388
    genome graphs ... 386–387
    localized view of ... 387
    materials and online resources ... 403–406
        Mauve alignment, of *Shigella* and *E. coli* genome ... 406
        Mauve 1.3.0, for analysis ... 404–405
    methods
        adjacency graph ... 409–410
        black-white genome graph ... 406–407
        distance without linear chromosomes ... 410
        distance with/without linear chromosomes ... 410–411
        edge graph construction ... 408, 409
        genome graph ... 407–408
        sorting with/without linear chromosomes ... 411–412

Doyle, J., ... 525
DpsTe family of proteins ... 86
DreLG3, zebrafish chromosome ... 376
Dress, A., ... 515
DSSP, predicted secondary structure information ... 150
Duplication-degeneration-complementation model (DDC) ... 273
DUST filtering programs ... 248
Dynamic programming (DP) algorithm ... 144–145

E

ECgene databases ... 184
Efficient Evaluation of Edit Paths (EEEP) ... 463
Efron, B., ... 319
Eichler, E., ... 444
Elutrap Electroelution System ... 37
EMBL ... 4, 8, 11–13, 17, 21, 23, 24, 130
    web-based submission tool, Webin ... 11
EMBOSS open source analysis package ... 187
Empirical models of nucleotide substitution ... 260
ENCODE Genome Annotation Assessment Project (EGASP) ... 173
ENCyclopedia Of DNA Elements (ENCODE) project ... 128, 173
Ensembl ... 93, 105–107, 129, 133, 137, 171, 186, 377, 434, 443
    database ... 128, 130
    genome browser ... 134
    pipeline for eukaryotic gene prediction ... 171–172
    software project ... 93
Entrez
    gene database ... 5–6
    nucleotide sequences ... 22
    protein sequences ... 22
Entrez PubMed ... 14
Environmental sampling sequences (ENV), GenBank division ... 16
ERPIN program ... 193
EST databases ... 184
European Bioinformatics Institute (EBI) ... 93
Evolution models for genome elements
    amino acid mutation models ... 263–265
    amino acid models incorporating correlation ... 269–270
    amino acid models incorporating features ... 267–269
    generating mutation matrices ... 265–267
    Codon models ... 262–263
    for M7 and M8 hypotheses ... 263
    rate matrix **Q** ... 262
    DNA substitution models ... 259–261
    modeling rate heterogeneity along sequence ... 261
    models based on nearest neighbor dependence ... 261–262
    functional divergence after duplication ... 275–276

microsatellite and ALU repeats
    evolution of 280–282
Markov models of DNA evolution 257–259
    instantaneous rate matrix of transition
        probabilities 258–259
    properties of 258
models of DNA segment evolution 271–273
protein domain evolution 276–279
    introns evolution of 279–280
RNA model of evolution 271
sub-functionalization model 273–275
sub-neo-functionalization 274
Ewald sphere for x-ray diffraction
    pattern for protein 66, 83–84
Expectation-Maximization algorithm 202
Expectation maximization (EM) 262
Expressed sequence tags (EST) 20, 25,
    163, 172, 181, 184, 185, 187, 188, 192,
    200, 204
Expression Set (exprSet) normalization 98
expresso command 99
ExprSet normalized expression data 99

F

Family-Wise Error Rate (FWER) 323–324
FASTA format 12
Fast Fourier Transformation (FFT),
    homologous segments detection 153
FEATURES field in GenBank flat file 15
Felsenstein, J., ... 259, 260, 261, 268, 269, 280,
    353, 449, 495
Felsenstein's model 260
Fernández-Baca, D., ... 514, 515, 516, 519
Fernau, H., ... 523
FFT-NS-1 and FFT-NS-2 protocol 153
Fgenesh++, transcript sequences 184
Fixation indices, for genetic variation 494
2Fo-Fc and Fo-Fc map, 71. *See also* Proteins
Foster, P. G., ... 357
French, G. S., ... 77
FUGUE default pairwise structural
    alignment methods 153
Full-length open reading frame (FL-ORF) clones 7
Functional Annotation Of Mammalian
    (FANTOM) 128

G

Galtier, N., ... 351
Gaps in alignments 268
Gcrma bioconductor packages 96
GenBank 13, 17, 20, 25, 130,
    185, 404, 459
    annotator 18
    flat file 13–14
    records, division 16
    sequence record 13
    Web-based submission tool, BankIt 11
genbiomultialign.exe program 213
genchangepts. exe program 213
GeneMark.hmm 165
Genemark programs 172
Gene Ontology (GO) 111
    development and 113–118
    online resources 112–113
    principles of 118–121
General Feature Format (GFF) 130, 133–136, 168
General time-reversible (GTR) 462
Generic genome browser 131, 133, 136
Genes
    database 7
    duplication 272, 334
    finding
        in environmental sequence samples 169–170
        eukaryotes in 170–173
        prokaryotes in 164–169
        software for 166
    in GC-rich genomes 174
    orders, conservation of
        for DreLG3 and Tni3 376
        and inference of ancestral gene orders 368
    products, functional annotations of 459
    recombination 471
        gag region of HIV-1, use of PhylPro 473
        dataset containing all sequences,
            creation of 474–475
        first phylogenetic profile, creation of 475
        import of sequences in VOP 474
        PhylPro application window 480
        profile interpretation 475–477
        profiles, comparing of 478
        program availability and data sources
            and format 473
        recombinant sequences, ... 478–479, 481
        sequence set, refinement of 477
    structure and splice graph 201
    transfer, ... 334
Genome annotation 125
    materials for
        databases 128–130
        nucleotides and nucleosomes 127–128
        transcripts 126–127
    methods
        public annotation 131–133
Genome duplication
    base of radiation of teleost fish 368
    beginning of teleost radiation 370
    DreLG3 and DreLG12 379
    event, in teleost lineage 370

events in chordate evolution and R3
    in ray-fin lineage 367
    production of DreLG3 and DreLG12 379
    R1 and R2, early vertebrate 367, 369
GenomeScan, transcript sequences............................... 184
Genome survey sequences (GSS),
    GenBank division 7, 16
GenScan, transcript sequences 184
getLayout() function... 101
GIGO rule..242–243
GISMO
    program for encoding of genes 168
    SVM-based gene finder... 175
GLAM motif discovery algorithms............................... 242
Glimmer programs 174–175
Global alignment strategy 147
Glycoprotein (GP120)... 321
Goldberg, L. A., ... 516
GOLD database... 426
Goldman, N., ... 262, 267, 268, 306, 325, 348, 356
Gonnett, G. H., ... 264
GO protein domains 96, 108, 112, 114,
    119–122, 245, 302
Gordon, A., ... 525
GO slim documentation on ontology............................ 122
G-protein coupled receptor activity (GO:0004930)...... 121
G-protein coupled receptor (GPCR) 121
GRAMENE, data resource for comparative genome
    analysis in grasses..................................... 93
Gramm, J., ... 531
Grantham, R., ... 262
GRIMM-Synteny..404, 431, 433
    identifying homologous blocks across
        multiple genomes 437
    forming synteny blocks from sequence-based
        local alignments 439–441
    GRIMM-anchors: filtering out
        alignments with 438–439
    preparing the input for................................. 437–438
    sample run 443–445
    usage and options, for forming synteny
        blocks.. 441–443
    mgr_macro.txt for input to GRIMM
        or MGR.. 435–436
    vs. GRIMM 433
GS-Finder, programs for start site identification 175
Gusfield, D., ...514, 530

H

Hampton Research crystallization and
    cryoprotection kits 72
Hanging drop method for protein solution 73
Hannenhalli, S.,..445, 447

HAPMART, data resource for results
    of HAPMAP project 93
Harpending, H., ... 496
Hasegawa, M.,...260, 261
has_participant relationship.. 120
Haussler, D., ... 261
HCP-TOCSY (Total Correlation Spectroscopy),
    RNA resonances 39
Heatmap plots.. 108
Heaviest Bundling, dynamic programming
    algorithm 202–203
Heber, S., ... 190, 200, 201, 202
help.search() function.. 95
Henikoff, S., ... 266
Hesper, B., ... 146
Heteronuclear Single Quantum Coherences (HSQCs) .. 43
Heterozygosity calculation 494
He, X., ... 274
hgnc_symbol as attribute 107
Hidden Markov models (HMMs) 144,
    165, 276, 321
    based EasyGene... 175
    based GeneWise program 171
    transition probabilities............................... 268
Hierarchical likelihood ratio tests (hLRT).................... 499
High throughput cDNA sequences (HTC),
    GenBank division 16
High throughput genomic (HTG) sequence.................... 8
HKL2000 suite component, Denzo and Scalepack......... 77
hmm ES-3.0 programs... 172
HNNCOSY (Correlation Spectroscopy) experiment...... 45
Hochberg, Y.,...323, 325
Hogeweg, P.,... 146
Ho, J. W. K., ... 340
Holder, M., ... 301
Holm's, S.,... 323
Homogeneous Markov process......................... 337, 352
Homolog..437
Homoplasy..488, 500, 503
hoxaa cluster of zebrafish... 368
HOX clusters..271, 368, 369
HP-HETCOR (Heteronuclear Correlation).................... 39
Hs.2012(TCN1) gene structure by
    splice graph representation..................... 202
Hudson, R. R.,.. 496
Huelsenbeck, J. P., ... 301
Human chromosome 17 (Hsa17)................................... 369
Human genome repetitious organization 257
Human immunodeficiency virus (HIV)-2
    transactivation response element (TAR)
    RNA, sequence and structure............. 31–32
Hyman, I. T., ... 350
Hypervariable V1/V2 loop regions of protein 321

I

Ideal Mismatch value 91
Iisoform problem... 199–200
Iisoform reconstruction and heaviest
    bundling.. 202–203
import all .spot files.. 100
Importance quartet puzzling (IQP)............................. 295
importing .spot data 100
Improbizer motif discovery algorithms......................... 242
Indels mutations 299
Inferring patterns of migration..................................... 501
    direct and indirect evidence 502–503
".info" for statistics.. 186
Insensitive Nuclei Enhancement by Polarization
    Transfer (INEPT) delays......................... 45
International Nucleotide Sequence Database
    (INSD) 3, 5, 6, 8, 11, 13, 17, 21, 23, 24, 130
    accessing of
        Entrez retrieval system................................. 22
        ftp sites 21
    archival primary sequence database
        annotation and propagation........................... 19–20
        vector contamination 20
        wrong strand annotation............................. 20–21
    sequence data submission in
        Web-based submission tools........................... 11–12
    sequence processing
        batch submissions processing of........................... 18
        submissions.. 16–18
        updates and maintenance of.......................... 18–19
International Table of Crystallography................................64
Internet Explorer... 130
INTERPRO protein domains...................................... 108
Invertebrate sequences (INV), GenBank
    division.. 16
is_a complete relationship.. 120
is_a relationship... 116

J

Jalview protein multiple sequence
    alignment editor..................................... 159
JASPAR
    "*broad-complex 1*" motif.. 234
    motif databases 232
Jayaswal, V.,..344, 357, 359
Jermann, T. M.,.. 331
Jim Kent web site 186
Jones, D., ... 265
Jones, D. T., ...264, 268
J scalar coupling constants and RNA structures 40
Jukes-Cantor model 260
Jukes, T. H., ...259, 260, 269, 353

K

Kannan, S., ... 515
Karev, G. P.,... 277
Karplus, K.,..44, 48
Karplus parameterization 44
KEYWORDS field in GenBank flat file 13
Kimura, M.,...259, 353
Kishino-Hasegawa (KH) test 298
Koonin, E. V.,... 277
Kreitman, M.,........................... 312, 313, 314, 317, 325

L

Lagergren, J.,..516, 519, 525
LAMA program.. 244
Lanave, C., ... 261
Lateral genetic transfer (LGT)...................................... 457
    data sources....................................... 458–459
    inference, by topological comparisons
        of trees 463–464
        EEEP method and edit operations.................... 463
        unrooted protein tree, reconciling...................... 464
    limitations and pitfalls, of surrogate
        methods 467
    phylogenetic inference and supertree
        construction 462–463
        CLANN and MRP, use of.................... 463
        Nexus format creation and burn-in
            phase identification................................ 462
    program availability 459
    putative homologs and orthologs, clustering of proteins
        All-*versus*-all BLASTP 460
        hybrid protein clustering strategy 461
    sequence alignment, validation,
        and trimming................................ 460–461
        alignment algorithms, in each MRC 460
        TEIRESIAS pattern-detection
            algorithm 460–461
        WOOF and GBLOCKS, use of......................... 461
Lewis, P. O.,... 301
Lewontin, R. C.,.. 274
".lib.info.gz" for EST library information...................... 186
Likelihood ratio test (LRT)...............................275, 320
Limma package 92
Lineage through time (LTT)... 496
Linkage disequilibrium...270, 503
Liò, P., ... 268
Lipman, D. J.,.. 145
listAttributes function.. 107
Listeria monocytogenes MolRep program after rotation
    and translation searches 79
listFilters function 107
list-Marts() function 105

Li, W.-H., ... 261
Local alignment strategy 147
Locally collinear blocks (LCBs) 404, 405
Lockhart, P. J., ... 351
LOCUS field in GenBank flat file 13
Log likelihood ratio (LLR) of PWM 237

M

maBoxplot() functions 109
MAFFT, multiple sequence alignment package 153–154
MAList object 102
maNorm() and *maPlot*() functions 109
MapViewer 7
Markov chain Monte Carlo (MCMC) 210, 217, 302–304
Markov process 257, 258, 268, 269, 333, 334, 336, 337
Marray package 100
MAS5.0 normalization procedure 91
Massingham, T., ... 325
Matched-pairs tests, of symmetry 340, 343, 345
    and complex alignments 345
    with more than two sequences 342–345
    with two sequences 340–342
matrix() function 94
Matrix representation with parsimony (MRP) 463, 525
Mauve 402, 404–406
Maximum a posteriori probability (MAP) 301
Maximum likelihood (ML) 289, 499
Maximum parsimony, of DNA sequences 499
McDonald, J., ... 312, 313, 314, 317
McDonald-Krietman test, selection on proteins ... 315–320
    adh locus, encoding alcohol dehydrogenase 312–313
    Fisher's exact test's *p* values, application of 314
    limitation 314
    for polarized variant 313
MDscan algorithms 241
Membrane-bound organelle (GO:0043227) 122
Methylation Sensitive Representational Difference
    Analysis (MS-RDA), sequencing 127
Metropolis-Hastings Monte Carlo simulation 495
MGED Ontology Resources 113
Microarray data
    annotation of 93
    data filtering 92
    differentially expressed genes detection of 92
    methods
        Affymetrix data normalization of 96–99
        cDNA microarray data 100–104
        differential expression detection
            by limma 104–105
        R essentials 93–95
        R packages and bioconductor 95

noisy signal 89
pre-processing
    Affymetrix data of 90–91
    two-color data of 91–92
quality assessment 90
Microsatellites
    codominant markers 492
    length, distribution in human genome 281
Migration history and patterns, determination of
    data types of 488, 489, 492, 493
    methods
        Bayesian-based population genetics 497
        classical population genetics 494–495
        coalescent-based population genetics 495–497
        genetic variation in European
            humans, components of 498
        gene trees in inferring migrations, use of 500
        inference structure 501–503
        multivariate statistical analysis 498
        NCPA, distribution of related haplotypes 501
        network construction 499–501
        phylogenetic methods 498–499
        structure output demonstration 497
    organism with wind-borne spore dispersal 486
    RFLP, RAPD and SSCP data 493
    sampling scheme designing 486–488
Migration rates theta estimation 495
Miller indices 83
MITRA motif discovery algorithms 242
Model Based Expression Index (MBEI) 91
Molecular phylogenetics 331
MolRep
    programs 72
    solutions 85
Monte Carlo
    simulations 344
    studies on matrices 266
Moran, S., ... 516, 518
Motif databases with web servers 244
Motif discovery
    algorithms with web servers 241
    and limitations 246–247
MotifSampler motif discovery algorithms 242
Mozilla Firefox 130
Multiple alignment formats (MAF) files 186
Multiple gene trees 504
Multiple sequence alignment (MSA) 12, 143, 159, 188, 208, 472, 475
    dynamic programming 144–145
    iteration of 146
    materials for
        global and local 147
        sequences selection of 146–147

methods
  MAFFT 153–154
  MUSCLE 150–152
  PRALINE 148–150
  ProbCons 154–155
  SPEM 155
  T-Coffee program 152–153
  protocol for 145–146
  reliability and evolutionary hypothesis 144
  types of 147–148
Murphy, W. J., ... 447
MUSCLE, multiple alignment software 149–151, 150–151
Mutations
  advantages 272
  rate of 503
MySQL database 186

N

Nadeau, J. H., ... 272
National Center for Biotechnology
  Information's (NCBI) 7, 22, 24, 185, 377, 405, 426, 434, 459
NCBI RefSeq project 20
Nearest neighbor interchange (NNI) 293, 294
Neighbor-Joining technique 152
Nested Clade Phylogeographical Analysis (NCPA) 501
Network visualization 418, 423, 426
Nicolas, F., ... 527
Nielsen, R., ... 263
NJ, tree calculation 159
Noisy signal 89
Non-canonical G:U pair 271
Non-coding DNA 164–165
Non-membrane-bound organelle
  (GO:0043228) 123
Non-parametric bootstrap 296–298
normalizeBetweenArrays() function 104
normalizeWithinArrays() function 102
Nuclear Overhauser effect spectroscopy (NOESY),
  RNA resonances 38
Nucleotide sequences
  alignments of 334, 339, 342
  equilibrium frequency of 259
  and haplotype diversity 494
  international collaboration 4–5
  NCBI RefSeq project 5–6
Nucleotide substitution, probabilistic model 314–319
  assumptions violated 315–316
  gap occurrence and strength of selection,
    relationship 319
  likelihood ratio test statistic, Δ 317
  log-likelihood of parameters 316
  missing data 316
  non-parametric bootstrap, gor
    diversifying selection 319
  random-site methods of detecting selection 317
  synonymous and non-synonymous
    mutations ratio ω 315
  use of parametric bootstrap techniques 318
Nucleotide triphosphates (NTPs) 29–30

O

OBO-Edit biomedical ontologies 112
OBO_REL relationship 120
Ohno, S.,.. 445
Olfactory receptor genes 256
Omit map .... 82. *See also* Proteins
Open biomedical ontologies (OBO)
  collection and principles of 113
  foundry for 115
Open reading frame (ORF) in genome sequence 164
ORGANISM field in GenBank flat file 14
Orphan sequences 147
Ortholog identification pipeline (ORTH) 374
Orthologs 370, 374, 376, 436, 460
  and paralogs, evolutionary relationships 371
O to build the model programs 72
Ozery-Flato, M., ... 445

P

Pagel, M., ... 270
Paired-mismatch (MM) probes 90
pairs command 98
PAM, amino acid exchange matrices 144
PAM matrices 265–267
Paralog 274, 275, 374–376
Paralog Identification pipeline (PARA) 374
Parametric bootstrap 296, 319, 333, 348, 349, 352, 357, 359
part_of relationship 116
Patent sequences (PAT), GenBank division 16
Patterson functions 68
Pedersen, A.-M. K., ... 263
Perfect match (PM) probes 90
Perfect phylogenies
  construction of
    formalization of problems 511–513
    parameter 514–515
  deviation measures of 515
    modification of input phylogenies 517–518
    relaxed evolutionary models 516–517
  minimizing of
    modification of input data 520
    penalty, phylogenetic number and number
      of bad state 519
Pf1-phage solution 55
Phobius prediction techniques 157

Bioinformatics Index

Phosphorus-fitting of doublets from singlets (P-FIDS) 43
Phylogenetic inference 267, 332
    applications of phylogenies 528
        haplotypes, prediction of 529–531
        phylogenetic footprinting, for
            regulatory elements 528–529
    combining phylogenies, approaches to 523–525
        agreement subtrees 527
        strict consensus supertrees 525–527
    computational problems in 508–510
    computing distances, between
        phylogenies 520–523
        crossing number, visualization of 523
        subtree prune and regraft operation 521
        tree bisection and reconnection 522
    parametrization and 510
Phylogenetics model, evaluation
    modeling process
        single site in a sequence 336–337
        single site in two sequences 337–338
    phylogenetic assumptions 334–336
        conditions relating to phylogenetic assumption 335–336
        sites evolved independently,
            Markov process 334–335
        violation of 334
Phylogenetic trees 288
    common forms of bifurcating tree 288
    effective and accurate estimation of trees 290–291
    footprinting and 528
    and phylogenetic number 516
    relationships, analysis of 255
        sequence selection and alignment 256
    scoring trees 289–290
    sequences for 145
Phylogeny reconstruction method 256
Phyme motif discovery algorithms 242
Pipeline interface program (PIP) 373
Plant, fungal, and algal sequences (PLN),
    GenBank division 16
Plastid DNA 489
plotMA() function 101
plotPrintTipLoess() function 102
Point estimation of trees
    approaches of 295–296
    initial tree proposal 291–292
        star-decomposition and stepwise addition 292
    refining of 292–294
    resampling from tree space 295
    stopping criteria 294–295
Point mutations 333, 334
poli-Ala search probe 85
Pollock, D. D., ... 269

Polyploidization 272
Population recombination rate 504
Population size 272
Posada, D., ... 355
Position-specific count matrix (PSCM) 235
Position weight matrices (PWMs) 232
Potential ancestral protein interaction networks 418
Power law distributions 276
PRALINE
    MSA toolkit for protein sequences 148
    standard web interface 149
PREDICTED RefSeq records 6
Pre-profile processing, optimization technique 148
Primate sequences (PRI), GenBank division 16
PRINTS motif databases 232
Print-tip lowess normalization 91
ProbCons progressive alignment algorithm for protein
    sequences 154
PROFsec prediction methods 150
Programs for filtering "noise" in DNA and protein
    sequences 240
Progressive alignment protocol 145
PROSA II potential 269
PROSITE motif databases 232
Protégé biomedical ontologies 112
Protégé Ontologies Library 113
Protein Data Bank 68
Protein interaction data, comparative
    genomics techniques
    BioLayout Java, visualization of interaction data
        input data 423
        protein interaction data, loading
            and visualization 423–424
        yeast protein interaction network, inferring
            potential ancestral state of 424–425
    building a PostgreSQL database 420
    input data
        sequence data 420
        yeast protein interaction data 419
    parsing BLAST outputs and sequence
        similarity searches 420
    querying database and building
        phylogenetic profiles 420–422
    sequence conservation, analysis of 422–423
Protein polyubiquitination (GO:0000209) 120
Proteins 311
    analogous and homologous, distinguishment 270
    coding genes 164
    coding sequences 256
    correcting for multiple tests 323–325
        Hochberg's FWER method 325
        Holm's method and Bonferroni
            procedure 323–324
    crystallization 63

crystal preparation and data collection................ 64
    data measurement and processing................ 66–67
    X-ray scattering/diffraction 65–66
domains...276
    ligand interactions 311
    location of diversifying selection..................... 319–320
    McDonald-Kreitman test.. 312
    mutations, in *Drosophila* subgroup.......................... 313
    preparation of materials................................. 320–322
        estimating confidence in regions
            of alignment............................. 321
        profile HMM 321
        variation in posterior probability,
            correct alignment 321–322
    probabilistic models of codon
        evolution 314–319
    protein interactions.. 311, 417
    solvent interactions... 269
    structure determination
        crystal cryoprotection.. 73
        crystallization for 72–73
        data measurement for 74
        data processing.. 74–75, 77
        materials for... 72
        model building.. 71, 81–83
        molecular replacement.......................67–68, 77–78
        refinement... 69–71
        rotation function... 69
        structure refinement...................................... 79–80
    transmembrane (TM) regions of 157
    x-ray crystallography, refinement techniques in......... 70
ProtKB/Swiss-Prot Release.. 184
PROVISIONAL RefSeq records..................................... 6
Pseudogenes... 256, 272
PSI-BLAST homology search engine........................... 150
PSIPRED prediction methods...................................... 150
Putative isoforms filtration 203–204

Q

QUANTA programs 72
Quantitative comparative analysis, of frequency
        distributions of proteins 276
QuickScore motif discovery algorithms........................ 242

R

Ragan, M.,... 525
Randomly amplified polymorphic DNA (RAPD)......... 493
Rate matrix Q.. 262
RCurl and XML packages 109
RData normalized data... 99
read.table() function 109
Reciprocal best hits (RBH) 366, 372
Recombination, homologous chromosomes 334

REFERENCE field in GenBank flat file........................ 14
Reference Sequence (RefSeq) database 5
REFMAC5 (CCP4i suite) program.......................... 79–80
Refmac5 programs.. 72
RefSeq, transcript sequences... 184
Reganor gene finder 174
RegulonDB motif databases... 232
Repbase packages 187
RepeatMasker filtering programs 248
Residual dipolar couplings (RDC) data.......................... 40
Resource Description Framework 118
Restriction fragment length polymorphisms (RFLP) ... 493
Restriction Landmark Genomic Scanning (RLGS)...... 127
Reticulation... 500, 503
".retired.lst.gz" for retired sequences information
        of previous version 186
Reversible Markov process 335, 339, 344, 352, 353,
        355–357
REVIEWED RefSeq records 6
RGList object.. 100
Ribose nuclear Overhauser effect (NOE) patterns........... 38
Ribosomal RNA genes... 271
R methods.. 93
RNA
    degradation plot by function plotAffyRNAdeg 97
    ligand recognition 29
    polymerase III-derived transcript 282
    purification by anion-exchange
        chromatography 37–38
    purification by denaturing polyacrylamide
        electrophoresis 36–37
    samples desaltation by CentriPrep concentrator........ 37
T7 RNA polymerase 30
RNA structure determination by NMR
        techniques 29–30
    materials for
        anion-exchange chromatography........................ 34
        polyacrylamide gel electrophoresis...................... 33
        in vitro transcription 32–33
    methods for
        NMR resonance assignment and restraint
            collection............................. 38–47
        RNA sample preparation and purification...... 34–38
        structure calculation.. 47–51
    TAR RNA hairpin loop with Tat 32
Robertson, D.,... 472
Robertson, D. L.,... 331
Robust local regression based
        normalization.. 91
Robust Multi-array Average (RMA).............................. 91
Rodent sequences (ROD), GenBank division 16
Rogers, A. R.,.. 496
Rotation function 69

RSA Tools.. 241
Rzhetsky, A.,.. 267, 271, 279

S

Saccharomyces cerevisiae genome, BLAST study............. 418
SAKURA, Web-based submission tools 11
SAP default pairwise structural alignment
 methods 153
Scaffold records 9
Scalepack for macromolecular crystallography 72
Scalepack, HKL2000 suite component 77
SCPD motif databases 232
SCPE simulation.. 269
SeaView, graphical editor 159
Secretory granule (GO:0030141) 121
SEG filtering programs 248
Seipel, A., ... 261
Self-organizing maps (SOMs).................................... 165
separate .gal file 101
".seq.all.gz" for all sequences.................................... 186
Sequence
 alignment of,..12, 146, 181,
  186, 189, 268, 319, 460, 475, 477,
  498, 528
 and annotation data 17
 based clustering algorithms 292
 motifs and biological features 231–232
 records, INSD types
  direct submissions in............................. 6–7
  EST/STS/GSS...................................... 7
  high throughput genomic (HTG) sequence 8
  mammalian gene collection 7
  microbial genome 7–8
  third-party annotation 9–11
  whole genome shotgun (WGS)
   sequencing 8–9
 subfamilies of.. 148
Sequence data, computational tools, for. *See also*
 Softwares, for bioinformaticians
 GRIMM
  data for input to 435–436
  Hannenhalli-Pevzner methodology.............. 432
  identifying rearrangements in two genomes 445
  input format and output 446
  sample run 448
  usage and options........................... 446–448
   (*see also* Unsigned genomes)
 MGR
  data for input to 435–436
  graphical visualization modern genome
   and ancestral permutation..................... 452
  input format................................ 446, 449
  mgr_macro.txt file 442–443
  Newick Format and ASCII representation........ 449

 reconstructs rearrangement
  multiple genomes............431–433, 448–451
  sample run 450–451
  usage and options.......................... 449–450
Sequence Revision History tool 18
Sequencestructure distance score, S_{dist}............................ 269
Sequence Tagged Sites (STSs) 7
Sequin, annotation and analysis of nucleotide
 sequences 12
".seq.uniq.gz" for representative
 sequences for each cluster 186
SeSiMCMC motif discovery algorithms....................... 242
Shamir, R., ... 445
Shimodaira, H., ...298, 504
Short and long interspersed elements
 (SINEs and LINEs) 256
Siggenes package 92
Signal sensors methods for genomic sequence....... 170–171
Significance Analysis of Microarray
 data (SAM).. 92
Simes, R. J., ... 325
Similarity (distance) matrix 145
Simulated annealing (SA)................................... 295, 296
Sing, C. F., ... 501
Single gene, duplication of 272
Single nucleotide polymorphisms (SNPs) 489, 492, 529
Single-stranded conformation polymorphism
 (SSCP)... 493
Singlet-double-triplet (SDT) 262
Site-wise Likelihood Ratio (SLR).................................. 320
Sitting drop method for protein solution......................... 73
skip argument.. 109.
  See also Sequence
Skyline plots, for changes in demographic
 parameters...................................... 496–497
Slatkin, M.,... 496
Sliding window analysis.................................... 207–209
Smith, V. S.,..509, 512
SNAP programs 172
Snir, S., ...516, 518
Softwares, for bioinformaticians. *See also* BLAST
 BAMBE and MrBayes 301
 CLANN... 463
 estimating parameters in Bayesian
  framework 497
 GRIMM-Synteny, GRIMM, and MGR 433
 Mauve genome alignment 404, 406
 MUSCLE.. 150
 phylogenetic packages,........................288, 291,
  294, 295, 306
 ProbCons.. 154
 in public genome annotation databases 136
 publicly available prokaryotic gene-finding 166
 R statistical 325

SeaView, editing tools 159
SPEM.. 155
SOURCE field in GenBank flat file 14
Species- or taxa-specific interactions 418
SPEM-protocol.. 155
Splice Graph for gene structure and
    alternative splicing 201
"splice_obs" table for observation table......................... 190
Splice site,..170, 172, 181, 187,
    189, 193, 201
"splice" table for interpretation table............................. 190
STACKdb databases... 184
Stationary Markov process 258, 335, 339, 347,
    352, 357, 359
Steel, M., ...515, 522, 525
Stoye, J., ... 145
STRAP editor program.. 159
Structurally constrained protein evolution (SCPE)........ 269
Structural replacement matrices, comparison of........... 265
Stuart, A., ... 341
Substitution model, phylogenetic studies 332, 338
    assumption of independent and identical
        processes testing............................ 348–352
            likelihood-ratio test 348
            permutation tests 349
            protein-coding genes, models for relationship
                among sites 349–351
            RNA-coding genes, models
                for relationship among sites 351–352
    bias... 338
    model selection, general approaches to 355–358
        parametric bootstrap and non-nested
            models.. 356–358
    signal.. 339
        compositional and covarion signal 339
        historical signal 338
        phylogenetic signal and non-historical
            signals 339
        stationary, reversible, and homogeneous condition
            testing 339–340
            Matched-Pairs tests 340–345
    time-reversible substitution model 352–355
        hierarchical likelihood-ratio test,
            problems encountered............................. 354
        time-reversible Markov models 353
    visual assessment of compositional
        heterogeneity 345–348
        tetrahedral plot for..................................... 346–347
Subtree pruning and regrafting (SPR)............293, 294, 464
SwissProt identifiers 419
Swofford-Olsen-Waddell-Hillis (SOWH) 299
Synteny........................14, 366, 369, 404, 406, 431, 437, 439,
    443. *See also* GRIMM-Synteny
Synthetic sequences (SYN), GenBank division 16

T

tab-delimited five-column table files 12
Tagle, D. A., ... 528
TAMO motif discovery algorithms 241–242
Tang, H., ... 274
"T-A-T-A-A-T" DNA regular expression............. 233–236
Tavaré, S., ... 334
Taylor, W., ... 265
Tbl2asn command-line program 12
T-Coffee..152–157, 459, 460
Telford, M. J., ... 358
Templeton, A. R., ... 501
Tesler, G., ... 448
N,N,N',N'-Tetramethyl-ethylenediamine (TEMED) 52
Thermosynechococcus elongatus, electronic
    density map of Dps from 81
Third-party annotation 9–11
Third Party Annotation (TPA), GenBank division......... 16
Thorne, J. L., ... 268
Thornton, J., ... 265
Three-group EX run. changept 217
Tibshirani, R. J., ... 319
TICO, programs for start site identification................. 175
TIGR
    database... 128, 130
    Gene Indices... 184
TMHMM prediction techniques................................. 157
Tni3, pufferfish chromosome 376
Tompa, M., ... 529
Torsion angle dynamics (TAD) 50
TPA dataset, mRNA sequence 10–11
Transcription factor binding sites (TFBSs) 127, 231
TRANSFAC motif databases.. 232
Translation function 69
Trans-membrane helix... 267
Transposons... 288
Transverse Relaxation-Optimized Spectroscopy
    (TROSY) versions of HCN experiments 46
Tree bisection and reconnection (TBR) 293, 294
T-test and ANOVA... 92
Ttranslation start site... 165
Tukey Biweight algorithm 91
Twinscan, transcript sequences 184
Two-color microarray experiments 91

U

UCSC Genome Browser..125, 128, 129,
    131, 132, 135, 136, 187, 192, 211, 222, 239, 246
Unannotated sequences (UNA), GenBank division 16
UniGene databases... 184
UniVec database 11, 20
Unpolymerized acrylamide/bisacrylamide, neurotoxin 52
Unsigned genomes 447–448, 453

UPGMA, tree calculation 159
use-Dataset() function 107
useMart() function 106

V

Vagility, migration patterns...................................... 502, 503
VALIDATED RefSeq records 6
Variance Stabilizing Normalization (VSN) method........ 91
VARSPLIC feature annotation 184
VecScreen BLAST analysis, graphical
    display of... 20
VERSION field in GenBank flat file............................. 13
Viral sequences (VRL), GenBank division...................... 16
Vi- RLGS computational method for
    mapping spot in genome........................ 127
Volcano plot... 105

W

Walsh, J. B., ... 272
Warnow, T., ...514, 515
WATERGATE-2D NOESY spectra
    of TAR RNA 41
Watson-Crick A:U, G:C pairs.. 271
W3C's Web Ontology Language 112
Web-based submission tools...................................... 11–12
Webin, EMBL Web-based submission tools................... 11
Web Ontology Language (OWL)................................. 117
Web resources
    full-length isoforms for.. 204
    multiple alignment programs.................................... 149
Weeder algorithms 241
Weighted sum of pairs (WSP) score 153–154
WGS project accession number 9

Whelan, S.,..262, 306, 348
which *is_a* relationship 116
Whole genome shotgun (WGS) sequencing 8–9
Whole human sorcin
    diffraction oscillation image with program
        Xdisp... 75
    output of Denzo autoindexing routine 76
Wilson, K. S., ... 77
World Wide Web Consortium (W3C)........................ 117
Wormbase, data resource for Caenorhabditis biology
    and genomics 93
Wright, S., ... 494
write.table() function.. 109
WUBlast packages... 187
WU-BLAST program 374

X

XDisplay for macromolecularcrystallography 72
Xfit programs... 72
XNU filtering programs 248
Xplor programs... 72

Y

Yang, Z., ...261, 262, 263
YASPIN prediction methods... 150
Yeast protein interaction network.......................... 418, 423
    ancestral state inference of............................... 424–426

Z

Zakharov, E. V.,.. 346
Zdobnov, E.,.. 443
Zhang, J.,... 274
Zharkikh, A.,.. 261

INDEX

A

ACCESSION field in GenBank flat file 13
affy bioconductor packages 96
affy_hg_u95av2 as filter 107–108
Affymetrix GeneChip™ chips 90
Akaike Information Criterion (AIC) 354, 355, 499
AlignACE algorithms 241
Alignment editors.. 158–159
Alignment of sequence 12, 146, 181, 186, 189,
268, 319, 460, 475, 477, 498, 528
Alternative splicing... 179–180
    methods
        alignment filtering 189
        and available public data 181, 184–186
        candidates for... 189–190
        data mining.. 190–192
        pre-processing and.. 187–188
        transcript sequences alignment of....................... 188
    and splice graph 201
    transcript, types of... 180–181
AMBER module... 50
Amino acid replacement models, comparison of........... 264
Among-site rate variation (ASRV) 462
AMoRe programs... 72
Amplified Fragment Length Polymorphisms
        (AFLPs)... 493
Analysis of molecular variance (AMOVA) 495
Ancestral interaction networks 418
Anchor optimization 151–152
Ancient outgroup 372
annaffy and *annBuilder* annotation package..................... 93
ANN-Spec motif discovery algorithms.......................... 242
Annotation data for sequence 17
ANOVA normalization methods 92
Aquifex aeolicus 458, 464
ArrayExpress microarray datasets..................................... 95
ASAP database... 191
ATP citrate synthase activity (GO:0003878) 117

B

Bacteriophage (PHG) and bacterial (BCT) sequences,
        GenBank division 16
BAliBASE MSA benchmark database 147
BankIt, GenBank's web-based submission tool 11
Bayes factors (BF) 354, 355
Bayesian Information Criterion (BIC).................... 354, 355
Bayesian phylogenetics 301, 499
Bayesian schema design... 190
Bayesian tree inference 300–302
    estimation of.. 301–302
    MCMC algorithm and
        chain convergence and
            mixing.. 303–304
        sampling 302–303
    priors specification of.. 304–305
Bioconductor
    packages... 93
    software project for analysis and comprehension
        of genomic data... 95
Bio-dictionary gene finder (BDGF)............................... 170
BioMart
    generic data management system 93
    package 105
biomaRt annotation package... 93
BioProspector algorithms 241
BLAST
    all-*vs*-all BLASTP 460
    BLASTP threshold for cluster existence 461
    conversion output into GFF 135
    databases, for genome 374
    general platform and computing/programming
        tools 419
    graphical display VecScreen analysis 21
    -m9 option for a tab-delimited
        format 133–134
    output and GFF format for human
        genome 134
    protein BLAST bit score 422
    PSI-BLAST homology search engine..................... 150
    for *Saccharomyces cerevisiae*
        genome 418, 420
    searches.. 17
    tool 137
    vs. SIM4 188
    WU-BLAST program... 374
BlockMaker algorithms.. 241
BLOCKS motif databases... 232
BLOSUM
    amino acid exchange matrices................................. 144
    matrix 154–155
    series of matrices... 265, 266

Bootstrap
- procedure 172
- techniques, for phylogeny 320
- values 298

Bowker matched-pairs test of symmetry 341
Boxplot comparing intensities for arrays
after normalization 98

C

CCP4i package programs 72
CC-TOCSY transfer 44
Change point analysis 209–210
Changept 209, 210, 212, 213–215, 217, 219, 222, 226
Character-state matrix, for phylogeny .... 509, 513. *See also* Perfect phylogenies
ChEBI, GO biological process ontology 119
Chemokine receptor genes 256
Chromosomes, ancestral and gene orders reconstruction
- hypotheses, DreLG3 and DreLG12
  - distribution 379
  - pufferfish fusion hypothesis 379–380
  - zebrafish fission hypothesis 379–380
- for linkage Group 3 (DreLG3) and Group 12 (DreLG12) 377
- Tni2 and Tni3 relation 377–379
- Tni2, regions of conserved gene order 381
- in Zebrafish and pufferfish 379

CINEMA, Java interactive tool 159
Circular and linear chromosomes, DCJ on
- adjacency graph, applications of 399
  - circular intermediates, restriction of 403
  - distance formula 402–403
  - DCJ distances, genome of *Shigella* and *E. coli* 402
  - 0 labels 400–401
  - paths and cycles, odd and even paths 401–402
  - procedures 403
  - telomeres 399–400
- edge graph, applications of
  - on capped lines 395–396
  - caps and null chromosomes 395
  - closure and distance 397, 399
  - menu of operations 396–397
  - paths and cycles 397

Clone-based sequencing technology 8–9
ClustalW 152, 159, 187, 382, 460, 481
- multiple alignment program 159

CNS programs 72
CNS/Xplor syntax 48–49
3D-Coffee protocol 152–153
Cold Spring Harbor Laboratory (CSHL) 93
COMMENT field in GenBank flat file 14
Compare Prospector algorithms 241

CON division 9
Confidence in phylogenetic trees,
bootstrapping 297
Confidence intervals on point estimates 296–297
- limitations of 300
- non-parametric bootstrap 297–298
- parametric bootstrap 298–299
Consensus motif discovery algorithms 242
Constructed entries (CON), GenBank
division 16
Content sensors for genomic sequence 170–171
Contig sequences 9
Coot programs 72
CRAN mirror website 95
Critica programs 174–175
Cross_Match packages 187

D

Danio rerio 370, 377
- Linkage Group 3 (DreLG3) 377
Database of Interacting Proteins (DIP) 419
data.frame 95
Dayhoff matrices, in database search
methods 263–264
DBTBS motif databases 232
DDBJ's webbased submission tool SAKURA 11
Decision theory *(DT)* 354, 355
DEFINITION field in GenBank flat file 13
Degenerative mutations, in regulatory elements 273
Deleterious mutations 273
de novo gene-finding accuracy 173
de novo motif discovery 236–237
Denzo
- HKL2000 suite component 77
- macromolecular crystallography 72
develops_from relationship 116
Dialign 156, 157, 481
Directed acyclic graph (DAG) 201
Diversifying selection, location of 319–320
- likelihood ratio test and SLR approach 320
- posterior probability calculation 319
D. melanogaster and *D. simulans* in axt format 211–217
- attribution of function 226
- conservation levels for 220
- convergence assessing 217–218
- genomic regions, enrichment and
  depletion of 224–225
- number of groups and degree of pooling 219
- profile generation and 221–223
- segment generation 225–226
cDNA microarrays 92
Double cut and join (DCJ), circular chromosomes 386
- adjacency graph 390–393
- on circular and linear chromosomes 395–403
- on circular chromosomes 386–395

DCJ operation 387–389
    block interchange 389
    on genome 388
    localized view of 387
distance
    DCJ acting on cycle 393
    formula 393
    removal of 1-cycle 395
    sorting of edge graph 394
distance formula 393
edge graph 389–390
genome graphs 386–387
materials and online resources 403–406
    Mauve alignment, of *Shigella* and
        E. coli genome 406
    Mauve 1.3.0, for analysis 404–405
methods
    adjacency graph 409–410
    black-white genome graph 406–407
    distance with or without linear
        chromosomes 410–411
    distance without linear chromosomes 410
    edge graph construction 408, 409
    genome graph 407–408
    sorting with/without linear
        chromosomes 411–412
DpsTe family of proteins 86
DreLG3, zebrafish chromosome 376
DSSP, predicted secondary structure information 150
Duplication-degeneration-complementation
    model (DDC) 273
DUST filtering programs 248
Dynamic programming (DP) algorithm 144–145

E

ECgene databases 184
Efficient Evaluation of Edit Paths (EEEP) 463
Elutrap Electroelution System 37
EMBL 4, 8, 11–13, 17, 21, 23, 24, 130
    web-based submission tool, Webin 11
EMBOSS open source analysis package 187
Empirical models of nucleotide
    substitution 260
ENCODE Genome Annotation Assessment Project
    (EGASP) 173
ENCyclopedia Of DNA Elements (ENCODE)
    project 128, 173
Ensembl 93, 105–107, 129, 133, 137, 171,
    186, 377, 434, 443
    database 128, 130
    genome browser 134
    pipeline for eukaryotic gene prediction 171–172
    software project 93
Entrez
    gene database 5–6
    nucleotide sequences 22
    protein sequences 22
Entrez PubMed 14
Environmental sampling sequences (ENV),
    GenBank division 16
ERPIN program 193
EST databases 184
European Bioinformatics Institute (EBI) 93
Evolution models for genome elements
    amino acid mutation models 263–265
        amino acid models incorporating
            correlation 269–270
        amino acid models incorporating
            features 267–269
        generating mutation matrices 265–267
    Codon models 262–263
        for M7 and M8 hypotheses 263
        rate matrix Q 262
    DNA substitution models 259–261
        modeling rate heterogeneity along
            sequence 261
        models based on nearest neighbor
            dependence 261–262
    evolution of microsatellite and ALU
        repeats 280–282
    Markov models of DNA evolution 257–259
        instantaneous rate matrix of transition
            probabilities 258–259
        properties of 258
    models of DNA segment evolution 271–273
    protein domain evolution 276–279
        evolution of introns 279–280
    RNA model of evolution 271
    sub-functionalization model 273–274
    sub-neo-functionalization 274
    tests for functional divergence after
        duplication 275–276
    tests for sub-functionalization 274–275
Ewald sphere for x-ray diffraction pattern
    for protein 66, 83–84
Expectation-Maximization algorithm 202
Expectation maximization (EM) 262
Expressed sequence tags (ESTs) 7, 20, 25, 163,
    172, 181, 184, 185, 187, 188, 192, 200, 204
Expression Set (exprSet) normalization 98
expresso command 99
ExprSet normalized expression data 99

F

Family-Wise Error Rate (FWER) 323–324
FASTA format 12
Fast Fourier Transformation (FFT), homologous
    segments detection 153
FEATURES field in GenBank flat file 15
Felsenstein's model 260

FFT-NS-1 and FFT-NS-2 protocol 153
Fgenesh++, transcript sequences 184
Fixation indices, for genetic variation 494
2Fo-Fc and Fo-Fc map.. 71.
    See also Protein structure
FUGUE default pairwise structural alignment
        methods 153
Full-length open reading frame (FL-ORF)
        clones 7
Functional Annotation Of Mammalian
        (FANTOM) 128

G

gag proteins.. 474
Gaps in alignments...................................... 143, 268
gcrma bioconductor packages... 96
GenBank 13, 17, 20, 25, 130, 185, 404, 459
    annotator 18
    flat file.. 13–14
    records, division 16
    sequence record.. 13
    web-based submission tool, BankIt 11
genbiomultialign.exe program 213
genchangepts.exe program 213
GeneMark.hmm.. 165
Genemark programs.. 172
Gene Ontology (GO)... 111
    development and 113–118
    online resources 112–113
    principles of 118–121
General Feature Format (GFF)............... 130, 133–136, 168
General time-reversible (GTR)....................................... 462
Generic genome browser..................................... 131, 133, 136
Genes
    database 7
    duplication.. 272, 334
    finding
        in environmental sequence samples 169–170
        in eukaryotes.. 170–173
        in prokaryotes... 164–169
        software for prokaryotes................................. 166
    in GC-rich genomes.. 174
    orders, conservation of
        for DreLG3 and Tni3..................................... 376
        and inference of ancestral gene orders 368
    products, functional annotations of......................... 459
    recombination.. 471
        gag region of HIV-1, use of PhylPro................. 473
            dataset containing all sequences,
                creation of 474–475
            first phylogenetic profile,
                creation of 475
            import of sequences in VOP....................... 474
            PhylPro application window 480

profile interpretation............................... 475–477
profile parameters, refinement of................ 477
profiles, comparing of 478
program availability and data sources
        and format 473
recombinant sequences,
        determination................................ 478–479
removal of recombinant sequences......... 479, 481
sequence set, refinement of 477
structure and splice graph 201
transfer.. 334
Genome annotation... 125
    materials for
        databases 128–130
        nucleotides and nucleosomes 127–128
        transcripts 126–127
    methods
        public annotation 131–133
Genome duplication
    at base of radiation of teleost fish 368
    at beginning of teleost radiation 370
    DreLG3 and DreLG12.. 379
    event, in teleost lineage....................................... 370
    events in chordate evolution and R3 in
            ray-fin lineage 367
    production of DreLG3 and DreLG12 379
    R1 and R2, early vertebrate 367, 369
GenomeScan, transcript sequences................................ 184
Genome survey sequences (GSS), GenBank
        division.. 7, 16
GenScan, transcript sequences 184
getLayout() function.. 101
GIGO rule 242–243
GISMO
    program for encoding of genes 168
    SVM-based gene finder.. 175
GLAM motif discovery algorithms................................. 242
Glimmer programs 174–175
Global alignment strategy 147
Glycoprotein (GP120)... 321
GOLD database... 426
GO protein domains 96, 108, 112, 114,
        119–122, 245, 302
GO slim documentation on ontology............................. 122
G-protein coupled receptor activity
        (GO:0004930) 121
G-protein coupled receptor (GPCR) 121
GRAMENE, data resource for comparative genome
        analysis in grasses....................................... 93
GRIMM
    data for input to 435–436
    Hannenhalli-Pevzner methodology......................... 432
    identifying rearrangements in two genomes 445
    input format and output 446

sample run 448
usage and options 446–448
(see also Unsigned genomes)
GRIMM-Synteny 404, 431, 433
identifying homologous blocks across multiple
genomes 437
forming synteny blocks from sequence-based
local alignments 439–441
GRIMM-anchors: filtering out alignments
with 438–439
preparing the input for 437–438
sample run 443–445
usage and options, for forming synteny
blocks 441–443
mgr_macro.txt for input to GRIMM
or MGR 435–436
vs. GRIMM 433
GS-Finder, programs for start site identification 175

H

Hampton Research crystallization and cryoprotection
kits 72
Hanging drop method for protein solution 73
HAPMART, data resource for results of HAPMAP
project 93
has_participant relationship 120
HCP-TOCSY (Total Correlation Spectroscopy),
RNA resonances 39
Heatmap plots 108
Heaviest Bundling, dynamic programming
algorithm 202–203
Helicobacter pylori 503
help.search() function 95
Heteronuclear Single Quantum Coherences
(HSQCs) 43
Heterozygosity calculation 494
hgnc_symbol as attribute 107
Hidden Markov models (HMMs) 144, 165, 276, 321
based EasyGene 175
based GeneWise program 171
transition probabilities 268
Hierarchical likelihood ratio tests (hLRT) 499
High throughput cDNA sequences (HTC), GenBank
division 16
High throughput genomic (HTG) sequence 8
HKL2000 suite component, Denzo and Scalepack 77
hmm ES-3.0 programs 172
HNNCOSY (Correlation Spectroscopy)
experiment 45
Homogeneous Markov process 337, 352
Homolog 437
Homoplasy 488, 500, 503
hoxaa cluster of zebrafish 368
HOX clusters 368, 369

HOX, or globin genes 271
HP-HETCOR (Heteronuclear Correlation) 39
Hs.2012(TCN1) gene structure by splice graph
representation 202
Human chromosome 17 (Hsa17) 369
Human immunodeficiency virus (HIV)-2
transactivation response element (TAR)
RNA, sequence and structure 31–32
Hypervariable V1/V2 loop regions
of protein 321

I

Ideal Mismatch value 91
Iisoform problem 199–200
Iisoform reconstruction and heaviest
bundling 202–203
import all .spot files 100
Importance quartet puzzling (IQP) 295
importing .spot data 100
Improbizer motif discovery algorithms 242
Indels mutations 299
Inferring patterns of migration 501
direct evidence 502
indirect evidence 502–503
Insensitive Nuclei Enhancement by Polarization
Transfer (INEPT) delays 45
International Nucleotide Sequence
Database (INSD) 3, 5, 6, 8, 11,
13, 17, 21, 23, 24, 130
accessing of
Entrez retrieval system 22
ftp sites 21
archival primary sequence database
annotation and propagation 19–20
vector contamination 20
wrong strand annotation 20–21
sequence data submission in
Web-based submission tools 11–12
sequence processing
batch submissions processing of 18
submissions 16–18
updates and maintenance of 18–19
International Table of Crystallography 64
INTERPRO protein domains 108
Invertebrate sequences (INV), GenBank
division 16
is_a complete relationship 120
is_a relationship 116

J

jalview protein multiple sequence alignment
editor 159
JASPAR 234
motif databases 232

Jim Kent web site 186
J scalar coupling constants and RNA structures 40
Jukes-Cantor model 260

K

Karplus parameterization 44
KEYWORDS field in GenBank
    flat file 13
Kishino-Hasegawa (KH) test 298

L

LAMA program... 244
Lateral genetic transfer (LGT)...................................... 457
    data sources... 458–459
    inference, by topological comparisons
        of trees 463–464
        EEEP method and edit operations.................... 463
        unrooted protein tree, reconciling..................... 464
    limitations and pitfalls, of surrogate methods.......... 467
    phylogenetic inference and supertree
        construction 462–463
        CLANN and MRP, use of................................ 463
        Nexus format creation and burn-in phase
        identification.. 462
    program availability 459
    putative homologs and orthologs, clustering
        of proteins
        All-*versus*-all BLASTP 460
        hybrid protein clustering strategy 461
    sequence alignment, validation,
        and trimming 460–461
        alignment algorithms, in each MRC 460
        TEIRESIAS pattern-detection
        algorithm 460–461
        WOOF and GBLOCKS, use of........................ 461
Likelihood ratio test (LRT)................................... 275, 320
limma package 92
Lineage through time (LTT)... 496
Linkage disequilibrium... 270, 503
listAttributes function... 107
Listeria monocytogenes MolRep program after
    rotation and translation searches.............. 79
listFilters function 107
list-Marts() function 105
Local alignment strategy 147
Locally collinear blocks (LCBs) 404, 405
LOCUS field in GenBank flat file 13
Log likelihood ratio (LLR) of PWM............................ 237

M

maBoxplot() functions 109
MAFFT, multiple sequence alignment
    package 153–154

MAList object 102
maNorm() and *maPlot*() functions................................. 109
MapViewer... 7
Markov chain Monte Carlo
    (MCMC)........................210, 217, 302–304
Markov process........257, 258, 268, 269, 333, 334, 336, 337
marray package 100
MAS5.0 normalization procedure 91
Matched-pairs tests, of symmetry...................340, 343, 345
    to analyze complex alignments 345
    with more than two sequences......................... 342–345
    with two sequences 340–342
matrix() function 94
Matrix representation with parsimony
    (MRP) 463, 525
Mauve..402, 404–406
Maximum a posteriori probability (MAP) 301
Maximum likelihood (ML) 289, 499
Maximum parsimony, of DNA sequences 499
McDonald-Krietman test, selection
    on proteins 312–320
    adh locus, encoding alcohol dehydrogenase 312–313
    Fisher's exact test's *p* values, application of 314
limitation 314
    for polarized variant... 313
MDscan algorithms... 241
Membrane-bound organelle (GO:0043227) 122
Methylation Sensitive Representational Difference
    Analysis (MS-RDA),
    sequencing 127
Metropolis-Hastings Monte Carlo simulation.............. 495
MGED Ontology Resources.. 113
MGR
    data for input to.. 435–436
    graphical visualization modern genome
        and ancestral permutation....................... 452
    input format... 446, 449
    mgr_macro.txt file 442–443
    Newick Format and ASCII representation............. 449
    reconstructs rearrangement multiple
        genomes 431–433, 448–451
    sample run 450–451
    usage and options 449–450
Microarray data
    annotation of 93
    data filtering 92
    differentially expressed genes detection of................ 92
    methods
        Affymetrix data normalization of 96–99
        cDNA microarray data 100–104
        differential expression detection
        by limma 104–105
        R essentials 93–95
        R packages and bioconductor 95

noisy signal 89
pre-processing
    Affymetrix data of 90–91
    two-color data of 91–92
    quality assessment 90
Microarrays (ChIP/chip) and sequencing paired-end
    di-tags (ChIP/PETs) 127
Microsatellites
    codominant markers 492
    length, distribution in human genome 281
Migration history and patterns, determination of
    methods
        Bayesian-based population genetics 497
        classical population genetics 494–495
        coalescent-based population genetics 495–497
        genetic variation in European humans,
            components of 498
        gene trees in inferring migrations,
            use of 500
        inference structure 501–503
        multivariate statistical analysis 498
        NCPA, distribution of related
            haplotypes 501
        network construction 499–501
        phylogenetic methods 498–499
        structure output demonstration 497
    organism with wind-borne spore dispersal 486
    RFLP, RAPD and SSCP data 493
    sampling scheme designing 486–488
    types of data 488, 489, 492, 493
Migration rates theta estimation 495
Miller indices 83
MITRA motif discovery algorithms 242
Model Based Expression Index (MBEI) 91
Molecular phylogenetics 331
MolRep programs 72
MolRep solutions 85
Monte Carlo simulations 344
Monte Carlo, studies on matrices 266
Motifs and biological features 231–232
Motif databases with web servers 244
Motif discovery algorithms with web servers 241
Motif discovery and limitations 246–247
MotifSampler motif discovery algorithms 242
Multiple alignment formats (MAF) files 186
Multiple gene trees 504
Multiple sequence alignment (MSA) 12, 143, 159, 188,
    208, 472, 475
    dynamic programming 144–145
    iteration of 146
    materials for
        global and local 147
        sequences selection of 146–147
    methods
        MAFFT 153–154

MUSCLE 150–152
PRALINE 148–150
ProbCons 154–155
SPEM 155
T-Coffee program 152–153
protocol for 145–146
reliability and evolutionary hypothesis 144
types of 147–148
MUSCLE, multiple alignment software 149–151,
    150–151
Mutations
    advantages 272
rate of 503
MySQL database 186

N

National Center for Biotechnology Information
    (NCBI) 7, 22, 24, 185, 377,
    405, 426, 434, 459
NCBI Ref Seq project 20
Nearest neighbor interchange (NNI) 293, 294
Neighbor-Joining technique 152
Nested Clade Phylogeographical Analysis (NCPA) 501
Network visualization 418, 423, 426
NJ, tree calculation 159
Noisy signal 89
Non-canonical G:U pair 271
Non-coding DNA 164–165
Non-membrane-bound organelle (GO:0043228) 123
Non-parametric bootstrap 296–298
normalizeBetweenArrays() function 104
normalizeWithinArrays() function 102
Nuclear DNA 489
Nuclear Overhauser effect spectroscopy (NOESY),
    RNA resonances 38
Nucleotide sequence
    alignments of 334, 339, 342
    equilibrium frequency of 259
    and haplotype diversity 494
    international collaboration 4–5
    NCBI RefSeq project 5–6
Nucleotide substitution, probabilistic model 314–319
    assumptions violated 315–316
    gap occurrence and strength of selection,
        relationship 319
    likelihood ratio test statistic, Δ 317
    log-likelihood of parameters 316
    missing data 316
    non-parametric bootstrap, gor diversifying
        selection 319
    random-site methods of detecting selection 317
    synonymous and non-synonymous mutations
        ratio ω 315
    use of parametric bootstrap techniques 318
Nucleotide triphosphates (NTPs) 29–30

O

OBO-Edit biomedical ontologies 112
OBO_REL relationship 120
Olfactory receptor genes 256
Omit map ....82. *See also* Protein structure
Open Biomedical Ontologies (OBO)
    collection and principles of 113
    foundry for 115
Open reading frame (ORF) in genome sequence 164
ORGANISM field in GenBank flat file 14
Orphan sequences 147
Ortholog identification pipeline (ORTH) 374
Orthologs 370, 374, 376, 436, 460
    and paralogs, evolutionary relationships 371
O program 72

P

Paired-mismatch (MM) probes 90
pairs command 98
PAM, amino acid exchange matrices 144
PAM matrices 265–267
Paralog 274, 275, 374–376
Paralog heterogeneity test 274
Paralog Identification pipeline (PARA) 374
Parametric bootstrap 296, 319, 333, 348, 349, 352, 357, 359
part_of relationship 116
Patent sequences (PAT), GenBank division 16
Patterson functions 68
Perfect match (PM) probes 90
Perfect phylogenies
    construction of
        formalization of problems 511–513
        parameter 514–515
    measures of deviation 515
        modification of input phylogenies 517–518
        relaxed evolutionary models 516–517
    minimizing of
        modification of input data 520
        penalty, phylogenetic number and number of bad state 519
Pf1-phage solution 55
Phobius prediction techniques 157
Phosphorus-fitting of doublets from singlets (P-FIDS) 43
Phylogenetic footprinting 528
Phylogenetic inference 267, 332
    applications of phylogenies 528
        haplotypes, prediction of 529–531
        phylogenetic footprinting, for regulatory elements 528–529
    combining phylogenies, approaches to 523–525
        agreement subtrees 527
        strict consensus supertrees 525–527
    computational problems in 508–510
    computing distances, between phylogenies 520–523
        crossing number, visualization of 523
        subtree prune and regraft operation 521
        tree bisection and reconnection 522
    parametrization and 510
Phylogenetic number 516
Phylogenetic relationships, analysis of 255
    sequence selection and alignment 256
Phylogenetics model, evaluation
    modeling process
        single site in a sequence 336–337
        single site in two sequences 337–338
        phylogenetic assumptions 334–336
            conditions relating to phylogenetic assumption 335–336
            sites evolved independently, Markov process 334–335
            violation of 334
Phylogenetic tree for sequences 145
Phylogenetic trees 288
    common forms of bifurcating tree 288
    effective and accurate estimation of trees 290–291
    scoring trees 289–290
Phylogeny reconstruction method 256
Phyme motif discovery algorithms 242
Pipeline interface program (PIP) 373
Plant, fungal, and algal sequences (PLN), GenBank division 16
Plastid DNA 489
plotMA() function 101
plotPrintTipLoess() function 102
Point estimation of trees
    approaches of 295–296
    initial tree proposal 291–292
        star-decomposition and stepwise addition 292
    refining of 292–294
    resampling from tree space 295
    stopping criteria 294–295
Point mutations 333, 334
poli-Ala search probe 85
Polyploidization 272
Population recombination rate 504
Population size 272
Position-specific count matrix (PSCM) 235
Position-specific scoring matrix (PSSM) 234, 236
Position weight matrices (PWMs) 232
Power law distributions 276
PRALINE
    MSA toolkit for protein sequences 148
    standard web interface 149
PREDICTED RefSeq records 6
Pre-profile processing, optimization technique 148

Primate sequences (PRI), GenBank division 16
PRINTS motif databases 232
Print-tip lowess normalization 91
ProbCons progressive alignment algorithm for
                protein sequences 154
PROFsec prediction methods 150
Programs for filtering "noise" in DNA and protein
                sequences 240
Progressive alignment protocol...................... 145
PROSA II potential 269
PROSITE motif databases............................... 232
Protégé
    biomedical ontologies 112
    OWL editor 112
Protégé Ontologies Library............................. 113
Protein crystallization................................. 63–64
    crystal preparation and data collection......... 64
        data measurement and processing.......... 66–67
        X-ray scattering/diffraction 65–66
Protein Data Bank.. 68
Protein interaction data, comparative genomics techniques
    BioLayout Java, visualization of interaction data
        input data 423
            protein interaction data, loading and
                visualization 423–424
            yeast protein interaction network, inferring
                potential ancestral state of.............. 424–425
    building a PostgreSQL database............................... 420
    Input data
        sequence data.. 420
        yeast protein interaction data............................ 419
    parsing BLAST outputs and sequence similarity
        searches 420
    querying database and building phylogenetic
        profiles 420–422
    sequence conservation, analysis of 422–423
Protein polyubiquitination (GO:0000209).................... 120
Proteins
    analogous and homologous, distinguishment 270
    coding genes 164
    coding sequences 256
    domains 276
        ligand interactions 311
        protein interactions............................ 311, 417
        solvent interactions 269
        x-ray crystallography, refinement techniques in 70
Proteins, action of selection on 311
    correcting for multiple tests 323–325
        Hochberg's FWER method............................... 325
        Holm's method and Bonferroni
            procedure 323–324
    location of diversifying selection...................... 319–320
    McDonald-Kreitman test................................ 312
    mutations, in *Drosophila* subgroup........................ 313

preparation of materials 320–322
    estimating confidence in regions
        of alignment 321
    profile HMM 321
    variation in posterior probability, correct
        alignment 321–322
    probabilistic models of codon evolution 314–319
Protein structure
    determination
        molecular replacement 67–68
        rotation function 69
    materials for determination of 72
    methods
        crystal cryoprotection............................ 73
        crystallization for 72–73
        data measurement for 74
        data processing............................ 74–75, 77
        model building 81–83
        molecular replacement 77–78
        structure refinement............................ 79–80
    model building... 71
    refinement 69–71
ProtKB/Swiss-Prot Release 184
PROVISIONAL RefSeq records............................ 6
Pseudogenes 256, 272
PSI-BLAST homology search engine 150
PSIPRED prediction methods......................... 150
Pufferfish fusion hypothesis 379–380
Putative isoforms filteration 203–204

Q

QUANTA programs 72
Quantitative comparative analysis, of frequency
        distributions of proteins 276
QuickScore motif discovery algorithms.................... 242

R

Randomly amplified polymorphic DNA
        (RAPD) 493
Rate matrix **Q** 262
RCurl and XML packages 109
RData normalized data..................................... 99
read.table() function 109
Reciprocal best hits (RBH) 366, 372
Recombination, homologous chromosomes 334
Records, INSD types
    direct submissions in................................. 6–7
    EST/STS/GSS 7
    high throughput genomic (HTG) sequence 8
    mammalian gene collection 7
    microbial genome 7–8
    third-party annotation 9–11
    whole genome shotgun (WGS) sequencing 8–9
REFERENCE field in GenBank flat file...................... 14

Reference Sequence (RefSeq) database 5
REFMAC5 (CCP4i suite) program 79–80
Refmac5 programs 72
RefSeq, transcript sequences 184
Reganor gene finder 174
RegulonDB motif databases 232
Repbase packages 187
RepeatMasker filtering programs 248
Repeats, in human genome 256
Repetitious organization of human genome 257
Residual dipolar couplings (RDC) data 40
Resource Description Framework 118
Restriction fragment length polymorphisms
 (RFLP) 493
Restriction Landmark Genomic
 Scanning (RLGS) 127
Reticulation 500, 503
Reversible Markov process 335, 339, 344, 352, 353,
 355–357
REVIEWED RefSeq records 6
RGList object 100
Ribose nuclear Overhauser effect
 (NOE) patterns 38
Ribosomal RNA genes 271
R methods 93
RNA
 degradation plot by function plotAffyRNAdeg 97
 ligand recognition 29
 polymerase III-derived transcript 282
 purification by anion-exchange
  chromatography 37–38
 purification by denaturing polyacrylamide
  electrophoresis 36–37
 samples desaltation by CentriPrep concentrator 37
T7 RNA polymerase 30
RNA structure determination by
 NMR techniques 29–30
 materials for
  anion-exchange chromatography 34
  polyacrylamide gel electrophoresis 33
  in vitro transcription 32–33
 methods for
  NMR resonance assignment and restraint
   collection 38–47
  RNA sample preparation and
   purification 34–38
  structure calculation 47–51
  TAR RNA hairpin loop with Tat 32
Robust local regression based normalization 91
Robust Multi-array Average (RMA) 91
Rodent sequences (ROD), GenBank division 16
Rotation function 69
RSA Tools 241

S

Saccharomyces cerevisiae genome, BLAST study 418
SAKURA, Web-based submission tools 11
SAP default pairwise structural
 alignment methods 153
Scaffold records 9
Scalepack for macromolecular crystallography 72
Scalepack, HKL2000 suite component 77
Schwikowski, B 529
SCPD motif databases 232
SCPE simulation 269
SeaView, graphical editor 159
Secondary structure 268
secretory granule (GO:0030141) 121
SEG filtering programs 248
Selection 164, 226, 263, 271,
 313, 316, 368, 500, 504
Self-organizing maps (SOMs) 165
Sequence-based clustering algorithms 292
Sequence Revision History tool 18
Sequencestructure distance score, S_{dist} 269
Sequence Tagged Sites (STSs) 7
Sequin, annotation and analysis of nucleotide
 sequences 12
 tbl2asn 12
SeSiMCMC motif discovery algorithms 242
Short and long interspersed elements
 (SINEs and LINEs) 256
siggenes package 92
Signal sensors methods for
 genomic sequence 170–171
Significance Analysis of Microarray data (SAM) 92
Similarity (distance) matrix 145
Simulated annealing (SA) 295, 296
Single gene, duplication of 272
Single nucleotide polymorphisms (SNPs) 489, 492, 529
Single-stranded conformation polymorphism
 (SSCP) 493
Singlet-double-triplet (SDT) 262
Site-wise Likelihood Ratio (SLR) 320
Sitting drop method for protein solution 73
skip argument 109
Skyline plots, for changes in demographic
 parameters 496–497
Sliding window analysis 207–209
SNAP programs 172
Software, for bioinformaticians. *See also* BLAST
 BAMBE and MrBayes 301
 CLANN 463
 estimating parameters in Bayesian framework 497
 GRIMM-Synteny, GRIMM, and MGR 433
 Mauve genome alignment 404, 406

MUSCLE 150
   phylogenetic packages 288, 291, 294, 295, 306
   ProbCons 154
   in public genome annotation databases 136
   publicly available prokaryotic gene-finding 166
   R statistical 325
   SeaView, editing tools 159
   SPEM 155
SOURCE field in GenBank flat file 14
Species- or taxa-specific interactions 418
SPEM-protocol 155
Splice Graph for gene structure and alternative
   splicing 201
"splice_obs" table for observation table 190
Splice site 170, 172, 181, 187, 189, 193, 201
"splice" table for interpretation table 190
STACKdb databases 184
Stationary Markov process 258, 335, 339, 347, 352, 357, 359
STRAP editor program 159
Structurally constrained protein evolution
   (SCPE) 269
Structural replacement matrices, comparison of 265
Subfamilies of sequences 148
Substitution model, phylogenetic studies 332, 338
   assumption of independent and identical
      processes testing 348–352
      likelihood-ratio test 348
      permutation tests 349
      protein-coding genes, models for relationship
        among sites 349–351
      RNA-coding genes, models for relationship
        among sites 351–352
   bias 338
   model selection, general approaches to 355–358
      parametric bootstrap and non-nested
        models 356–358
   signal 339
      compositional and covarion signal 339
      historical signal 338
      phylogenetic signal and non-historical signals ... 339
   stationary, reversible, and homogeneous
      condition testing 339–340
      Matched-Pairs tests 340–345
   time-reversible substitution model 352–355
      hierarchical likelihood-ratio test, problems
        encountered 354
      time-reversible Markov models 353
   visual assessment of compositional
      heterogeneity 345–348
      tetrahedral plot for 346–347
Subtree pruning and regrafting (SPR) 293, 294, 464
SwissProt identifiers 419
Swofford-Olsen-Waddell-Hillis (SOWH) 299

Synteny .... 14, 366, 369, 404, 406, 431, 437, 439, 443. *See also* GRIMM-Synteny
Synthetic sequences (SYN), GenBank division 16

T

TAMO motif discovery algorithms 241–242
Tbl2asn command-line program 12
T-Coffee 152–157, 459, 460
Template sequences for *in vitro* transcription of HIV-2
   TAR RNA 35–36
N,N,N,N-Tetramethyl-ethylenediamine (TEMED) 52
Thermosynechococcus elongatus, electronic density
   map of Dps from 81
Third-party annotation 9–11
Third Party Annotation (TPA), GenBank division 16
TICO, programs for start site identification 175
TIGR
   database 128, 130
   Gene Indices 184
TMHMM prediction techniques 157
Tni3, pufferfish chromosome 376
Torsion angle dynamics (TAD) 50
TPA dataset, mRNA sequence 10–11
Transcription factor binding sites (TFBSs) 127, 231
TRANSFAC motif databases 232
Translation function 69
Trans-membrane helix 267
Transmembrane (TM) regions of proteins 157
Transposons 256, 288
Transverse Relaxation-Optimized Spectroscopy (TROSY)
   versions of HCN experiments 46
Tree bisection and reconnection (TBR) 293, 294
T-test and ANOVA 92
Ttranslation start site 165
Tukey Biweight algorithm 91
Twinscan, transcript sequences 184
Two-color microarray experiments 91

U

UCSC Genome Browser 125, 128, 129, 131, 132, 135, 136, 187, 192, 211, 222, 239, 246
Unannotated sequences (UNA), GenBank division 16
UniGene databases 184
UniVec database 11, 20
Unpolymerized acrylamide/bisacrylamide, neurotoxin 52
Unsigned genomes 447–448, 453
UPGMA, tree calculation 159
use-Dataset() function 107
useMart() function 106

V

Vagility 502, 503
VALIDATED RefSeq records 6

Variance Stabilizing Normalization (VSN) method........ 91
VARSPLIC feature annotation 184
VecScreen BLAST analysis, graphical display of............. 20
VERSION field in GenBank flat file................................ 13
Viral sequences (VRL), GenBank division....................... 16
Vi- RLGS computational method for mapping
    spot in genome... 127
volcano plot, ... 105

W

WATERGATE-2D NOESY spectra
    of TAR RNA 41
Watson-Crick A:U,G:C pairs... 271
W3C's Web Ontology Language 112
Web-based submission tools.................................... 11–12
Webin, EMBL Web-based submission tools................... 11
Web Ontology Language (OWL)................................. 117
Web resources for full-length isoforms.......................... 204
Web sites of multiple alignment programs 149
Weeder algorithms 241
Weighted sum of pairs (WSP) score 153–154
WGS project accession number 9
which *is_a* relationship 116
Whole genome shotgun (WGS) sequencing 8–9
Whole human sorcin
    diffraction oscillation image with program
        Xdisp... 75
    output of Denzo autoindexing routine 76
World Wide Web Consortium (W3C)......................... 117
Wormbase, data resource for Caenorhabditis biology
    and genomics 93
write.table() function.. 109
WUBlast packages 187
WU-BLAST program 374

X

XDisplay for macromolecular crystallography 72
Xfit programs 72
XNU filtering programs 248
Xplor programs 72

Y

YASPIN prediction methods 150
Yeast protein interaction network 418, 423
    ancestral state inference of 424–426

Z

Zebrafish fission hypothesis 379–380

MUSCLE 150
   phylogenetic packages 288, 291, 294, 295, 306
   ProbCons 154
   in public genome annotation databases 136
   publicly available prokaryotic gene-finding 166
   R statistical 325
   SeaView, editing tools 159
   SPEM 155
SOURCE field in GenBank flat file 14
Species- or taxa-specific interactions 418
SPEM-protocol 155
Splice Graph for gene structure and alternative
   splicing 201
"splice_obs" table for observation table 190
Splice site 170, 172, 181, 187, 189, 193, 201
"splice" table for interpretation table 190
STACKdb databases 184
Stationary Markov process 258, 335, 339, 347, 352,
   357, 359
STRAP editor program 159
Structurally constrained protein evolution
   (SCPE) 269
Structural replacement matrices, comparison of 265
Subfamilies of sequences 148
Substitution model, phylogenetic studies 332, 338
   assumption of independent and identical
      processes testing 348–352
      likelihood-ratio test 348
      permutation tests 349
      protein-coding genes, models for relationship
         among sites 349–351
      RNA-coding genes, models for relationship
         among sites 351–352
   bias 338
   model selection, general approaches to 355–358
      parametric bootstrap and non-nested
         models 356–358
   signal 339
      compositional and covarion signal 339
      historical signal 338
      phylogenetic signal and non-historical signals ... 339
   stationary, reversible, and homogeneous
      condition testing 339–340
      Matched-Pairs tests 340–345
   time-reversible substitution model 352–355
      hierarchical likelihood-ratio test, problems
         encountered 354
      time-reversible Markov models 353
   visual assessment of compositional
      heterogeneity 345–348
      tetrahedral plot for 346–347
Subtree pruning and regrafting (SPR) 293, 294, 464
SwissProt identifiers 419
Swofford-Olsen-Waddell-Hillis (SOWH) 299

Synteny 14, 366, 369, 404, 406, 431,
   437, 439, 443. See also GRIMM-Synteny
Synthetic sequences (SYN), GenBank division 16

T

TAMO motif discovery algorithms 241–242
Tbl2asn command-line program 12
T-Coffee 152–157, 459, 460
Template sequences for *in vitro* transcription of HIV-2
   TAR RNA 35–36
N,N,N,N-Tetramethyl-ethylenediamine (TEMED) 52
Thermosynechococcus elongatus, electronic density
   map of Dps from 81
Third-party annotation 9–11
Third Party Annotation (TPA), GenBank division 16
TICO, programs for start site identification 175
TIGR
   database 128, 130
   Gene Indices 184
TMHMM prediction techniques 157
Tni3, pufferfish chromosome 376
Torsion angle dynamics (TAD) 50
TPA dataset, mRNA sequence 10–11
Transcription factor binding sites (TFBSs) 127, 231
TRANSFAC motif databases 232
Translation function 69
Trans-membrane helix 267
Transmembrane (TM) regions of proteins 157
Transposons 256, 288
Transverse Relaxation-Optimized Spectroscopy (TROSY)
   versions of HCN experiments 46
Tree bisection and reconnection (TBR) 293, 294
T-test and ANOVA 92
Ttranslation start site 165
Tukey Biweight algorithm 91
Twinscan, transcript sequences 184
Two-color microarray experiments 91

U

UCSC Genome Browser 125, 128, 129, 131, 132,
   135, 136, 187, 192, 211, 222, 239, 246
Unannotated sequences (UNA), GenBank division 16
UniGene databases 184
UniVec database 11, 20
Unpolymerized acrylamide/bisacrylamide, neurotoxin 52
Unsigned genomes 447–448, 453
UPGMA, tree calculation 159
use-Dataset() function 107
useMart() function 106

V

Vagility 502, 503
VALIDATED RefSeq records 6

Variance Stabilizing Normalization (VSN) method 91
VARSPLIC feature annotation 184
VecScreen BLAST analysis, graphical display of 20
VERSION field in GenBank flat file 13
Viral sequences (VRL), GenBank division 16
Vi- RLGS computational method for mapping
        spot in genome 127
volcano plot, ... 105

W

WATERGATE-2D NOESY spectra
        of TAR RNA 41
Watson-Crick A:U,G:C pairs 271
W3C's Web Ontology Language 112
Web-based submission tools 11–12
Webin, EMBL Web-based submission tools 11
Web Ontology Language (OWL) 117
Web resources for full-length isoforms 204
Web sites of multiple alignment programs 149
Weeder algorithms 241
Weighted sum of pairs (WSP) score 153–154
WGS project accession number 9
which *is_a* relationship 116
Whole genome shotgun (WGS) sequencing 8–9

Whole human sorcin
    diffraction oscillation image with program
        Xdisp 75
    output of Denzo autoindexing routine 76
World Wide Web Consortium (W3C) 117
Wormbase, data resource for Caenorhabditis biology
        and genomics 93
write.table() function 109
WUBlast packages 187
WU-BLAST program 374

X

XDisplay for macromolecular crystallography 72
Xfit programs 72
XNU filtering programs 248
Xplor programs 72

Y

YASPIN prediction methods 150
Yeast protein interaction network 418, 423
    ancestral state inference of 424–426

Z

Zebrafish fission hypothesis 379–380

Printed in the United States of America